Drug Delivery of siRNA Therapeutics

Drug Delivery of siRNA Therapeutics

Special Issue Editors

Gaetano Lamberti
Anna Angela Barba

MDPI • Basel • Beijing • Wuhan • Barcelona • Belgrade • Manchester • Tokyo • Cluj • Tianjin

Special Issue Editors
Gaetano Lamberti
Department of Industrial Engineering, University of Salerno
Italy

Anna Angela Barba
Department of Pharmacy, University of Salerno
Italy

Editorial Office
MDPI
St. Alban-Anlage 66
4052 Basel, Switzerland

This is a reprint of articles from the Special Issue published online in the open access journal *Pharmaceutics* (ISSN 1999-4923) (available at: https://www.mdpi.com/journal/pharmaceutics/special_issues/drug_delivery_of_siRNA_therapeutics).

For citation purposes, cite each article independently as indicated on the article page online and as indicated below:

LastName, A.A.; LastName, B.B.; LastName, C.C. Article Title. *Journal Name* **Year**, *Article Number*, Page Range.

ISBN 978-3-03936-200-4 (Hbk)
ISBN 978-3-03936-201-1 (PDF)

Contents

About the Special Issue Editors

Anna Angela Barba, Ph.D., is a chemical engineer. She teaches Industrial Pharmaceutical Plants at the Department of Pharmacy, University of Salerno, Italy. Her research activities focus on the development of non-conventional techniques and intensified processes involving microwave energy and ultrasonic energy in the production of active molecule delivery systems and in agro-food treatments. The results of her scientific activities are reported in numerous articles published in international journals, national technical journals, communications at international and national conferences, and books/monographs. An overview of her research group is available on the website https://gruppotpp.unisa.it.

Gaetano Lamberti, Ph.D., is a chemical engineer, and he teaches Transport Phenomena at the Department of Industrial Engineering, University of Salerno, Italy. His research interests focus on the applications of transport phenomena in pharmaceutical/biomedical sciences, as well as in food science, and on polymer transformation processes, with special emphasis on flow-induced crystallization. The results of his research and an overview on his research group are summarized on the website https://gruppotpp.unisa.it. Gaetano Lamberti is the single author or a co-author of more than 130 papers published in international journals on these subjects.

Editorial

Drug Delivery of siRNA Therapeutics

Gaetano Lamberti [1,2] and Anna Angela Barba [1,3,*]

1 Eng4Life Srl, Spin-off Accademico, Via Fiorentino, 32, 83100 Avellino, Italy; glamberti@unisa.it
2 Dipartimento di Ingegneria Industriale; Università degli Studi di Salerno, via Giovanni Paolo II, 132 84084 Fisciano (SA), Italy
3 Dipartimento di Farmacia; Università degli Studi di Salerno, via Giovanni Paolo II, 132 84084 Fisciano (SA), Italy
* Correspondence: aabarba@unisa.it

Received: 14 February 2020; Accepted: 16 February 2020; Published: 20 February 2020

Small interfering RNA (siRNA) is a class of nucleic acid-based drugs (NABDs) able to block gene expression by interaction with mRNA before its translation. Small interfering RNAs (siRNAs) therefore present extraordinary potential due to their ability to silence the expression of disease-causing genes. Even if the mechanism of action has been successfully investigated (Nobel Prize in Physiology or Medicine 2006 to Andrew Z. Fire and Craig C. Mello "for their discovery of RNA interference – gene silencing by double-stranded RNA") and siRNA drugs can be candidates to fight, in principle, any diseases. However, the practice of siRNA-based therapies is restricted because of relevant inconveniences. SiRNAs are negatively charged large macromolecules and this entails difficult crossing of cell membranes; they undergo rapid degradation by plasma enzymes and are easily subjected to fast hepatic/renal clearance sequestration. These features seriously hinder siRNAs' usability in therapeutics. Currently, the scientific community focused on gene therapy research is developing studies to overcome the obstacles related to siRNA's features.

This Special Issue of *Pharmaceutics* titled "Drug Delivery of siRNA Therapeutics" aims to present the state of the art of siRNA delivery, embracing investigation strategies of international research groups with different experiences and skills. The Special Issue will thus be devoted to presenting the current connections between experimental and in silico approaches for therapies based on siRNA delivery, accounting for all the most promising techniques based on liposomes, polymeric and inorganic nanoparticles, aptamers, chemical modification of siRNAs, and so on.

Reviews (five) and research papers (eight) constitute this Special Issue. A representative international scientific community focused on gene-therapies researches is represented by 12 different countries involving 75 scientists with multidisciplinary skills.

In the reviews, different research activities cover several disciplines of investigation mainly focused on approaches of siRNA therapies to combat several kinds of cancer in laboratory conditions and the current state of siRNA–lipid delivery systems in clinical trials.

In Marson et al., [1] studies on poly(amidoamine)-based dendrimers as attractive nanovectors for siRNA delivery into cells, were presented. In particular, an introduction to RNAi-based therapeutics and the advantages offered by dendrimers as siRNA nanocarriers were discussed. Subsequent linked studies reported in Laurini et al., [2] present the development of poly(amidoamine)-based amphiphilic dendrons—structures able to auto-organize themselves into nanosized micelles which ultimately outperform their covalent dendrimer counterparts in in vitro and in vivo gene silencing. In Barba et al., [3] the current status of siRNA-lipid delivery systems in clinical trials was addressed, offering an updated overview on the clinical goals and the next challenges of this new class of therapeutics which will soon replace traditional drugs. Farra et al., [4] focused their studies on the description of the therapeutic potential of siRNAs and polymer-/lipid-based delivery systems for ovarian cancer. After a brief description of ovarian cancer and siRNA features, they summarized the strategies employed to minimize siRNA delivery problems, the targeting strategies to ovarian cancer and the preclinical models available. They also discussed the most interesting works published in the last three years about polymer-/lipid-based materials for siRNA delivery. In Dinis Ano Bom et al., [5] attention was devoted to the use of aptamers as delivery agents of siRNA in nanoparticle formulations in cancer treatments, alone or in combination with chemotherapy.

Research papers deal with experimental new strategies to design and develop innovative suitable and effective vectors for siRNA delivery such as liposomes, dendrimers, aptamers, polymer–lipid systems, polymeric, co-polymeric and magnetic nanoparticles.

Stiina Kontturi et al., [6] aimed their studies at the development of efficient and safe administration systems devoted to the delivery of oligonucleotide-based drugs. In particular, they produced a light-triggered liposomal delivery system for oligonucleotide delivery based on a non-cationic and thermosensitive liposome with indocyanine green as a photosensitizer ingredient. Hao et al., [7] focused their studies on the combination of chemotherapeutic drugs and siRNA as an emerging modality for cancer therapy. They developed a functionalized mixed micelle-based delivery system for targeted co-delivery of methotrexate and survivin siRNA. Hattori et al., [8] presented studies on three types of cationic liposomes/siRNA complexes (siRNA lipoplexes) on gene-silencing actions in tumor cells. They used three types of cationic cholesterol derivatives to investigate an optimal formulation to achieve the best performance in terms of gene-silencing and cellular uptake effects. In Egorova et al., [9] researches on modular peptide carriers for the delivery of siRNAs to therapeutic angiogenesis inhibition were performed. In particular, the transfection properties of siRNA as polyplexes were studied in breast cancer cells and endothelial cells. Fatemian et al., [10] investigated the use of inorganic pH-dependent carbonate apatite nanoparticles to efficiently deliver various classes of therapeutics into cancer cells. Co-delivery of drugs and genetic materials (siRNAs) was studied in in vivo research. Ewe et al., [11] presented a research on the chemical modifications of polyethylenimines used to produce polymeric nanoparticles, promising structures towards the development of more efficient non-viral delivery systems. In particular, they concentrated their attention on tyrosine-modified polyethylenimines with low or very low molecular weight for siRNA delivery. Jin et al., [12] focused their work on polyethyleneimine-modified magnetic Fe_3O_4 nanoparticles prepared for the delivery of therapeutic siRNAs to contrast oral cancer cells' growth. Craparo et al., [13] studied the formulation and properties of a novel protonable copolymer, based on polyaspartamide, able to form polyplex structures with siRNA to be used in antiasthmatic therapy.

Conflicts of Interest: The authors declare no conflict of interest.

References

1. Marson, D.; Laurini, E.; Aulic, S.; Fermeglia, M.; Pricl, S. Evolution from Covalent to Self-Assembled PAMAM-Based Dendrimers as Nanovectors for siRNA Delivery in Cancer by Coupled in Silico-Experimental Studies. Part I: Covalent siRNA Nanocarriers. *Pharmaceutics* **2019**, *11*, 351. [CrossRef] [PubMed]
2. Laurini, E.; Marson, D.; Aulic, S.; Fermeglia, M.; Pricl, S. Evolution from Covalent to Self-Assembled PAMAM-Based Dendrimers as Nanovectors for siRNA Delivery in Cancer by Coupled in Silico-Experimental Studies. Part II: Self-Assembled siRNA Nanocarriers. *Pharmaceutics* **2019**, *11*, 324. [CrossRef] [PubMed]
3. Barba, A.A.; Bochicchio, S.; Dalmoro, A.; Lamberti, G. Lipid Delivery Systems for Nucleic-Acid-Based-Drugs: From Production to Clinical Applications. *Pharmaceutics* **2019**, *11*, 360. [CrossRef] [PubMed]
4. Farra, R.; Maruna, M.; Perrone, F.; Grassi, M.; Benedetti, F.; Maddaloni, M.; Boustani, M.E.; Parisi, S.; Rizzolio, F.; Forte, G.; et al. Strategies for Delivery of siRNAs to Ovarian Cancer Cells. *Pharmaceutics* **2019**, *11*, 547. [CrossRef] [PubMed]
5. Dinis Ano Bom, A.P.; da Costa Neves, P.C.; de Almeida, C.E.B.; Silva, D.; Missailidis, S. Aptamers as Delivery Agents of siRNA and Chimeric Formulations for the Treatment of Cancer. *Pharmaceutics* **2019**, *11*, 684. [CrossRef] [PubMed]
6. Kontturi, L.-S.; van den Dikkenberg, J.; Urtti, A.; Hennink, W.; Mastrobattista, E. Light-Triggered Cellular Delivery of Oligonucleotides. *Pharmaceutics* **2019**, *11*, 90. [CrossRef] [PubMed]

7. Hao, F.; Lee, R.; Yang, C.; Zhong, L.; Sun, Y.; Dong, S.; Cheng, Z.; Teng, L.; Meng, Q.; Lu, J.; et al. Targeted Co-Delivery of siRNA and Methotrexate for Tumor Therapy via Mixed Micelles. *Pharmaceutics* **2019**, *11*, 92. [CrossRef] [PubMed]
8. Hattori, Y.; Shimizu, S.; Ozaki, K.; Onishi, H. Effect of Cationic Lipid Type in Folate-PEG-Modified Cationic Liposomes on Folate Receptor-Mediated siRNA Transfection in Tumor Cells. *Pharmaceutics* **2019**, *11*, 181. [CrossRef] [PubMed]
9. Egorova, A.A.; Shtykalova, S.V.; Maretina, M.A.; Sokolov, D.I.; Selkov, S.A.; Baranov, V.S.; Kiselev, A.V. Synergistic Anti-Angiogenic Effects Using Peptide-Based Combinatorial Delivery of siRNAs Targeting VEGFA, VEGFR1, and Endoglin Genes. *Pharmaceutics* **2019**, *11*, 261. [CrossRef] [PubMed]
10. Fatemian, T.; Moghimi, H.R.; Chowdhury, E.H. Intracellular Delivery of siRNAs Targeting AKT and ERBB2 Genes Enhances Chemosensitization of Breast Cancer Cells in a Culture and Animal Model. *Pharmaceutics* **2019**, *11*, 458. [CrossRef] [PubMed]
11. Ewe, A.; Noske, S.; Karimov, M.; Aigner, A. Polymeric Nanoparticles Based on Tyrosine-Modified, Low Molecular Weight Polyethylenimines for siRNA Delivery. *Pharmaceutics* **2019**, *11*, 600. [CrossRef] [PubMed]
12. Jin, L.; Wang, Q.; Chen, J.; Wang, Z.; Xin, H.; Zhang, D. Efficient Delivery of Therapeutic siRNA by Fe3O4 Magnetic Nanoparticles into Oral Cancer Cells. *Pharmaceutics* **2019**, *11*, 615. [CrossRef] [PubMed]
13. Craparo, E.F.; Drago, S.E.; Mauro, N.; Giammona, G.; Cavallaro, G. Design of New Polyaspartamide Copolymers for siRNA Delivery in Antiasthmatic Therapy. *Pharmaceutics* **2020**, *12*, 89. [CrossRef] [PubMed]

Review

Evolution from Covalent to Self-Assembled PAMAM-Based Dendrimers as Nanovectors for siRNA Delivery in Cancer by Coupled In Silico-Experimental Studies. Part I: Covalent siRNA Nanocarriers

Domenico Marson, Erik Laurini *, Suzana Aulic, Maurizio Fermeglia and Sabrina Pricl

Molecular Biology and Nanotechnology Laboratory (MolBNL@UniTS), Department of Engineering and Architecture, University of Trieste, 34127 Trieste, Italy

* Correspondence: erik.laurini@dia.units.it; Tel.: +39-040-558-3432

Received: 21 June 2019; Accepted: 16 July 2019; Published: 18 July 2019

Abstract: Small interfering RNAs (siRNAs) represent a new approach towards the inhibition of gene expression; as such, they have rapidly emerged as promising therapeutics for a plethora of important human pathologies including cancer, cardiovascular diseases, and other disorders of a genetic etiology. However, the clinical translation of RNA interference (RNAi) requires safe and efficient vectors for siRNA delivery into cells. Dendrimers are attractive nanovectors to serve this purpose, as they present a unique, well-defined architecture and exhibit cooperative and multivalent effects at the nanoscale. This short review presents a brief introduction to RNAi-based therapeutics, the advantages offered by dendrimers as siRNA nanocarriers, and the remarkable results we achieved with bio-inspired, structurally flexible covalent dendrimers. In the companion paper, we next report our recent efforts in designing, characterizing and testing a series of self-assembled amphiphilic dendrimers and their related structural alterations to achieve unprecedented efficient siRNA delivery both in vitro and in vivo.

Keywords: RNAi therapeutics; siRNA delivery; covalent dendrimers; PAMAM dendrimers; nanovectors; gene silencing

---

1. RNA Interference and Challenges in Small Interference RNA Therapeutics

Discovered in 1986 by the Nobel laureates Fire and Mello [1], RNA interference (RNAi)—also known as post-transcriptional gene silencing (PTGS)—is a compendium of mechanisms involving small RNAs that regulate the expression of genes in a variety of eukaryotic organisms. In simple terms, the RNAi process implies the cleavage of endogenous long double-stranded RNAs into short ribonucleic acid sequences (the so called small interfering RNAs or siRNAs, usually 21–23 bases long) by the action of the Dicer endonuclease [2]. Upon incorporation into the multiprotein RNA-induced silencing complex (RISC), the double helical siRNAs are unwound into two strands: The sense (or passenger) strand is discarded, and the antisense (or guide) strand is paired to a complementary mRNA sequence via the RISC complex. Upon binding, the targeted mRNA is in turn degraded by a RISC subunit (known as Argonaute 2) endowed with endonuclease activity. This last step ultimately results in the prevention of mRNA translation into the corresponding protein or, in other words, in gene silencing. Finally, once the target mRNA degradation is accomplished, the RISC complex can be recycled to digest other targets on the mRNA, greatly improving inhibition efficiency [3].

In the post-genomic era, siRNAs of a synthetic nature can be designed and synthesized under good manufacturing practice (GMP) to target complementary regions on any gene of a known sequence to achieve its downregulation for curative purposes [4–6]. However, key to the therapeutic utility of these RNAi triggers is the ability to introduce them into their target cells of the human body. Indeed, naked siRNAs are not amenable to therapeutic administration. For instance, when a siRNA is administered intravenously, it is readily digested by nucleases and largely cleared from the kidney glomeruli before reaching the diseased organs. Moreover, the negative charge and large size of a naked siRNA make it difficult to pass through the plasma membrane of a target cell. Even if the siRNA molecules could reach the target tissue and be taken up by target cells, they must avoid degradation in lysosomes via endosomal escape, a process in which the efficiency of these nucleic acid fragments is notoriously low. As a consequence, one of the major challenges in successful RNAi therapeutics is the discovery of safe, efficient and effective siRNA delivery vectors. Such delivery vehicles must at least protect each siRNA from nucleases in the serum or extracellular media, enhance siRNA transport across the cell membrane, and guide the siRNA to its proper location through interactions with the intracellular trafficking machinery. Ideally, both viral and non-viral (nano)vectors can deliver siRNA into cells [7], although, despite impressive transfection efficiency, the use of the former is limited by safety concerns, including genotoxic and immunogenicity-mediated adverse events [8]. On the contrary, non-viral delivery systems have great potential for the safer delivery of siRNA therapeutics, although so far their performance in transfection efficiency has not reached the level requested for full clinical exploitation [7,9–11].

2. Role of Dendrimers as siRNA Nanocarriers

In the variegated scenario of non-viral (nano)materials for siRNA delivery, dendrimers have quickly grown as a family of synthetic nano-sized, radially symmetric molecules with fine-defined, homogeneous and monodisperse composition endowed with enormous potential as gene therapy nanovectors [12–14]. Structurally, dendrimers are constituted by three distinct domains (Figure 1a): (1) A fundamental atom or, most frequently, a group of atoms defined as the core; (2) the branching units, which, emanating from the core through diverse chemical reactions, allow the dendrimeric molecule to grow in geometrically organized radial layers known as generations (G); and (3) an exponentially increasing number of peripheral surface groups which constitute a multivalent nanoscale array and can therefore form high-affinity interactions with a variety of biological targets [15].

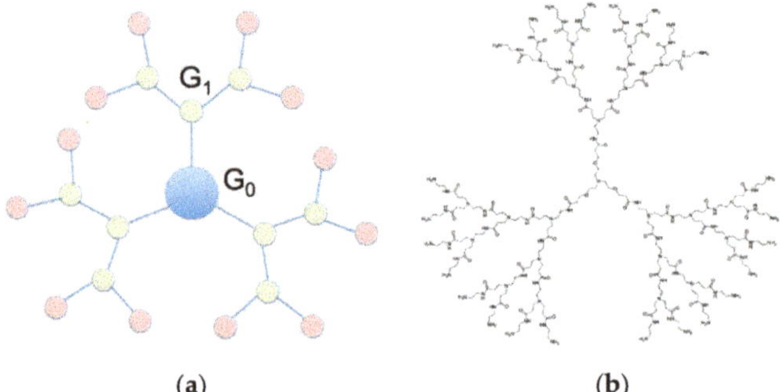

Figure 1. (**a**) Cartoon representation of a dendrimer structure highlighting its three, distinct structural motifs: The core (in light blue), the branching units forming the different G generations (in light green), and the terminal groups (in light pink). According to a consolidate dendrimer nomenclature, the core constitutes Generation 0 (G_0). Therefore, the subsequent Generations G_1, G_2, … G_n refer to the corresponding level of branching, as annotated in panel (**a**). (**b**) Molecular structure of the triethanolamine (TEA)-core poly(amidoamine) dendrimers. For clarity, in panel (**b**), only a dendrimer of Generation 4 (G_4) is shown.

From the synthetic viewpoint, both divergent and convergent pathways (or combinations of thereof) can be adopted to prepare dendrimers with different generations with precisely defined, regular structures [16,17]. To date, more than fifty families of dendrimers each with unique chemistry and properties have been produced and are under investigation in a diversity of different biomedical applications [18]. Among these, poly(amidoamine) (aka PAMAM) dendrimers undoubtedly constitute the molecules most widely explored as nanocarriers for both drug and gene delivery [13,19–21]. In the specific field of nucleic acid delivery and release, this popularity can be ascribed to several beneficial features of PAMAMs, including (i) the chemical nature of their terminal groups, which, being primary amines, are fully protonated at the physiological pH of 7.4. This entails extremely favorable electrostatic (Coulombic) interactions of these dendrimers with the negatively charged nucleic acid fragments and the subsequent mutual condensation into nanoscopic particles, often called dendriplexes; and (ii) the presence of tertiary amines within the dendritic branched structure which, becoming protonated at lower pH values pertaining to endosomes and lysosomes, mediate the osmotic swelling and subsequent disruption of the membranes of these vesicles, ultimately promoting the intracellular release of the siRNA cargo. This mechanism, known as the proton-sponge hypothesis, relies on the assumption (under debate; [22]) that the unprotonated amines of PAMAMs can absorb protons as they are pumped into the lysosome/endosome, resulting in more protons being pumped in, thus leading to an increased influx of Cl^- ions and water. A combination of the osmotic swelling and a swelling of the dendrimers themselves because of the repulsion between protonated amine groups causes the rupture of the lysosomal/endosomal membranes, resulting in the subsequent release of its contents into the cytoplasm [23].

3. Structurally Flexible PAMAM Dendrimers for Safe, Efficient and Effective siRNA Delivery

Despite the wealth of studies dating back to the early 90s yielding highly promising results for PAMAM-based dendrimers as DNA nanovectors [24–28], only the last 10 years have witnessed systematic investigations of this class of molecules in siRNA delivery (see Table S1 in Supplementary Materials) [19,29–32]. In this arena, successive rounds of structure optimization led us to the design of PAMAM dendrimers with a triethanolamine (TEA) core (Figure 1b) [33]. The rationale behind the conception and synthesis of this new dendrimer family was that the TEA-core molecules, having the branching units starting away from the central amine with a distance of 10 successive bonds, should feature an extended core. As such, they were expected to be less congested in space with respect to the prototypical NH_3-core PAMAM dendrimers, in which the branches sprout directly from the small ammonia focal point. As a consequence, the TEA-core dendrimers—with less densely packed branches and terminal units—should be endowed with an enhanced flexibility of their arms and, as such, should perform better as siRNA nanocarriers than their NH_3-core counterparts. In essence, the hypothesis of a greater flexibility translating into the more effective enwrapping of the nucleic acid fragment was inspired by the behavior of histones, whose structure dynamics allows for conformational changes related to DNA binding (required for post-translational modification) and unbinding (required to prevent transcription) [34].

3.1. *Prediction of Enhanced Flexibility and siRNA Interactions of TEA-Core Dendrimers by Computer Simulations*

The hypothesis of diverse flexibility between TEA- and NH_3-core based PAMAMs was verified by atomistic molecular dynamics (MD) simulations [35–37]. Figure 2a,b shows two MD equilibrated structures of the G_5 TEA-core and NH_3-core PAMAM dendrimers at a physiological pH (7.4), respectively. As intuitively perceived from these images, the TEA-core molecule is characterized by a more open conformation, with a uniform void distribution within its interior, whilst its ammonia-core counterpart is remarkably more compact, featuring a non-homogeneous, restricted void spacing. Thus, the conformation of the TEA-core PAMAM dendrimers is such that the outer branches can freely move and adjust to optimize binding with its siRNA cargo (Figure 2a). On the contrary, the more rigid and compact structure of NH_3-core PAMAMs prevents these molecules from any induced-fit conformational readjustment, and, consequently, not all of the terminal groups are available to self-orient for optimal nucleic acid binding (Figure 2b).

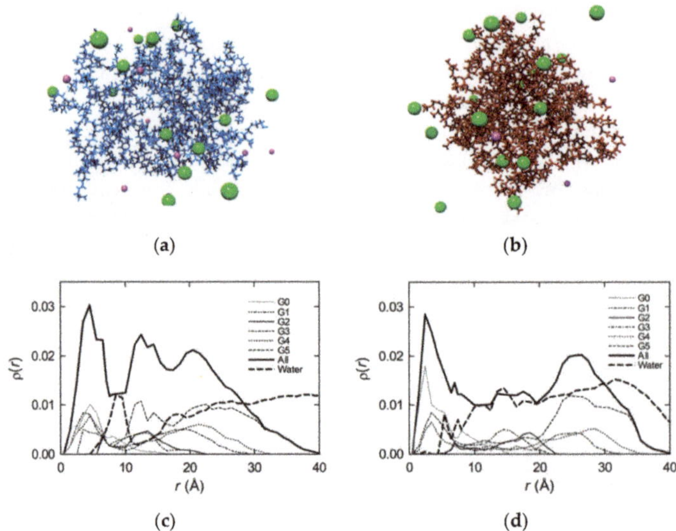

Figure 2. Equilibrated molecular dynamics conformations of G_5 TEA-core (**a**) and NH_3-core (**b**) poly(amidoamine) (PAMAM) dendrimers in physiological solution (pH = 7.4, ionic strength = 0.15 M NaCl). Each dendrimer molecule is represented as colored sticks, some ions and counterions are visualized as purple (Na^+) and green (Cl^-) spheres, and water molecules are not shown for clarity. Average radial monomer density $\rho(r)$ for subsequent generations (from G_0 to G_5) of the TEA-core (**c**) and the NH_3-core PAMAMs (**d**). In all cases, the origin was set at the molecular center of mass. Adapted from [37], published by RSC, 2013.

Quantitative substantiation was obtained from calculating the average radial monomer density $\rho(r)$ for each dendrimer type (shown in Figure 2c,d for subsequent dendrimer generations up to G_5), a quantity defined as the number of atoms whose centers of mass locate within a spherical shell of radius r and thickness Δr. Accordingly, the integration of $\rho(r)$ over r yields the total number of dendrimer monomers as:

$$N(r) = 4\pi \int_0^R r^2 \rho(r) dr. \quad (1)$$

Focusing attention on G_5 as an example, in the case of the TEA-core molecule, the whole dendrimer $\rho(r)$ curve (shown as a thick continuous line in Figure 2c) is characterized by the presence of two minima—the first (more pronounced) located approximately 10 Å away from the core and the second at around 17 Å—each followed by two relative maxima, at about 13 and 21 Å, respectively. These features of $\rho(r)$ constitute a clear indication that the dendrimer core region is denser with respect to the middle–outer molecular portions (which are fairly hollow) and that the higher sub-generation monomers generate a crowded layer at the dendrimer periphery. This, in turn, accounts for the presence of a uniform distribution of hollow spaces in the central dendrimer structure, which can be filled up by a significant number of solvent molecules, as testified by the corresponding thick dashed line in Figure 2c. Considering the same data for the alternative G_5 NH_3-core dendrimer, the $\rho(r)$ profile for the whole molecule (thick continuous line; Figure 2d) is representative of a complete different trend: Indeed, after the core peak, the curve quickly reaches a plateau value that spans the entire central dendrimer region and finally increases again at the molecular periphery. In other words, all dendrimer sub-generations afford a substantial contribution to the whole density curve, supporting the visual evidence (Figure 2b) of a more uniform monomer distribution within the dendrimeric structure. In line with this observation, the corresponding water density profile (thick dashed line; Figure 2d) does not feature any pronounced maximum in any specific region of the molecule but rather exhibits a uniform distribution throughout the dendrimer interior.

The postulated enhanced ability of the more flexible, extended-core (TEA) PAMAMs in interacting with siRNA molecules with respect to smaller core (NH_3) dendrimers was next predicted by computer simulations based on the so-called molecular mechanics/Poisson-Boltzmann surface area (MM/PBSA) methodology [35–40] (see Supporting Information for detailed explanation). To this purpose, the free energy of binding normalized by the total number of charged dendrimer terminal groups ($\Delta G_{bind}/N$) between successive generations of the two different PAMAM-based molecules towards the siRNA sequence directed against the mRNA coding for the heat shock protein 27 (Hsp27)—a small molecular chaperone which is a vital regulator of cell survival and a major player in drug resistance—was calculated [35]. This normalization procedure was required to compare the affinity of the different dendrimer generations towards the double-stranded (ds) RNA fragment. As can be seen in Table 1, $\Delta G_{bind}/N$ is negative for all systems considered, indicating that, under *in silico* physiological conditions (pH 7.4 and 0.15 M NaCl), the association of both dendrimeric nanovectors with their nucleic acid payloads is a thermodynamically favorable and spontaneous process. However, for each dendrimer generation, the TEA-core PAMAMs show a superior affinity for the ds-RNA sequence (i.e., $\Delta G_{bind}/N$ more negative) with respect to their NH_3-core counterparts. Additionally, there is a notable increase in binding strength in passing from G_4 to G_5, substantially ascribable to an enhanced favorable enthalpic component $\Delta H_{bind}/N$. This aspect accounts for the general trend of better binding and, hence, better properties as nanocarriers of high generation dendrimers, which is in agreement with experimental evidence [32] Contextually, the entropic contribution is less unfavorable (i.e., smaller) in the case of the TEA-core molecules. This lower value of $-T\Delta S_{bind}/N$ can be connected again to the enhanced flexibility and, consequently, the greater capacity of the conformational adaptation of all generations of the enlarged-core dendrimers in enwrapping the ds-RNA molecule, followed by an enhanced productive binding of the nucleic acid.

Table 1. *In silico* normalized free energy of binding ($\Delta G_{bind}/N$) and its major components (binding enthalpy $\Delta H_{bind}/N$ and entropy variation $-T\Delta S_{bind}/N$) for G_4–G_6 TEA-core and NH_3-core PAMAMs in complex with heat shock protein 27 (Hsp27) small interfering RNA (siRNA) at pH 7.4 and 0.15 M NaCl. The normalization factor *N* is the generation-specific total number of charged dendrimer terminal groups (*N*). All values are expressed in kcal/mol[1]. Adapted from [35] with permission of John Wiley and Sons.

	TEA-Core PAMAMs			**NH_3-Core PAMAMs**		
G	**$\Delta G_{bind}/N$**	**$\Delta H_{bind}/N$**	**$-T\Delta S_{bind}/N$**	**$\Delta G_{bind}/N$**	**$\Delta H_{bind}/N$**	**$-T\Delta S_{bind}/N$**
4	−7.57 [1]	−9.82	2.25	−4.57	−8.02	3.45
5	−14.9	−17.9	3.02	−11.5	−16.0	4.43
6	−17.0	−20.5	3.55	−14.1	−18.8	4.77

[1] Standard deviation for all data in Table 1 is less that 1%.

The equilibrated MD snapshots of these two G_5 dendrimer series in complex with the Hsp27 siRNA shown in Figure 3 offer additional insightful structural information. In fact, as can be inferred from Figure 3a, the conformation of the TEA-core dendrimers is such that its outer branches can readily move towards the phosphate backbone of the siRNA during complex formation so that its charged amine groups can arrange themselves via induced-fit for optimal binding with the nucleic acid. On the contrary, the more rigid and compact structure of the alternative PAMAM molecule prevents it from undergoing a significant conformational readjustment required by the induced-fit (Figure 3b); as a consequence, a smaller number of amine groups are available for optimal siRNA binding.

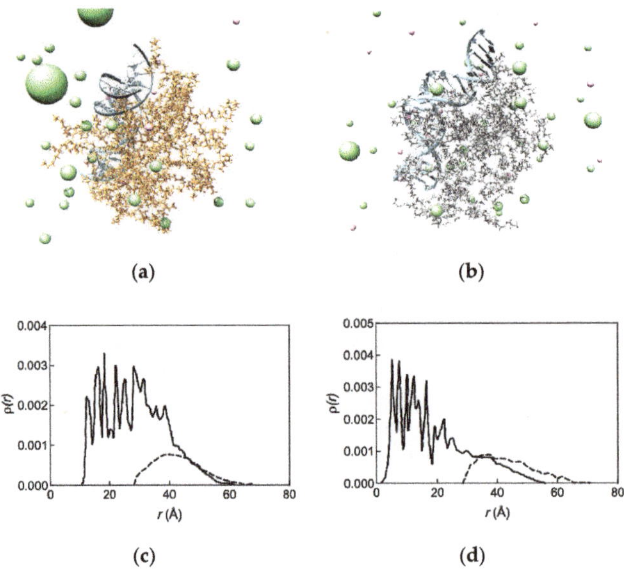

Figure 3. Equilibrated molecular dynamics conformations of G_5 TEA-core (**a**) and NH_3-core (**b**) PAMAM dendrimers in complex with Hsp27 siRNA in physiological solution (pH = 7.4, ionic strength = 0.15 M NaCl). Each dendrimer molecule is represented as colored sticks, some ions and counterions are visualized as light pink (Na^+) and light green (Cl^-) spheres, the siRNAs are portrayed as light blue ribbons, and water molecules are not shown for clarity. Radial density distribution $\rho(r)$ of the dendrimer terminal nitrogen atoms in G_5 TEA-core (**c**) and NH3-core (**d**) PAMAMs in complex with Hsp27 siRNA (continuous lines). The corresponding distributions of the siRNA phosphorous atoms in each siRNA/dendrimer complex are shown as dashed lines. Adapted from [35,36] with permission of John Wiley and Sons and Bentham Science Publishers.

This differential behavior in siRNA binding is more evident when analyzing the radial density distributions of the relevant nucleic acid/dendrimer complexes reported in Figure 3c,d. For the G_5 TEA-core dendrimer/siRNA complex, the density profiles of the primary, positively charged nitrogen atoms stretches further out towards the molecule periphery due to the electrostatic attraction of the siRNA negatively charged phosphate moieties (Figure 3c). Contextually, the density distribution of the phosphorous siRNA atoms reveals a good penetration of the nucleic acid fragment within the dendrimer outer shell. Contrarily, even in complex with siRNA the NH_3-core G_5, PAMAM maintains a more compact conformation that is characterized by a high degree of branch back-folding; as a result, the density of the terminal amines on the dendrimer surface is lower, and the corresponding siRNA phosphorous density distribution curve shows only a partial penetration of the nucleic acid within the dendrimer molecular structure (Figure 3d).

3.2. High-Generation TEA-Core PAMAM Dendrimers as Effective In Vitro and In Vivo siRNA Nanocarriers

3.2.1. In Vitro Data

The predicted ability of TEA-core PAMAMs to generate nanoscale siRNA/dendrimer complexes (aka dendriplexes) was experimentally verified and characterized in vitro. Figure 4a,b shows the atomic force microscopy (AFM) images of the Hsp27/dendrimer nanoparticles obtained using TEA-core dendrimers from G_1 to G_7 [41].

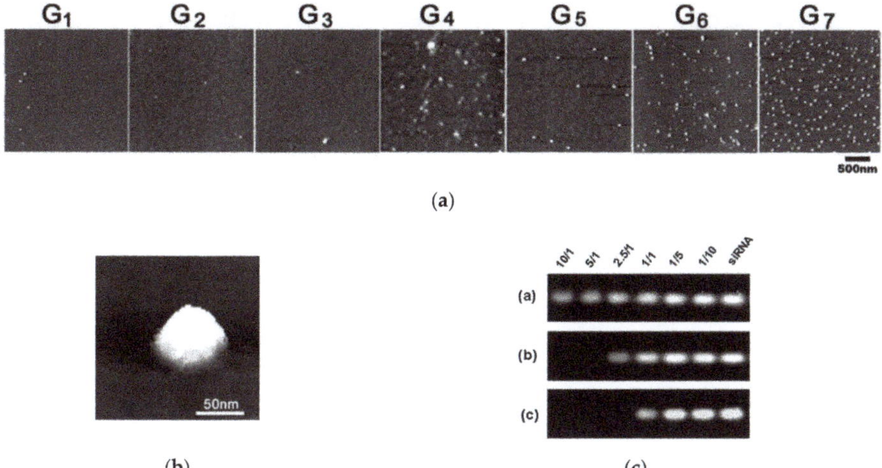

Figure 4. Three-dimensional atomic force microscopy (AFM) images of (**a**) Hsp27 siRNA in complex with TEA-core PAMAM dendrimers of increasing generation (**G_1–G_7**) and (**b**) a single spherical siRNA/TEA-core dendrimer (**G_7**) complex at a final siRNA concentration of 0.0125 mg/L and at a dendrimer-to-siRNA charge ratio (N/P) of 10. (**c**) Gel retardation of Hsp27 siRNA with three different TEA-core PAMAMs ((**G_1**) (**a**), (**G_4**) (**b**) and (**G_7**) (**c**)) as a function of the N/P ratio (from 10/1 to 1/10 from left to right; last lane: Naked siRNA). Adapted from [41] with the permission of the RSC.

While only a few siRNA/dendrimer assemblies can be observed for smaller dendrimers (G_1 and G_3), an increasing number of nanoscale particles are formed with increasing dendrimer generation (Figure 4); in particular, starting from G_4, the siRNA/dendrimer nanocomplexes progressively become more uniform, well-defined, and compact. This suggests that G_4 is the lowest threshold TEA-core dendrimer generation for effective siRNA complexation. This assertion is substantiated by the results of the RNA mobility assay illustrated in Figure 4c: While no gel retardation effect is observed with the G_1 dendrimer even at the maximum dendrimer-to-siRNA charge (N/P) ratio adopted (10/1), the siRNA mobility is considerably retarded by the G_4 molecule for $N/P \geq 2.5$, and, for the same N/P value, the siRNA shift in the gels is completely almost prevented by the G_7 dendrimer [41]. The complexes formed between siRNA and high generation ($\geq G_4$) TEA-core PAMAMs are completely stable in physiological conditions, require strong ionic detergents such as Sodium Dodecyl Sulfate (SDS) to be disrupted and, contrarily to naked siRNA, to show considerable resistance to RNase degradation [33].

The rapid and efficient cellular uptake of siRNA/TEA-core dendrimer nanoassembly was demonstrated using the human castration-resistant prostate cancer (CRPC) PC-3 cell line by confocal fluorescence imaging. In these experiments, a non-silencing (i.e., scrambled) Hsp27 siRNA sequence labeled with the green fluorescent dye Alexa 488 was employed [42]. Figure 5a shows that, 4 h after treatment, cells are populated by the green fluorescent siRNA–dendrimer nanoparticles which reside exclusively in the cell cytoplasm (Figure 5b,c). Contextually, after cell treatment with a similar amount of Alexa 488-labelled naked siRNA, no green fluorescent nanoparticles are detected inside the cells.

The successful siRNA delivery and specific gene silencing of these flexible TEA-core PAMAMs was next validated in the first proof-of-concept (POC) study targeting again Hsp27 PC-3 cells [42]. Prostate cancer is the second most commonly occurring cancer in men and the fourth most commonly occurring cancer overall; although most patients initially respond well to first-line hormone-based therapies associated with androgen ablation, after a very short time period (≈2 years), they unfortunately relapse [43], and no effective therapeutic regimen is available to treat this progressive condition. Therefore, anticancer treatments based on targeting survival genes (e.g., Hsp27) using RNAi constitute an interesting option in contrasting CRPCs [44]. The results from the POC study (Figure 6) show that transfecting PC-3 cells with TEA-core G_7/Hsp27 siRNA complexes yields the potent, specific, and long-lasting downregulation (>50% silencing after five days) of both targeted mRNA and protein [42].

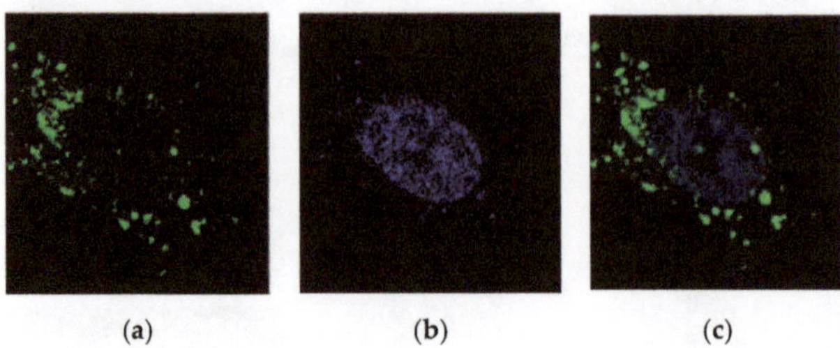

Figure 5. (**a**) Confocal fluorescence imaging of G_7 TEA-core dendrimer-mediated cellular uptake of siRNA using a non-silencing siRNA sequence labeled with the green fluorescent dye Alexa 488. (**b**) Blue fluorescence images of a nucleus of the same cells stained by TOPRO-3. (**c**) Merged fluorescent images of a and b confirming the exclusive cytoplasm localization of the dendrimer/siRNA complexes. Adapted from [42] with permission of John Wiley and Sons.

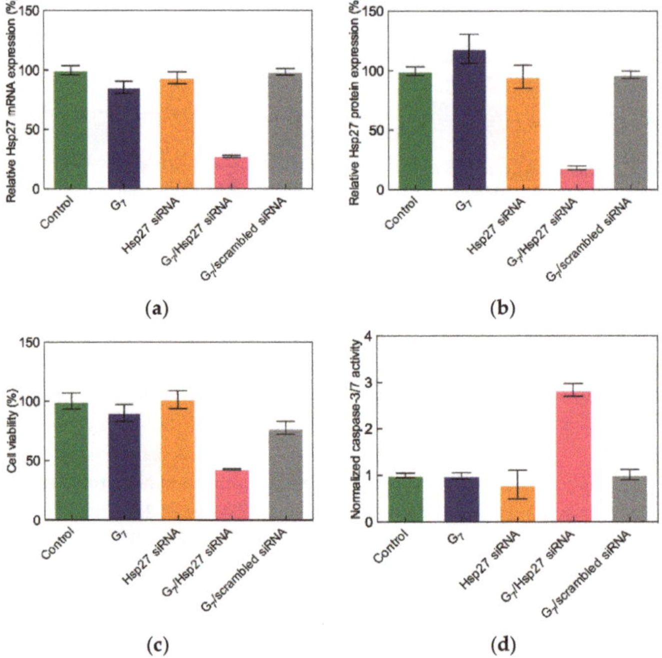

Figure 6. Quantitative analysis of (**a**) Hsp27 mRNA levels (determined by quantitative reverse transcription polymerase chain reaction (RT-qPCR)), (**b**) Hsp27 protein expression (determined by western blot analysis), (**c**) cell proliferation (determined by the 3-(4,5-dimethylthiazol-2-yl)-2,5-diphenyl tetrazolium bromide (MTT) assay), and (**d**) caspase-3/7 activity (measured by a colorimetric assay) in prostate cancer-3 (PC-3) cells three days after the TEA-core G_7-mediated delivery of 50 nM Hsp27-targeting siRNA (N/P = 10) (hot pink bars in all panels). Data for dendrimer G_7 alone (blue bars), naked Hsp27 siRNA (orange bars), and scrambled siRNA–G_7 complexes (gray bars) are shown for comparison. Values are expressed as % relative to control (non-treated cells, green bars). Adapted from [42] with permission of John Wiley and Sons.

Following the knockdown of Hsp27 in PC-3 cells by the G_7 TEA-core PAMAM dendrimer/siRNA complex, a remarkable anti-proliferative effect is observed (Figure 6c), in agreement with previous results from Rocchi et al. [45], thus showing that the inhibition of Hsp27 protein expression negatively affects PC-3 cell survival.

The main mechanism beyond this cellular effect was next demonstrated to proceed via caspase-dependent induced apoptosis. Apoptosis is programmed cell death that involves the controlled dismantling of intracellular components while avoiding inflammation and damage to surrounding cells. Apoptotic caspases are a family of endoproteases that provide critical links in cell regulatory networks controlling cell death. Specifically, the activation of these enzymes results in the inactivation/activation of substrates, and the generation of a cascade of signaling events permitting such controlled demolition of cellular components. As seen in Figure 6d, a three-fold increase in caspase-3/7 activity after Hsp27 siRNA/G_7 TEA-core dendrimer complex treatment relative to controls is detected, ultimately confirming that Hsp27 siRNA delivered by the G_7 TEA-core dendrimer is very effective in inhibiting Hsp27 expression and thereby inducing caspase-dependent anticancer activity in human CRPC PC-3 cells.

The dependence of the TEA-core PAMAM dendrimer-mediated siRNA silencing effect on siRNA concentration, dendrimer generation, N/P ratio and incubation time for transfection was also investigated [42]. A clear dose-dependent gene silencing was observed using the G_7 TEA-core PAMAM nanocarrier, in line with a RNAi-mediated gene silencing mechanism. Moreover, in agreement with the molecular simulation results (see Table 1), the gene silencing efficacy was seen to increase with dendrimer generation, the G_7 molecule being the best delivery nanovector for Hsp27 siRNA in cell-based assays. With this high generation dendrimer, optimal gene silencing was achieved at an N/P value of 10, this siRNA/dendrimer charge ratio likely providing not only the best nucleic acid degradation protection but also the most effective siRNA deployment into the cellular cytoplasm by virtue of a very efficient endosomal escape mechanism. Finally, an incubation time for transfection of 24 h was determined as the optimal condition for G_7 TEA-core PAMAM dendrimer-mediated siRNA delivery to CRPC PC-3 cells for Hsp27 gene silencing.

Sensitivity to serum proteins constitutes an Achilles' heel of cationic nanovectors in siRNA delivery. In fact, since high N/P ratios are always required for efficient nucleic acid transfection (as discussed above), the resulting overall charge of the delivery complexes is positive. As such, electrostatic forces may drive their interaction with the negatively charged serum proteins, this process eventually resulting in the disintegration of the nanoassemblies with consequent premature nucleic acid release and degradation. The lead G_7 TEA-core PAMAM dendrimer was indeed not exempt from this drawback, as no Hsp27 gene silencing was observed in the presence of 10% serum with G_7 nanovectors loaded with 50 nM Hsp27 siRNA at the optimal, serum-free N/P ratio of 10. In trying to overcome these negative results, gene silencing experiments were performed at progressively increasing dendrimer-to-siRNA ratios. Potent Hsp27 gene silencing is indeed reached in PC-3 cells at N/P = 40, supported by a strong inhibition of cell proliferation and high caspase-3/7 activation, as shown in Figure 7. Pleasingly, the potent in vitro anticancer activity of the G_7 TEA-core PAMAM-based nanovectors was paralleled by the complete absence of toxicity, as assayed by 3-(4,5-dimethylthiazol-2-yl)-2,5-diphenyl tetrazolium bromide (MTT) tests for cell viability and by lactate dehydrogenase (LDH) release for cell membrane damage [42]. Finally, no acute toxicity was observed in preliminary tests conducted by treating healthy mice via a tail vein injection with scrambled siRNA–G_7 complexes, naked Hsp27 siRNA, and the G_7 TEA-core dendrimer alone, as well as a glucose solution as control.

In aggregate, these findings supported the concept that the flexible, high-generation TEA-core PAMAM dendrimers could be effective nanovectors for in vitro siRNA delivery.

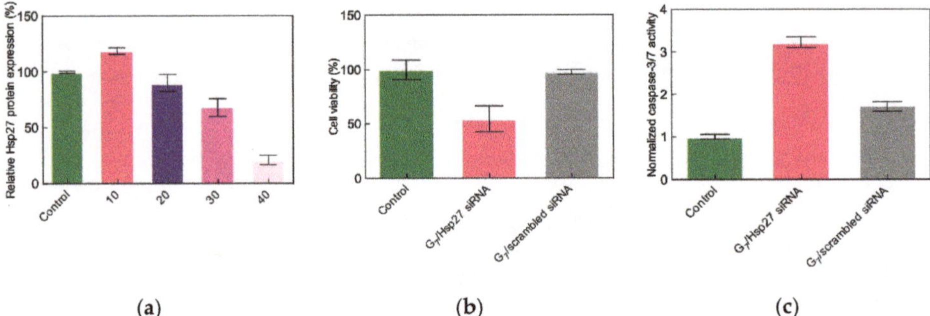

(a) (b) (c)

Figure 7. (**a**) Effect of N/P ratio on Hsp27 gene silencing, (**b**) inhibition of cell growth, and (**c**) caspase-3/7 activity in PC-3 cells three days after TEA-core G_7-mediated delivery of 50 nM Hsp27-targeting siRNA in the presence of 10% serum (hot pink bars in all panels). Data for scrambled siRNA–G_7 complexes (gray bars) are shown for comparison. Values are expressed as % relative to control (non-treated cells, green bars). Adapted from [42], with permission of John Wiley and Sons.

3.2.2. In Vivo Data

Encouraged by the results obtained in vitro, high-generation TEA-core dendrimers were tested for in vivo gene silencing performance in various cancer models, including prostate [46], ovarian [47], liver [48], and glioblastoma [49,50]. The application to ovarian cancer treatment represents a particularly interesting example, since, beside the remarkable silencing of AKT—a gene activated in 36% of primary tumors, with the activation being associated with high-grade cancer and aggressive clinical behavior, protection to induced apoptosis, drug resistance and disease relapse [51]—the combined administration of the anticancer drug paclitaxel with AKT siRNA delivered using G_6 TEA-core PAMAM nanovectors resulted in a synergistic therapeutic effect in vivo.

In this study, the effective AKT gene-silencing mediated by the G_6 dendrimer nanocarriers (Figure 8a,b) was first verified in SKOV-3 cells, a gold standard model for drug-resistant ovarian cancer, as shown in Figure 8c–f. A non-specific (NS) siRNA sequence was also used in all experiments for comparison.

As seen from Figure 8c, the G_6 TEA-core PAMAM-mediated gene silencing effect is dependent on siRNA concentration, with approximately 70% inhibition achieved using 50 nM siRNA. Interestingly, AKT knockdown negatively affects the expression of $p70^{S6K}$, an AKT effector protein involved in cell survival within the phosphatidylinositol 3-kinase (PI3K)/AKT pathway-driven ovarian cancer development [51]. With the same nanovector/siRNA system, the inhibition of cancer cell growth is evident after 72 h post transfection (Figure 8e), and the loss of cell viability can be ascribed to induced apoptosis, as monitored by the significant increase of caspase-3 activity (Figure 8f) [47].

Current chemotherapeutic regimens unfortunately have limited efficacy in patients with advanced ovarian cancer [52]. Paclitaxel is the current first-line drug for treating this malignancy in the clinics, and induced apoptosis has been established as one of the main mechanisms of action of this molecule. Therefore, we speculated that, when administered together, AKT-targeting siRNA and paclitaxel might act synergistically in promoting specific cancer cell death. This hypothesis was initially verified in in vitro experiments, showing how the combined treatment of AKT siRNA/G6 TEA-core dendrimer nanoparticles substantially enhanced SKOV-3 cell growth inhibition via induced apoptosis, as illustrated in Figure 9a,b.

These promising results prompted us to proceed with in vivo experiments. Accordingly, SKOV-3 cells were subcutaneously injected in nude mice, and tumors were allowed to grow until they reached a volume of approximately 30 mm^3. The xenografts were then first treated with intratumoral injections of either AKT siRNA/ or non-specific siRNA/G_6 TEA-core dendriplexes in the absence of paclitaxel. The differential reduction of tumor volume is well evident in the upper panel of Figure 9c:

A tumor shrinkage of ≈50% in mice administered with the targeted siRNA with respect to those receiving the non-specific treatment can be readily appreciated. Concomitantly, immunohistochemistry confirmed the drastic decrease of AKT expression in the AKT knockdown mice (Figure 9d, right panel), the corresponding tumors showing signs of necrosis (Figure 9d, left panel).

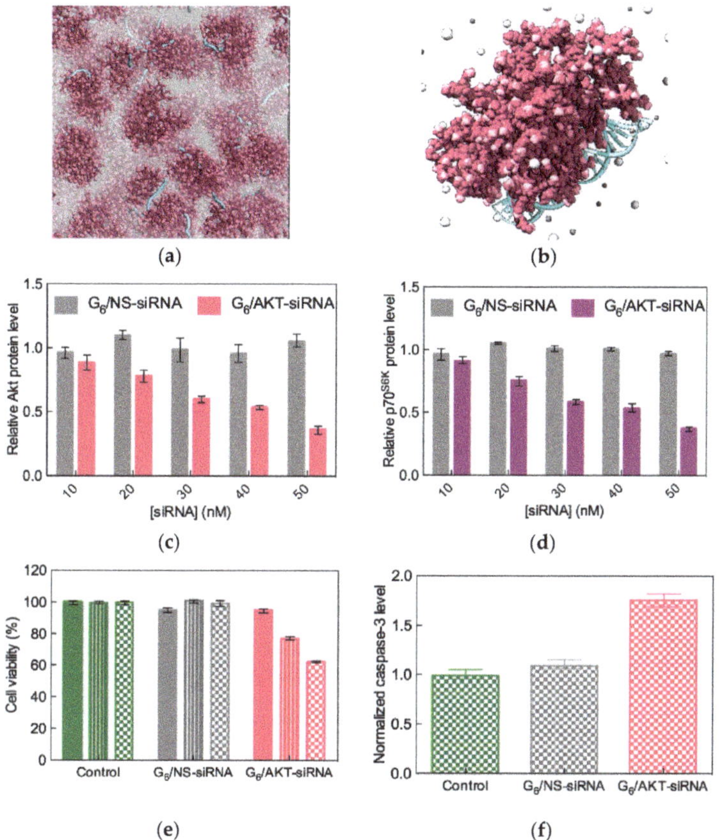

Figure 8. (**a**) Simulation of G_6 TEA-core PAMAM dendrimers in complex with AKT siRNA at N/P = 5. The dendrimers are portrayed as wine-colored spheres, with charged amine groups depicted in pink. siRNA molecules are shown as turquoise sticks. A transparent gray field is used to represent the solvent environment. (**b**) Zoomed view of one single G_6 TEA-core PAMAM molecule in complex with one AKT siRNA (colors as in panel a). Some Cl^- and Na^+ counterions are shown as white and light gray spheres, respectively; water molecules are not shown for clarity. siRNA concentration-dependent inhibition of AKT (**c**) and its downstream effector $p70^{6SK}$ (**d**) in SKOV-3 cells three days after TEA-core G_6-mediated delivery (N/P = 5). Data for non-specific (NS) siRNA–G_6 complexes are shown for comparison. Protein expression levels were determined by western blotting, quantified by densitometry, and are expressed as fold-change normalized to β-actin. (**e**) Time-dependent growth inhibition of SKOV-3 cells transfected with G_6 TEA-core PAMAM dendrimers (N/P = 5) and with non-specific (NS) siRNA–G_6 complexes, as determined by the MTT assay. Values are expressed as % relative to control (non-treated cells). Filled bars: 24 h post transfection (p.t.), striped bars: 48 p.t., checked bars: 72 h p.t. (**f**) Caspase-3 activation in SKOV-3 cells determined 72 p.t. with G_6 TEA-core PAMAM dendrimers at N/P = 5. Data for non-specific (NS) siRNA–G_6 complexes are shown for comparison. Values are expressed as fold change normalized to β-actin used as control. Adapted from [47], which is an open access article published under an ACS AuthorChoice License.

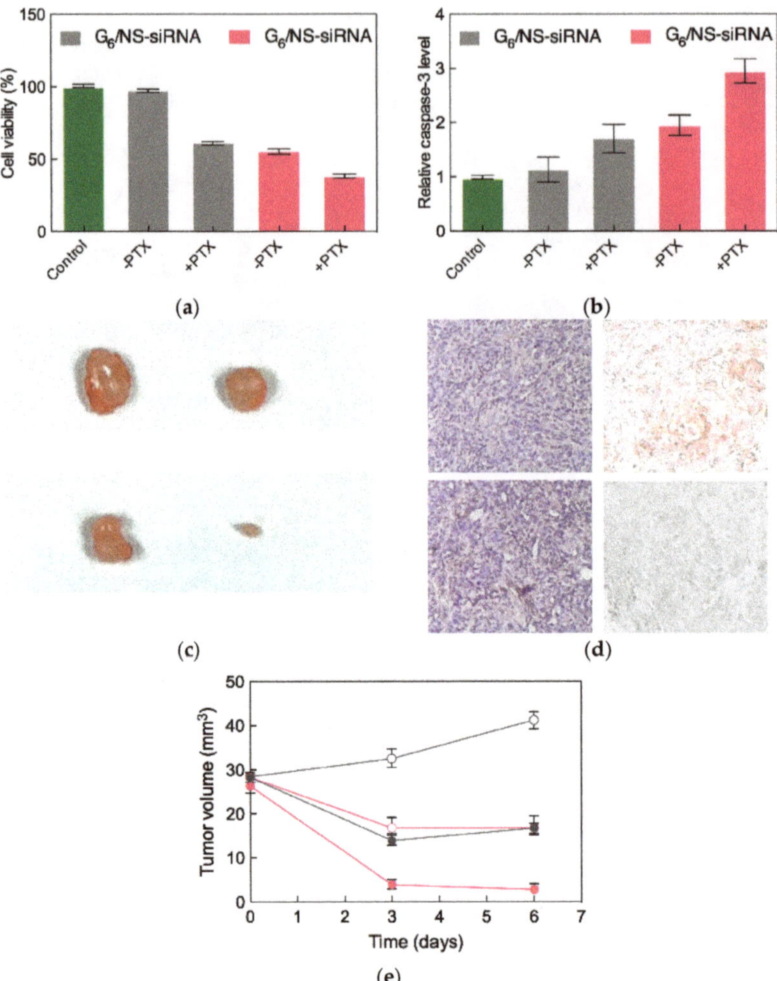

Figure 9. SKOV-3 cell viability (**a**) and relative caspase-3 activation (**b**) after transfection with non-specific (NS) siRNA or AKT siRNA delivered by the G_6 TEA-core dendrimer nanovectors (N/P = 5) alone or in combination with paclitaxel (100 nM). Non-treated cells and β-actin were used as respective controls. (**c**) Tumor volumes from SKOV-3 xenografted mice treated with non-specific (NS) siRNA (top panel, left) or AKT siRNA delivered by the G_6 TEA-core dendrimer nanovectors (N/P = 5) alone (top panel, right) or in combination with paclitaxel (100 nM, bottom panel). (**d**) Drastic reduction of AKT levels in tumor xenografts injected with AKT siRNA G_6 TEA-core dendriplexes (**bottom, left**), and the corresponding histological sample showing sign of necrosis (**bottom, right**), compared with xenografts treated with NS siRNA delivered with the same nanovectors showing no reduction of AKT levels (**top, left**) and no necrosis (top, right). (**e**) Tumor volume during combined treatment of AKT siRNA G_6 TEA-core dendriplexes/paclitaxel (filled hot pink circles) of SKOV-3 mice xenografts, compared with nanodelivered AKT siRNA (open hot pink circles) or paclitaxel alone (filled gray circles). Data for nanodelivered NS siRNA are shown for control (open gray circles). Adapted from [47], which is an open access article published under an ACS AuthorChoice License.

Finally, the same assays were performed in tandem with paclitaxel administration via mice intraperitoneal injection. The corresponding xenografts reveal a remarkable cooperative action of the two treatments with respect to the chemotherapeutic drug per se (Figure 9c, lower panel), with a reduction of tumor growth of 85% compared to the treatment with paclitaxel alone (Figure 9f) while the G_6 TEA-core dendrimer or the AKT siRNA alone have no effect (data not shown). To the best of our knowledge, this constitutes the first study documenting a potentiated anticancer effect of paclitaxel while co-administered with AKT siRNA mediated by dendrimer nanovectors to ovarian cancer both in vitro and in vivo.

3.3. Low-Generation TEA-core PAMAM Dendrimers as Effective In Vitro and In Vivo siRNA Nanocarriers

Though the high-Generation G_6 and G_7 TEA-core dendrimers were very effective in delivering siRNA molecules to various cancer models in vitro and in vivo, the large-scale chemical synthesis of these molecules required for their clinical applications is laborious and very time/resource consuming (e.g., extensive dendrimer purification is intrinsically difficult, being hampered by the presence of highly similar side-products) [53]. These issues imply that the good manufacturing practice (GMP) claimed for products expected to undergo clinical trials is technically very challenging. Therefore, finding a way to endow lower generation dendrimers with effective and efficient siRNA delivery could be a worthy goal per se, as discussed below.

3.3.1. Functional Delivery of Sticky siRNA

In 2007, Jean Paul Behr and his group showed that small interfering RNAs bearing short complementary A_n/T_n (n = 5–8) sticky overhangs delivered using polyethyleneimine (PEI) result in enhanced gene silencing with respect to standard siRNAs [54]. Since PEI is one of the best non-viral DNA carriers but its efficiency drops dramatically during siRNA transfection [55], its ability to quantitatively deliver sticky siRNAs could be attributed only to the presence of the complementary A_n/T_n overhangs. According to Behr's explanation, the latter can self-assemble into gene-like, long double-stranded RNA, and this in turn allows for the successful cellular delivery by PEI by virtue of a mechanism utterly similar to that governing plasmid DNA transfection. On the other hand, we reasoned that an additional contributing factor to the enhanced gene silencing of nano-delivered ssiRNAs could also be related to the inherent flexibility of the terminal, single-strand nucleic acid fragments; this might allow them to behave as clamps that, just like protruding molecular arms, can better enwrap and tightly hold the nanovector, thereby enhancing binding and, ultimately, delivery. Accordingly, we set on to verify these concepts with the ultimate purpose of exploiting low generation TEA-core PAMAM dendrimers as efficient nanocarriers in RNAi.

3.3.2. In Vitro Preliminary Data of Sticky siRNA Delivery by Lower Generation TEA-Core PAMAMs

Under this perspective, we initially constructed two sticky siRNA molecules with complementary A_5/T_5 and A_7/T_7 3′-overhangs to target Hsp27 and TCTP (a highly conserved protein present in all eukaryotic organisms that regulate cell survival in human tumors) in prostate and breast cancer models [56,57]. In parallel, we also synthesized four additional non-complementary ssiRNAs (i.e., A_5/A_5, A_7/A_7, T_5/T_5, and T_7/T_7) to investigate the effect of chemistry, length and non-complementarity on the relevant gene silencing potential with respect to the standard (A_2/T_2) siRNAs discussed above [56,57].

Next, we preliminarily tested our TEA-core dendrimers from Generation 4 to Generation 7 for their ability to deliver complementary ssiRNAs in PC-3 cells. Figure 10a shows that, contrarily to conventional A_2/T_2 siRNA delivery, for which Generation 6 or (better) 7 dendrimers were requested to achieve biological effects (see § 3.2, [42,47]), a significant gene silencing can be achieved in PC-3 cells starting from the G_5 nanocarriers, the A_7/T_7 ssiRNA being somewhat more effective than the A_5/T_5 counterpart for all nanovector generations (Figure 10a). Since G_5 was the lowest and most effective dendrimer generation for the in vitro delivery of ssiRNAs, further investigations were carried out using this nanovector. In particular, the best, long-term gene silencing was obtained with the G_5 TEA-core assisted delivery of 50 nM ssiRNAs at N/P ratio = 10, as highlighted in Figure 10b.

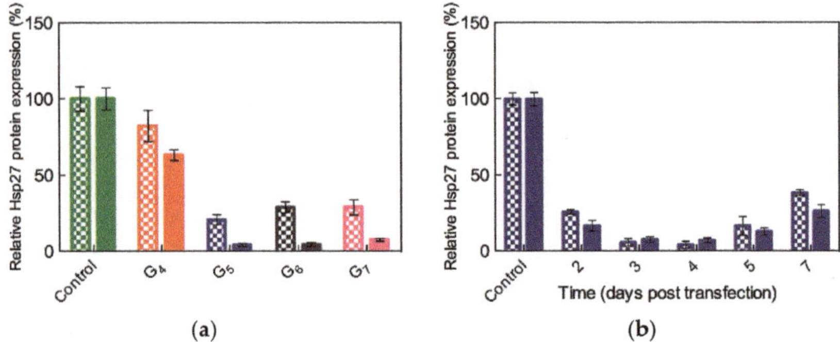

Figure 10. (**a**) Hsp27 gene silencing upon delivery of complementary ssiRNAs mediated by different generations of TEA-core PMAMAM to PC-3 cells. Checked bars: Data for A_5/T_5 ssiRNA; solid bars: Data for A_7/T_7 ssiRNA. In these experiments, vinculin was used as reference and non-treated cells were used for control. (**b**) Long-term Hsp27 silencing achieved with A_5/T_5 and A_7/T_7 ssiRNAs (50 nM) delivered by G_5 TEA-core PAMAMs at N/P = 10. Redrawn from [56], with permission of the American Chemical Society.

3.3.3. *In Silico* Binding Affinity of ssiRNAs with G_5 TEA-Core Dendrimer Nanovectors

Before embarking in further time- and resource-consuming experimental investigations, we performed MD simulations to predict and understand if and how the different ssiRNA overhangs could impact G_5 TEA-core mediated delivery. The *in silico* investigation started by verifying our own hypothesis, according to which the protruding, flexible overhangs could promote a better interaction and stronger binding of monomeric ssiRNAs to their dendrimeric nanocarriers via molecular dynamics simulations. The MD results are summarized in Figure 11a (see Table A1 in Appendix A for full MD results), while some exemplificative images extracted from the corresponding equilibrated MD trajectories are shown in Figure 12.

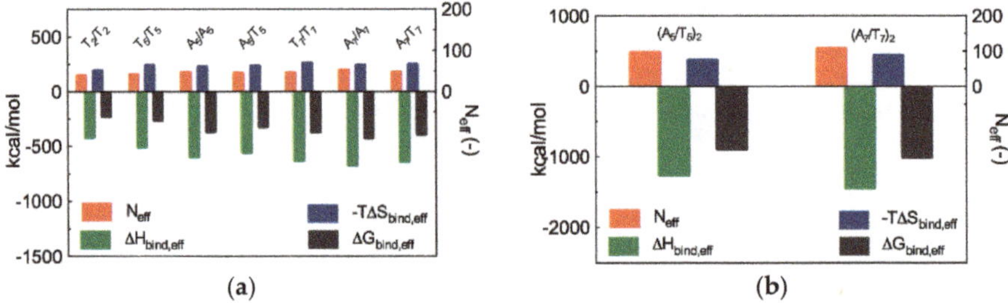

Figure 11. Total effective free energy ($\Delta G_{bind,eff} = \Delta H_{bind,eff} - T\Delta S_{bind,eff}$), enthalpic ($\Delta H_{bind,eff}$), and entropic ($-T\Delta S_{bind,eff}$) components for the binding of (**a**) ssiRNAs featuring complementary and non-complementary overhangs of different length and (**b**) dimeric ssiRNAs with the G_5 TEA-core PAMAM dendrimer. N_{eff} is the number of effective dendrimer positive charges involved in nucleic acid binding (see Tables A1 and A2 in Appendix A and text for more details). (**b**) Redrawn from [57], with permission of the American Chemical Society.

The computer simulations reveal that both the nature of the overhangs and their length influence the interaction of the relevant ssiRNAs with the G_5 dendrimer nanocarrier. For the first aspect, the first three columns in Table A1 show that ssiRNAs with A_n/A_n overhangs are characterized by the most favorable free energy of binding values (ΔG_{bind} = −409.9 kcal/mol for A_5/A_5 and −447.9 kcal/mol for A_7/A_7), followed by the ssiRNAs bearing complementary overhangs (ΔG_{bind} = -387.4 kcal/mol for A_5/T_5 and -422.9 kcal/mol for A_7/T_7), and, last, by the ssiRNAs with T_n/T_n overhangs, which are characterized by the lowest nanovector affinity (ΔG_{bind} = −316.8 kcal/mol for T_2/T_2, −344.8 kcal/mol for T_5/T_5, and −402.2 kcal/mol for T_7/T_7, respectively). Concerning the second aspect, these data clearly indicate that the presence of longer overhangs enhances nanovector/nucleic acid binding, the ssiRNA with the shorted overhangs T_2/T_2 being the one with the lowest ΔG_{bind} value in the entire series.

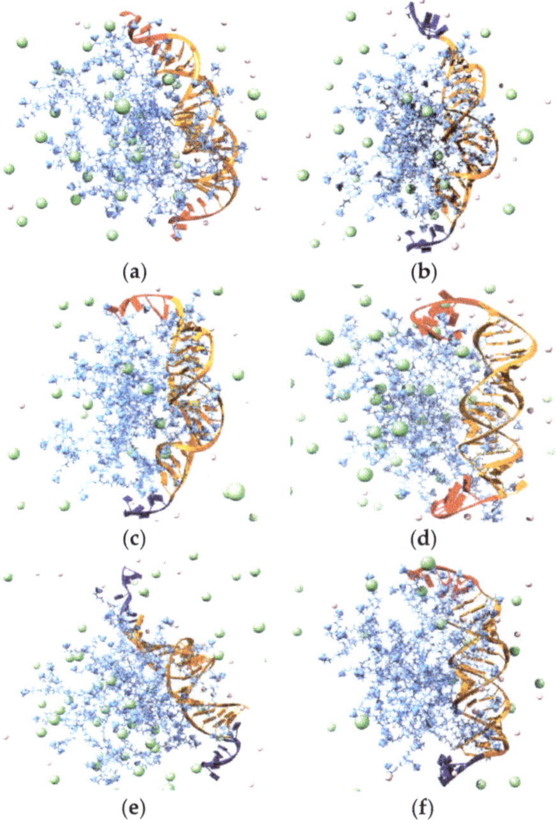

Figure 12. Examples of equilibrated molecular dynamics (MD) snapshots of G_5 TEA-core dendrimer in complex with A_5/A_5 (**a**), T_5/T_5 (**b**), A_5/T_5 (**c**), A_7/A_7 (**d**), T_7/T_7 (**e**), and A_7/T_7 (**f**) ssiRNAs at pH 7.4 and in the presence of 0.15 M NaCl. In all panels, the dendrimer is shown as cornflower blue sticks, and the terminal charged amine groups are highlighted as sticks-and-balls. The ssiRNA is portrayed as an orange ribbon, with the two overhangs (A_n) and (T_n) colored in red and navy blue, respectively. Some Cl^- and Na^+ ions and counterions are shown as light green and light pink spheres, respectively. Water molecules are not shown for clarity. Redrawn from [57], with permission of the American Chemical Society.

Further interesting data were obtained by calculating the effective free energy of binding ($\Delta G_{bind,eff}$), i.e., the specific energetic contribution to ssiRNA/nanocarrier complex formation afforded only by those dendrimer branches in constant and productive interaction with its nucleic acid cargo. An analysis of each ssiRNA/dendrimer nanoassembly MD simulation allowed us to precisely identify and quantify these dendrimer residues (N_{eff}); next, a per residue decomposition of the total binding free energy (see Supporting Information for details) led to the corresponding values of $\Delta G_{bind,eff}$ (Figure 11a and Table A1). The first notable result regards N_{eff}: Indeed, not only the smallest number of dendrimer branches involved in nanovector binding (38) pertains to the nucleic acid fragment with the shortest overhangs (T_2/T_2), but also the values of N_{eff} follow a clear increasing trend from T_n/T_n to A_n/A_n to A_n/T_n ssiRNAs. Concomitantly, the corresponding $\Delta G_{bind,eff}$ values become more favorable (i.e., more negative) in the same order.

The deconvolution of $\Delta G_{bind,eff}$ in its enthalpic ($\Delta H_{bind,eff}$) and entropic ($-T\Delta S_{bind,eff}$) components (Figure 11a and Table A1) reveals that the nanovector/siRNA interaction is prevalently enthalpic in nature, although entropic effects linked to the released of ions, counterions, and water molecules in the bulk solvent upon complex formation also contribute in modulating the individual intermolecular affinities. Thus, taking the ssiRNAs series bearing five-nucleotide long overhangs as an example, it is easily seen that the A_5/A_5 ssiRNA has both the most favorable enthalpic contribution ($\Delta H_{bind,eff} = -592.0$ kcal/mol) and the least unfavorable entropic term ($-T\Delta S_{bind,eff} = 227.2$ kcal/mol) in the homologous series. The best results for the A_5/A_5 ssiRNA/G_5 TEA core PAMAM complex can be rationalized, from the enthalpic viewpoint, by taking into account the high number of favorable electrostatic interactions, supported by the greatest value of N_{eff} for this homologous series (46), along with other non-bonded, stabilizing contacts between the nanovector and the nucleic acid, including its overhangs (see Figure 12a). From the entropic perspective, the more rigid nature and the enhanced clamping propensity of the A-based overhangs with respect to the T-based ones translate into more permanent and effective contacts between the full nucleic acid fragments (overhangs included) and the positively charged dendrimer terminal groups (Figure 12a). The least performing nanoassembly in this series, i.e., the one involving the T_5/T_5 ssiRNA, is characterized by the smaller value of N_{eff} (41), the lowest enthalpic variation ($\Delta H_{bind,eff} = -503.1$ kcal/mol) and the most unfavorable entropic component ($-T\Delta S_{bind,eff} = 241.1$ kcal/mol). This latter term is quickly understood, considering the remarkable flexibility of the T_n protruding overhangs, which fluctuate in the solvent for most of the time of the corresponding MD trajectory (Figure 12b). When interacting with the nanovector terminal groups, these overhang molecular movements are frozen, resulting in a considerable loss of degrees of freedom and, hence, entropic penalty. As a further effect, the corresponding ssiRNA/nanocarrier opposite-charge contacts are suboptimal, and this directly reflects in a decrease of $\Delta H_{bind,eff}$. The ssiRNA with complementary overhangs A_5/T_5 exhibits an intermediate behavior ($-\Delta H_{bind,eff} = -554.9$ kcal/mol and $-T\Delta S_{bind,eff} = 233.4$ kcal/mol), stemming from a compensatory effect between the rigid A_n arm (with high binding tendency) and the springy T_n arm (endowed with less efficient dendrimer clasping propensity), as illustrated in Figure 12c. An utterly analogous situation—governed by the same molecular factors described above—is observed for the homologous ssiRNA series bearing longer overhangs, for which the affinity towards the G_5 TEA-core dendrimer increases in the order $A_7/A_7 > A_7/T_7 > T_7/T_7$ (Figures 11 and 12d–f, Table A1). These results allowed us to draw some general considerations about the effect of nature and lengths of the ssiRNA overhangs on their interaction with the G_5 TEA-core dendrimer, as follows. First, longer overhangs are more beneficial to nanovector/ssiRNA interactions than shorter ones by virtue of the higher number of dendrimer residues (N_{eff}) in permanent and efficient contact with the nucleic acid fragment. In addition, for a given length of non-complementary overhangs, the more flexible nature of the T_n sequence is detrimental to nanoassembly formation with respect to the alternative A_n strand, since the relevant less-optimized nanoparticle structure and the larger entropic penalty paid upon dendrimer/ssiRNA complex formation result in a lower affinity of the nucleic acid fragment for its nanovector.

The next part of the *in silico* investigation was devoted to verify the second hypothesis, according to which ssiRNAs could self-assemble into gene-like structures via the formation of hybrid bridges between the complementary overhang sequences and, in doing so, enhance their affinity for nanovectors. In his original work [54], Behr already showed that this oligomerization or concatenation process enhanced cooperative and multivalent PEI/A_8/T_8 ssiRNA interactions, thereby leading to better delivery efficiency. Most importantly, however, since no concatemers were detected in the absence of nanovectors (i.e., PEI or G_5 TEA-core PAMAM dendrimers), we further reasoned that the nanocarriers themselves must play an active role in directing encounters between individual ssiRNA/nanovector complexes, thus promoting complementary overhang concatemerization. To assess these concepts, we performed further MD simulations on G_5 TEA-core PAMAMs in complex with two dimeric ssiRNAs, $(A_5/T_5)_2$ and $(A_7/T_7)_2$ (see Figure 13a,b). The computational results are graphically reported in Figure 11b (and numerically listed in Table A2). When comparing these data with those relative to the monomeric ssiRNAs (i.e., A_5/T_5 and A_7/T_7, Figure 11a and Table A1), some important information can be immediately appreciated. First, the number of nanovector-charged branches in productive contact with the nucleic acid are larger than twice the sum of the value predicted for the analogous monomeric nanoassemblies (i.e., N_{eff} = 96 for $(A_5/T_5)_2$ and (2 × 44) = 88 for A_5/T_5, and N_{eff} = 107 for $(A_7/T_7)_2$ and (2 × 47) = 94 for A_7/T_7, respectively, Table A3). This enhancement of stabilizing intermolecular contacts for the concatenated systems can be ascribed to the presence of the extra double-stranded portion of the hybridized ssiRNAs (Figure 13a,b), which, being more rigid and globally more negatively charged than the single stranded overhangs, induces a further conformational adaptation of the dendrimer terminal units to accommodate a larger number of favorable electrostatic interactions.

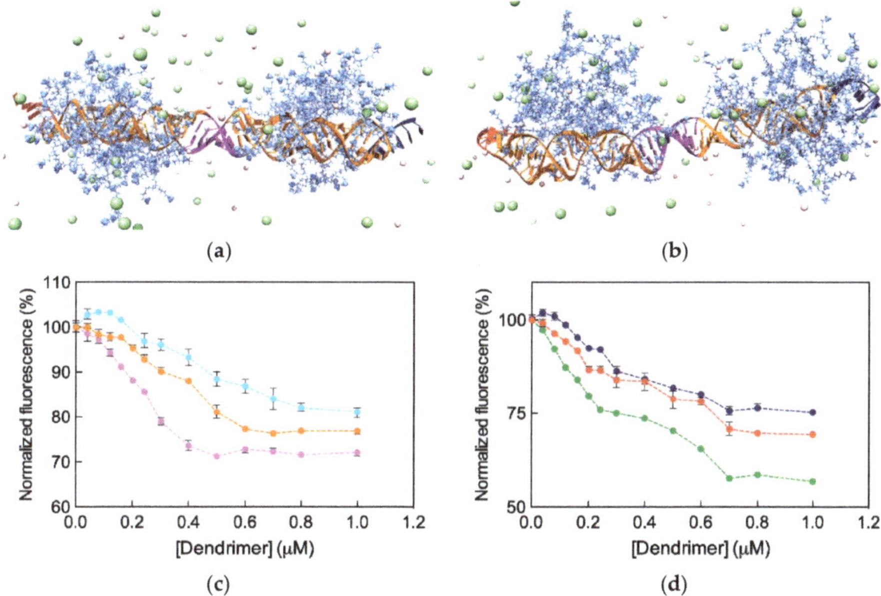

Figure 13. Equilibrated MD snapshots of the $(A_5/T_5)_2$ (**a**) and $(A_7/T_7)_2$ (**b**) dimeric ssiRNAs in complex with the G_5 TEA-core dendrimer pH 7.4 and in the presence of 0.15 M NaCl. Molecule representations and color scheme as in Figure 12. The double-stranded portion of the concatenated (hybridized) ssiRNAs is highlighted in purple. (**c**) and (**d**) Experimental binding of ssiRNAs bearing complementary and non-complementary overhangs with the G5 TEA-core dendrimer by ethidium bromide (EB) displacement assay. Color legend: (**c**) Light blue, T_5/T_5 ssiRNA; orange, A_5/A_5 ssiRNA; light purple, A_5/T_5 ssiRNA; dark blue; (**d**) T_7/T_7 ssiRNA; red, A_7/A_7 ssiRNA; light green, A_7/T_7 ssiRNA. Adapted from [57], with permission of the American Chemical Society.

The synergistic effect of ssiRNA concatemerization is also evident in the corresponding binding thermodynamics. Indeed, both $\Delta G_{bind,eff}$ and $\Delta H_{bind,eff}$ for the hybridized ssiRNAs are more favorable than two times the corresponding values for the monomeric ssiRNAs, while the decrease in the entropic contributions ($-T\Delta S_{bind,eff}$) for the dimeric ssiRNAs is less disfavoring for the former systems with respect to twice the values for the latter ones, as summarized in Table A3.

All these data indeed provide a computational support to the idea that dimeric ssiRNAs generated by nucleic acid fragments bearing complementary overhangs which might hybridize into a central $(A_n/T_n)_2$ double-stranded portion result in a synergistic binding with the G_5 TEA-core dendrimer nanovector with respect to siRNAs characterized by both short and/or non-complementary overhangs. Pleasingly, these theoretical predictions were confirmed by ethidium bromide (EB) displacement fluorescence spectroscopy assays (Figure 13c,d), according to which the experimental binding affinity of the different ssiRNAs for the dendrimer follows exactly the same order anticipated by simulations, that is: $A_n/T_n > A_n/A_n > T_n/T_n$.

The final step of the *in silico* study concerned another fundamental aspect in nanovector-assisted effective siRNA delivery and gene silencing—the disassembly of the nanocomplexes in the cellular cytoplasm to make the nucleic acid cargo available to the RNAi machinery. To the purpose, advanced computational techniques based on steered molecular dynamics (SMD) simulations were applied to the four complexes formed by ssiRNAs bearing non-complementary overhangs, as well as by the standard (T_2/T_2) siRNA and the G_5 TEA-core dendrimer (see Supplementary Materials)). Briefly, during SMD runs, each siRNA molecule was drifted away from its nanocarrier using a constant pulling speed, and the behavior of the force require to break the corresponding complexes was recorded as a function of time. The results from SMD simulations are shown in Figure 14a, from which it is seen that the peak force that needs to be exerted to dissociate the nucleic acids fragments from their nanocarrier increases in the order: 730 pN for T_2/T_2, 753 pN for T_5/T_5, 794 pN for A_5/A_5, 824 pN for T_7/T_7, and 862 pN for A_7/A_7. If the nanovector/ssiRNA disassembly force is plotted against the corresponding effective formation free energy ($\Delta G_{binf,eff}$) value (Figure 11a and Table A1), a linear relationship is obtained (Figure 14b, $R^2 = 0.95$), indicating that the tighter the ssiRNA/dendrimer binding, the stronger the force required to disassemble the corresponding complex. In other words, a very high affinity between nanocarrier and cargo, although useful for protection and transport, will ultimately be detrimental to the final step, i.e., efficient release; accordingly, the ideal system must represent the best compromise among these counteracting effects.

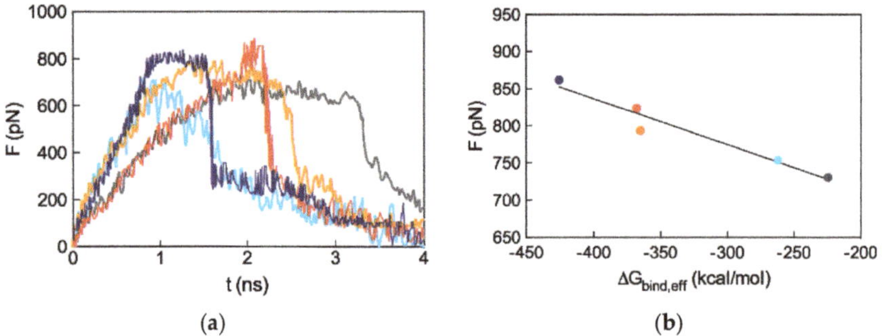

Figure 14. (**a**) Profiles of the average force required to unbind ssiRNAs from their G_5 TEA-core dendrimer nanovectors as obtained from steered molecular dynamics (SMD) simulations. Color legend: Dark blue, (T_7/T_7) ssiRNA; light blue, (T_5/T_5) ssiRNA; gray, (T_2/T_2) (i.e., non-sticky) siRNA; red, (A_7/A_7) ssiRNA; orange, (A_5/A_5) ssiRNA. (**b**) Relationship between the SMD peak force and the corresponding effective free energy of binding $\Delta G_{bind,eff}$ for the corresponding ssiRNA and the G_5 dendrimers. Redrawn from [57], with permission of the American Chemical Society.

3.3.4. In Vitro Delivery of ssiRNAs with G_5 TEA-Core Dendrimer Nanovectors

The uptake of ssiRNA/G_5 TEA-core PAMAM dendriplexes by PC-3 cells was first verified using live-cell confocal microscopy that confirmed both efficient internalization and cytoplasmic localization of the nucleic acid-loaded nanocarriers. Since the mechanism presiding cellular uptake of nanoparticles can involve several pathways, including macropinocytosis, clathrin-mediated endocytosis, and caveolae-mediated endocytosis [58], the mechanism of uptake was investigated using specific inhibitors and biomarkers of various endocytic pathways. As an example, Figure 15a shows that a meaningful reduction of G_5 TEA core dendrimer/A_5/T_5 ssiRNA complexes was achieved only in the presence of the macropinocytosis inhibitor cytochalasin D, while only very weak effect was obtained in the presence of the two alternative inhibitors, that is genistein (an inhibitor of caveolae-mediated endocytosis) and chlorpromazine, a clathrin-mediated uptake specific blocker. In addition, a robust colocalization of the Alexa 647-labelled nanoparticles with dextran (a prototypic macropinocytosis biomarker) was observed, while minor-to-moderate colocalization was evidenced using either the transferrin of cholera toxin B (biomarkers for clathrin- and caveolae-mediated endocytosis) (Figure 15b).

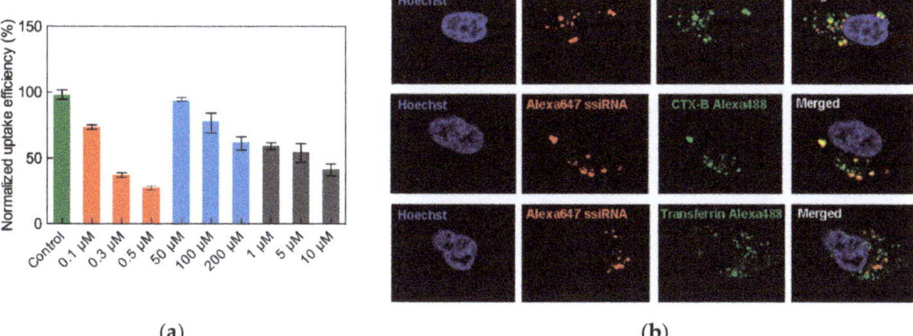

Figure 15. (**a**) Effect of cytochalasin D (a macropinocytosis inhibitor, red bars), genistein (a caveolae-mediated endocytosis inhibitor, light blue bars), and chlorpromazine (a clathrin-mediated endocytosis inhibitor, gray bars) on the uptake of Alexa 647-labelled A_5/T_5 ssiRNA/G_5 TEA-core dendrimer nanoparticles by PC-3 cells. Values are normalized to Alexa 647-labeled ssiRNA/G_5 TEA-core dendrimer nanoparticles uptake in the absence of any inhibitor. (**b**) Colocalization of the Alexa 647-labelled A_5/T_5 ssiRNA/G_5 TEA-core dendrimer nanoparticles with different endocytosis biomarkers: Top panel, dextran (macropinocytosis biomarker); middle panel, cholera toxin B (caveolae-mediated endocytosis biomarker); bottom panel, transferrin (clathrin-mediated endocytosis biomarker).

Based on all results discussed above, the ssiRNAs A_n/T_n, A_n/A_n, and T_n/T_n (n = 5 or 7) were selected for further in vitro experiments. Taking again the Hsp27 as the target gene in different cancer cell lines, the efficiency and specificity of the gene silencing effect were evaluated both at the mRNA and protein levels [57]. Figure 16 illustrates some of the results obtained in these tests.

As seen in this figure, the G_5 TEA-core dendrimer/non-sticky siRNA (A_2/T_2) assembly confirmed its inability to elicit gene silencing in all cell lines, while, in agreement with computational predictions, a substantial effect (approximately 90%) was achieved with ssiRNAs bearing complementary overhangs (i.e., A_n/T_n). Moreover, again in line with *in silico* results, ssiRNAs ending with non-complementary sequences can induce up to 70% gene silencing, as also seen in Figure 16. Specifically, for these systems, as revealed by simulations, if, on the one side, a higher overhang rigidity leads to stronger nanovector/ssiRNA interaction, on the other side, longer and more flexible overhangs are more beneficial for the subsequent nucleic acid delivery. The results from all *in silico* investigations described above unambiguously supported the conclusion that, among all ssiRNAs with non-complementary overhangs we synthesized to empower low-generation TEA-core PAMAMs with effective delivery capacity, those bearing A_7/T_7 terminals represent the best compromise in terms of both nanovector binding and unbinding ability.

Before moving the A_7/T_7 ssiRNA/G_5 TEA-core dendrimer complexes to in vivo tests, we further investigated their anticancer effects resulting from Hsp27 silencing in PC-3 cells. Giving the previous results obtained with the delivery of siRNA by the TEA-core dendrimer of Generation 6 and 7 discussed above, we expected this smaller PAMAM to be devoid of toxicity and its complexes with the Hsp27-targeting ssiRNAs to significantly suppress cell growth via a caspase-induced apoptosis. Indeed, as shown in Figure 17a, a notable inhibition of cell proliferation was detected after delivering

A_7/T_7 ssiRNA with the G_5 TEA-core nanovector with respect to non-treated cells, cells treated with the G_5 dendrimer alone, or with a scrambled (non-silencing) ssiRNA.

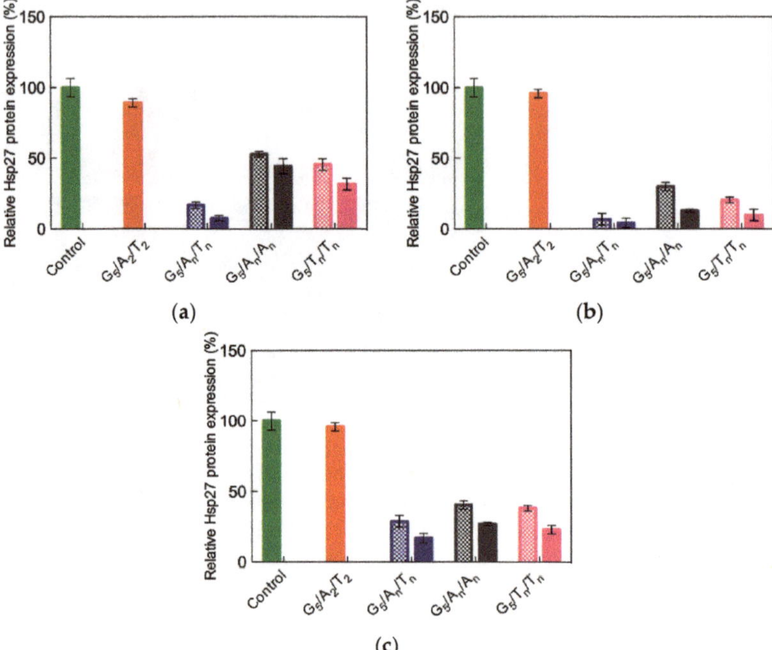

Figure 16. Hsp27 gene silencing upon delivery of different ssiRNAs (50 nM) mediated by G_5 TEA-core PAMAM (N/P = 10) to PC-3 cells (**a**), MDA-MB-231 cells (**b**), and MCF-7 cells (**c**). Checked bars: Data for ssiRNA with n = 5; solid bars: Data for ssiRNA with n = 7. In these experiments, vinculin was used as reference, and non-treated cells were used for control (green solid bar). Data for non-sticky siRNA (A_2/T_2) are also shown for comparison (red solid bar). MDA-MB-231 and MCF-7 are two different breast cancer cell lines. Redrawn from [57], with permission of the American Chemical Society.

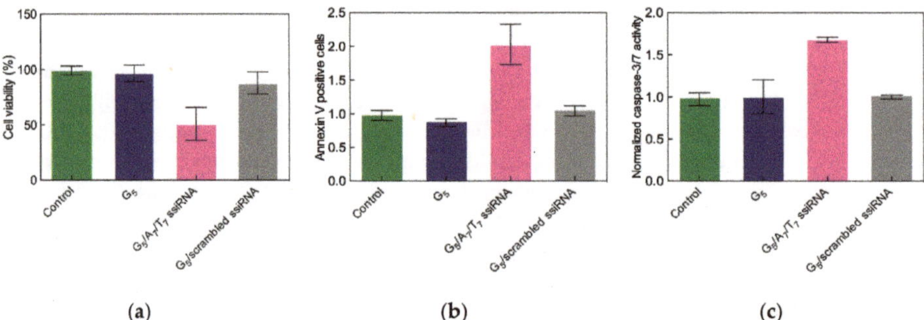

Figure 17. (**a**) Cell proliferation, (**b**) apoptosis, and (**c**) caspase 3/7 activity in PC-3 cells treated with A_7/T_7 ssiRNAs (50 nM) delivered by G_5 TEA-core PAMAM (N/P = 10). Non-treated cells, the G_5 dendrimer alone and a scrambled (non-silencing) ssiRNA sequence were used for control. Data in panels (**a**) and (**c**) were measured as described in Figure 6. The apoptotic index was measured with fluorescence-activated cell sorting (FACS) flow cytometry by the annexin V assay four days after treatment. Redrawn from [56], with permission of the American Chemical Society.

Concomitantly, fluorescence-activated cell sorting (FACS) flow cytometry revealed a considerable increase of annex V-positive apoptotic cells paralleled with the relevant activation of the apoptotic Caspases 3 and 7. Finally, no toxicity mediated by the ssiRNA/dendrimer complexes was observed via further MTT and lactate dehydrogenase assays, supporting the potential for in vivo experiments with the A_7/T_7 ssiRNA/G_5 TEA-core dendrimer nanoparticles.

3.3.5. In Vivo Delivery of ssiRNAs with G_5 TEA-Core Dendrimer Nanovectors

The final part of the study concerned the evaluation of in vivo gene silencing by the in vitro most efficient ssiRNA/nanovector system. Accordingly, a prostate cancer PC-3 xenografted mouse model was adopted, to which the $A_7/T_7/G_5$ TEA core dendrimer nanoparticles were slowly administered via slow intratumoral injection. Treatment lasted one week, during which the mice survived well, showing no sign of induced toxicity or weight loss. After mice sacrifice, the expression of Hsp27 in the tumors was measured, as shown in Figure 18a,b. A significant downregulation of Hsp27 at both the mRNA and protein levels was observed, compared to all controls, confirming that the ssiRNA delivered by the dendrimer nanovector was able to elicit potent and specific RNAi also in vivo.

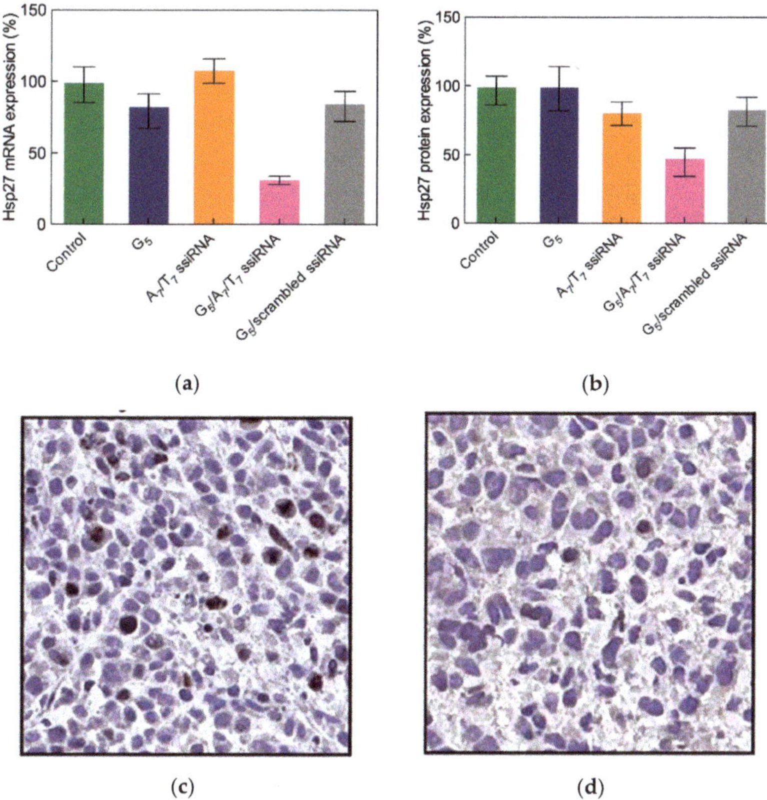

Figure 18. In vivo downregulation of Hsp27 at both mRNA (**a**) and protein (**b**) levels achieved after treating PC-3 cell xenografted nude mice with intratumoral injection of Hsp27 A_7/T_7 ssiRNA/G_5 complex, buffer solution (control), the dendrimer G5 alone, the A7/T7 ssiRNA alone and a scrambled (non-silencing) ssiRNA sequence/G_5 complex (all used as negative controls). (**c**) Evaluation of tumor cell proliferation via immunohistochemistry using Ki-67 staining after treatment with a scrambled ssiRNA sequence/G_5 (left) and the Hsp27 A_7/T_7 ssiRNA/G_5 complexes (right). Adapted from [56], with permission of the American Chemical Society.

Apoptotic caspase activation was also detected only in mice treated with the nanodelivered ssiRNA, and immunohistochemistry images obtained with Ki-67 antibody staining finally confirmed the remarkable inhibition of cell proliferation in the treated animals (Figure 18c,d).

4. Conclusions

In the last ten years, the number of studies involving dendrimers as safe, efficient and effective nanovectors for drug and nucleic acid delivery have increased exponentially. This is mainly due to the exquisite properties of these hyperbranched molecules which, by virtue of their nanoscale size, regularly repeating structure and functional surface groups, make them an ideal drug delivery platform. PAMAM dendrimers in particular bear primary amine groups on their periphery which, being positively charged at physiological pH (7.4), can aptly condense negatively charged nucleic acids for efficient gene or siRNA delivery. In addition, this class of dendrimers features tertiary amines in their interior which become protonated at endosomal pH (5.5), thereby promoting the so-called proton sponge effect and the subsequent release of their DNA/siRNA cargo in the cell cytoplasm.

During the same decade, our group has been particularly active in the field of design and optimization of PAMAM-based dendrimers for siRNA delivery. In particular, we designed, synthesized, and tested highly flexible triethanolamine-core PAMAM dendrimers which proved to be highly effective for siRNA delivery in cancer therapeutics both in vitro and in vivo, as discussed in this brief review. Based on its successful performance, the G_5 TEA-core PAMAM dendrimer was scheduled to enter clinical trials for siRNA-based cancer therapy in 2014; unfortunately, however, due to the unavailability of GMP dendrimer material, the foreseen clinical trial was delayed and ultimately replaced by the use of Smarticles® for the delivery of siRNA therapeutics [59].

Since the GPM production of dendrimer is quite a challenging process, we decided to exploit the quintessence of nanotechnology, i.e., the controlled self-assembly of small, synthetically amenable building blocks to generate nanosystems for siRNA delivery. Accordingly, we designed, synthesized, and tested amphiphilic dendrons which, upon auto-organization into micelles, were able to mimic the covalent, high generation dendrimers in size, structure and function—in particular for siRNA delivery. These exciting self-assembled nanovectors will be the subject of the second part of this review work.

Supplementary Materials: The following are available online at http://www.mdpi.com/1999-4923/11/7/351/s1, Table S1, Computational details.

Funding: This research was funded by the Italian Association for Cancer Research (AIRC), grant IG17413 to SP. The assistant professor position (RTDa) of SA is fully supported by the University of Trieste, in agreement with the actuation of the strategic planning financed by the Italian Ministry for University and Research (MIUR, triennial program 2016–2018) and the Regione Friuli Venezia Giulia (REFVG, strategic planning 2016-18), assigned to SP. This award is deeply acknowledged.

Acknowledgments: Authors wish to thank Ling Peng and her group for the longstanding, fruitful collaboration, the challenges in siRNA delivery nanovector design and optimization, the inspiring discussions and, above all, the personal friendship.

Conflicts of Interest: The authors declare no conflict of interest.

Appendix A

Table A1. Total and effective enthalpy (ΔH_{bind}, $\Delta H_{bind,eff}$), entropy ($-T\Delta S_{bind}$, $-T\Delta S_{bind,eff}$), and free energy of binding ($\Delta G_{bind} = \Delta H_{bind} - T\Delta S_{bind}$, $\Delta G_{bind,eff} = \Delta H_{bind,eff} - T\Delta S_{bind,eff}$) for ssiRNAs featuring complementary and non-complementary overhangs of different length with G_5 TEA-core PAMAM dendrimers. All values are in kcal/mol (standard deviation for all data in Table A1 is less that 3%). N_{eff} is the number of positive dendrimer charges effectively involved in ssiRNA binding. Adapted from [57] with permission of the American Society of Chemistry.

Overhangs	ΔH_{bind}	$-T\Delta S_{bind}$	ΔG_{bind}	N_{eff}	$\Delta H_{bind,eff}$	$-T\Delta S_{bind,eff}$	$\Delta G_{bind,eff}$
T_2/T_2	−571.1	254.3	−316.8	38	−415.6	190.7	−224.9
T_5/T_5	−609.9	265.1	−344.8	41	−503.1	241.1	−262.0
A_5/A_5	−659.7	249.8	−409.9	46	−592.0	227.2	−364.8
A_5/T_5	−637.4	250.0	−387.4	44	−554.9	233.4	−321.5
T_7/T_7	−678.4	276.2	−402.2	45	−626.7	258.9	−367.8
A_7/A_7	−714.8	266.9	−447.9	52	−669.1	243.6	−425.5
A_7/T_7	−690.2	267.3	−422.9	47	−637.3	248.2	−389.1

Table A2. Total and effective enthalpy (ΔH_{bind}, $\Delta H_{bind,eff}$), entropy ($-T\Delta S_{bind}$, $-T\Delta S_{bind,eff}$), and free energy of binding ($\Delta G_{bind} = \Delta H_{bind} - T\Delta S_{bind}$, $\Delta G_{bind,eff} = \Delta H_{bind,eff} - T\Delta S_{bind,eff}$) for the two dimeric ssiRNAs with G_5 TEA-core PAMAM dendrimers. All values are in kcal/mol (standard deviation for all data in Table A2 is less that 3%). N_{eff} is the number of positive dendrimer charges effectively involve in ssiRNA binding. Adapted from [57] with permission of the American Society of Chemistry.

Overhangs	ΔH_{bind}	$-T\Delta S_{bind}$	ΔG_{bind}		$\Delta H_{bind,eff}$	$-T\Delta S_{bind,eff}$	$\Delta G_{bind,eff}$
$(A_5/A_5)_2$	−1382.7	407.3	−975.4	96	−1260.3	372.7	−887.6
$(A_7/T_7)_2$	−1480.4	455.9	−1024.5	107	−1441.2	437.2	−1004.0

Table A3. Synergistic effect of the dimeric ssiRNA concatemerization on their G_5 TEA-core PAMAM dendrimer effective binding thermodynamics with respect to the corresponding monomeric ssiRNAs. Data from Tables A1 and A2. All values are in kcal/mol.

	$(A_5/T_5)_2$	$2 \times (A_5/T_5)$	$(A_7/T_7)_2$	$2 \times (A_7/T_7)$
N_{eff}	96	88	107	94
$\Delta H_{bind,eff}$	−1260.3	−1109.8	−1441.2	−1274.6
$-T\Delta S_{bind,eff}$	372.7	446.8	437.2	496.4
$\Delta G_{bind,eff}$	−887.6	−663.0	−1004.0	−778.2

References

1. Fire, A.; Xu, S.; Montgomery, M.K.; Kostas, S.A.; Driver, S.E.; Mello, C.C. Potent and specific genetic interference by double-stranded RNA in *Caenorhabditis elegans*. *Nature* **2005**, *391*, 806–811. [CrossRef] [PubMed]
2. Bernstein, E.; Caudy, A.A.; Hammond, S.M.; Hannon, G.J. Role for a bidentate ribonuclease in the initiation step of RNA interference. *Nature* **2001**, *409*, 363–366. [CrossRef] [PubMed]
3. Ameres, S.L.; Martinez, J.; Schroeder, R. Molecular basis for target RNA recognition and cleavage by RISC. *Cell* **2007**, *131*, 101–112. [CrossRef] [PubMed]
4. Bobbin, M.L.; Rossi, J.J. RNA interference (RNAi)-based therapeutics: Delivery on the promise? *Annu. Rev. Pharmacol. Toxicol.* **2016**, *56*, 103–122. [CrossRef] [PubMed]
5. Pecot, C.V.; Calin, G.A.; Coleman, R.L.; Lopez-Berestein, G.; Sood, A.K. RNA interference in the clinic: Challenges and future directions. *Nat. Rev. Cancer* **2011**, *11*, 59–67. [CrossRef] [PubMed]
6. Castanotto, D.; Rossi, J.J. The promises and pitfalls of RNA-interference-based therapeutics. *Nature* **2009**, *457*, 426–433. [CrossRef] [PubMed]

7. Khalil, I.A.; Yamada, Y.; Harashima, H. Optimization of siRNA delivery to target sites: Issues and future directions. *Expert Opin. Drug Deliv.* **2018**, *15*, 1053–1065. [CrossRef] [PubMed]
8. Ledford, H. Gene-silencing technology gets first drug approval after 20-year wait. *Nature* **2018**, *560*, 291–292. [CrossRef] [PubMed]
9. Liu, F.; Wang, C.; Gao, Y.; Li, X.; Tian, F.; Zhang, Y.; Fu, M.; Li, P.; Wang, Y.; Wang, F. Current transport systems and clinical applications for small interfering RNA (siRNA) drugs. *Mol. Diagn. Ther.* **2018**, *22*, 551–569. [CrossRef] [PubMed]
10. Durymanov, M.; Reineke, J. Non-viral delivery of nucleic acids: Insight into mechanisms of overcoming intracellular barriers. *Front. Pharmacol.* **2018**, *9*, 971. [CrossRef] [PubMed]
11. Yin, H.; Kanasty, R.L.; Eltoukhy, A.A.; Vegas, A.J.; Dorkin, J.R.; Anderson, D.G. Non-viral vectors for gene-based therapy. *Nat. Rev. Genet.* **2014**, *15*, 541–555. [CrossRef] [PubMed]
12. Mignani, S.; Rodrigues, J.; Roy, R.; Shi, X.; Ceña, V.; El Kazzouli, S.; Majoral, J.P. Exploration of biomedical dendrimer space based on in-vivo physicochemical parameters: Key factor analysis. (Part 2). *Drug Discov. Today* **2019**. [CrossRef] [PubMed]
13. Araújo, R.V.; Santos, S.D.S.; Igne Ferreira, E.; Giarolla, J. New advances in general biomedical applications of PAMAM dendrimers. *Molecules* **2018**, *23*, 2849. [CrossRef] [PubMed]
14. Leiro, V.; Santos, S.D.; Pego, A.P. Delivering siRNA with dendrimers: In vivo applications. *Curr. Gene Ther.* **2017**, *17*, 105–119. [CrossRef] [PubMed]
15. Kannan, R.M.; Nance, E.; Kannan, S.; Tomalia, D.A. Emerging concepts in dendrimer-based nanomedicine: From design principles to clinical applications. *J. Intern. Med.* **2014**, *276*, 579–617. [CrossRef] [PubMed]
16. Tomalia, D.A.; Christensen, J.B.; Boas, U. *Dendrimers, Dendrons and Dendritic Polymers: Discovery, Applications and the Future*; Cambridge University Press: London, UK, 2012.
17. Walter, M.V.; Malkoch, M. Simplifying the synthesis of dendrimers: Accelerated approaches. *Chem. Soc. Rev.* **2012**, *41*, 4593–4609. [CrossRef] [PubMed]
18. Kim, Y.; Park, E.J.; Na, D.H. Recent progress in dendrimer-based nanomedicine development. *Arch. Pharm. Res.* **2018**, *41*, 571–582. [CrossRef] [PubMed]
19. Li, J.; Liang, H.; Liu, J.; Wang, Z. Poly (amidoamine) (PAMAM) dendrimer mediated delivery of drug and pDNA/siRNA for cancer therapy. *Int. J. Pharm.* **2018**, *546*, 215–225. [CrossRef]
20. Luo, K.; He, B.; Wu, Y.; Shen, Y.; Gu, Z. Functional and biodegradable dendritic macromolecules with controlled architectures as nontoxic and efficient nanoscale gene vectors. *Biotechnol. Adv.* **2014**, *32*, 818–830. [CrossRef]
21. Jędrych, M.; Borowska, K.; Galus, R.; Jodłowska-Jędrych, B. The evaluation of the biomedical effectiveness of poly(amido)amine dendrimers generation 4.0 as a drug and as drug carriers: A systematic review and meta-analysis. *Int. J. Pharm.* **2014**, *462*, 38–43. [CrossRef]
22. Benjaminsen, R.V.; Mattebjerg, M.A.; Henriksen, J.R.; Moghimi, S.M.; Andresen, T.L. The possible "proton sponge" effect of polyethylenimine (PEI) does not include change in lysosomal pH. *Mol. Ther.* **2013**, *21*, 149–157. [CrossRef] [PubMed]
23. Behr, J.P. The proton sponge: A trick to enter cells viruses did not exploit. *Chimia* **1997**, *51*, 34–36.
24. Haensler, J.; Szoka, F.C. Polyamidoamine cascade polymers mediate efficient transfection of cells in culture. *Bioconjug. Chem.* **1993**, *4*, 372–379. [CrossRef] [PubMed]
25. Kukowska-Latallo, J.F.; Bielinska, A.U.; Johnson, J.; Spindler, R.; Tomalia, D.A.; Baker, J.R., Jr. Efficient transfer of genetic material into mammalian cells using Starburst polyamidoamine dendrimers. *Proc. Natl. Acad. Sci. USA* **1996**, *93*, 4897–4902. [CrossRef] [PubMed]
26. Eichman, J.D.; Bielinska, A.U.; Kukoswka-Latallo, J.F.; Baker, J.R., Jr. The use of PAMAM dendrimers in the efficient transfer of generic material into cells. *Pharm. Sci. Technol. Today* **2000**, *3*, 232–245. [CrossRef]
27. Guillot-Nieckowski, M.; Eisler, S.; Diederich, F. Dendritic vectors for gene transfection. *New J. Chem.* **2007**, *31*, 1111–1127. [CrossRef]
28. Mintzer, M.A.; Simanek, E.E. Non viral vectors for gene delivery. *Chem. Rev.* **2009**, *109*, 259–302. [CrossRef]
29. Cao, Y.; Liu, X.; Peng, L. Molecular engineering of dendrimer nanovectors for siRNA delivery and gene silencing. *Front. Chem. Sci. Eng.* **2017**, *11*, 663–675. [CrossRef]
30. Palmerston Mendes, L.; Pan, J.; Torchilin, V.P. Dendrimers as nanocarriers for nucleic acid and drug delivery in cancer therapy. *Molecules* **2017**, *22*, 1401. [CrossRef]

31. Kesharwani, P.; Benerjee, S.; Gupta, U.; Amin, M.C.I.M.; Padhye, S.; Sarkar, F.H.; Iyer, A.K. PAMAM dendrimers as promising nanocarriers for RNAi therapeutics. *Mater. Today* **2015**, *18*, 565–572. [CrossRef]
32. Liu, X.; Rocchi, P.; Peng, L. Dendrimers as non-viral vectors for siRNA delivery. *New J. Chem.* **2012**, *36*, 256–263. [CrossRef]
33. Zhou, J.; Wu, J.; Hafdi, N.; Behr, J.P.; Erbacher, P.; Peng, L. PAMAM dendrimers for efficient siRNA delivery and potent gene silencing. *Chem. Commun.* **2006**, *22*, 2362–2364. [CrossRef]
34. Venkatesh, S.; Workman, J.L. Histone exchange, chromatin structure and the regulation of transcription. *Nat. Rev. Mol. Cell. Biol.* **2015**, *16*, 178–189. [CrossRef]
35. Karatasos, K.; Posocco, P.; Laurini, E.; Pricl, S. Poly(amidoamine)-based dendrimer/siRNA complexation studied by computer simulations: Effects of pH and generation on dendrimer structure and siRNA binding. *Macromol. Biosci.* **2012**, *12*, 225–240. [CrossRef]
36. Posocco, P.; Laurini, E.; Dal Col, V.; Marson, D.; Karatasos, K.; Fermeglia, M.; Pricl, S. Tell me something that I do not know. Multiscale molecular modeling of dendrimer/dendron organization and self-assembly in gene therapy. *Curr. Med. Chem.* **2012**, *19*, 5062–5087. [CrossRef]
37. Posocco, P.; Laurini, E.; Dal Col, V.; Marson, D.; Peng, L.; Smith, D.K.; Klajnert, B.; Bryszewska, M.; Caminade, A.-M.; Majoral, J.P.; et al. Multiscale modeling of dendrimers and dendrons for drug and nucleic acid delivery. In *Dendrimers in Biomedical Applications*; Klajnert, B., Peng, L., Ceña, V., Eds.; RSC Publishing: Cambrige, UK, 2013; pp. 148–166.
38. Pavan, G.M.; Posocco, P.; Tagliabue, A.; Maly, M.; Malek, A.; Danani, A.; Ragg, E.; Catapano, C.V.; Pricl, S. PAMAM dendrimers for siRNA delivery: Computational and experimental insights. *Chem. Eur. J.* **2010**, *16*, 7781–7795. [CrossRef]
39. Marson, D.; Laurini, E.; Posocco, P.; Fermeglia, M.; Pricl, S. Cationic carbosilane dendrimers and oligonucleotide binding: An energetic affair. *Nanoscale* **2015**, *7*, 3876–3887. [CrossRef]
40. Mehrabadi, F.S.; Hirsch, O.; Zeisig, R.; Posocco, P.; Laurini, E.; Pricl, S.; Haag, R.; Kemmner, W.; Calderón, M. Structure–activity relationship study of dendritic polyglycerolamines for efficient siRNA transfection. *RSC Adv.* **2015**, *5*, 78760–78770. [CrossRef]
41. Shen, X.-C.; Zhou, J.; Liu, X.; Wu, J.; Qu, F.; Zhang, Z.-L.; Pang, D.-W.; Quélèver, G.; Zhang, C.-C.; Peng, L. Importance of size-to-charge ratio in construction of stable and uniform nanoscale RNA/dendrimer complexes. *Org. Biomol. Chem.* **2007**, *5*, 3674–3681. [CrossRef]
42. Liu, X.-X.; Rocchi, P.; Qu, F.; Zheng, S.-Q.; Liang, Z.; Gleave, M.; Iovanna, J.; Peng, L. PAMAM dendrimers mediate siRNA delivery to target Hsp27 and produce potent antiproliferative effects on prostate cancer cells. *ChemMedChem* **2009**, *4*, 1302–1310. [CrossRef]
43. Siegel, R.L.; Miller, K.D.; Jemal, A. Cancer statistics. *CA Cancer J. Clin.* **2016**, *66*, 7–30. [CrossRef]
44. Huang, Y.; Jiang, X.; Liang, X.; Jiang, G. Molecular and cellular mechanisms of castration resistant prostate cancer. *Oncol. Lett.* **2018**, *15*, 6063–6076. [CrossRef]
45. Rocchi, P.; So, A.; Kojima, S.; Signaevsky, M.; Beraldi, E.; Fazli, L.; Hurtado-Coll, A.; Yamanaka, K.; Gleave, M. Heat shock protein 27 increases after androgen ablation and plays a cytoprotective role in hormone-refractory prostate cancer. *Cancer Res.* **2004**, *64*, 6595–6602. [CrossRef]
46. Liu, X.; Liu, C.; Catapano, C.V.; Peng, L.; Zhou, J.; Rocchi, P. Structurally flexible triethanolamine-core dendrimers as effective nanovectors to deliver RNAi-based therapeutics. *Biotechnol. Adv.* **2014**, *32*, 844–852. [CrossRef]
47. Kala, S.; Mak, A.S.C.; Liu, X.; Posocco, P.; Pricl, S.; Peng, L.; Wong, A.S.T. Combination of dendrimer-nanovector-mediated small interfering RNA delivery to target AKT with the clinical anticancer drug paclitaxel for effective and potent anticancer activity in treating ovarian cancer. *J. Med. Chem.* **2014**, *57*, 2634–2642. [CrossRef]
48. Reebye, V.; Sætrom, P.; Mintz, P.J.; Huang, K.W.; Swiderski, P.; Peng, L.; Liu, C.; Liu, X.; Lindkaer-Jensen, S.; Zacharoulis, D.; et al. Novel RNA oligonucleotide improves liver function and inhibits liver carcinogenesis in vivo. *Hepatology* **2014**, *59*, 216–227. [CrossRef]
49. Cui, Q.; Yang, S.; Ye, P.; Tian, E.; Sun, G.; Zhou, J.; Sun, G.; Liu, X.; Chen, C.; Murai, K.; et al. Downregulation of TLX induces TET3 expression and inhibits glioblastoma stem cell self-renewal and tumorigenesis. *Nat. Commun.* **2016**, *7*, 10637–10651. [CrossRef]

50. Lang, M.F.; Yang, S.; Zhao, C.; Sun, G.; Murai, K.; Wu, X.; Wang, J.; Gao, H.; Brown, C.E.; Liu, X.; et al. Genome-wide profiling identified a set of miRNAs that are differentially expressed in glioblastoma stem cells and normal neural stem cells. *PLoS ONE* **2012**, *7*, e36248–e36251. [CrossRef]
51. Song, M.; Bode, A.M.; Dong, Z.; Lee, M.H. AKT as a therapeutic target for cancer. *Cancer Res.* **2019**, *79*, 1019–1031. [CrossRef]
52. Christie, E.L.; Bowtell, D.D.L. Acquired chemotherapy resistance in ovarian cancer. *Ann. Oncol.* **2017**, *28* (Suppl. 8), viii13–viii15. [CrossRef]
53. Svenson, S. The dendrimer paradox—Highly medical expectations but poor clinical translation. *Chem. Soc. Rev.* **2015**, *44*, 2228–2238. [CrossRef]
54. Bolcato-Bellemin, A.L.; Bonnet, M.E.; Creusat, G.; Erbacher, P.; Behr, J.P. Sticky overhangs enhance siRNA-mediated gene silencing. *Proc. Natl. Acad. Sci. USA* **2007**, *104*, 16050–16055. [CrossRef]
55. Boussif, O.; Lezoualc'h, F.; Zanta, M.; Mergny, M.D.; Scherman, D.; Demeneix, B.; Behr, J.P. A versatile vector for gene and oligonucleotide transfer into cells in culture and in vivo: Polyethyleneimine. *Proc. Natl. Acad. Sci. USA* **1995**, *92*, 7297–7301. [CrossRef]
56. Liu, X.; Liu, C.; Laurini, E.; Posocco, P.; Pricl, S.; Qu, F.; Rocchi, P.; Peng, L. Efficient delivery of sticky siRNA and potent gene silencing in a prostate cancer model using a generation 5 triethanolamine-core PAMAM dendrimer. *Mol. Pharm.* **2012**, *9*, 470–481. [CrossRef]
57. Posocco, P.; Liu, X.; Laurini, E.; Marson, D.; Chen, C.; Liu, C.; Fermeglia, M.; Rocchi, P.; Pricl, S.; Peng, L. Impact of siRNA overhang for dendrimer-mediated siRNA delivery and gene silencing. *Mol. Pharm.* **2013**, *10*, 3262–3273. [CrossRef]
58. Behzadi, S.; Serpooshan, V.; Tao, W.; Hamaly, M.A.; Alkawareek, M.Y.; Dreaden, E.C.; Brown, D.; Alkilany, A.M.; Farokhzad, O.C.; Mahmoudi, M. Cellular uptake of nanoparticles: Journey inside the cell. *Chem. Soc. Rev.* **2017**, *46*, 4218–4244. [CrossRef]
59. First-in-Human Safety and Tolerability Study of MTL-CEBPA in Patients with Advanced Liver Cancer (OUTREACH). Available online: https://clinicaltrials.gov/ct2/show/NCT02716012 (accessed on 17 July 2019).

Review

Evolution from Covalent to Self-Assembled PAMAM-Based Dendrimers as Nanovectors for siRNA Delivery in Cancer by Coupled in Silico-Experimental Studies. Part II: Self-Assembled siRNA Nanocarriers

Erik Laurini, Domenico Marson *, Suzana Aulic, Maurizio Fermeglia and Sabrina Pricl

Molecular Biology and Nanotechnology Laboratory (MolBNL@UniTS), Department of Engineering and Architecture, University of Trieste, 34127 Trieste, Italy

* Correspondence: domenico.marson@dia.units.it; Tel.: +39-040-558-3750

Received: 21 June 2019; Accepted: 8 July 2019; Published: 10 July 2019

Abstract: In part I of this review, the authors showed how poly(amidoamine) (PAMAM)-based dendrimers can be considered as promising delivering platforms for siRNA therapeutics. This is by virtue of their precise and unique multivalent molecular architecture, characterized by uniform branching units and a plethora of surface groups amenable to effective siRNA binding and delivery to e.g., cancer cells. However, the successful clinical translation of dendrimer-based nanovectors requires considerable amounts of good manufacturing practice (GMP) compounds in order to conform to the guidelines recommended by the relevant authorizing agencies. Large-scale GMP-standard high-generation dendrimer production is technically very challenging. Therefore, in this second part of the review, the authors present the development of PAMAM-based amphiphilic dendrons, that are able to auto-organize themselves into nanosized micelles which ultimately outperform their covalent dendrimer counterparts in in vitro and in vivo gene silencing.

Keywords: RNAi therapeutics; siRNA delivery; amphiphilic dendrons; PAMAM dendrimers; self-assembling; nanovectors; gene silencing

1. Self-Assembling PAMAM-Based Amphiphilic Dendrons: A New Paradigm for siRNA Delivery

In its broadest sense, self-assembly describes the natural tendency of physical systems to exchange energy with their surroundings and assume patterns or structures of low free energy. Random thermal motions bring constituent particles together in various configurations, and only those with significantly favorable interaction energy forming, tend to persist, and eventually become predominant. The information on the shape and size of the ultimate self-assembled entity is embodied in the structures of the individual components. A system slowly approaching equilibrium assumes a simple repetitive structure, while a dynamic system may generate structures of great complexity. For example, molecules in a cooling glass of water self-assemble as simple ice crystals, while the same molecules in a turbulent cloud with temperature and humidity gradients self-assemble as complex snowflakes of enormous variety. In relation to chemistry, self-assembly constitutes the quintessence of nanotechnology-based techniques leading to the design of novel materials [1–4]. It relies on the cumulative effects of multiple non-covalent interactions to assemble molecular building blocks into supramolecular entities in a reversible, controllable, and specific way, yet with relatively little synthetic effort. In particular is the ability of self-assembled structures to behave as more than the sum of their individual parts, and exhibit completely new properties [5].

With these concepts in mind, the authors envisaged the idea of creating small amphiphilic poly(amidoamine) (PAMAM)-based dendrons which, upon auto-organization into nanosized micelles, could mimic the covalent, high-generation dendrimer counterparts in size, shape and function, in particular for in vitro and in vivo siRNA delivery [6–8], as set out below.

2. siRNA Delivery by Single-Tail Self-Assembling Amphiphilic Dendrons

2.1. Design, Optimization and Chemico-Physical Characterization of Single-Tail Self-Assembling Amphiphilic Dendrons

Our work in this field started with the design, optimization and synthesis of a series of amphiphilic dendrons (**1–6**) characterized by a single hydrophobic alkyl chain of variable length and a hydrophilic PAMAM head with 8 primary amine terminal groups [7], as shown in Scheme 1. In addition, two non-amphiphilic dendrons characterized by a hydrophilic pentaethylene glycol (PEG) chain (**7**) and by the presence of the sole PAMAM head (**8**) were also produced as negative reference compounds.

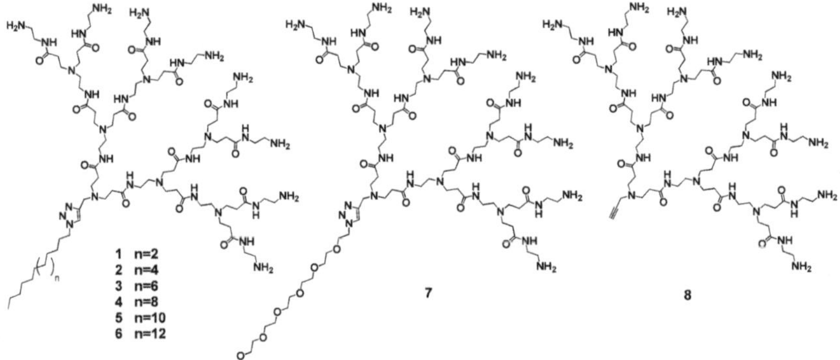

Scheme 1. The chemical structure of the PAMAM-based self-assembling amphiphilic dendrons bearing a single alkyl chain of variable length (**1–6**). The two non-amphiphilic dendrons characterized by a hydrophilic pentaethylene glycol (PEG) chain (**7**) and by the presence of the sole PAMAM head (**8**) used as negative reference compounds are also shown (see text for details). Adapted from [7] with the permission of John Wiley and Sons, 2016.

Mesoscale simulations (see Figure S1, Table S1, and text in Supporting Information for details) were initially employed to predict the self-assembly of all dendrons **1–8**. According to the in silico results, the amphiphilic dendrons **1–6** were able to auto-organize into spherical nanosized micelles (shown in Figure 1 for dendrons **4**, **3** and **2** as an example), with an average diameter D and aggregation number N_{agg} ranging from 5.6 nm and 5 for micelles. This was generated by dendron **1** to 8.1 nm and 13 for those originated by the self-assembling of dendron **6** (Table 1). Contextually, no stable nanostructure formation was observed for the negative controls **7** and **8**, as shown in the left panel of Figure 1d for dendron **8**.

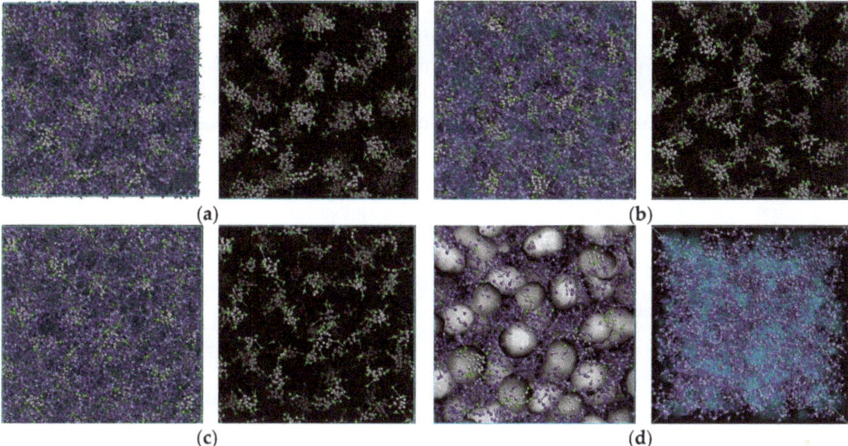

Figure 1. The computer simulation of the self-assembled micelles generated by the amphiphilic dendrons **4** (**a**), **3** (**b**), and **2** (**c**). In all panels, the hydrophilic PAMAM head is portrayed as purple and lilac beads, the hydrophobic tail is represented by light gray beads while the linker between the hydrophilic and the hydrophobic portion of the molecule is shown as green beads. Water and counterions are portrayed as a gray field for clarity. The right images in each panel show only the micellar cores. (**d, left**) Zoomed view of the micelles formed by dendron **4**. The micellar cores are highlighted by a light gray surface. (**d, right**) Computer simulation of the non-amphiphilic dendron **8** selected as negative control, showing the non-self-assembling characteristics of this system.

Table 1. Micelle aggregation number (N_{agg}), surface charge density (σ_m, e/nm^2), diameter (D, nm), critical micelle concentration (CMC, μM) and free energy of micellization (ΔG_m, kJ/mol) for dendrons **1–6** as predicted by computer simulations. The corresponding experimental values of the micelle diameter (D_{exp}, nm) and critical micelle concentration (CMC_{exp}, μM) are also reported. Adapted from [7] with the permission of John Wiley and Sons.

Dendron	N_{agg}	D	D_{exp}	CMC	CMC_{exp}	ΔG_m
1	5	5.6 ± 0.1	-	415	398 ± 22	−38.7
2	6	5.9 ± 0.1	-	127	116 ± 11	−44.4
3	7	6.2 ± 0.1	6.6 ± 0.1	37.2	49.9 ± 3.3	−50.5
4	8	6.4 ± 0.1	6.8 ± 0.1	11.4	15.6 ± 1.1	−56.4
5	10	7.1 ± 0.2	7.3 ± 0.1	9.13	11.2 ± 0.6	−57.5
6	13	8.1 ± 0.2	7.8 ± 0.2	8.50	8.24 ± 3.8	−57.8

The values of the critical micelle concentration (CMC) for the amphiphilic dendrons **1–6** (see Supporting Information for details) were further investigated from which a clear inverse relationship between the hydrophobic tail length and the CMC was obtained (Table 1). In other words, the amphiphilic dendrons bearing a longer hydrophobic component are able to self-assemble and pack more efficiently than those characterized by shorter alkyl tails. In addition, the associated values of the free energy of micellization (ΔG_m) predicted by simulation (see Supporting Information for details) were all largely negative, supporting the spontaneous and thermodynamically favored micelle formation for all dendrons **1–6** (Table 1). As the hydrophilic portion was the same in all amphiphilic dendrons, the differential contribution to ΔG_m (and hence to CMC) is obviously related to the length of the hydrophobic chain.

Experimental confirmation of in silico predictions were first performed by dynamic light scattering (DLS) and transmission electron microscopy (TEM). Both these techniques confirmed that the amphiphilic dendrons predominantly formed nanosized spherical micelles, as shown in Figure 2 for dendrons **3–6**. In particular, data analysis confirmed that an increase of the hydrophobic tail length from C_{14} (**3**) to C_{22} (**6**) was paralleled by an increase of the micellar diameter D_{exp} from 6.6 nm to 7.8 nm, in full agreement with the computer-based predictions (Table 1). Quite interestingly, the overall dimensions of these micellar nano-objects are indeed very close to those characterizing the high-generation structurally flexible triethanolamine (TEA)-core covalent dendrimers discussed in detail in the companion paper [9] (e.g., the average molecular diameters for generation 4, 5 and 6 TEA-core dendrimers are equal to 5.2, 7.7 and 10.1 nm, respectively). Therefore, at least from the standpoint of structure and dimensions, the self-assembled nanoparticles obtained from the amphiphilic dendrons can be considered as supramolecular dendrimer mimics.

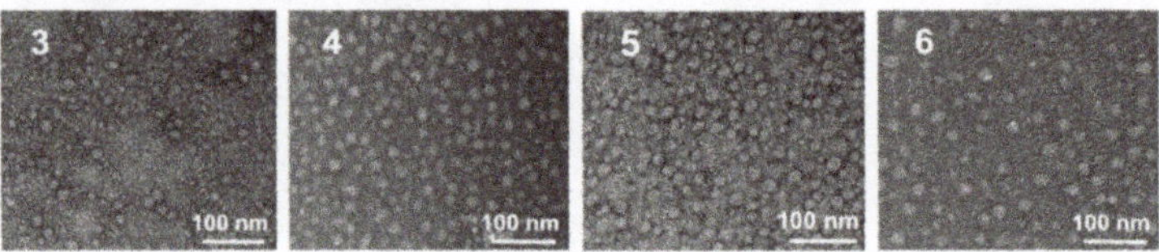

Figure 2. TEM images of spherical micelles formed by the amphiphilic dendrons **3–6**. Adapted from [7] with the permission of John Wiley and Sons, 2016.

The predicted CMC values were also confirmed by experiments using pyrene as a hydrophobic fluorescent probe (Table 1). The progressive increase in alkyl chain length resulted in a drop of CMC_{exp} from 398 μM for dendron **1** to 8.24 μM for dendron **6**. The high values of the computational/experimental CMC for systems **1** and **2** are the likely reason why micelle formation by these two dendrons was not observed under the conditions adopted for DLS/TEM experiments. Further, in agreement with the simulation, no CMC value could be experimentally determined for **7** and **8**, confirming the non-self-assembling properties of these two negative-control systems.

2.2. *In Silico/Experimental Interaction of Single Tail Self-Assembling Dendrons and siRNA*

Mesoscale simulations were again exploited for predicting the interactions between the amphiphilic dendrons **1–8** and siRNA molecules, as illustrated in Figure 3.

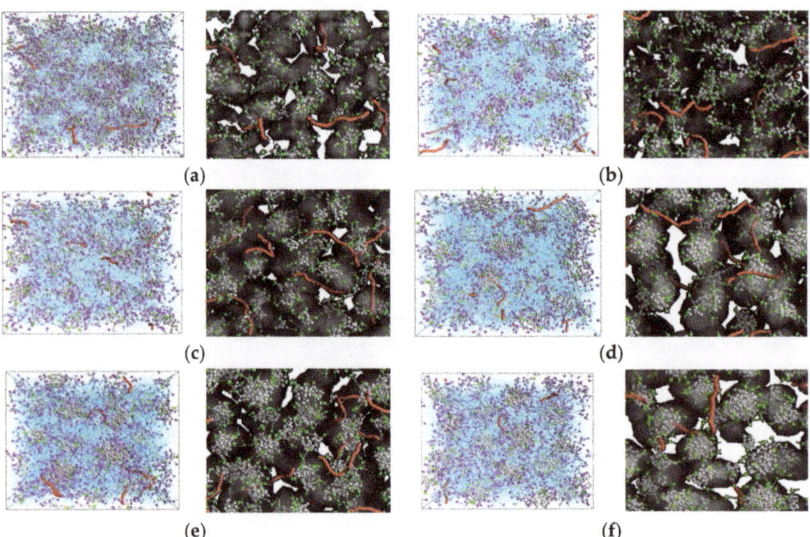

Figure 3. *Cont.*

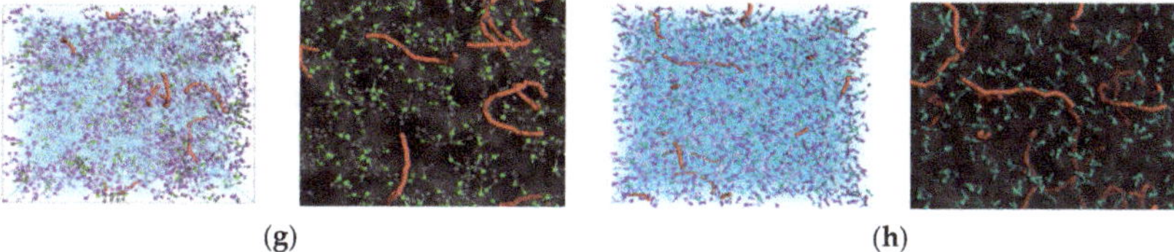

(g) (h)

Figure 3. Morphologies of the systems formed at a dendrimer-to-siRNA charge ratio (N/P) = 10 by the siRNA molecules and the amphiphilic dendrons **1** (**a**), **2** (**b**), **3** (**c**), **4** (**d**), **5** (**e**), **6** (**f**), **7** (**g**) and **8** (**h**) as obtained from molecular simulations. Left panels are zoomed images of the right panels. Colors as in Figure 2, except for the siRNA molecules, shown as red sticks and the water and counterions, portrayed in the left panels as light blue field for clarity. Adapted from [7] with the permission of John Wiley and Sons, 2016.

The computer images showed that, at the same dendrimer-to-siRNA charge ratio (N/P) of 10, the nucleic acid fragments are well encased within the micellar network formed by all self-assembling amphiphilic dendrimers **1–6** and so are efficiently protected from the surrounding environment. For the non-self-assembling systems **7** and **8**, the siRNA molecules are interspersed within an unorganized solution of single dendrons and water, and therefore in principle more susceptible to ultimate degradation by e.g., RNAses.

The in silico quantitative characterization of the siRNA/amphiphilic dendron interactions was next carried out by atomistic molecular dynamics (MD) simulations performed in the framework of the so-called molecular mechanics/Poisson-Boltzmann surface area (MM/PBSA) methodology [9–15] (the methodology is described in detail in the Supporting Information in the companion paper [9]). The values of free energy of binding (ΔG_{bind}) between the different self-assembling dendrons **1–6** towards the siRNA sequence directed against the mRNA coding for the heat shock protein 27 (Hsp27)—a small molecular chaperone which is a vital regulator of cell survival and a major player in drug resistance—were initially calculated, as listed in the first column of Table 2. However, in order to perform a rigorous discussion about these aspects, the concept of effective free energy of binding ($\Delta G_{bind,eff}$) was introduced, as follows. From an extensive analysis of each dendron micelle/siRNA complex MD trajectory, the number N_{eff} of dendron branches (each bearing a positive charge) in permanent contact with the siRNA was determined, leading to the values listed in the second columns of Table 2. The precise identification of each dendrimer branch involved in the siRNA binding, and the corresponding individual contribution afforded to the overall binding energy, was then carried out by a per-residue free energy decomposition technique (described in the Supporting Information in the companion paper [9]). The cumulative results of this analysis are reported in the third column of Table 2, showing the effective contribution to the total free energy of binding ($\Delta G_{bind,eff}$). However, to be able to compare simulation data among themselves (and with experimental data, see below) the effective-charge-normalized values of $\Delta G_{bind,eff}$, i.e., $\Delta G_{bind,eff}/N_{eff}$, were considered, as shown in the fifth column of Table 2.

Table 2. The free energy of micellization (ΔG_{bind}, kcal/mol), number of effective charges (N_{eff}), effective free energy of binding ($\Delta G_{bind,eff}$, kcal/mol), and effective-charge-normalized free energy of binding ($\Delta G_{bind,eff}/N_{eff}$) between siRNA and dendrons **1–6** as derived from molecular simulations. The last column reports the siRNA/dendron binding data (C_{50}, μM) obtained from ethidium bromide (EB) displacement assays. Adapted from [7] with the permission of John Wiley and Sons, 2016.

Dendron	ΔG_{bind}	N_{eff}	$\Delta G_{bind,eff}$	$\Delta G_{bind,eff}/N_{eff}$	C_{50}
1	−8.23 ± 0.91	13 ± 1	−2.47 ± 0.32	−0.19 ± 0.03	-
2	−13.4 ± 1.1	18 ± 1	−4.30 ± 0.45	−0.24 ± 0.03	42.0 ± 2.0
3	−17.1 ± 1.3	22 ± 1	−6.00 ± 0.36	−0.27 ± 0.02	16.9 ± 0.5
4	−22.6 ± 1.2	31 ± 1	−12.7 ± 0.41	−0.41 ± 0.04	7.0 ± 0.3
5	−46.6 ± 1.6	49 ± 1	−28.9 ± 0.70	−0.59 ± 0.04	5.3 ± 0.4
6	−57.8 ± 2.3	62 ± 1	−34.7 ± 0.62	−0.56 ± 0.05	4.8 ± 0.1

From the values reported in Table 2, the effective interaction between the micelles formed by the different dendrons with siRNA ($\Delta G_{bind,eff}$) increases with increasing alkyl chain length, the last three dendrons definitely being the strongest siRNA binders in the order: **4** < **5** < **6**. Limiting the discussion to these three self-assembling systems, the micelles formed by dendron **4** were characterized by a total number of charges equal to +64, the 48% of which (N_{eff} = 31 are involved in productive siRNA binding. This translates into the corresponding $\Delta G_{bind,eff}/N_{eff}$ value of −0.41 kcal/mol. In contrast, the other two efficient self-assembled micelles generated by dendrons **5** and **6** were not only both able to exploit 61% of their positive charges (49/80 and 62/102 for **5** and **6**, respectively) but they also did it more efficiently ($\Delta G_{bind,eff}/N_{eff}$ = −0.59 kcal/mol for **5** and −0.56 kcal/mol for **6**, Table 1), ultimately characterizing the corresponding micelles as the best siRNA binders of the whole series.

In silico data were experimentally confirmed by fluorescent ethidium bromide (EB) displacement assays (see Figure A1 in Appendix A). According to these tests, the value of C_{50} (i.e., the dendron concentration at which the fluorescence of siRNA-intercalated EB is reduced by 50%), is an indication of siRNA binding efficiency: The lower the C_{50} value, the stronger the siRNA/micelle binding. As seen in Figure A1, the negative-control (i.e., non-self-assembling) dendrons **7** and **8** were not able to displace EB from the nucleic acid. In line, also for the self-assembling dendron **1**, bearing the shortest alkyl tail (C_{14}), the C_{50} value was not determined, due to the high CMC value required for its self-assembly. On the other hand, for the dendron series **2**–**6**, progressively decreasing C_{50} values were obtained (Figure A1 and Table 2, last column), with dendron **6** being the most effective siRNA binder, in full agreement with computer-based predictions.

2.3. *In Vitro siRNA Delivery Performance of Single-Tail Self-Assembling Dendrons*

Based on the results described above, dendrons **4**–**6** were selected for in vitro delivery of siRNA molecules targeting both the Hsp27 and the translationally controlled tumor protein (TCTP, a highly conserved protein present in all eukaryotic organisms that regulate cell survival in human tumors) in human castration-resistant prostate cancer (HCPC) PC-3 and C4-2 cell lines (Figure 4a,b). In these experiments, all other dendrons (**1**, **3**, **7** and **8**) were also tested for comparison and/or negative controls.

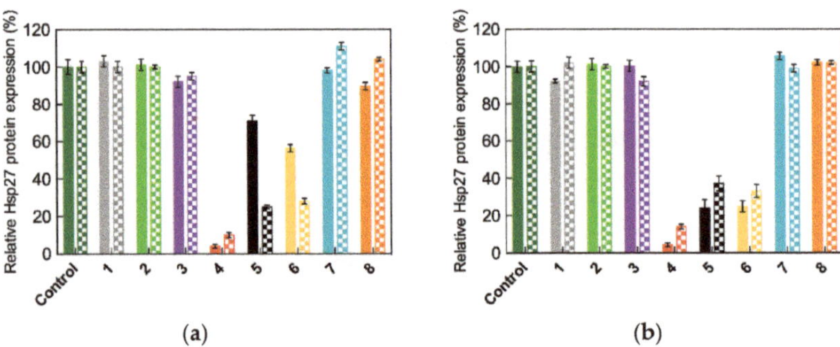

Figure 4. *Cont.*

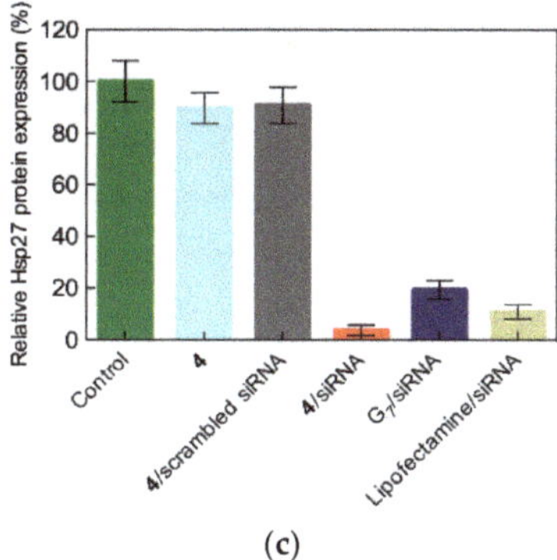

(c)

Figure 4. Self-assembled dendron-mediated siRNA delivery and gene silencing of Hsp27 (**a**) and TCTP proteins (**b**) in human prostate cancer PC-3 (solid bars) and C4-2 (patterned bars) cell lines (N/P = 10). Control: Untreated cells. The results obtained from non-self-assembling dendrons **1**, **7** and **8** are also shown for negative-control purposes. (**c**) Expression of Hsp27 in PC-3 cells treated with 50 nM siRNA delivered by self-assembled dendron **4** (N/P = 10), by Lipofectamine, and by the high-generation covalent triethanolamine (TEA)-core PAMAM-based dendrimer G_7. Control: Untreated cells. Negative controls: Self-assembled dendron **4** alone and a scrambled (i.e., non-silencing) siRNA sequence. The expression levels of Hsp27 and TCTP proteins were quantified by western blots 72 h post treatment using vinculin as the control. Adapted from [7] with the permission of John Wiley and Sons, 2016.

As shown in this Figure, the best siRNA delivery capacity was obtained with the nanovector based on the self-assembled form of dendron **4**, whose related gene silencing effect was even superior to that obtained both with the gold standard commercial vector Lipofectamine and with the high-generation (G_7) covalent PAMAM-based dendrimer. This featured an extended triethanolamine (TEA) core developed by our group and discussed in detail in the companion paper [9]. As anticipated by in silico predictions, nanocarriers formed by dendrons bearing shorter aliphatic (**1–3**), hydrophilic (**7**) or no chains (**8**) fail to elicit meaningful gene silencing in all cases. Indeed, while **7** and **8** are intrinsically unable to self-assembly (negative controls), the former dendron set (**1–3**) likely cannot generate stable micelles and/or siRNA complexes at the experimental conditions employed for transfection, due to their high CMC and C_{50} values (Tables 1 and 2). On the other hand, nanomicelles from dendrons with longer alkyl tails (**5** and **6**) were also able to achieve significant siRNA delivery and silenced the target Hsp27 and TCTP genes, yet with performance lower than that obtained with dendron **4**.

These last results seem to be in slight contradiction with the computer-based prediction reported in Table 2, according to which the best (i.e., most favorable) values of $\Delta G_{bind,eff}/N_{eff}$ and C_{50} are indeed obtained for self-assembled dendrons **5** and **6**. This ranks these nanovectors as the tightest siRNA binders. To investigate whether this disagreement was real or only apparent, cellular uptake experiments based on flow cytometry were performed using siRNA in complex with micelles from **4**, **5**, and **6** (Figure 5a). These data clearly show that this step is not governing the differential behavior of these nanovector/siRNA performance along the in vitro gene silencing pathway, as all three systems are characterized by comparable cellular uptake efficiency.

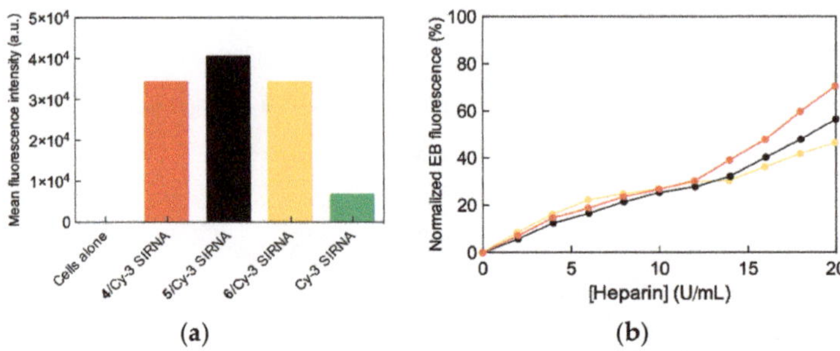

Figure 5. (**a**) The mean fluorescence intensity of PC-3 cells treated with siRNA labeled with the fluorescent dye Cy-3 in complex with self-assembled nanovectors **4**, **5**, and **6** (50 nM siRNA, N/P = 10). Non-treated cells and cells treated with labeled siRNA alone were used as control. (**b**) siRNA release from their complexes with micelles of dendrons **4** (red), **5** (black), and **6** (yellow) as assessed using EB displacement assay performed in the presence of heparin. Adapted from [7] with the permission of John Wiley and Sons, 2016.

With this aspect ruled out, it was reasoned that the siRNA release of its nanovector could be the factor underlying the best performance in gene silencing of the **4**/siRNA complex. Thus, it was determined that it was the dissociation of the nucleic acid fragments from their **4**, **5**, and **6** self-assembled nanovectors via the EB displacement assay in the presence of heparin, a highly negatively charged polysaccharide that can compete with siRNA in nanocarrier binding. As can be seen in Figure 5b, by increasing heparin concentration the siRNA molecules were more efficiently displaced from their complex with **4** than from the two alternative dendron micelles. This led us to the conclusion that the siRNA release process might be more effective from the self-assembled dendron **4** than from the **5** and **6** complexes, respectively.

Collectively, these results led to the conclusion that, in agreement with computer predictions, the supermolecular assemblies formed by **5** and **6** were actually too stable to dissemble in physiological conditions. This enhanced stability negatively affects the siRNA release in the cellular environment, thereby impairing the relevant gene silencing effect. At the same time, the complexes of micelles originated by dendron **4** and siRNA represent the best compromise in term of nucleic acid fragment binding and release efficiency, and these properties ultimately translate into the most potent biological performance.

The final step of the in vitro characterization concerned the investigation of the cellular toxicity of our single-tail self-assembled dendrons. As shown in Figure 6 for the PC-3 cell line as an example, none of the self- and non-self-assembling dendrons were endowed with cytotoxic properties. Analogous results were obtained with the alternative prostate cancer cell line, C4-2.

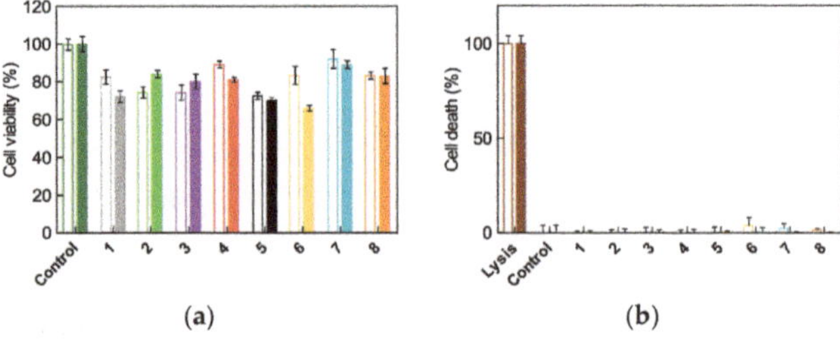

Figure 6. Metabolite toxicity (**a**) and membrane damaging activity (**b**) of self- and non-self-assembling dendrons **1–8** and their complexes with a scrambled (i.e., non-silencing) siRNA sequence (50 nM siRNA, N/P = 10) in PC-3 cells using the 3-(4,5-dimethylthiazol-2-yl)-2,5-diphenyl tetrazolium bromide (MTT) assay and the lactate dehydrogenase (LDH) assay, respectively. Untreated cells were used for control. Adapted from [7] with the permission of John Wiley and Sons, 2016.

In essence, all in silico and in vitro results concurred to indicate **4** as the amphiphilic dendron endowed with the best features for efficient self-assembly, and easy and stable formation of complexes with the siRNA molecules for delivery. Once inside the cell, there is effective disassembly during endosomal release for subsequent potent gene silencing.

*2.4. In Vivo siRNA Delivery Performance of Single-Tail Self-Assembling Dendron **4***

The in vitro best performing nanomicelles generated by the dendron **4** were finally challenged for in vivo gene silencing in a prostate cancer xenografted nude mouse model. The results obtained are gathered in Figure 7.

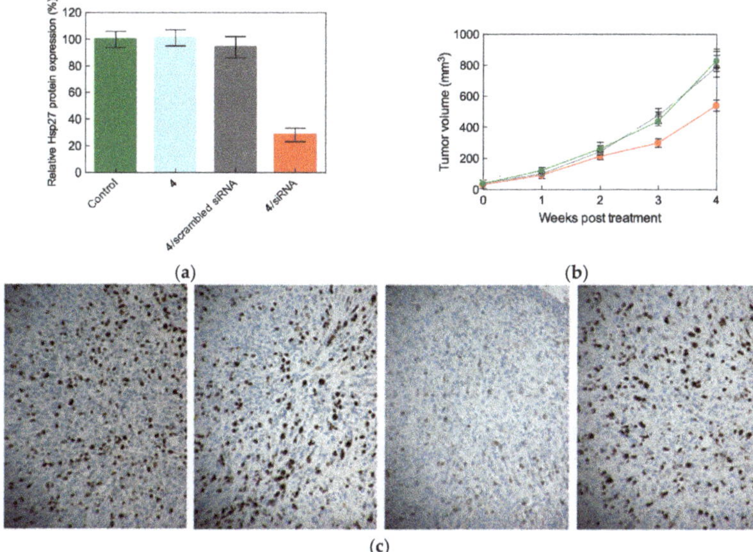

Figure 7. (**a**) In vivo gene silencing achieved by treating nude mice bearing prostate cancer (PC-3) tumors (30–50 mm^3) with Hsp27 siRNA delivered by the self-assembling dendron **4**, buffer solution (control), the dendron **4** alone, and a scrambled siRNA/**4** complex (negative controls). Treatments (3 mg/kg siRNA, N/P = 5) were administered for a period of 4 weeks via intraperitoneal injections (twice per week). (**b**) In vivo inhibition of tumor growth quantified by tumor size and (**c**) antiproliferation activity in tumors assessed by immunohistochemistry in mice treated as described for panel (**a**). Panel (**c**), from left to right: control, **4** alone, **4**/siRNA, **4**/scrambled siRNA. Adapted from [7] with the permission of John Wiley and Sons, 2016.

Images in Figure 7 support the effective and specific in vivo gene silencing induced upon administration of the Hsp27-directed siRNA delivered by the self-assembled nanovector **4**. In particular, an impressive reduction of ~80% in Hsp27 expression and of ~65% in tumor size after 4 weeks of treatment were achieved (Figure 7a,b).

These data, coupled with no discernable signs of in vivo toxicity (Figure 8) and the absence of mice weight loss, support the potential utilization of the amphiphilic compound **4** as a siRNA vector for future therapeutic implementations.

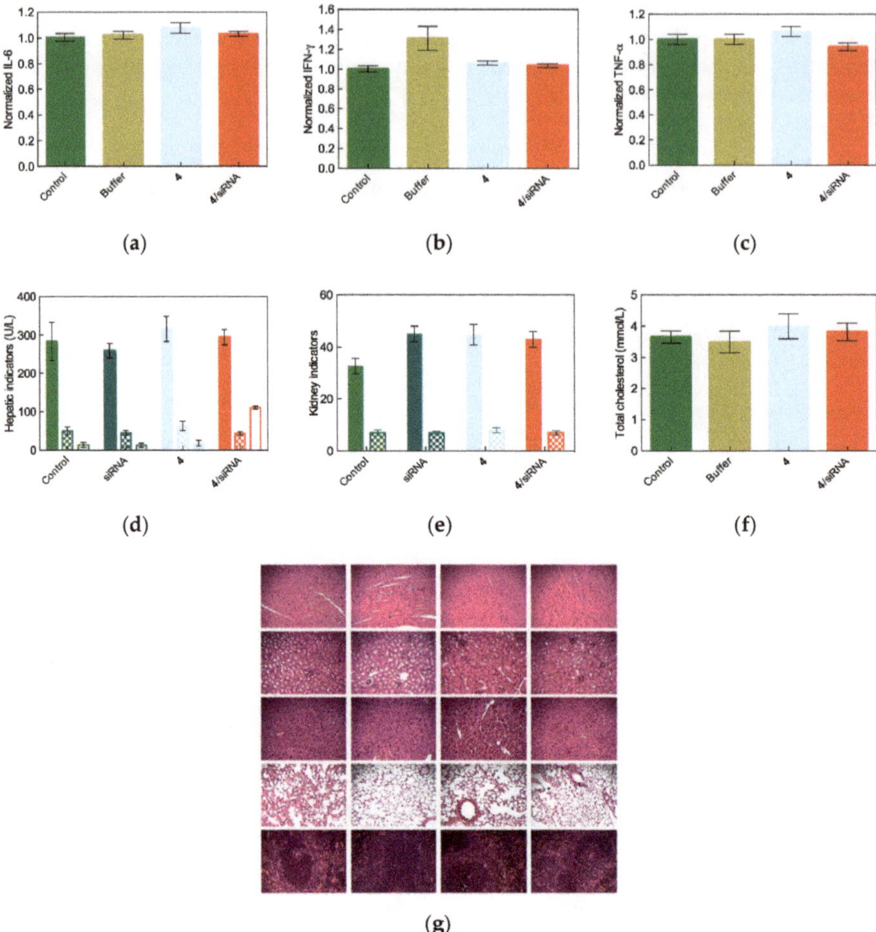

Figure 8. In vivo levels of inflammatory cytokines interleukine-6 (IL-6, **a**), interferon-gamma (INF-γ, **b**), and tumor necrosis factor-alpha (TNF-α, **c**), observed in male C57BL/6 mice treated with **4** and its siRNA complex (3.0 mg/kg siRNA, N/P = 5) by intravenous injection. Non-treated mice and animals treated with buffer solutions were used for control. Mice were sacrificed 24 h after treatment and the ELISA assay was performed on the collected serum. (**d**) Hepatic enzyme levels (aspartic aminotransferase AST (solid bars), alanine transferase ALT (patterned bars), gamma-glutamine transferase γ-GLT (empty bars), (**e**) kidney functions (creatinine CRE (mmol/L, solid bars), blood urea nitrogen BUN (μmol/L, patterned bars), and (**f**) cholesterol level in the blood measured in the treated mice after sacrifice. (**g**) Pathology of major organs from treated mice (HES staining). Columns (from left to right): Control, **4**, **4**/siRNA, and **4**/scrambled siRNA. Rows (from top to bottom): Heart, kidney, liver, lung, spleen. Adapted from [7] with the permission of John Wiley and Sons, 2016.

3. siRNA Delivery by Double-Tail Self-Assembling Amphiphilic Dendrons

3.1. Design, Optimization and Chemico-Physical Charactrization of Double-Tail Self-Assembling Amphiphilic Dendrons

With the haste of further improving the siRNA delivery and related gene silencing performance of the lead self-assembling amphiphilic dendron **4**, it was questioned whether increasing its hydrophobic portion would render this molecule even more efficient and effective. According to the devised strategy, it was ultimately decided to decorate **4** with a second C_{18}-long alkyl chain. In order to do so, a series of computer-based molecular design and optimization studies were carried out, finally leading to the new double-tail amphiphilic dendron structure **AD** (where **AD** stands for Amphiphilic Dendron) shown in Figure 9a [6]. In buffer solution per se, molecular simulations predicted **AD** to self-assemble into large (~200 nm) vesicle-like structures (Figure 9b) called dendrimersomes [16]. After synthesis of **AD** by click-chemistry, dynamic light scattering (DLS) and transmission electron microscopy (TEM) indeed confirmed both the size and the shape of the supermolecular entities formed by **AD** upon self-assembly (Figure 9c). Notably, the vesicle corona structure closely matched a bilayer, characterized by a thickness of 7 nm—approximately equal to twice the molecular length of **AD** (~3.5 nm).

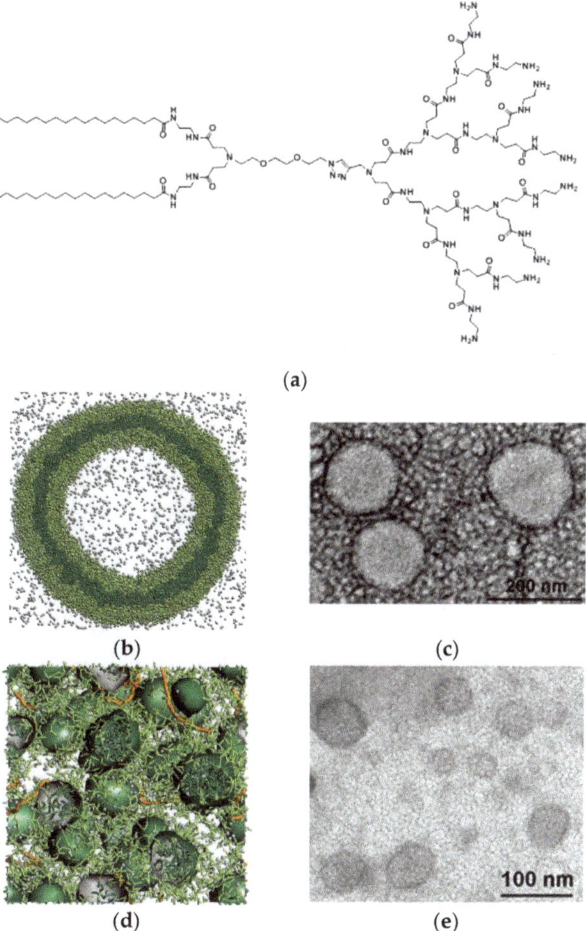

Figure 9. (**a**) The structure of the double-tail amphiphilic dendron simulated (**b**) and experimental TEM images (**c**) of the vesicle-like dendrimersomes formed by **AD** alone upon self-assembling in buffer solution. In panel b, the hydrophilic units of **AD** are portrayed in light green, the hydrophobic units are depicted in dark green, while light grey spheres are used to show some representative water molecules. Simulated (**d**) and experimental TEM images (**e**) of the nanosized spherical micelles formed by **AD** in the presence of siRNA molecules in solution. In panel (**d**), colors as in panel (**b**), while the nucleic acid fragments are portrayed as orange sticks. Adapted from [6] with permission of John Wiley and Sons, 2014.

Quite surprisingly, however, when simulations of the self-assembly of **AD** were performed in the presence of siRNA many highly ordered smaller spherical micelles of 6–8 nm diameter were obtained, with the siRNA fragments nicely interspersed within them (Figure 9d). The explanation for this vesicular to micellar structural transition likely resides in the greater positively charged surface area exposed by the micelles with respect to the dendrimersomes, which provides more efficient electrostatic interactions with the negatively charged siRNA. This, in turn, leads to the formation of more stable and compact complexes between the self-assembled nanocarriers and the nucleic acids molecules. The computer predictions were supported by TEM imaging (Figure 9e), confirming that **AD** is able to dynamically self-assemble into responsive and adaptive supramolecular assemblies in the presence of external stimuli, i.e., negatively-charged small polyanions.

3.2. In Vitro siRNA Delivery Performance of the Double-Tail Self-Assembling Dendron AD

The nanosized micelles formed by self-assembly of the double-tail amphiphilic dendron **AD** were able to effectively protect siRNA from degradation (e.g., by RNAses) and to promote fast and quantitative uptake by castrate-resistant prostate cancer PC-3 cells—almost 100% internalization of the nanovector/siRNA complexes was reached within 30 min. As concerns the mechanism of cell internalization of the **AD**/siRNA complexes, macropinocytosis was found to be the leading cellular entry mechanism for this nanovector/siRNA system, in utter analogy with the G_5 covalent PAMAM dendrimer counterpart discussed in the companion paper (Figure 15a in [9]). Indeed, neither chlorpromazine (an inhibitor of clathrin-mediated endocytosis) nor genistein (a caveolae-mediated uptake specific blocker) were effective while cytochalasin D (a specific macropinocytosis inhibitor that, by binding to actin filaments, interferes with actin polymerization and assembly) inhibited cellular uptake in a dose-dependent manner, as presented in Figure 10b. Again in analogy with what observed for the G_5 TEA-core PAMAM covalent dendrimer (Figure 15b in [9]), confocal microscopy provided images showing significant co-localization of the siRNA/**AD** complexes with the macropinocytosis biomarker dextran. Whereas, negligible co-localization was observed in cells treated with the two alternative biomarkers transferrin (clathrin-mediated endocytosis) and CTX-B (caveolae-mediated endocytosis). Thus, these experiments further corroborate the concept that these self-assembling PAMAM-based amphiphilic dendrons behave very similarly to their covalent high generation counterpart.

The gene silencing effect induced by Hsp-27 targeting siRNA delivered by **AD** nanomicelles was next investigated both at the mRNA and protein levels using prostate (PC-3) and breast (MCF-7 and MDA-MB231) cancer cell lines. As seen in Figure 10, a remarkable attenuation of the Hsp27 expression was achieved in all cases. This effect was comparable, if not greater, than that achieved with Lipofectamine, the reference standard reagent for siRNA/DNA transfection (e.g., **AD** was 6 times more efficient than Lipofectamine in downregulating Hsp27 protein expression in PC-3 cells, Figure 10b).

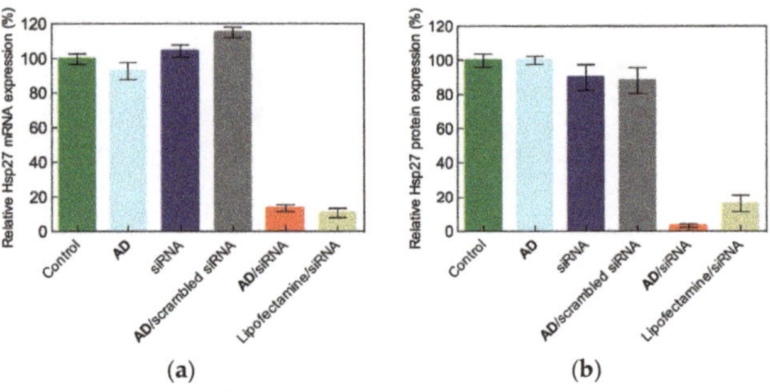

Figure 10. *Cont.*

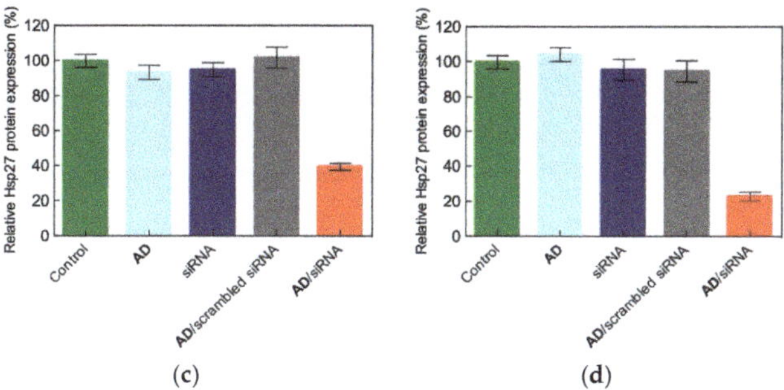

Figure 10. Downregulation of Hsp27 at the mRNA (**a**) and protein (b) level in prostate cancer PC-3 cells treated with siRNA (50 nM) delivered by the self-assembled double-tail dendron **AD** (N/P = 5). Inhibition of Hsp27 expression by siRNA/**AD** complexes (50 nM siRNA, N/P = 5) in breast cancer MCF-7 (**c**) and MDA-MB231 (**d**). Adapted from [6] with permission of John Wiley and Sons, 2014.

The knockdown of Hsp27 in these cancer cells by siRNA delivered by **AD** micelles was paralleled by significant anti-proliferative effects (~75% of cancer cell growth inhibition), in agreement with the evidence that Hsp27 silencing negatively affects PC-3 cell survival [9,17]. Similarly to what was observed for the high generation covalent PAMAM dendrimers [9], the mechanism leading to cell death relied on caspase-dependent induced apoptosis, that is, the programmed cell death was promoted by apoptotic caspases, a family of endoproteases that provide critical links in cell regulatory networks controlling cell death. Accordingly, a 3-fold increase in caspase-3/7 activation was observed on average, confirming that siRNA delivered by the nanovectors based on self-assembled **AD** is very effective in silencing Hsp27 and thereby inducing caspase-dependent anticancer activity in vitro. A remarkable additional feature of this biological outcome is that the silencing effect was completely maintained for one week, and was also effective when transfection was performed in the presence of serum, a prototypical Achilles' heel of cationic nanovectors in siRNA delivery. In fact, the negatively charged serum proteins can not only compete with siRNA binding to the nanocarrier, but also may, by virtue of strong electrostatic forces, lead to the the disintegration of the nanoassemblies with consequent premature nucleic acid release and degradation [9].

From another perspective, when confronted with its single-tail analogue **4,** the efficiency of in vitro silencing induced by the double-tailed **AD** dendron/siRNA complexes was practically identical (e.g., 97% for **AD** and 96% for **4** in PC-3 cells, Figure 10b,c, respectively). Most importantly, however, remarkably good results were obtained with **AD** when challenged against cancer stem cells (notoriously the cancer cell population most refractory to treatments), and also stem and primary cells that constitute the target of deadly viruses such as HIV-1. As shown in Figure 11, **AD** was indeed able to elicit significant RNAi interference (RNAi) in glioblastoma stem cells (PBT003), and in HIV-1 infected human primary peripheral blood mononuclear cells (PBMC-CD4$^+$), and hematopoietic stem cells (HSC-CD34$^+$). Specifically, upon treatment of the hardly tractable PBT003 cells with **AD** carrying the siRNA was directed against the signal transducer and activator of transcription 3 (STAT3), which is a key player protein in glioma-initiating cells, thought to be responsible for glioblastoma induction, progression and recurrence. A 35% reduction in the corresponding mRNA level was observed, while only 24% mRNA reduction was achieved upon transfection with Lipofectamine (Figure 11a). The data relative to the two HIV-infected cell lines (PBMC-CD4$^+$ and HSC-CD34$^+$) transfected with the siRNA targeting the HIV-1 Tat/Rev gene are more striking. The Tat and Rev HIV-1 proteins are essential positive regulators of gene expression through interaction with RNA target elements present within the 5′ translated leader sequence and envelope gene, respectively. Importantly, the genes encode the Rev and Tat overlap, with each being produced from a different reading frame. As seen from Figure 11b,c, in both cases a remarkable 50% reduction in the Tat/Rev mRNA expression was observed, while the prototypical delivery vector Lipofectamine (aka RNAiMAX) failed to induce any gene silencing in these cells infected by the acquired immunodeficiency syndrome (AIDS) virus. Fundamentally, this translated into a 50% reduction of the viral infection (Figure 11d,e) and, to the best of the authors' knowledge, constitutes the first report on the successful and safe delivery of siRNA into primary and stem cells.

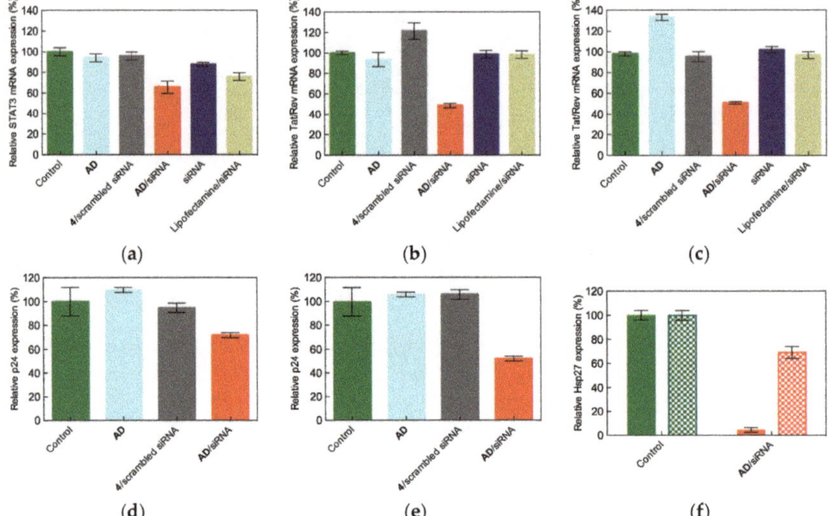

Figure 11. (**a**) STAT3 knockdown in glioblastoma stem cells (PBT003) treated with siRNA/**AD** complexes (50 nM siRNA, N/P = 5). The knockdown of Tat/Rev in primary human peripheral blood mononuclear cells (PBMC-CD4$^+$) (**b**) and hematopoietic CD34$^+$ stem cells (HSC-CD4$^+$) (**c**) treated as in (**a**). Effective inhibition of HIV replication in PBMC-CD4$^+$ (**d**) and HSC-CD4$^+$ (**e**) after treatment with the anti-Tat/Rev siRNA delivered by **AD** (50 nM siRNA, N/P = 5). Viral loading was assessed using HIV-1 p24 antigen ELISA 3 days post-treatment. PBMC-CD4$^+$ cells and HSC-CD34$^+$ cells were infected by NL4-3 virus at a multiplicity of infection (MOI) = 0.001 and by JR-FL virus (MOI = 0.005), respectively, for 5 days before transfection. (**f**) Proton-sponge effect in PC-3 cells transfected with siRNA/**AD** in the absence (solid bars) and presence (patterned bars) of the proton pump inhibitor Bafilomycin A1. Untreated cells were used as the control. Adapted from [6] with permission of John Wiley and Sons, 2014.

As for the high generation covalent PAMAM dendrimer analogues presented in the companion paper [9], it was reasoned that also in the case of the **AD** micelles, the so-called proton sponge effect [18] could be invoked to explain nucleic acid release from the self-assembled double-tail nanovectors and, therefore, potent siRNA delivery within the cell cytoplasm. According to the debated proton sponge concept, when cationic nanoparticles such as those formed by **AD** and its siRNA cargo enter acidic endosomal (and lysosomal) vesicles, unsaturated amino groups sequester protons supplied by the proton pump v-ATPase. These sequestered protons cause the pump to continue functioning, leading to the retention of chloride ions and water molecules. Eventually, osmotic swelling causes the rupture of the vesicle, allowing the cationic nanoparticles to enter the cytoplasm. As the **AD** amphiphiles feature tertiary amines in the interior of their PAMAM heads which become protonated at the endosomal acidic pH (5.5), the authors verified whether the proton sponge effect was effectively at play also during their mediated siRNA delivery. Indeed, as illustrated in Figure 11f, Hsp27 silencing in PC-3 cells was significantly reduced in the presence of Bafilomycin A1 (a proton pump inhibitor that prevents endosome acidification). This implies that the **AD**-mediated siRNA delivery was indeed dependent on the endosomal acidification process and that the proton sponge effect played a role in the endosomal escape and cytoplasmic release of the nucleic acid.

Finally, the same protocol adopted for investigating the cellular toxicity of the single-tail self-assembled dendrons described in Section 2.3 was applied to **AD**. As expected, no discernible toxicity was observed in all cell lines (PC-3, PBMC-CD4$^+$, HSC-CD34$^+$, and PBT003 transfected with 50 nM siRNA and **AD** at N/P = 5 and 10) using both the MTT and LDH assays, supporting further in vivo gene silencing experiments with this nanocarrier.

*3.3. In Vivo siRNA Delivery Performance of the Double-Tail Self-Assembling Dendron **AD***

The final steps to assess the therapeutic potential of siRNA delivery based on the double-tail self-assembled amphiphilic dendron **AD** were constituted by in vivo experiments using prostate cancer (PC-3) xenograft mouse models and Hsp27 as the target gene for knockdown. Figure 12a,b reveal that, following a 5-week animal treatment by intraperitoneal injections (2/week) of siRNA/**AD** complexes (3 mg/kg Hsp27 siRNA, **AD** at N/P = 5), there was a remarkable downregulation of Hsp27 at both mRNA (62%) and protein (73%) levels. This was paralleled by an effective inhibition of tumor growth, as shown in Figure 12c. Further, as for the case of the single-tail amphiphilic analogue **4**, no major organ injury or histopathological changes were evidenced in the animal subjected to **AD**/siRNA treatments (Figures 12d and A2).

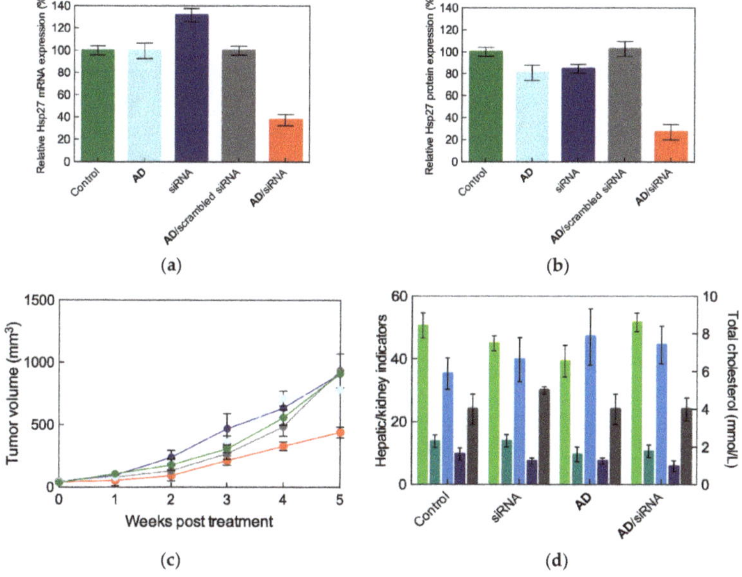

Figure 12. In vivo effective gene silencing of Hsp27 at the mRNA (**a**) and protein (**b**) levels observed in prostate cancer (PC-3) xenograft mice after treatment with Hsp27 siRNA delivered by **AD** (3 mg/kg siRNA and **AD** at N/P = 5, injection twice a week for 5 weeks). Hsp27 mRNA and protein levels were quantified by quantitative reverse transcription polymerase chain reaction (qRT-PCR) and western blotting, respectively. Non-treated mice were used for the control. **AD** alone, siRNA alone and a scrambled (non-silencing) siRNA sequence delivered by **AD** were used as negative controls. (**c**) Tumor volume during treatment of PC-3 mice xenografts with siRNA/**AD** complexes (red symbols), compared with non-treated animals (green), and animals treated by AD alone (light blue), siRNA alone (dark blue), and a scrambled (non-silencing) siRNA sequence delivered by **AD** (gray symbol) used as negative controls. (**d**) Hepatic enzyme levels (alanine transferase ALT (light green bars), gamma-glutamine transferase γ-GLT (teal bars), (kidney functions (creatinine CRE (mmol/L, sky blue bars), blood urea nitrogen BUN (μmol/L, dark blue bars), and cholesterol level in the blood (gray bars, values on the right axis) measured in male C57BL/6 mice treated with intravenous administration of siRNA/AD (3 mg/kg siRNA and **AD** at N/P = 5). Untreated mice were used for control. Adapted from [6] with permission of John Wiley and Sons, 2014.

In summary, the nanosized micelles generated upon self-assembly of the double-tail amphiphilic PAMAM-based dendron **AD** were not able to deliver siRNA to a different panel of cancer cell lines with efficiency in vitro and in vivo comparable to those formed by its single-tail precursor **4**. However, they were also capable of eliciting gene silencing effects in the highly challenging and treatment-refectory human primary cells and stem cells. Accordingly, these versatile and safe delivery properties of **AD**, coupled with easy formulation, render AD a promising nanocarrier for highly efficient, cell-type-independent RNAi therapeutics.

4. siRNA Delivery by the Double-Tail, Dual Targeting Self-Assembling Amphiphilic Dendron AD/$E_{16}G_6$RGDK

4.1. Design, Optimization and Chemico-Physical Charactrization of the Double-Tail, Dual Targeting Self-Assembling Amphiphilic Dendron **AD**/*$E_{16}G_6$RGDK*

In the general design strategy of nanocarriers to achieve effective siRNA (and in general drug) delivery, two well-known criteria must be satisfied: (i) The delivered therapeutic is be able to reach the desired disease (e.g., tumor) site(s) after administration with minimal loss in their quantity and activity during their journey in the blood stream; and (ii) they only affect diseased cells without exerting harmful effects to healthy organs and tissues. These requirements may be enabled using two strategies: Passive and active drug targeting of drugs. The former approach relies on the (still controversial) intrinsic enhanced permeability and retention effect (EPR) [19,20], according to which molecules with size in the nanometer range tend to accumulate in diseased (especially tumor) tissues much more than they do in normal tissues, thereby leaving little space for specific molecular design and optimization. Active targeting employs directed intermolecular interactions (e.g., ligand binding to disease-overexpressed receptors or other molecular recognition processes) to confer more specificity to the delivery system. In the end, besides enhanced therapeutic action, active targeting can result in reduced non-specific interactions and localization of the drug in peripheral tissues, thus minimizing unwanted side-effects. One of the biggest challenges in active targeting consists in designing new nanovectors (or modifying existing performing ones) with targeting moieties that do not interfere with its cargo loading capacity and its final mission purposes. Obviously, these chemical entities must be fully available to interact with their cellular counterparts and be endowed with strong affinity towards their targets (e.g., surface receptors) in order to trigger endocytosis.

With the goal of empowering the high-performance double-tail self-assembling amphiphilic dendron **AD** with active targeting ability for siRNA specific delivery to cancer cells, the authors considered the RGDK peptide as the starting point for molecular design and optimization. The RGDK peptide is an ideal moiety for cancer targeted nanomedicine in that it features a dual targeting capacity within a single, small molecular scaffold. On the one side, the RGD sequence is known to be able to target tumor endothelium via interaction with $\alpha v \beta_3$-integrin (a receptor overexpressed in tumor vasculature [21]); while on the other side, the RGDK sequence has been shown to bind the neutropilin-1 (Nrp-1, a transmembrane hub receptor with multiple ligands that is abundant on the surface of cancer cells [22]), and in so doing, promoted cell uptake and penetration.

Several sequence rounds of computer optimization led to the final design of the $E_{16}G_6$RGDK peptide shown in Figure 13a [8]. This new chemical entity features three distinct segments serving different functions, as follows: (1) A highly negatively charged tail composed of 16 glutamic acid residues (E_{16}), for anchoring the peptide to the positively charged surface of the self-assembled **AD** micelles via strong, favorable electrostatic interactions; (2) the positively charged RGDK warhead for fostering cancer cell targeting and homing by virtue of the dual interaction with $\alpha v \beta_3$-integrin and Nrp-1 receptors; and (3) a neutral spacer of optimal length composed of 6 glycine residues (G_6) separating the oppositely charged segments, (1) and (2).

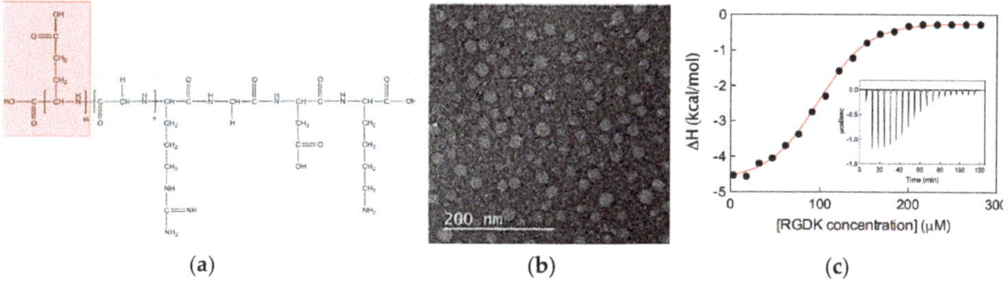

(a) (b) (c)

Figure 13. (**a**) Chemical structure of the $E_{16}G_6$RGDK peptide designed as a double targeting moiety for enhancing the siRNA delivery to cancer cells mediated by the double-tail self-assembling **AD** dendron. The light red box encases the E_{16} fragment, the light green box marks the G_6 spacer while the light blue box highlights the RGDK warhead. (**b**) TEM imaging of the siRNA/**AD**/ $E_{16}G_6$RGDK nanosized micelles formed upon simple addition of the targeting peptide to pre-formed siRNA/**AD** complexes. (**c**) Integrated ITC profile for the titration of siRNA/**AD** complexes with the $E_{16}G_6$RGDK peptide. The solid red line represents data fitting with a sigmoidal function. The insert shows the corresponding ITC raw data. Adapted from [8] with permission of the American Chemical Society, 2018.

With the optimized new peptide structure at hand, and the previous knowledge about the self-assembling process of **AD** and its related performance as siRNA nanocarrier (Section 3), the authors initially verified the interaction of the $E_{16}G_6$RGDK molecules with pre-formed siRNA/**AD** complexes in solution by means of different experimental techniques. Both DLS and TEM confirmed the formation of nanosized (30–45 nm in diameter) spherical micelles (shown in Figure 13b), characterized by a ζ-potential value of +15 mV, approximately equal to half of the value measured for the siRNA/**AD** assemblies (+32 mV). This substantial decrease in positive surface charges further supported the interaction of the negatively charged (−15) peptide with the siRNA/**AD** complexes.

However, the ultimate confirmation of the effective binding of the $E_{16}G_6$RGDK targeting peptides to the siRNA/AD nanomicelles was obtained by isothermal titration calorimetry (ITC). ITC is a straightforward and noninvasive titration-based method for binding interaction analysis performed in solution at constant temperature [23]. The analysis of ITC curves directly yields the binding enthalpy (ΔH_{bind}) and, upon data fitting with a suitable thermodynamic model, the binding constant K_b can be obtained. Once K_b is known, the free energy of binding ΔG_{bind} is calculated using the relationship $\Delta G_{bind} = -RT \ln K_{bind}$, where R is the universal gas constant and T is the absolute temperature. Finally, the variation in entropy upon binding is derived from the fundamental Gibbs equation $\Delta G_{bind} = \Delta H_{bind} - T\Delta S_{bind}$. The binding thermodynamics between the targeting peptides and the siRNA/AD micelles was found to be characterized by a favorable (i.e., negative) enthalpic variation ($\Delta H_{bind} = -5.0 \pm 0.2$ kcal/mol), as evidenced by the exothermic peaks in the corresponding thermogram (Figure 13c). Also, a small, favorable (i.e., positive) entropic change ($T\Delta S = 1.9$ kcal/mol) was estimated, leading to an overall favorable (i.e., negative) ΔG_{bind} value of −6.8 kcal/mol. The substantial enthalpic nature of the binding between the RGKD peptide and the siRNA/**AD** micelles is the result of electrostatic forces as the main supramolecular interaction drivers. However, the favorable entropic contribution was ascribed to the stabilizing peptide/micelle hydrophobic interactions and the concomitant release of ions and water molecules into the bulk solvent [24–26].

The structure of the siRNA/**AD**/$E_{16}G_6$RGDK was next investigated at the molecular level via computer simulations. Extensive (1 μs) atomistic molecular dynamics (MD) simulations revealed the formation of stable, Janus-like nanoparticles featuring a siRNA molecule bound on one face of each **AD** micelle and 4 peptides adsorbed on the opposite face (Figure 14a). Further, MD simulations showed that the negatively charged peptide tails ($E_{16}G_6$) were adsorbed onto the positively charged surface micelle while, most importantly, the 4 RGDK warhead was free from sterical hindrance and protruded deeply into the solvent (Figure 14b).

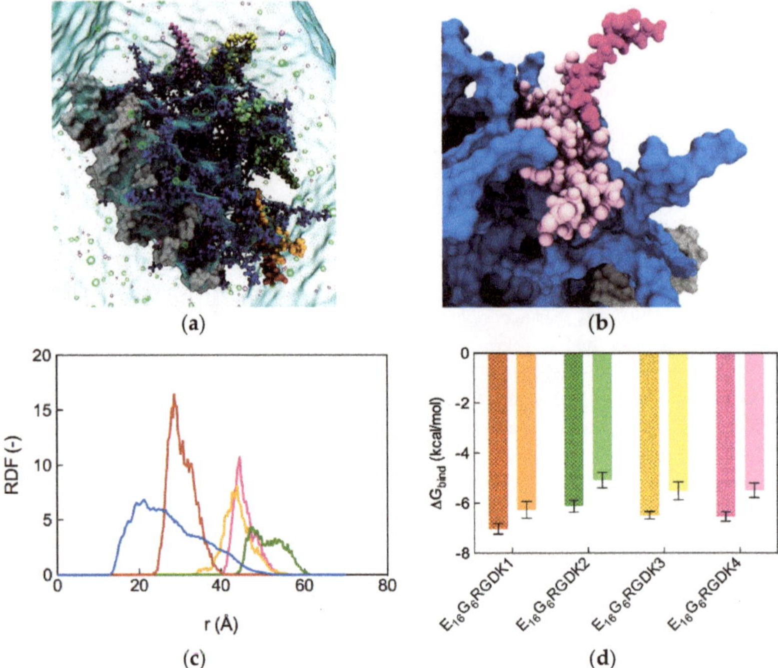

Figure 14. (**a**) Image extracted from the equilibrated portion of the MD simulation of a siRNA/**AD** micelle in complex with 4 $E_{16}G_6$RGDK peptides performed in 150 nM NaCl water solution. The 4 peptides are numbered and colored as follows: $E_{16}G_6$RGDK1, tail ($E_{16}G_6$) sandy brown, head (RDGK) firebrick; $E_{16}G_6$RGDK2, tail light green, head olive drab; $E_{16}G_6$RGDK3, tail khaki, head golden rod; $E_{16}G_6$RGDK4, tail plum, head magenta. The **AD** micelle is in a dark slate blue sticks-and-balls representation, the siRNA molecule is highlighted by its gray van der Waals surface, the peptides are shown as colored spheres, Na^+ and Cl^- ions are depicted as purple and green hollow spheres, while a light cyan transparent surface is used to represent the water solvent. (**b**) Zoomed view of the image in panel (**a**), showing the $E_{16}G_6$RGDK4 peptide with its tail (plum) attached to the micelle surface and its RGDK warhead (magenta) protruding into the solvent. In this panel, water and ions are omitted for clarity. (**c**) Radial distribution function (RDF) from the **AD** center of mass for the positively charged **AD** terminal groups (dark slate blue), and for each of the four RGDK warhead groups (colors as in panel (**a**)). (**d**) Free energy of binding (ΔG_{bind}) of each $E_{16}G_6$RGDK peptide on the siRNA/**AD** micelle. The patterned bars represent the ΔG_{bind} values for the entire peptide molecule, while the solid bars show the contribution to ΔG_{bind} afforded by the negatively charged ($E_{16}G_6$) tails only. Adapted from [8] with permission of the American Chemical Society, 2018.

These pictorial views were quantified first by calculating the radial distribution function (RDF) from the **AD** micelle center of mass for the positively charged **AD** terminal groups and the 4 targeting peptide RGDK warheads, as shown in Figure 14c. As seen from this Figure, all 4 RGDK moieties are positioned far away from the nanomicelle periphery. Accordingly, they are potentially available for interaction with their cellular target receptors. Next, the free energy of binding ΔG_{bind} of each single $E_{16}G_6$RGDK peptide with the siRNA/**AD** nanomicelle was estimated by processing the MD data according to the MM/PBSA methodology [9–15]. Figure 14d shows not only that all peptides are characterized by favorable ΔG_{bind} values (from −7.0 to −6.2 kcal/mol), but also that the major contribution to binding is afforded by the strong electrostatic interaction between the negatively charged peptide tails ($E_{16}G_6$) and the positively charged micellar surface, as expected by the original molecular design.

4.2. *In Vitro Targeted siRNA Delivery Performance of Double-Tail, Dual Targeting Self-Assembling Amphiphilic Dendron AD/E$_{16}$G$_6$RGDK*

The authors verified that the presence of the E$_{16}$G$_6$RGDK peptides on the surface of the siRNA/**AD** micelles was indeed effective in specifically targeting cancer cells via dual binding with the designated target receptors $\alpha v\beta_3$integrin and Nrp-1. The cellular uptake of the siRNA/**AD**/E$_{16}$G$_6$RGDK complexes was investigated again with PC-3 cells, in which both receptors were highly expressed [27,28]. As seen in Figure 15a, the amount of siRNA/**AD**/E$_{16}$G$_6$RGDK complexes internalized by PC-3 cells was three times higher than that measured for the siRNA/**AD** micelle. Further, the cellular uptake of the nanocarrier modified with the targeting moieties was inhibited in the presence of a cyclic RGD peptide. This clearly indicates that this last molecule competed with the peptide-coated nanovector for integrin binding via the RDG motif.

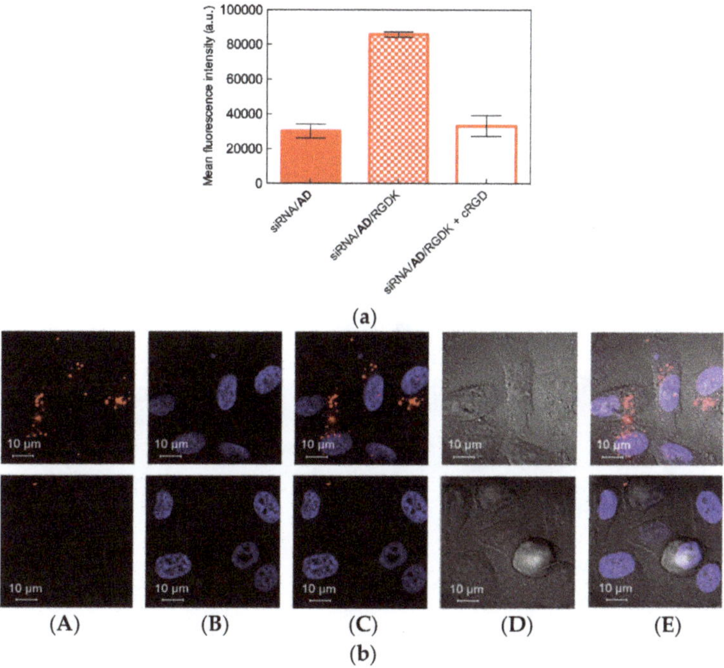

Figure 15. (**a**) Flow cytometry analysis of cellular uptake of siRNA/**AD** and siRNA/**AD**/E$_{16}$G$_6$RGDK nanoparticles in PC-3 cells using a siRNA labeled with the fluorescent Cy5 die. The last column shows the cell uptake results for the siRNA/**AD**/E$_{16}$G$_6$RGDK obtained in the presence of the cyclic peptide c-RGD, an $\alpha v\beta_3$-integrin integrin binder. (**b**) Cellular uptake visualized by confocal microscopy of the siRNA/**AD**/E$_{16}$G$_6$RGDK complexes in the absence (top panel) and in the presence (bottom panel) of a Nrp-1 antibody. Images from left to right: (**A**) cy5-labelled siRNA; (**B**) nucleus (4′,6-diamidino-2-phenylindole (DAPI) staining); (**C**) merged (**A**,**B**). The last two images (**D**,**E**) are the bright field version of (**B**,**C**). In all these experiments, the siRNA/**AD**/E$_{16}$G$_6$RGDK nanoassemblies were prepared using 50 nM siRNA, siRNA/**AD** N/P = 10, and **AD**/E$_{16}$G$_6$RGDK molar ratio = 5. Adapted from [8] with permission of the American Chemical Society, 2018.

Next, PC-3 cells pretreated with a monoclonal antibody (mAb) directed against Nrp-1 were used to assay the internalization of the siRNA/**AD**/E$_{16}$G$_6$RGDK nanoassemblies in comparison with untreated cells. As can be easily seen in the corresponding confocal microscopy images (Figure 15b), in the presence of the Nrp-1 mAb (bottom panel), the uptake of thc siRNA-loaded peptide-modified **AD** micelles was drastically reduced when the mAb was absent (upper panel). This indicates that the targeting peptide-decorated siRNA delivery system was effectively cell internalized also via its interaction of its RGDK warhead with the Nrp-1 receptor.

Following verification of the successful attribution of the desired targeting properties to our double-tail self-assembling **AD** nanovector, the authors checked whether its excellent endosomal escape features, required for efficient siRNA delivery and release, was preserved. By confocal microscopy, it was observed that 2 h after transfection, almost all of the siRNA molecules were localized in the cell cytoplasm (Figure A3). This confirms that the peptide-decorated **AD** nanovectors were also able to effectively circumvent endosomal trapping and release their nucleic acid cargo for subsequent RNAi.

The authors further assessed whether the endowment of the double-tail self-assembling dendron **AD** with targeting properties eventually conferred by the RGDK-based peptides rendered this nanovector more effective in siRNA delivery to cancer cells with respect to the undecorated **AD** micelles. Their RNAi performance in silencing Hsp27 in PC-3 cancer cells was compared. The relevant results showed that a remarkable Hsp27 knockdown was achieved both at the mRNA (56%, Figure 16a) and the protein (90%) levels with a very low (20 nM) siRNA concentration (for comparison, a 50 nM siRNA concentration was required to produce similar results either with the covalent G_5 PAMAM dendrimer [9] or with the **AD** micelles without targeting peptide decoration (Figure 10)). Moreover, the same experiments performed in the presence of either the c-RGD peptide or the Nrp-1 directed mAb resulted in a decreased gene silencing effect (as shown by the last two columns in Figure 16a). This confirms that the siRNA/AD/$E_{16}G_6$RGDK nanoparticles indeed interact with the two target receptors.

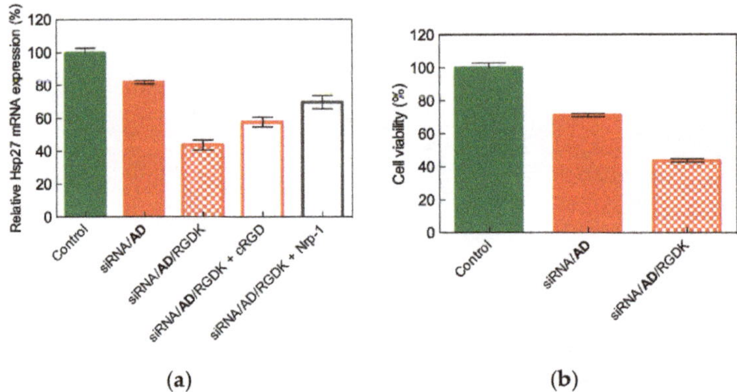

Figure 16. (**a**) mRNA levels following knockdown of Hsp27 in PC-3 cells with siRNA/**AD** and siRNA/**AD**/$E_{16}G_6$RGDK nanoparticles. The last two columns show the results of the same experiments performed in the presence of the integrin binder c-RGD and of the mAb directed toward the Nrp-1 receptor, respectively. (**b**) Comparison of the antiproliferative activity exerted on PC-3 cells by siRNA/**AD** and siRNA/**AD**/$E_{16}G_6$RGDK nanoparticles. In both panels, non-treated cells were used for control, siRNA concentration = 20 nM, siRNA/**AD** N/P = 10, and **AD**/$E_{16}G_6$RGDK molar ratio = 5. Adapted from [8] with permission of the American Chemical Society, 2018.

Following siRNA delivery to PC-3 cells with the peptide-decorated **AD** nanomicelles, a drastic anticancer activity was observed again using only a 20 nM siRNA concentration (Figure 16b). This corresponds to an effect of 30% higher than that achieved with the pristine **AD** nanovectors. Concomitantly, the **AD**/$E_{16}G_6$RGDK nanoassemblies exhibited a safe toxicity profile, similar to that presented by the **AD** micelles (Figure A4), and in addition to its notable improvement in hemolytic effects (Figure A4c). All these data led the authors to perform the final testing of the siRNA/AD/E16G6RGDK nanoparticles in vivo.

4.3. *In Vivo Targeted siRNA Delivery Performance of Double-Tail, Dual Targeting Self-Assembling Amphiphilic Dendron AD/$E_{16}G_6$RGDK*

The in vivo gene silencing and anticancer effects of the peptide-decorated **AD** nanomicelles was investigated again using PC-3 xenograft nude mice. The first analysis was conducted to verify their enhanced tumor targeting ability with respect to the undecorated **AD** nanovectors using a fluorescent-labeled Hsp27 siRNA. Accordingly, ex vivo imaging of isolated tumors and the quantification of the relevant mean fluorescence intensity were performed (Figure 17a,b), from which the substantially higher tumor accumulation of the targeted nanoparticles was substantiated.

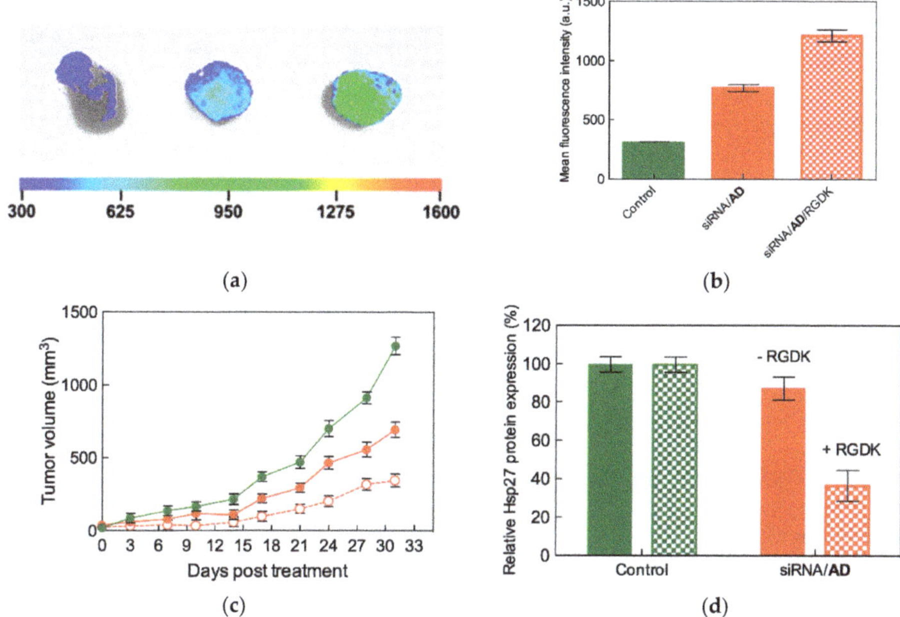

Figure 17. (**a**) Ex vivo fluorescence images obtained from tumor isolated from PC-3 xenograft mice treated with PBS buffer (delivery medium used as control, left), siRNA/**AD** (center), and siRNA/**AD**/$E_{16}G_6$RGDK nanoparticles (right). (**b**) Quantification of the mean fluorescence intensity for the images in panel **a**. PBS buffer (**c**) Tumor growth in untreated animals (green filled symbols), and in animals treated with siRNA/**AD** (red filled symbols). (**d**) In vivo knockdown of Hsp27 with siRNA/**AD** (-RGKD) and siRNA/**AD**/$E_{16}G_6$RGDK (+ RGDK) nanoparticles. Untreated animals were used as the control. The treatments (0.25 mg/kg siRNA, siRNA/**AD** N/P = 10, and **AD**/$E_{16}G_6$RGDK molar ratio = 5) were administered by intravenous injection twice per week. Adapted from [8] with permission of the American Chemical Society, 2018.

After 4 weeks of mice treatment with a very low siRNA concentration (0.25 mg/kg)—12-fold less than the conventional concentration used for siRNA delivery in mice—both the tumor volume and the corresponding levels of Hsp27 protein were drastically reduced when siRNA was administered via the peptide-decorate **AD** nanovectors, as shown in Figure 17c,d. In particular, at day 31 after treatment with the siRNA/**AD**/$E_{16}G_6$RGDK nanoparticles, tumor growth was reduced by 73% with respect to the control and by 50% relative to siRNA/**AD** treatment (Figure 17c). Consistently, Hsp27 expression in tumors was reduced by 65% and 41% upon siRNA treatment with the targeted nanovectors with respect to the control and **AD**-delivered siRNA, respectively (Figure 17d).

These highly promising results, coupled with the excellent in vivo toxicity profile exhibited by the **AD**/$E_{16}G_6$RGDK nanocarriers confirmed the high potential for these targeting nanosystems as safe and efficient siRNA delivery, gene silencing and consequent anticancer nanotherapeutics.

5. Conclusions

As defined by Nature.com [29], "self-assembly is the process by which an organized structure spontaneously forms from individual components, as a result of specific, local interactions among the components. When the constitutive components are molecules, the process is termed molecular self-assembly". A key feature of molecular self-assembly is the multivalent, cooperative and synergistic nature of the intermolecular interactions leading to the organization of individual molecular entities into well-defined nanosized structures. This approach presents at least two major conceptual advantages. The first relies on the evidence that the driving forces governing self-assembly lead to the formation of the nano-objects virtually with no flaws, as interactions between the nanomicelle building blocks are mediated by specific molecular recognition ultimately resulting in complex and ordered nanoscale structures. The second, by no means of less important benefit to self-assembly is that very small amounts of material are required to accomplish the process.

In the companion paper [9], the authors presented the design, synthesis and gene silencing activity of modified PAMAM-based covalent nanovectors for siRNA delivery in cancer therapeutics. However, high generation dendrimer synthesis is extremely laborious and time-consuming, since the final product purification is difficult and hampered by the presence of highly similar side products. Thus, notwithstanding the highly promising results achieved with these molecules, the difficulties inherent in large-scale good manufacturing practice (GMP) production of high generation dendrimers led the authors to explore the potential of self-assembly in the design of new efficient siRNA nanocarriers.

Thus, as presented in this second short review, the small PAMAM-based amphiphilic dendron **4**, composed by a PAMAM head and a C_{18}-long hydrocarbon tail, was initially developed. This was safe and effective in delivering siRNA both in vitro and in vivo. The natural evolution of **4** was its double-tail counterpart **AD**. The in vivo gene silencing effect obtained with siRNA delivered by **AD** nanomicelles in cancer treatment was comparable to that obtained with its precursor dendron **4**. However, **AD** nanovectors were also capable of eliciting gene silencing effects in the highly challenging and treatment-refectory human primary cells and stem cells.

The RNAi results of both the amphiphilic dendrons **4** and **AD** could be mainly attributed to passive targeting via the EPR effect in addition to their excellent siRNA delivery and endosomal escape capabilities. The authors focused on equipping **AD** with the dual $E_{16}G_6$RGDK targeting peptide, with the final goal of endowing the resulting nanovectors with cancer cell-directing specificity and, accordingly, higher anticancer activity. Compared to **AD**, the peptide-decorated **AD** nanomicelles exhibited more than 10-fold greater in vivo RNAi at significantly lower siRNA doses with respect to both covalent PAMAM dendrimers [9] and the non-targeted **AD** nanovectors. Our current activity in the field is progressing along this line, with the computer-assisted design of new targeting moieties for decorating self-assembling nanovectors in the development of gene silencing-based personalized medicine against different highly challenging and deadly (e.g., glioblastoma and pancreatic) human cancers.

Supplementary Materials: The following are available online at http://www.mdpi.com/1999-4923/11/7/324/s1, Figure S1. Schematic representation of the coarse-grained DPD model of the AD dendron. The different bead types are colored as follows: RC, dark magenta; R, plum; L, dark turquoise; G, chartreuse; C, light gray. Table S1. Example of DPD interaction parameters used to simulate the self-assembling of the amphiphilic dendron AD per se and in the presence of siRNA molecules.

Funding: This research was funded by the Italian Association for Cancer Research (AIRC), grant IG17413 to SP. The assistant position (RTDa) of SA is fully supported by the University of Trieste, in agreement with the actuation of the strategic planning financed by the Italian Ministry for University and Research (MIUR, triennial program 2016–2018) and the Regione Friuli Venezia Giulia (REFVG, strategic planning 2016–18), assigned to SP. This award is deeply acknowledged.

Acknowledgments: Authors wish to thank Ling Peng and her group for the longstanding, fruitful collaboration, the challenges in siRNA delivery nanovector design and optimization, the inspiring discussions and, above all, the personal friendship.

Conflicts of Interest: The authors declare no conflict of interest.

Appendix A

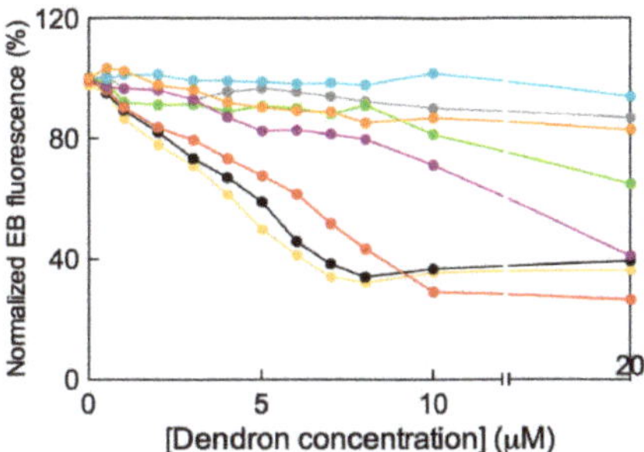

Figure A1. Experimental determination of siRNA binding to self-assembling dendrons **1** (gray), **2** (light green), **3** (purple), **4** (red), **5** (black), and **6** (yellow) by EB assay. Non-self-assembling dendrons **7** (cyan) and **8** (orange) were used as negative controls. Adapted from [7] with the permission of John Wiley and Sons, 2016.

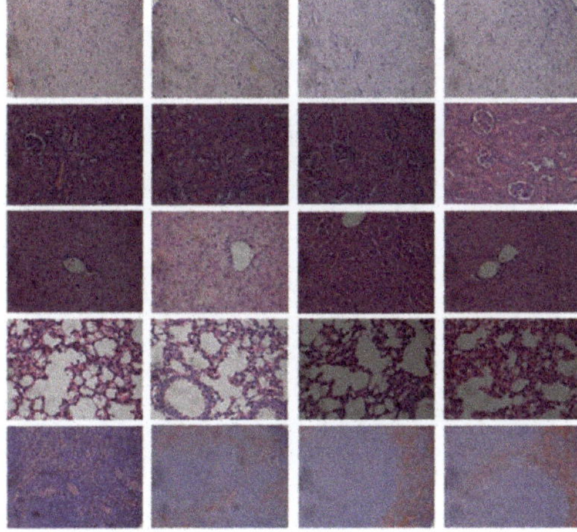

Figure A2. Pathology of major organs from treated mice (HES staining). Columns (from left to right): control, siRNA, **AD**, **AD**/siRNA. Rows (from to bottom): heart, kidney, liver, lung, spleen. Adapted from [6] with the permission of John Wiley and Sons, 2014.

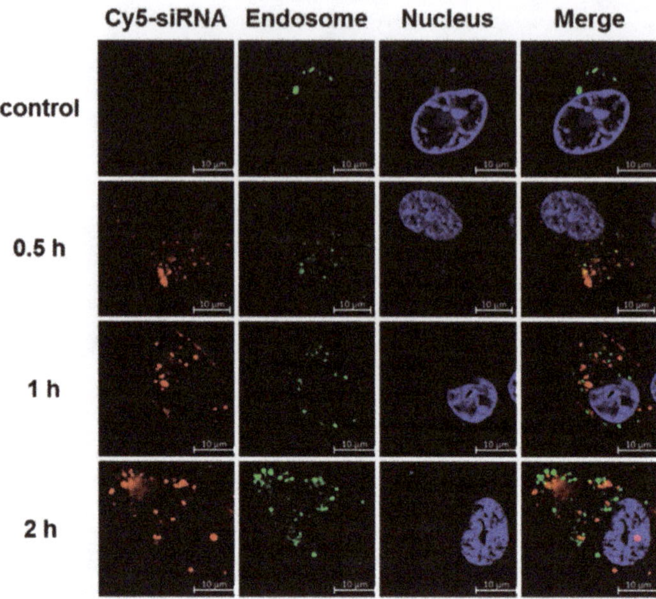

Figure A3. Confocal microscopy images of the successful endosomal escape of the siRNA/**AD**/$E_{16}G_6$RGDK nanoparticles in PC-3 cells after incubation times 0 (control), 0.5, 1, and 2 h Red channel images show the cy5-labeled siRNA/AD/$E_{16}G_6$RGDK nanoparticles, the green channel images reveal the Lyso Tracker red-marked endosomes, and the blue channel images show the cell nuclei stained with Hoechst 33342. In all these experiments, the siRNA/**AD**/$E_{16}G_6$RGDK nanoassemblies were prepared using 50 nM siRNA, siRNA/**AD** N/P = 10, and **AD**/$E_{16}G_6$RGDK molar ratio = 5. Adapted from [8] with the permission of the American Chemical Society, 2018.

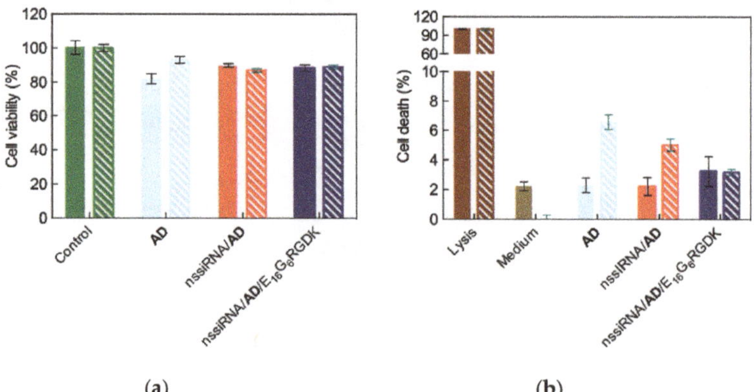

(**a**) (**b**)

Figure A4. *Cont.*

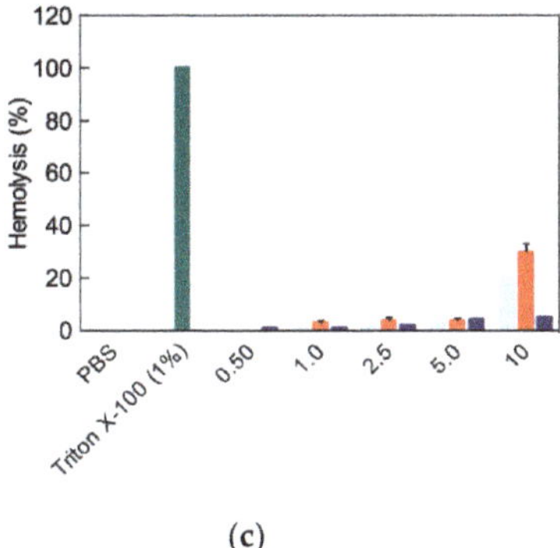

(c)

Figure A4. (**a**) Toxicity assessment using the MTT assay in prostate cancer PC-3 cells (solid bars) and in human embryonic kidney (HEK293) cells (patterned bars) of the **AD** micelles alone, **AD** in complex with a non-silencing siRNA (nssiRNA/**AD**) and the targeting peptide-decorated **AD** in complex with a non-silencing siRNA (nssiRNA/**AD**/$E_{16}G_6$RGDK). (**b**) Toxicity assessment using the LDH assay in prostate cancer PC-3 cells (solid bars) and in human embryonic kidney (HEK293) cells (patterned bars) of the **AD** micelles alone, **AD** in complex with a non-silencing siRNA (nssiRNA/**AD**) and the targeting peptide-decorated **AD** in complex with a non-silencing siRNA (nssiRNA/**AD**/$E_{16}G_6$RGDK). (**c**) Hemolysis effect measured via UV absorption at 540 nm for the **AD** micelles alone, **AD** in complex with a non-silencing siRNA (nssiRNA/**AD**) and the targeting peptide-decorated **AD** in complex with a non-silencing siRNA (nssiRNA/**AD**/$E_{16}G_6$RGDK) at different **AD** concentrations (corresponding to nssiRNA concentrations of 10, 20, 50, 100, and 200 nM, respectively). In all these experiments, the siRNA/**AD**/$E_{16}G_6$RGDK nanoassemblies were prepared using 50 nM siRNA, siRNA/**AD** N/P = 10, and **AD**/$E_{16}G_6$RGDK molar ratio = 5. Adapted from [8] with the permission of the American Chemical Society, 2018.

References

1. Lehn, J.-M. Toward self-organization and complex matter. *Science* **2002**, *295*, 2400–2403. [CrossRef] [PubMed]
2. Whitesides, G.M.; Grzybowski, B. Self-Assembly at all scales. *Science* **2002**, *295*, 2418–2421. [CrossRef] [PubMed]
3. Mattia, E.; Otto, S. Supramolecular systems chemistry. *Nat. Nanotechnol.* **2015**, *10*, 111–119. [CrossRef] [PubMed]
4. Ariga, K.; Nishikawa, M.; Mori, T.; Takeya, J.; Shrestha, L.K.; Hill, J.P. Self-assembly as a key player for materials nanoarchitectonics. *Sci. Technol. Adv. Mater.* **2019**, *20*, 51–95. [CrossRef] [PubMed]
5. Whitty, A. Cooperativity and biological complexity. *Nat. Chem. Biol.* **2008**, *4*, 435–439. [CrossRef]
6. Liu, X.; Zhou, J.; Yu, T.; Chen, C.; Cheng, Q.; Sengupt, K.; Huang, Y.; Li, H.; Liu, C.; Wang, Y.; et al. Adaptive amphiphilic dendrimer-based nanoassemblies as robust and versatile siRNA delivery systems. *Angew. Chem. Int. Ed. Engl.* **2014**, *53*, 11822–11827. [CrossRef] [PubMed]
7. Chen, C.; Posocco, P.; Liu, X.; Cheng, Q.; Laurini, E.; Zhou, J.; Liu, C.; Wang, Y.; Tang, J.; Dal Col, V.; et al. Mastering dendrimer self-assembly for efficient siRNA delivery: From conceptual design to in vivo efficient gene silencing. *Small* **2016**, *12*, 3667–3676. [CrossRef]
8. Dong, Y.; Yu, T.; Ding, L.; Laurini, E.; Huang, Y.; Zhang, M.; Weng, Y.; Lin, S.; Chen, P.; Marson, D.; et al. A dual targeting dendrimer-mediated siRNA delivery system for effective gene silencing in cancer therapy. *J. Am. Chem. Soc.* **2018**, *140*, 16264–16274. [CrossRef]
9. Marson, D.; Laurini, E.; Aulic, S.; Fermeglia, M.; Pricl, S. Evolution from covalent to self-assembled PAMAM-based dendrimers as nanovectors for siRNA delivery in cancer by coupled in silico-experimental studies. Part I: Covalent siRNA nanocarriers. *Pharmaceutics* **2019**, in press.
10. Karatasos, K.; Posocco, P.; Laurini, E.; Pricl, S. Poly(amidoamine)-based dendrimer/siRNA complexation studied by computer simulations: Effects of pH and generation on dendrimer structure and siRNA binding. *Macromol. Biosci.* **2012**, *12*, 225–240. [CrossRef]

11. Posocco, P.; Laurini, E.; Dal Col, V.; Marson, D.; Karatasos, K.; Fermeglia, M.; Pricl, S. Tell me something that I do not know. Multiscale molecular modeling of dendrimer/dendron organization and self-assembly in gene therapy. *Curr. Med. Chem.* **2012**, *19*, 5062–5087. [CrossRef] [PubMed]
12. Posocco, P.; Laurini, E.; Dal Col, V.; Marson, D.; Peng, L.; Smith, D.K.; Klajnert, B.; Bryszewska, M.; Caminade, A.-M.; Majoral, J.P.; et al. Multiscale modeling of dendrimers and dendrons for drug and nucleic acid delivery. In *Dendrimers in Biomedical Applications*; Klajnert, B., Peng, L., Ceña, V., Eds.; RSC Publishing: Cambrige, UK, 2013; pp. 148–166.
13. Pavan, G.M.; Posocco, P.; Tagliabue, A.; Maly, M.; Malek, A.; Danani, A.; Ragg, E.; Catapano, C.V.; Pricl, S. PAMAM dendrimers for siRNA delivery: Computational and experimental insights. *Chem. Eur. J.* **2010**, *16*, 7781–7795. [CrossRef] [PubMed]
14. Marson, D.; Laurini, E.; Posocco, P.; Fermeglia, M.; Pricl, S. Cationic carbosilane dendrimers and oligonucleotide binding: An energetic affair. *Nanoscale* **2015**, *7*, 3876–3887. [CrossRef] [PubMed]
15. Mehrabadi, F.S.; Hirsch, O.; Zeisig, R.; Posocco, P.; Laurini, E.; Pricl, S.; Haag, R.; Kemmner, W.; Calderón, M. Structure–activity relationship study of dendritic polyglycerolamines for efficient siRNA transfection. *RSC Adv.* **2015**, *5*, 78760–78770. [CrossRef]
16. Percec, V.; Wilson, D.A.; Leowanawat, P.; Wilson, C.J.; Huges, A.D.; Kaucher, M.S.; Hammer, D.A.; Levine, D.H.; Kim, A.J.; Bates, F.S.; et al. Self-assembly of Janus dendrimers into uniform dendrimersomes and other complex architectures. *Science* **2010**, *328*, 1009–1014. [CrossRef] [PubMed]
17. Rocchi, P.; So, A.; Kojima, S.; Signaevsky, M.; Beraldi, E.; Fazli, L.; Hurtado-Coll, A.; Yamanaka, K.; Gleave, M. Heat shock protein 27 increases after androgen ablation and plays a cytoprotective role in hormone-refractory prostate cancer. *Cancer Res.* **2004**, *64*, 6595–6602. [CrossRef] [PubMed]
18. Behr, J.P. The proton sponge: A trick to enter cells viruses did not exploit. *Chimia* **1997**, *51*, 34–36.
19. Nichols, J.W.; Bae, Y.H. EPR: Evidence and fallacy. *J. Control. Release* **2014**, *190*, 451–464. [CrossRef]
20. Danhier, F. To exploit the tumor microenvironment: Since the EPR effect fails in the clinic, what is the future of nanomedicine? *J. Control. Release* **2016**, *244*, 108–121. [CrossRef]
21. Desgrosellier, J.; Cheresh, D. Integrins in cancer: Biological implications and therapeutic opportunities. *Nat. Rev. Cancer* **2010**, *10*, 9–22. [CrossRef]
22. Teesalu, T.; Sugahara, K.; Kotamraju, V.; Ruoslahti, E. C-end rule peptides mediate neuropilin-1-dependent cell, vascular, and tissue penetration. *Proc. Natl. Acad. Sci. USA* **2009**, *106*, 16157–16162. [CrossRef] [PubMed]
23. Velázquez Campoy, A.; Freire, E. ITC in the post-genomic era...? Priceless. *Biophys. Chem.* **2005**, *115*, 115–124. [CrossRef] [PubMed]
24. Thornalley, K.A.; Laurini, E.; Pricl, S.; Smith, D.K. Enantiomeric and diastereomeric self-assembled multivalent nanostructures: Understanding the effects of chirality on binding to polyanionic heparin and DNA. *Angew. Chem. Int. Ed. Engl.* **2018**, *57*, 8530–8534. [CrossRef] [PubMed]
25. Rodrigo, A.C.; Laurini, E.; Vieira, V.M.P.; Pricl, S.; Smith, D.K. Effect of buffer at nanoscale molecular recognition interfaces - electrostatic binding of biological polyanions. *Chem. Commun.* **2017**, *53*, 11580–11583. [CrossRef] [PubMed]
26. Marson, D.; Laurini, E.; Fermeglia, M.; Smith, D.K.; Pricl, S. Mallard Blue binding to heparin, its SDS micelle-driven decomplexation, and interaction with human serum albumin: A combined experimental/modeling investigation. *Fluid Phase Equilib.* **2018**, *470*, 259–267. [CrossRef]
27. Goel, H.L.; Li, J.; Kogan, S.; Languino, L.R. Integrins in prostate cancer progression. *Endocr. Relat. Cancer* **2008**, *15*, 657–664. [CrossRef] [PubMed]
28. Soker, S.; Takashima, S.; Miao, H.Q.; Neufeld, G.; Klagsbrun, M. Neuropilin-1 is expressed by endothelial and tumor cells as an isoform-specific receptor for vascular endothelial growth factor. *Cell* **1998**, *92*, 735–745. [CrossRef]
29. Available online: https://www.nature.com/subjects/self-assembly (accessed on 4 July 2019).

 *pharmaceutics*

Review

Lipid Delivery Systems for Nucleic-Acid-Based-Drugs: From Production to Clinical Applications

Anna Angela Barba [1,2], Sabrina Bochicchio [1,*], Annalisa Dalmoro [1,2] and Gaetano Lamberti [1,3]

1 Eng4Life Srl, Spin-off Accademico, Via Fiorentino, 32, 83100 Avellino, Italy
2 Dipartimento di Farmacia; Università degli Studi di Salerno, via Giovanni Paolo II, 132 84084 Fisciano (SA), Italy
3 Dipartimento di Ingegneria Industriale; Università degli Studi di Salerno, via Giovanni Paolo II, 132 84084 Fisciano (SA), Italy
* Correspondence: sbochicchio@unisa.it; Tel.: +39-089968291

Received: 8 July 2019; Accepted: 23 July 2019; Published: 24 July 2019

Abstract: In the last years the rapid development of Nucleic Acid Based Drugs (NABDs) to be used in gene therapy has had a great impact in the medical field, holding enormous promise, becoming "the latest generation medicine" with the first ever siRNA-lipid based formulation approved by the United States Food and Drug Administration (FDA) for human use, and currently on the market under the trade name Onpattro™. The growth of such powerful biologic therapeutics has gone hand in hand with the progress in delivery systems technology, which is absolutely required to improve their safety and effectiveness. Lipid carrier systems, particularly liposomes, have been proven to be the most suitable vehicles meeting NABDs requirements in the medical healthcare framework, limiting their toxicity, and ensuring their delivery and expression into the target tissues. In this review, after a description of the several kinds of liposomes structures and formulations used for in vitro or in vivo NABDs delivery, the broad range of siRNA-liposomes production techniques are discussed in the light of the latest technological progresses. Then, the current status of siRNA-lipid delivery systems in clinical trials is addressed, offering an updated overview on the clinical goals and the next challenges of this new class of therapeutics which will soon replace traditional drugs.

Keywords: NABDs; siRNA; liposomes; clinical trials

1. Introduction

Gene therapy is the treatment of diseases at molecular level by "switching genes on or off" through the use of Nucleic Acid Based Drugs (NABDs) which, including oligodeoxynucleotides, plasmid DNA, ribozymes, siRNA, miRNA and related chemically-synthesized molecules, are highly degradable therapeutics that need to be loaded into a vehicle in order to reach the patient target cell/tissue where the defective or mutated gene has to be corrected [1,2]. Indeed, naked genetic materials cannot be easily systemically administered due to their toxicity, low stability in serum, rapid renal clearance, reduced uptake by target cells, phagocyte uptake and their ability in activating the immune response, all features that preclude their clinical development [3]. Therefore, choosing the right vector represents one the most critical steps in attaining a successful gene therapy, and carriers based on lipid materials seem to be the most promising, as they have similar structures to cell membranes [4]. To date, lipid-based NABDs delivery systems represent a novel approach for the treatment of different diseases, mostly including advanced cancers, amyloidosis, fibrosis, hypercholesterolemia and various virus infections, as evidenced by the numerous products under clinical trials, ongoing or planned [3,5–7].

One of the most significant advances in lipid-based NABDs therapies dates back only to August 2018, and is undoubtedly represented by Patisiran (ALN-TTR02), the first ever approved RNAi therapeutic by the Food and Drug Administration (FDA) and by the European Commission (EC). Produced by Alnylam Pharmaceuticals Corporation and sold under the trade name Onpattro™, ALN-TTR02 is an siRNA formulation based upon Stable Nucleic Acid Lipid Particle (SNALP) transfecting technology, and is indicated for the treatment of the Hereditary Transthyretin-mediated Amyloidosis (hATTR), a progressively incapacitating and often fatal genetic disorder. Thus, after 20 years from the first publication about the RNAi-mechanism discovery [8], in August 2018 the gene-silencing technology has had its first drug approval as stated in an article published by Nature and just entitled "Gene-silencing technology gets first drug approval after 20-year wait" [9]. It appears evident how the clinical translation of lipid-based NABDs delivery systems shelf has progressed [10–17], leading to the formulation of several lipid-based carriers (see schematization of Figure 1), including micelles, solid lipid nanoparticles and liposomes [18].

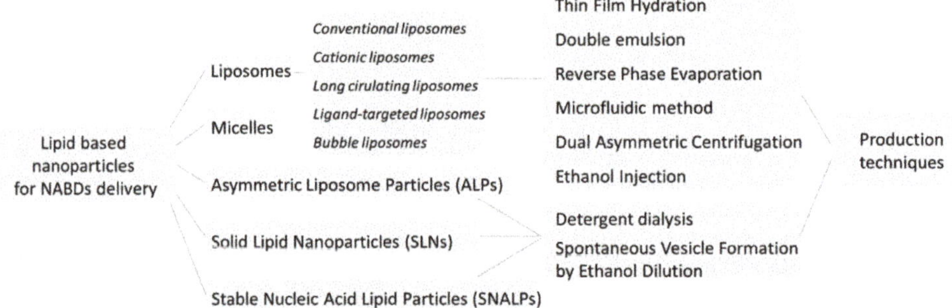

Figure 1. Schematization of lipid-based nanoparticles that are used for Nucleic Acid Based Drugs (NABDs) delivery and their production techniques.

Among these systems, liposomes are the most effective system. They are spherical particles composed of phospholipids, encapsulating a volume of aqueous medium showing several favorable features, such as an excellent biocompatibility, biodegradability, low toxicity, low immunogenicity, an ability to deliver a large piece of nucleic acids, structural flexibility and easiness of handling. Moreover, cationic liposomes, made by positively-charged lipids, are especially indicated for NABDs delivery, being this last characterized by negatively-charged phosphate groups. Their association gives compacted structures which can also electrostatically interact with the negatively-charged cell membrane, facilitating their cellular uptake [19]. In particular, in recent years, cationic liposomes have gained a lot of attention for siRNA delivery, improving their uptake into tumor tissues and their stability and bioavailability [20,21]. The siRNAs are 21–23 nucleotides long, double-stranded RNA, which, binding at specific sequences of messenger RNA (mRNA), "interfere" with the translation of proteins. Their therapeutic potential as a next generation medicine has been recently reviewed [22], together with the major delivery strategies currently adopted for their effective in vitro and in vivo delivery [23].

In this work an overview on the advances in NABDs-lipid systems delivery is presented. In particular, the attention is before focused upon liposomes as a lipid carrier system and on siRNA as NABDs. After, a discussion about liposomes structures, the formulations and production techniques suitable for NABDs delivery are reported. In the last part, the medical applications of siRNA-liposomes currently under investigation in clinical trials are then treated in detail.

2. Liposomes: Structures and Basic Formulations

Liposomes are vesicles characterized by a lipid bilayer that surrounds an aqueous core. They are mainly composed of phospholipids recognizable by a polar hydrophilic head and two apolar hydrophobic chains [24].

When dispersed in aqueous solutions, their polar heads interact with the aqueous environment, due to the hydrogen bonds and polar interactions, while their aliphatic chains interact with each other due to the van der Waals forces, leading to a lipid bilayers formation of which they constitute the lipophilic inner compartment [24,25]. During their formulation process, water-soluble drugs can be dissolved in the aqueous compartment while hydrophobic materials can be entrapped into the lipid bilayer [26–28]. Joined by being composed of biocompatible materials, liposomes show different structures, dimensions, lipid composition and surface charge. They can be composed by several concentric bilayers separated by aqueous compartments, with an external lipid bilayer containing other ever smaller bilayers separated by water cavities, like an onion structure. In this case liposomes are called Multilamellar Vesicles (MLVs) and show a size range of 500 nm to 5 μm, or by only one phospholipid bilayer surrounding an aqueous compartment. In this case liposomes can be differentiated in small, large and giant vesicles depending upon their dimension: They are called Small Unilamellar Vesicles (SUVs) if they have a 20 to 200 nm range size, Large Unilamellar Vesicles (LUVs) with a 200 to 1 μm range size and Giant Unilamellar Vesicles (GUVs) with a size larger than 1 μm. Finally, similar in dimension to MLVs there are multi-compartmental structures constituted by vesicles surrounded by other vesicles called Multi Vesicular Vesicles (MVVs) [19,29,30]. In Figure 2 a schematization of liposomes classification based on their structure, size and composition is presented.

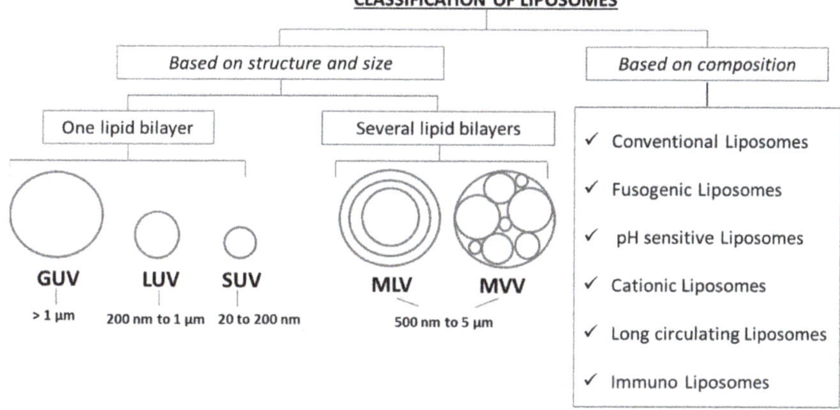

Figure 2. Liposomes classification by structure/size and lipid composition.

Liposomes are mainly made of phospholipids, which contain two major categories including glycerophospholipids and sphingomyelins. Composed by a hydrophilic head group, a hydrophobic side chain and a glycerol backbone, glycerophospholipids are those that characterized the eukaryotic cells, and are used in liposomes production by varying their head group i.e., phosphatidylcholine (PC), phosphatidylethanolamine (PE), phosphatidylserine (PS), phosphatidylglycerol (PG) [31].

2.1. Conventional Liposomes

Conventional liposomes are neutrally-charged vesicles made primarily from PC and cholesterol; the incorporation of this last into the lipid bilayer, with its hydroxyl group oriented towards the aqueous surface, and the aliphatic chain aligned parallel to the acyl chains [2], enhances the stability of the vesicles, reducing their permeability, and increasing their in vivo and in vitro performance [24,32–34].

Although reducing the in vivo drugs toxicity, being recognized as foreign substances by serum opsonines and destroyed by the reticuloendothelial system (RES), conventional liposomes are characterized by a short circulation time when intravenously administered [17,35].

To overcome this limit, the strategy of a vesicle surface modification by hydrophilic inert polymers has been adopted for the development of stealth liposomes, and the approach is discussed in the "Long circulating liposomes" paragraph. Moreover, the use of conventional liposomes for NABDs delivery usually results in a scarce encapsulation efficiency due to the lack of electrostatic interactions between the positive nucleic acids and the uncharged vesicles bilayer. To avoid this problem, cationic liposomes can be used, as discussed below.

2.2. Cationic Liposomes

Vesicles can be prepared by using positively-charged lipids, i.e., the 1,2-DiOleoyl-3-TrimethylAmmonium Propane (DOTAP), achieving cationic liposomes which can electrostatically attract the negative phosphate groups of NABDs, increasing their encapsulation efficiency. Moreover, the cationic liposomes also facilitate the intracellular uptake of NABDs and their endosomal escape, more efficiently if "helper" lipids are included in the formulation [19].

Indeed, neutral and zwitterionic lipids, mainly 1,2-dioleoylsn-glycero-3-phosphatidylcholine (DOPC), Di-Oleoyl-Phosphatidyl-Ethanoalamine (DOPE) and 1,2-DiStearoyl-sn-glycero-3-PhosphoCholine (DSPC), being more fusogenic than the cationic one, can affect the polymorphic features of the liposome–siRNA complexes, promoting the transition from a lamellar to a hexagonal phase, thus inducing fusion and a disruption of the membrane [36]. In particular the fusogenicity increases with the decrease of the degree of the saturation of the lipids hydrophobic tails, and with the decrease of their polar heads size (i.e., lipids with large polar heads and highly saturated tails prefer the lamellar phase, and thus are not fusogenic; lipids with small polar heads and highly unsaturated tails prefer the reverse hexagonal phase, and thus are fusogenic) [15]. DOPC lipid has been successfully used by the M.D. Anderson Cancer Center to produce siRNA-EphA2-DOPC, which is currently in a Phase I clinical trial for the treatment of advanced cancers, as discussed later in the "Current applications of lipid-based siRNA delivery" Section. Some of the most common cationic and helper lipids used for NABDs delivery are reported in Figure 3.

Cationic liposomes formulation is in constant improvement. In order to optimize siRNA encapsulation efficiency, Santel and collaborators have synthesized the cationic lipid β-l-Arginyl-2,3-l-diaminopropionic acid-N-palmityl-N-oleyl-amide trihydrochloride (AtuFECT), which when complexed with siRNA is termed AtuPLEX [37]. Atu027 is an AtuPLEX produced to treat advanced solid tumor [38] which has completed the Phase I and the Phase I/II of clinical trials (see Section 4).

Cationic liposomes can be also formulated by adding to the basic vesicles composition the lipidoids, lipid-like materials produced through the conjugate addition of an amine to an acrylate or acrylamide [4,39–43]. Akinc and collaborators have produced a complex called "LNP01" which is made by 98N12-5:cholesterol:PEG-lipid in a molar ratio of 42:48:10 (mol:mol:mol) respectively, with a total lipid:siRNA ratio of about 7.5:1 (wt:wt), a C14 alkyl chain length on the PEG lipid and a mean particle size of roughly 50–60 nm. The developed formulation is liver targeted (with a >90% injected dose distributed to the liver) and can induce, after repeated administrations, a long-duration gene silencing without any loss of activity [42,43].

Love and coworkers have also made complexes by combining lipidoid, cholesterol, and a polyethylene glycol modified lipid in a yield weight fractions of 52:20:28 (wt:wt:wt), obtaining an siRNA-directed liver gene silencing in mice at doses below 0.01 mg/kg [41]. A lipidoid nanoparticle which mediates potent gene knockdown in hepatocytes and immune cell populations after IV administration to mice, with siRNA EC50 values as low as 0.01 mg/kg, was successfully produced also by Whitehead and collaborator [39]. The lipidoids structures used in the cited works for siRNA-lipid complexes production are given in Figure 4.

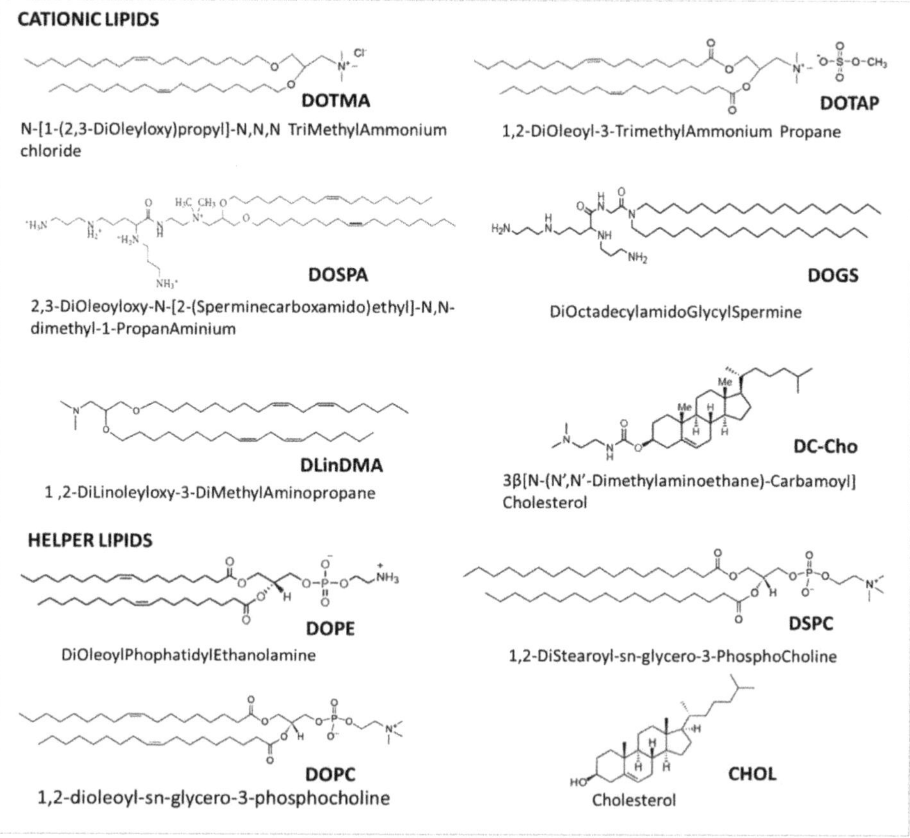

Figure 3. Structures of cationic and helper lipids used as components of various siRNA liposomal delivery systems.

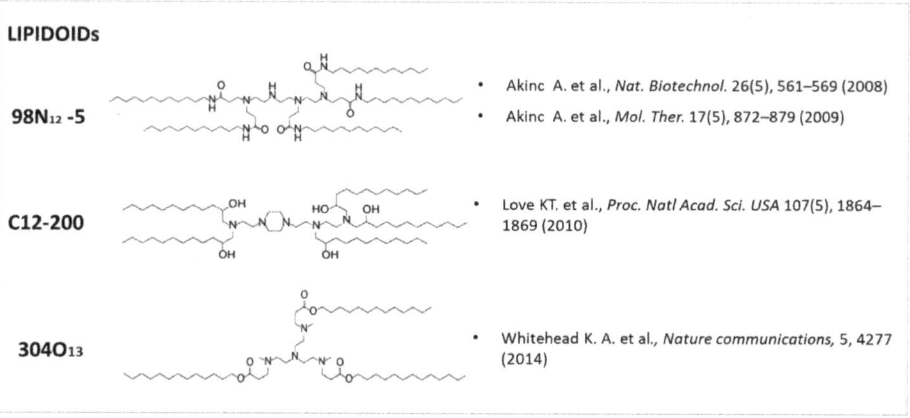

Figure 4. Lipidoids synthetized by different research groups and used in cationic liposomal formulations to increase siRNA encapsulation efficiency.

Finally, several commercial products based on cationic liposomes for NABDs transfection during in vitro experiments are available on the market, the most used ones are Lipofectine and LipofectAMINE, produced and sold by the Invitrogen Company.

Despite the advantages in strongly attracting NABDs, the use of cationic liposomes presents several drawbacks due to their instability, rapid systemic clearance, toxicity and their induction of immunostimulatory responses [44,45]. Alternatively, in order to ameliorate cationic liposomes performances in gene therapy, different techniques were developed for the production of relatively new cationic lipid structures, including Stable Nucleic Acid Lipid Particles (SNALP), as discussed in Section 3.

2.3. *Long Circulating Liposomes*

The rapid decrease of liposomal drug complexes in blood with the consequent accumulation in liver, spleen and other organs has led to the development of the long circulating liposomes, lipid vesicles characterized by a surface covered with inert polymeric molecules, such as oligosaccharides, glycoproteins, polysaccharides and synthetic polymers [28]. Among these, covering liposomes with the polyethylene glycol (PEG) represents an effective strategy to increase the repulsive forces between liposomes and serum-components, thus avoiding their elimination by the RES, while improving their stability and enhancing their circulation times in the blood [46–48]. Such surface modified liposomes have been called "PEGylated" or "stealth" liposomes, since they can evade recognition by T cells and macrophages, and avoid rapid clearance by the immune system [49,50]. In particular, the polymeric chain adsorbed on the liposomes surface consisting of tails, loops and trains, can form several structures on the bases of the polymeric layer size formed. Polymeric "pancake", "mushroom" or "brush" structures are formed with a PEG surface thickness of about 1.5 nm, 3.5 nm and 5 nm, respectively [18,51]. Doxil® represents the first FDA-approved drug carrier based on PEGylated liposome technology for the treatment of advanced ovarian cancer, multiple myeloma and HIV-associated Kaposi's sarcoma [52]. Moreover, several PEGylated liposomal products for NABDs delivery are in preclinical development, i.e., the lipidoid-based siRNA formulation 98N12-5 [53] or the DACC cationic lipoplex [54].

However, the long-circulating liposomes exhibit some drawbacks: (1) The PEGylation has to be optimized by modulating PEG length and density in order to control the interactions between the liposomal, the plasmatic and the endosomal membranes, which can prevent the vesicles' cellular uptake/escape from the endosome. This problem, known as "PEG dilemma", can be avoided by using cleavable PEG-lipids such as the "pH-responsive" or "enzyme-responsive" ones [47]; (2) covering with PEG does not guarantee the selectivity of the drug delivery to the target tissue. This problem can be overcome by using the targeting ligands conjugation approach [55] discussed below.

2.4. *Ligand-Targeted Liposomes*

Tumor tissues/blood vessels are characterized by specific and overexpressed receptors which can bind to antibodies, polysaccharides, proteins, polypeptides, aptamers and other molecules. The attachment of these specific targets to the NABDs-liposomes surface guarantees an active targeting of the carriers to the diseased tissue [28]. Recently, a Dual-Targeting Ligands approach was developed by Riaz and collaborators [49], and a combination of peptide and antibody ligands was used to fuctionalize a single liposomal formulation, obtaining an increased nanoparticles cell uptake [56].

In 1995 folic acid was conjugated for the first time onto the surface of nucleic acids liposomal carriers (folate-PEG-liposomes) by Wang and collaborators [57]. To date, the folate receptor is one of the most used targets for siRNA liposomal delivery, being overexpressed in several malignancies, such as ovarian and uterine cancer, osteosarcoma, meningioma and other cancer cells, but not in normal tissues [28].

The transferrin receptor-dependent mechanism is also used for liposomal NABDs internalization, i.e., Mendoça and coworkers have successfully developed a transferrin receptor-targeted liposome loaded with an anti-BCR-ABL siRNA or asODN for the treatment of chronic myeloid leukemia [58].

Recently, Zang and collaborators have developed a pH-sensitive cholesterol-Schiff base-polyethylene glycol (Chol-SIB-PEG)-modified cationic liposome–siRNA complex, conjugated with the recombinant humanized anti-EphA10 antibody (Eph). The Eph–PEG–SIB–Chol-modified liposome–siRNA complex (EPSLR) has shown a good endo-lysosomal escape, releasing siRNA into the cytoplasm after 4 h from in vitro transfection. Moreover, an in vivo study conducted in tumor-bearing mice shows that EPSLR can reach the diseased tissue more effectively [59].

Moreover, polysaccharides, i.e., galactose, mannose, dextran and hyaluronic acid can be also used as ligands to prepare glycosyl-liposomes targeting tissues overexpressing these receptors. The use of functionalization with polysaccharides and other targeting moieties for an effective NABDs delivery has been widely described in [18].

2.5. Bubble Liposomes

Lipid vesicles filled with gas are called "Bubble Liposomes". They are particles ranging from 1 to 10 micro meters, characterized by a gas core and a shell composed of several materials i.e., proteins, lipids, polymers, surfactants and galactose, used to deliver NABDs and other drugs into the cells and tissues [60]. In particular, the term "Bubble Liposomes" was coined by Suzuki, Maruyama and collaborators, who developed novel liposomes containing lipid nanobubbles loaded with perfluoropropane, used as ultrasound imaging agent. Briefly, "Bubble Liposomes" are prepared by producing at first polyethyleneglycol-modified liposomes (PEG-liposomes) through the reverse phase evaporation method, and placing them in vials with perfluoropropane gas, then sonicating in a bath sonicator. Upon exposure to ultrasound, these particles can induce cavitation, which supplies the energy required to deliver extracellular molecules into the cytosol, and this can thus be utilized as a gene delivery tool [61]. This phenomenon involves a greater cell/tissue permeability, improving NABDs penetration into the cells and ensuring, at the same time, their escape from the endosome with their consequent expression [62].

In that regard, it was demonstrated that bubble liposomes can deliver genes into cells even when the cells were exposed to ultrasound for only 1 s. They have also shown that this kind of liposome is more effective in delivering NABDs, i.e., the luciferase gene, into a tumor than are the conventional cationic liposomes used as transfection agents [63].

Recently, several research groups have opted for the use of bubble liposomes for NABDs delivery such as Negishi and collaborators, who have used bubble liposomes for a selective gene delivery to syndecan-2 overexpressing cancer cells [64], or also Endo-Takahashi and coworkers, in whose work bubble liposomes were successfully used to deliver miRNA (miR-126) for the cure of hindlimb ischemia by the systemic administration of miR-126-loaded bubble liposomes into mice coupled with US exposure [65].

3. Liposomes Preparation Techniques

There are many different methods for the preparation of liposomes [19,66–68], here the attention is focused on those available for NABDs encapsulation.

3.1. Thin Film Hydration

The most commonly used technique for liposomes preparation is the Thin Film Hydration (TFH) or the Bangham method, in which lipids are dissolved in an organic solvent, then evaporated through the use of a rotary evaporator leading to a thin lipid layer formation [69]. After the layer hydration by an aqueous buffer solution containing the hydrophilic drug to be loaded, Multilamellar Vesicles (MLVs) are formed, which can be reduced in size to produce Small or Large Unilamellar vesicles (LUV and SUV) by extrusion through membranes or by the sonication of the starting MLV [18,19,70].

Even if the method is not scalable, and toxic solvent traces could remain in the final formulation, its easiness to perform and low cost make this one of the most adopted techniques for liposomes preparation. The method is used also for the production of liposomes containing NABDs, whose encapsulation efficiency, however, is generally quite low, ranging from 3 to 45%, but can be increased by modulating the lipid mixtures [71,72]. In our previous works DOTAP cationic lipid was added to a cholesterol and phosphatidylcholine formulation in order to electrostatically attract siRNA molecules. Briefly, through the TFH method, followed by a duty cycle sonication, siCyD1-nanoliposomes [73] and siE2F1-nanoliposomes [74] with 100% siRNA encapsulation efficiency, able to reduce respectively colon CyD1 and E2F1 proteins expression after transfection in ex vivo human tissue cultures, were produced.

By modulating the lipid formulation adding DOTAP, Coated Cationic Liposomes (CCL) entrapping an antisense oligodeoxynucleotides (asODN), with a diameter of 188 nm and an encapsulation efficiency of 85–95% were also prepared through the TFH technique by Stuart and coworkers [75].

3.2. Double Emulsion

Liposomes can be also prepared through the Double Emulsion technique which involves lipids dissolution in a water/organic solvent mixture. The organic solution, containing water droplets, is mixed with an excess of aqueous medium, leading to a water-in-oil-in-water (W/O/W) double emulsion formation. After mechanical vigorous shaking, part of the water droplets collapse, giving Large Unilamellar Vesicles (LUVs) [76]. This method was used for NABDs encapsulation, but achieved a very low entrapment efficiency [77].

3.3. Reverse Phase Evaporation

The Reverse Phase Evaporation (REV) method also allows one to achieve LUVs loaded with NABDs [78]. In this technique a two-phase system is formed by phospholipids dissolution in organic solvents and aqueous buffer. The resulting suspension is then sonicated briefly until the mixture becomes a clear one-phase dispersion. The liposome formation is achieved after the organic solvent evaporation under reduced pressure. This technique has been used to encapsulate different large and small hydrophilic molecules included nucleic acids, i.e., in a work of Stuart and Allen, cationic vesicles loaded with nucleic acids and covered with neutral lipids, with significantly high incorporation efficiency (80–100%), and being stable in 50% human plasma at 37 °C, were produced [79]. Recently by means of a modified REV, Mokhtarieh and collaborators have produced Asymmetric Liposome Particles (ALPs) containing siRNA of about 200 nm in size with more than 90% encapsulation efficiency. The method provides the formation of two kinds of lipid inverted micelles composing the inner and outer lipid film. The siRNAs are entrapped in the inner one, made of ionizable cationic lipids, which is mixed with the outer lipid film made of conventional lipids. After a solvent evaporation and dialysis, siRNA-ALPs are achieved [80].

3.4. Microfluidic Method

A relatively new technology used for liposomes production is the Microfluidic method, unlike other bulk techniques, this one gives the possibility of controlling the lipid hydration process. The method can be classified in continuous-flow microfluidic and droplet-based microfluidic, according to the way in which the flow is manipulated [81]. In 2004, Jahn and collaborators described a microfluidic hydrodynamic focusing (MHF) method which operates in a continuous flow mode. Briefly lipids are dissolved in isopropyl alcohol which is hydrodynamically focused in a microchannel cross junction between two aqueous buffer streams. Vesicles size can be controlled by modulating the flow rates, thus controlling the lipids solution/buffer dilution process [82]. The method was successfully extended for producing oligonucleotides (ON) lipopolyplexes by using a microfluidic device consisting of three-inlet and one-outlet ports. Lipids ethanol solution and ON aqueous solutions are contained into sterile syringes connected to the inlet ports. Briefly, a fluid stream is split into two side streams at inlet port 1 or 2, while a fluid stream directly entered the center microchannel through inlet port 3. The resulting ON liposomes (about 115 nm in size) solution is collected at the outlet port [81].

3.5. Dual Asymmetric Centrifugation

Dual Asymmetric Centrifugation (DAC) is another method for the production of NABDs-liposomes [83]. This technique differs from the usual centrifugation because the sample is undergone, during the normal centrifugation process, to an additional rotation around its own vertical axis. By this way an efficient homogenization is achieved due to the two overlaying movements generated: The sample is pushed outwards, as in a normal centrifuge, and then it is pushed towards the center of the vial due to the additional rotation. Briefly, by mixing lipids and an NaCl-solution a viscous vesicular phospholipid gel (VPC) is achieved, which is then diluted to obtain liposomal dispersion. Liposome size can be regulated by optimizing DAC speed, lipid concentration and homogenization time. With this method Hirsch and coworkers have prepared siRNA-liposomes, about 109 nm in size, with high entrapping efficiency, ranging from 43 to 81%, depending upon batch size [84]. In 2011, also Adrian and collaborators have used the DAC method to produce liposomes containing siRNA, targeting the particles surface with an antibody for the selective interaction with neuroblastoma cells, achieving 190 to 240 nm particles with siRNA encapsulation efficiency of up to 50% [85]. By this technique, in only one step, sterile SUVs formulations in a highly reproducible manner can be prepared [86].

3.6. Ethanol Injection

Unilamellar liposomes can be prepared by the Ethanol Injection (EI) method which can be used for NABDs encapsulation. This method provides the rapid injection of an ethanolic solution, in which lipids are dissolved, into an aqueous medium containing nucleic acids to be encapsulated, through the use of a needle. Vesicles are spontaneously formed when the phospholipids are dispersed throughout the medium. By utilizing this method Stabilized Antisense-Lipid Particles (SALPs) can be obtained [87], i.e., Saffari and collaborators showed that by the Ethanol injection method DOTAP-liposomes loaded with an antisense oligonuclotide (AsODN) against protein kinase C alpha, with a size of 115 nm, not requiring downsizing with extrusion or other methods, can be achieved. They obtained an encapsulation efficiency of around 90% that was about seven times more than that of the TFH method, also tried out in their work [88]. In a work of Saad and collaborators cationic liposomes were produced by the EI method for the co-delivery of doxorubicin and siRNA targeted to MRP1 and BCL2 mRNA (suppressors of pump and nonpump cellular resistance) to multi-drug resistance (MDR) lung cancer cells, enhancing the chemotherapy efficiency [89].

3.7. Detergent Dialysis

The Detergent dialysis method can be used to encapsulate plasmid DNA. Briefly lipid and plasmid are solubilized in a detergent solution of appropriate ionic strength, after removing the detergent by dialysis, Small Stabilized Plasmid Lipid Particles (SPLPs) are formed. Unencapsulated DNA is then removed by ion-exchange chromatography and empty vesicles by sucrose density gradient centrifugation. The technique is highly sensitive to the cationic lipid content and to the salt concentration of the dialysis buffer, and the method is also difficult to scale [90]. Through this technique 100 nm liposomes loaded with nucleic acids with 60–70% encapsulation efficiency were achieved [91].

3.8. Spontaneous Vesicle Formation by Ethanol Dilution

SPLPs can also be produced through the Spontaneous Vesicle Formation by Ethanol Dilution method in which a stepwise or dropwise ethanol dilution provides the instantaneous formation of vesicles loaded with plasmid DNA by the controlled addition of lipid dissolved in ethanol to a rapidly mixing aqueous buffer containing DNA. In a work of Jeffs and collaborators, SPLP prepared by the stepwise approach had a DNA encapsulation efficiency of 81% and were monodisperse with a mean vesicle diameter of about 130 nm, while SPLP prepared by dropwise ethanol had a DNA encapsulation efficiency of 74% and mean vesicle size of about 115 nm [92]. Later the method has been applied successfully to NABDs smaller than plasmids, such as miRNAs and siRNAs, obtaining a Stable Nucleic Acid Lipid Particle (SNALP) of about 100 nm with a high nucleic acids encapsulation efficiency (>95%) [93,94]. In a work of Judje and collaborators 2′OMe-modified siRNA targeting apolipoprotein B (apoB) was successfully encapsulated inside 100 to 130 nm liposomes with an entrapment efficiency of 90–95% through the stepwise ethanol dilution method, obtaining a potent silencing of the endogenous gene target apoB when administered systemically in animals [95]. Recently, Wilner and Levy have described a method for targeting SNALP encapsulating siRNA with aptamers in order to obtain a selective targeted delivery along with an efficient siRNA-mediated gene knockdown [44,96].

Due to their stability, their ability in remain intact in circulation for many hours and accumulate at the target sites, the SNALPs systems have been successfully adopted in a variety of NABDs formulations which are currently in clinical development for the cure of several diseases i.e., solid tumors, Amyloidosis, Hepatitis B, Hypercholesterolemia and the Ebola Virus (see Section 4) [97–100].

3.9. siRNA Encapsulation in Preformed Liposomes

Finally, the entrapment of NABDs can be also obtained starting with preformed liposomes through two different methods: (1) A simple mixing of cationic liposomes with nucleic acids which gives electrostatic complexes called "lipoplexes" [101], where they can be successfully used to transfect cell cultures, but are characterized by their low encapsulation efficiency and poor performance in vivo [72,102]; and (2) a liposomes destabilization, slowly adding absolute ethanol to a cationic vesicles suspension up to a concentration of 40% v/v followed by the dropwise addition of nucleic acids achieving loaded vesicles; however, the two main steps characterizing the encapsulation process are too sensitive, and the particles have to be downsized [72].

4. Current Applications of Lipid-Based siRNA Delivery

During the last 20 years we assisted in a fast growth of the attention of the scientific community relative to the RNAi mechanism, and thus to NABDs new formulations, including small interfering RNA (siRNA), associated with lipid-based delivery strategy to be use for the cure of several diseases. As illustrated in Figure 5, starting from the 2001, year in which siRNA was successfully delivered into mammalian cells for the first time [103], until 2019, the numbers of publications related to "siRNA and liposome" in vitro, in vivo and in clinical trials applications, have increased dramatically, and different liposomal-based siRNA formulations have already shown good safety records in humans.

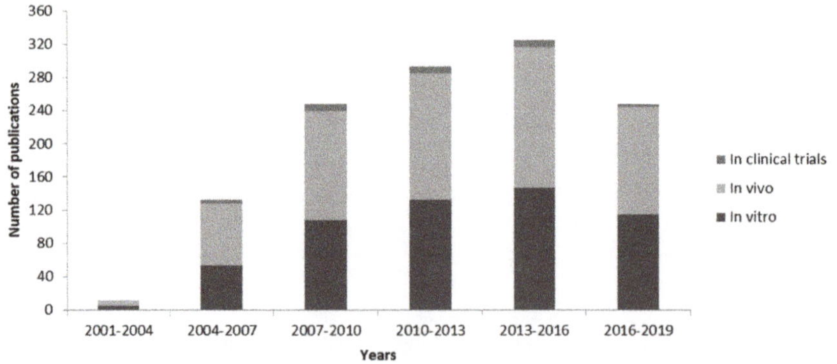

Figure 5. PubMed publications related to "siRNA liposome". Data recorded in the PubMed database were identified with the search terms "*siRNA liposome*" and "*in vitro*" or "*in vivo*" or "*in clinical trials*" used for the search query. Data were compiled and plotted as a bar graph for the number of publications since 2001 and every three years; data for 2019 are based on the first five months.

In particular, here the attention is focused on the applications of siRNAs delivered by lipid-based systems currently into the clinical trials stage. The clinical trials are summarized in Table 1, reporting the common name of the drug, the gene targeted, the disease which it is intended to cure, the company which is carrying out the experimentation, the identifier, the phase (phase I: Pharmacodynamics and Pharmacokinetics—sometimes noted as phase 0 and screening for safety; phase II: Establishing the efficacy of the drug; phase III: Final confirmation of safety and efficacy) and its status as obtained from [104]. In all the clinical trials, the pharmaceuticals were administered by intravenous injection (bolus and/or infusions), and the delivery system is a lipid-based one.

Table 1. The siRNA-lipid delivery systems used in clinical trials.

#	Drug	Target	Disease	Company	[104]		
					Identifier	**Phase**	**Status**
1.	ALN-VSP02	KSP and VEGF	Solid tumors	Alnylam Pharmaceuticals	NCT 00882180 NCT 01158079	I I	Completed Completed
2.	ALN-PCS02	PCSK9	Hypercholesterolemia	Alnylam Pharmaceuticals	NCT 01437059	I	Completed
3.	ALN-TTR01 ALN-TTR02	TTR	TransThyRetin (TTR)-mediated amyloidosis	Alnylam Pharmaceuticals	NCT 01148953 NCT 01559077 NCT 01617967 NCT 01961921 NCT 01960348 NCT02939820 NCT02510261 NCT03862807	I I II II III Approved for marketing III III	Completed Completed Completed Completed Completed Recruiting Recruiting
4.	siRNA-EphA2-DOPC	EPHA2	Advanced cancers	M.D. Anderson Cancer Center	NCT 01591356	I	Recruiting
5.	Atu027	PKN3	Advanced solid cancers	Silence Therapeutics	NCT 00938574 NCT 01808638	I I/II	Completed Completed
6.	ND-L02-s0201	HSP47	Fibrosis	Nitto Denko Corporation	NCT 01858935 NCT02227459 NCT03241264 NCT03538301	I I I II	Completed Completed Completed Recruiting
7.	TKM-ApoB	Apo B	Hypercholesterolemia	Tekmira Pharmaceuticals	NCT 00927459	I	Terminated
8.	TKM-080301	PLK1	Cancer (Polo-Like-Kinase 1)	Tekmira Pharmaceuticals	NCT 01437007 NCT 01262235 NCT 02191878	I I/II I/II	Completed Completed Completed
9.	TKM-100201	VP24, VP35, L-polymerase	Ebola Virus Infection	Tekmira Pharmaceuticals	NCT 01518881	I	Terminated
10.	TKM-100802	VP24, VP35, L-polymerase	Ebola Virus Infection	Tekmira Pharmaceuticals	NCT 02041715	I	Terminated
11.	DCR-MYC	MYC	Solid Tumors Multiple Myeloma Non-Hodgkins Lymphoma Hepatocellular Carcinoma	Dicerna Pharmaceuticals	NCT 02110563 NCT02314052	I I/II	Terminated Terminated
12.	ARB-001467	HBV proteins	Hepatitis B, Chronic	Arbutus Biopharma	NCT02631096	II	Completed

Alnylam Pharmaceuticals is the most active company in the field of siRNA therapeutics. Some of its drugs were delivered by the Stable Nucleic Acid Lipid Particles (SNALPs) technology of the Tekmira Pharmaceuticals Corporation. Several pharmaceuticals using this and other lipid-based delivery technologies involved in the clinical trials are listed below.

1. ALN-VSP02

The ALN-VSP02 is an Alnylam Pharmaceutics lipid nanoparticle formulation containing two small interfering RNAs (siRNAs) directed against the kinesin spindle protein (KSP) and vascular endothelial growth factor (VEGF) mRNAs, preventing their translation. Since their key role in tumor proliferation, VEGF and KSP downregulation can lead to a growth inhibition of the cells implicated in such disease. ALN-VSP02, produced through the Tekmira's SNALP technology, entered in a phase I dose-escalation trial (NCT 00882180, 2009, "A Multi-Center, Open Label, Phase 1 Dose-Escalation Trial to Evaluate the Safety, Tolerability, Pharmacokinetics and Pharmacodynamics of Intravenous ALN-VSP02 in Patients with Advanced Solid Tumors with Liver Involvement"). The study was completed, and a total of 41 patients were enrolled between Mar 2009 and Aug 2011, including 30 patients of the dose-escalation phase treated at 0.1 to 1.5 mg/kg (administered by IV infusion) and 11 of the expansion phase treated at 1.0 or 1.25 mg/kg. On the basis of the molecular analysis of biopsy samples from patients, the study confirmed the presence of residual siRNAs and mRNAs cleavage products at the site of actions. The drug was well tolerated at the highest dose, and then an expansion study has been initiated and completed (NCT01158079, 2010, "A Multi-center, Open-Label, Extension Study of ALN-VSP02 in Cancer Patients Who Have Responded to ALN-VSP02 Treatment"). Seven patients from the first study were treated for an average of 11.3 months (between Jul 2010 and Aug 2012), in order to further evaluate the safety, tolerability, pharmacokinetics and pharmacodynamics of the drug. Even if the efficacy of the drug was not tested during the phase I clinical trial, it is worth noting that one patient (of seven) with endometrial cancer shown a complete response (full tumor regression) and he has remained in remission and completed treatment after receiving 50 doses over 26 months. Three other patients have shown stable disease, having received 17–36 doses over 8–18 months. The results of these two studies are reported in [105].

2. ALN-PCS02

The ALN-PCS02 is a SNALP-formulated [106] RNAi therapeutic targeting PCSK9 mRNA, encoding for Convertase Subtilisin/Kexin type 9 (PCSK9) protein for the treatment of hypercholesterolemia. Binding to Low-Density Lipoprotein (LDL) receptors, PCSK9 leads to their degradation. Genetics studies have shown that reduced-plasma LDL cholesterol can be noticed as a consequence of the loss-of-function mutations in PCSK9 resulting in a decreased risk of coronary heart disease. Therefore, drugs that block PCSK9 can lower LDL.

Preclinical studies have given encouraging results in mice [107], thus a phase I clinical trial was started by Alnylam Pharmaceutics (NCT01437059, 2011, "A Phase 1, Randomized, Single-blind, Placebo-Controlled, Single Ascending Dose, Safety, Tolerability and Pharmacokinetics Study of ALN-PCS02 in Subjects With Elevated LDL-Cholesterol (LDL-C)"). The study, started in Sep 2011 and ended in Mar 2012, enrolled 32 patients with elevated LDL-Cholesterol levels. 24 over 32 patients received the siRNA-SNAPL pharmaceutical, dosed between 0.015 and 0.400 mg/kg, noticing an adverse effect (AE) level equal to that of the control group (eight patients treated with a placebo). Finally patients who received ALN-PCS at 0.400 mg/kg showed a 70% reduction in circulating PCSK9 plasma and a mean of 40% reduction in LDL-C with respect to those treated with the placebo [108].

3. ALN-TTR01 and ALN-TTR02

Two distinct siRNA-SNAPL formulations, the ALN-TTR01 and ALN-TTR02, targeting the TransThyRetin (TTR, TRANsports THYroxine and Retinol) gene encoding for the protein TransThyRetin (TTR), were developed by Alnylam Pharmaceuticals for the cure of amyloid diseases associated with a TTR overexpression. Both of the formulations have been investigated in phase I clinical trials (NCT 01148953, 2010, "A Phase 1, Randomized, Single-Blind, Placebo-Controlled, Dose Escalation Trial to Evaluate the Safety and Tolerability of a Single Dose of Intravenous ALN-TTR01 in Patients With TTR Amyloidosis") and (NCT 01559077, 2012, "A Phase 1, Randomized, Single-blind, Placebo-Controlled, Single Ascending Dose, Safety, Tolerability and Pharmacokinetics Study of ALN-TTR02 in Healthy Volunteers"). The results of these studies, both completed, were summarized by [109]. The two drugs were found to be safe and effective in reducing the TTR levels. In particular, ALN-TTR01, in which SNAPL are obtained by the Tekmira technique, and is administered to 32 patients with transthyretin amyloidosis, produced mild-to-moderate infusion-related reactions in 20.8% of patients, and at the highest dose of 1.0 mg/kg, a mean reduction at day seven of 38% for the TTR plasma level. ALN-TTR02, in which SNAPL are obtained by an LNP proprietary formulation [106], administered to 17 healthy volunteers, produced mild-to-moderate infusion-related reactions in 7.7% of patients, and at the highest dose of 0.3 mg/kg, reductions at day 28 of 56.6 to 67.1% for the TTR plasma level. Therefore, ALN-TTR02 was found to be more safe and effective than ALN-TTR01. Based on these results, Alnylam Pharmaceutics planned and carried out several more trials, a phase II (NCT 01617967, 2012, "A Phase 2, Open-Label, Multi-Dose, Dose Escalation Trial to Evaluate the Safety, Tolerability, Pharmacokinetics, and Pharmacodynamics of Intravenous Infusions of ALN-TTR02 in Patients With TTR Amyloidosis") was completed with 29 patients enrolled; an extended phase II (NCT 01961921, 2013, "A Phase 2, Multicenter, Open-Label, Extension Study to Evaluate the Long-Term Safety, Clinical Activity, and Pharmacokinetics of ALN-TTR02 in Patients With Familial Amyloidotic Polyneuropathy Who Have Previously Received ALN-TTR02") was also completed with 27 patients enrolled, followed by a phase III study (NCT 01960348, 2013, "APOLLO: A Phase 3 Multicenter, Multinational, Randomized, Double-blind, Placebo-controlled Study to Evaluate the Efficacy and Safety of ALN-TTR02 in Transthyretin (TTR)-Mediated Polyneuropathy (Familial Amyloidotic Polyneuropathy-FAP)"). In particular, in APOLLO study, which started in December 2013 and was completed in December 2018, 225 patients with hATTR amyloidosis polyneuropathy were enrolled and randomized 2:1 to receive 0.3 mg/kg Patisiran (name of the ALN-TTR02 formulation) or a placebo via intravenous infusion once every 3 weeks for 18 months. A consistent effect on the N-terminal prohormone of the brain natriuretic peptide at 18 months was observed in the overall APOLLO patients population (n = 225), showing good results also in the cardiac subpopulation (126 of 225 [56%]) patients treated with Patisiran compared with those treated with the placebo [110–112]. This trial represents the largest Phase 3 study of an RNAi strategy for the treatment of hATTR amyloidosis. Based on these results, an expansion study (NCT02939820 2016, "Expanded Access Protocol of Patisiran for Patients With Hereditary Transthyretin-Mediated Amyloidosis (hATTR Amyloidosis) With Polyneuropathy") was carried out, providing further data on the long-term safety and efficacy of ALN-TTR02 in patients with hATTR amyloidosis, and thus on August 2018 ALN-TTR02 received a marketing license in Europe for stage 1 and 2 neuropathy in hATTR adult patients.

Marketed as Onpattro™, Patisiran has become the first RNA interference lipid drug to win US Food and Drug Administration (FDA) approval, and is also the only specific medicinal treatment for this rare indication to gain approval in the USA. Two more studies of ALN-TTR02 are currently recruiting participants for a phase III clinical trials: NCT02510261, 2015, "A Multicenter, Open-Label, Extension Study to Evaluate the Long-term Safety and Efficacy of Patisiran in Patients With Familial Amyloidotic Polyneuropathy Who Have Completed a Prior Patisiran Clinical Study"—estimated enrollment 211 patients, starting date July 2015, estimated primary completion date Aug 2022; NCT03862807, 2019, "An Open-label Study to Evaluate Safety, Efficacy and Pharmacokinetics (PK) of Patisiran-LNP in Patients With Hereditary Transthyretin-mediated Amyloidosis (hATTR Amyloidosis) With Disease Progression Post-Orthotopic Liver Transplant"—estimated enrollment 20 participants, starting date Mar 2019, estimated primary completion date, Jan 2021.

4. siRNA-EphA2-DOPC

A neutral liposomes formulation loaded with a siRNA directed against the EPHA2 gene encoding for a tyrosine kinase which is overexpressed and implicated in tumor growth, was produced and tested in a preclinical study giving encouraging results, showing the good tolerability of nanoliposomal EphA2-targeted therapeutic (EPHARNA) in mice at different concentrations [113–116]. Therefore, a phase I dose-escalation clinical trial was started by the M.D. Anderson Cancer Center (NCT 01591356, 2012, "EphA2 Gene Targeting Using Neutral Liposomal Small Interfering RNA Delivery"). The study is currently recruiting participants, the estimated enrollment is of 40 subjects, the starting date was May 2012 and the estimated primary completion date is July 2020.

5. Atu027

AtuPLEX is a cationic LNP formulation produced by Silence Therapeutics containing siRNAs targeting the Cluster of Differentiation 31 (CD31) and the tyrosine kinase receptor for angiopoietins (TIE-2) mRNAs. The silencing of these endothelial cell–specific proteins can be a strategy to treat several kinds of advanced solid cancers. Obtaining a successful knockdown of the target genes into mouse vascular endothelium, an Atu027 formulation was produced using AtuPLEX technology to load siRNA, targeting the Protein Kinase N3 (PKN3), which is implicated in metastatic pancreatic tumor cell growth. Favorable preclinical data in animals [117,118] have led to a phase I clinical trial, based on Atu027 administered to patients with advanced solid tumors, by single and repeated intravenous infusion (NCT 00938574, 2009, "A Prospective, Open-label, Single Center, Dose Finding Phase I-study With Atu027 (an siRNA Formulation) in Subjects With Advanced Solid Cancer"). The study has been completed, the enrollment was of 34 subjects, the starting date was July 2009 and the study was concluded in April 2013. Atu027 was well tolerated up to dose levels of 0.336 mg/kg, showing low-grade toxicities (grade 1 or 2). The drug was safe in patients with advanced solid tumors, with 41% of patients having stable disease for at least 8 weeks [119,120]. In the light of these results, Silence Therapeutics began a new clinical trial, a phase I/II in which a combination of Atu027 and Gemcitabine was administered to patients with pancreatic carcinoma (APC) (NCT 01808638, 2013, "A phase Ib/IIa study of combination therapy with Gemcitabine and Atu027 in subjects with locally advanced or metastatic pancreatic adenocarcinoma"). The study has been completed, the enrollment was of 29 subjects with a starting date on Mar 2013 and a completion date on Jan 2016. The liposomal-formulated Atu027 in combination with Gemcitabine for the treatment of APC was safe and was well tolerated, a twice-weekly administration was better than a once-weekly regime, and the results are detailed in [121].

6. ND-L02-s0201

ND-L02-s0201 is a vitamin A–coupled cationic liposomal nanoparticle encapsulating an siRNA which inhibits the expression of Heat Shock Protein 47 (HSP47), a collagen-specific chaperone [122] involved in the fibrosis of liver and other organs. Vitamin A-moieties conjugated to the nanoparticles surface maximize the drug delivery to hepatic stellate cells.

Preclinical data have suggested that disrupting collagen synthesis via HSP47 may reverse fibrosis. In particular, five treatments with the siRNA-vitamin A-coupled liposomes almost completely resolved liver fibrosis and prolonged survival in rats [123,124], thus a phase I study of ND-L02-s0201 was started by the Nitto Denko Corporation delivering the pharmaceutical by IV single-dose injection in healthy patients (NCT 01858935, 2013, "A Phase 1, Randomized, Double-Blind, Placebo-Controlled, Escalating Single Dose Study to Evaluate the Safety, Tolerability and Pharmacokinetics of ND-L02-s0201 Injection, a Vitamin A-Coupled Lipid Nanoparticle Containing siRNA Against HSP47, in Healthy Normal Subjects"). The study was completed, the enrollment was of 56 subjects with a starting date on May 2013 and a completion date on May 2017. 90 mg IV of ND-L02-s0201 for three weeks was well tolerated in healthy Japanese and non-Japanese adults enrolled, with no clinically meaningful differences or intolerability [123].

In the light of the encouraging results, the Nitto Denko Corporation carried out three more trials: A phase I to evaluate the safety and the tolerability of multiple doses of ND-L02-s0201 in subjects with moderate to extensive hepatic fibrosis (NCT02227459, 2014, "A Phase 1b/2, Open Label, Randomized, Repeat Dose, Dose Escalation Study to Evaluate the Safety, Tolerability, Biological Activity, and Pharmacokinetics of ND-L02-s0201 Injection, A Vitamin A-coupled Lipid Nanoparticle Containing siRNA Against HSP47, in Subjects With Moderate to Extensive Hepatic Fibrosis (METAVIR F3-4)")—completed, 25 patients enrolled; an extended phase I to evaluate the safety, tolerability, and the effects of two ND-L02-s0201 (NCT03241264, 2017, "A Phase 1, Open-Label, Randomized-Sequence, Single-Crossover, Bridging Study to Evaluate the Single-Dose Pharmacokinetics, Safety, and Tolerability of Two ND-L02-s0201 Formulations, Frozen Versus Lyophilized, Administered by Intravenous Infusion to Healthy Male and Female Subjects")—completed, 12 patients enrolled; a phase II study (NCT03538301,

2018, "A Phase 2, Randomized, Double-Blind, Placebo-Controlled Study to Evaluate the Safety, Tolerability, Biological Activity, and PK of ND-L02-s0201 in Subjects With Idiopathic Pulmonary Fibrosis (IPF)")—currently recruiting participants, with an estimated enrollment of 120 participants. The initial date was May 2018, with an estimated primary completion date in Mar 2020.

7. TKM-ApoB

TKM-ApoB is a systemically-delivered RNAi therapeutic based on SNALP carrying an siRNA against the ApoB (Apolipoprotein B) mRNA in order to downregulate the protein which facilitates the uptake of LDL into various cell types and tissues, reducing the LDL cholesterol levels in hypercholesterolemia patients. In the light of the positive preclinical results [100] a phase I clinical trial has been initiated (NCT 00927459, 2009, "Study to Evaluate the Safety, Tolerability, Pharmacokinetics (PK), and Pharmacodynamics (PD) of Liposomal siRNA in Subjects With High Cholesterol"). The enrollment was of 23 subjects with a starting date on June 2009, but the study has been terminated [125]. Of the 23 subjects, 17 received treatment with a single dose IV infusion of the TKM-ApoB at seven different dosing levels, the other six subjects received the placebo control. The drug was well tolerated, with no evidence of liver toxicity, except for one of the two subjects treated with the highest drug dosage, who, although reporting an ApoB protein and LDL cholesterol reduction, has also reported symptoms of immune system stimulation [12]. Tekmira has declared it was not satisfied with the performance of its current TKM-ApoB LNP formulation, and thus they will be going to focus on selecting an alternative formulation with improved nanoparticle carriers and siRNA chemistry to recommence a clinical trial for the ApoB siRNA therapy.

8. TKM-080301

The TKM-080301 is a SNALP formulation of a siRNA against Polo-Like Kinase 1 (PLK1), a serine/threonine kinase that regulates key aspects of cell cycle progression and mitosis and is highly expressed in malignant cells. In preclinical models the anti-tumor activity of PLK1 inhibition through RNA interference have been demonstrated [126–128], thus several clinical trials based on the pharmaceutical TKM-080301 were carried out by Tekmira Pharmaceuticals.

At first, the feasibility of administering TKM-080301 via Hepatic Arterial Infusion (HAI) and its pharmacokinetics and pharmacodynamics in patients with unresectable primary liver cancer or liver metastases has been evaluated in a dose-escalation study (NCT 01437007, 2011, "A Phase 1 Dose Escalation Study of Hepatic Intra-Arterial Administration of TKM-080301 (Lipid Nanoparticles Containing siRNA Against the PLK1 Gene Product) in Patients With Colorectal, Pancreas, Gastric, Breast, Ovarian and Esophageal Cancers With Hepatic Metastases"). The study started in Aug 2011 and was completed in June 2012. Subsequently, to determine the safety and tolerability of TKM-080301 in adult patients with solid tumors or lymphomas that are refractory to standard therapy, or for whom there is no standard therapy, a phase I/II TKM-PLK1 clinical trial was carried out (NCT 01262235, 2010, "A Phase 1/2 Dose Escalation Study to Determine the Safety, Pharmacokinetics, and Pharmacodynamics of Intravenous TKM-080301 in Patients With Advanced Solid Tumors"). The study was completed, the enrollment was of 68 subjects, the starting date was Dec 2010 and the primary completion date was July 2015. Results of NCT 01437007 and NCT 01262235, reported in [129,130], indicate that TKM-080301 was generally well-tolerated by the majority of patients showing a preliminary antitumor efficacy, supporting PLK1 as a therapeutic target. Finally, to test the safety and tolerability of TKM-080301 in subjects with advanced hepatocellular carcinoma and to find the maximum tolerated dose (MTD) providing a preliminary assessment of anti-tumor activity of TKM-080301, an extended phase I/II TKM-PLK1 clinical trial for HCC was done (NCT 02191878, 2014, "Open-Label, Multi-Center, Phase 1, Dose Escalation Study With Phase 2 Expansion Cohort to Determine the Safety, Pharmacokinetics and Preliminary Anti-Tumor Activity of Intravenous TKM-080301 in Subjects With Advanced Hepatocellular Carcinoma"). The study was completed, the enrollment was of 43 participants, the starting date was Jun 2014 and the primary completion date was May 2016. TKM-080301 was generally well tolerated, with a starting dose of 0.3 mg/kg and MTD of 0.6 mg/kg. Four patients did not have any evaluable post baseline scan. Of the other 39 subjects who had received at least 0.3 mg/kg, 18 subjects (46.2%) had stable disease and eight subjects (23.1%) had a partial response. The median survival for the whole study population was 7.5 months. Results are reported in [131].

9. TKM-100201

TKM-100201 is a small interfering RNA lipid nanoparticle product which has been developed for the treatment of the Ebola Virus Disease (EVD), an infection characterized by an immune suppression and a systemic inflammatory response that causes the impairment of the vascular, coagulation and immune systems, leading to multiorgan failure [132]. The formulation is a mixture of three siRNAs that target the L, VP24, and VP35 proteins of the Zaire Ebola Virus strain based on SNALP technology. In preclinical studies the pharmaceutical efficiently provided a post-exposure protection in animal models [98,99,133], therefore, Tekmira Pharmaceutical initiated a phase I clinical trial based on IV infusion with dose-escalation (NCT01518881, 2011, "A Placebo-Controlled, Single-Blind, Single-Ascending Dose Study With Additional Multiple-Ascending Dose Cohorts to Evaluate the Safety, Tolerability and Pharmacokinetics of TKM-100201 in Healthy Human Volunteers"). The study started in Jan 2012, but was terminated because Tekmira Pharmaceutical has decided to reformulate the product, which presented a dangerous EBOV-induced inflammatory response [134].

10. TKM-100802

As a consequence of the previous clinical trial, Tekmira Pharmaceutical modified the drug TKM-100201, toward TKM-100802, a lyophilized formulation with an increased therapeutic index. The LNP and the pool of siRNAs remained unchanged, but the new formulation achieved 100% survival (6 primates over 6), using a 4-fold lower dose (0.5 mg/kg/dose), and a significant survival advantage (4 primates over 6) even at the 10-fold lower dose (0.2 mg/kg/dose) [135]. The new drug has undergone a phase I trial, administered by IV infusion to healthy patients, with dose-escalation (NCT02041715, 2014, "A Placebo-Controlled, Single-Blind, Single-Ascending Dose Study With Additional Multiple-Ascending Dose Cohorts to Evaluate the Safety, Tolerability, and Pharmacokinetics of TKM-100802 in Healthy Human Volunteers").

The Single Ascending Dose (SAD) part of the study has interested 14 patients, and Adverse Effects (AEs) have been observed at the highest dose (0.5 mg/kg/dose); thus also this study was terminated in Aug 2015 due to safety concerns arising from the development of flu-like symptoms in treated individuals, which were ultimately linked to cytokine release triggered by the action of the siRNA [133,136].

A TKM-130803 siRNA LNP-enhanced formulation in which the siRNA component has been adapted by two nucleotide substitutions in the VP35 siRNA and a single nucleotide substitution in the L-polymerase siRNA to ensure specificity to the West African Makona variant of Zaire Ebola Virus, has been developed for EVD therapy [137]. However the new product, administered at a dose of 0.3 mg/kg/d by intravenous infusion to adult patients with severe EVD, was not shown to improve survival when compared to historic controls. On 14 patients enrolled, 11 died and 3 survived, thus the trial was halted, and is not reported in *ClinicalTrials.gov* [138].

To date, the development of new approaches to fight EBOV are in progress, and a genome-wide siRNA screen to identify novel pro- and antiviral factors against the Ebola Virus life cycle is being performed [139].

11. DCR-MYC

DCR-MYC is a lipid nanoparticle-based formulation consisting of an siRNA directed against the mRNA of the proto-oncogene c-Myc, involved in cellular proliferation, differentiation and apoptosis, and overexpressed in a variety of cancers. The inhibition of c-Myc translation has a potential antineoplastic activity as demonstrated in preclinical studies [140]. Therefore, a phase I study of DCR-MYC in patients with solid tumors and hematological malignancies was started by Dicerna

Pharmaceuticals, delivering the drug by IV infusion (NCT 02110563, 2014, "Phase I, Multicenter, Dose Escalation Study of DCR-MYC in Patients With Solid Tumors, Multiple Myeloma, or Lymphoma"). This study, started in Apr 2014, was immediately followed by the subsequent phase I/II clinical trial (NCT 02314052, 2014, "Phase 1b/2, Multicenter, Dose Escalation Trial to Determine the Safety, Tolerance, Maximum Tolerated Dose and Recommended Phase 2 Dose of DCR-MYC, a Lipid Nanoparticle (LNP)-Formulated Small Inhibitory RNA (siRNA) Oligonucleotide Targeting MYC, in Patients With Hepatocellular Carcinoma (HCC)"). Unfortunately, despite showing promising initial clinical and metabolic responses across various dose levels [141], DCR-MYC clinical trials were terminated for "Sponsor decision" because preliminary results did not meet the Company's expectations for further developments.

12. ARB-001467

The ARB-001467 is an RNA interference agent comprised of three siRNAs delivered using lipid nanoparticle technology, in particular liposomes, targeting the Hepatitis B Virus (HBV) RNA, and inhibiting the production of all HBV proteins, reducing the cccDNA content [142]. Preclinical studies demonstrated that a single dose of ARB-001467 can reduce HBsAg in a chimeric mouse model while ARB-001467 multiple injections led to a 90% reduction of HBsAg levels and a 50% reduction of cccDNA within 28 days of treatment in chimpanzees [143]. Based on these results, the Arbutus Biopharma corporation carried out a phase II clinical trial (NCT02631096, 2015, "A Phase 2a Single-Blind, Randomized, Placebo-Controlled Study Evaluating the Safety, Anti-Viral Activity, and Pharmacokinetics of ARB-001467 in Non Cirrhotic, HBeAg Negative and Positive Subjects With Chronic HBV Infection Receiving Nucleos(t)Ide Analogue Therapy"). The study was completed, the enrollment was of 36 participants, the starting date was Dec 2015 and the completion date was May 2018. Subjects were enrolled in three cohorts: Cohort 1, HBeAg-negative (0.2 mg/kg); Cohort 2, HBeAg-negative (0.4 mg/kg); Cohort 3, HBeAg-positive (0.4 mg/kg) and ARB-001467 was given as an intravenous injection over two hours, monthly, for three months. The treatment was generally well tolerated, and all subjects receiving ARB-001467 experienced a reduction in HBsAg from the baseline; greater HBsAg reductions were observed with more frequent dosing (bi-weekly), and at the higher dose (0.4 mg/kg) [144].

Challenges and Limitations of Lipid-Based siRNA Delivery Systems

Good biocompatibility, biodegradability, low toxicity, structural variability, easiness of largescale production and the possibility of both hydrophilic and lipophilic drugs incorporation are all beneficial properties that make lipid-based systems advanced NABDs delivery vectors with respect to other carriers [145].

In fact, vectors based on polymers such as chitosan, cyclodextrin polyethyleneimine, poly lactic-*co*-glycolic acid, dendrimers (i.e., polycationic dendrimers as polyamidoamine (PAMAM) and polypropylenimine (PPI)), although being excellent candidates for the delivery of NABDs, constitute only a small number of gene vectors entered in clinical trials (i.e., siG12D LODER (Local Drug EluteR) for Pancreatic Cancer [146]) with respect to the lipid-based ones. Also considering carriers based on peptides, proteins or the so-called "next-generation nanoparticles" or rather viral capsids, as noninfectious protein-based nanoparticles, termed virus-like particles (VLPs) [147], the similarity of lipid based systems with biological membranes, with all the consequent advantages such as a facilitated uptake in the cells and a greater biocompatibility, makes them unique carriers increasingly promising in gene therapy.

Also siRNA chemical modifications have been rigorously investigated with the goal of prolonged half-life, increased cellular uptake and targeted delivery if the nucleic acid is conjugated with proteins, aptamers, cholesterol and its derivatives, cell penetrating peptides (CPPs) and carbohydrates. For instance, the delivery of RNAi therapeutics to liver hepatocytes by subcutaneous administration has been achieved by Alnylam Pharmaceuticals using the GalNAc-siRNA conjugate delivery platform. In this approach, the siRNAs are conjugated with multivalent N-acetylgalactosamine (GalNAc) residues that are recognized by the asialoglycoprotein receptor (ASGPR) expressed on human hepatocyte cell surfaces with the consequent endocytosis. GalNAc-siRNA conjugates are reactive in numerous animal models, and thus are significantly promising for targeting disease-causing genes produced in liver [148].

Anyhow, in recent years the attention of the scientific community has focused on the development of vectors for NABDs delivery and several companies, including Tekmira, Alnylam, Silence Therapeutics, and others, have introduced siRNA nanoparticle products in either the preclinical or clinical phases [13]. In particular, the use of lipid carriers systems for NABDs delivery has given excellent feedback in clinical applications, while the recent approval of ALN-TTR02 for marketing this product is certainly the proof, but it is not the only example. The favorable features of these vectors are also reflected in all those products that have successfully passed one or more phases of clinical trials, i.e., showing to be safe and well-tolerated, able to improve the pharmacokinetics and pharmacodynamics of siRNAs. However, despite the use of new synthesized lipids and technologies such as SNALP and other LNP-based products, toxicity and immunogenicity still represent a great hindrance for these novel class of therapeutics [149].

Moreover the successful use of a type of carrier system for a certain disease does not guarantee its efficiency for the treatment of another illness, i.e., the phase I clinical trial of TKM-ApoB formulation based on LNP technology, and containing an siRNA targeting ApoB to be used for the treatment of hypercholesterolemia, was terminated due to patients' immune systems stimulation, as well as another Tekmira Pharmaceutical product developed to combat the Ebola Virus (TKM-100802). On the contrary, the same LNP technology, in particular the SNALP subfamily, was successfully used for the cure of TransThyRetin (TTR)-mediated amyloidosis through the development of ALN-TTR02 or for the treatment of solid tumors, this through ALN-VSP02 formulation, which has passed the phase I of clinical trials.

Finally, the problem represented by the high manufacturing cost of these therapeutic products must be considered, i.e., Patisiran treatment is currently priced at $450,000 per patient per year.

5. Conclusions

The wide range of diseases that can be treated by NABDs which are being evaluated in clinical trials, ranging from tumors to viral infections, highlight the great therapeutic potential of this new class of biological therapeutics, whose limit is determined by the impossibility of being administered in their naked form. To date, engineered lipid delivery systems, as cationic liposomes used in the SNALP of Tekmira Pharmaceuticals, represent the approach of choice for a concrete NABDs delivery; however, although the technology has undergone rapid developments, becoming a reality with the marketing of the first RNA- lipid-based drug carrier, some barriers still need to be overcome.

Currently, the lack of a rapid, reliable and scalable technique, reflecting in NABDs-liposomes the high production cost for minimum amounts of product, and the impossibility of having a meticulous control over the possible biological responses following their administration, i.e., toxicity and immunogenicity, represent the major limits, thus their overcoming is the challenge to face in order to get, in the near future, more powerful alternatives to the common drugs for the cure of several disabling diseases.

Author Contributions: Conceptualization, G.L. and S.B.; Methodology, S.B. and A.D.; Formal Analysis, S.B. and A.D.; Investigation, S.B. and A.D.; Writing—Original Draft Preparation, S.B.; Writing—Review & Editing, S.B., A.A.B. and G.L.; Supervision, A.A.B. and G.L.

Funding: Part of the present work has been done within the funded project "Campania Oncoterapie" - POR FESR 2014-2020 - D.D. n. 4 22/01/2019. CUP B61418000470007.

Conflicts of Interest: The authors declare no conflict of interest.

References

1. Ramamoorth, M.; Narvekar, A. Non Viral Vectors in Gene Therapy—An Overview. *J. Clin. Diagn. Res.* **2015**, *9*, 1–6. [CrossRef] [PubMed]
2. Daraee, H.; Etemadi, A.; Kouhi, M.; Alimirzalu, S.; Akbarzadeh, A. Application of liposomes in medicine and drug delivery. *Artif. Cells Nanomed. Biotechnol.* **2016**, *44*, 381–391. [CrossRef] [PubMed]
3. Chen, J.; Guo, Z.; Tian, H.; Chen, X. Production and clinical development of nanoparticles for gene delivery. *Mol. Ther. Methods Clin. Dev.* **2016**, *3*, 16023. [CrossRef] [PubMed]
4. Zhang, S.; Zhi, D.; Huang, L. Lipid-based vectors for siRNA delivery. *J. Drug Target.* **2012**, *20*, 724–735. [CrossRef] [PubMed]
5. Sridharan, K.; Gogtay, N.J. Therapeutic nucleic acids: Current clinical status. *Br. J. Clin. Pharmacol.* **2016**, *82*, 659–672. [CrossRef] [PubMed]
6. Kaczmarek, J.C.; Kowalski, P.S.; Anderson, D.G. Advances in the delivery of RNA therapeutics: From concept to clinical reality. *Genome Med.* **2017**, *9*, 60. [CrossRef]
7. Davidson, B.L.; McCray, P.B., Jr. Current prospects for RNA interference-based therapies. *Nat. Rev. Genet.* **2011**, *12*, 329. [CrossRef] [PubMed]
8. Fire, A.; Xu, S.; Montgomery, M.K.; Kostas, S.A.; Driver, S.E.; Mello, C.C. Potent and specific genetic interference by double-stranded RNA in Caenorhabditis elegans. *Nature* **1998**, *391*, 806–811. [CrossRef] [PubMed]
9. Ledford, H. Gene-silencing drug approved. *Nature* **2018**, *560*, 291–292. [CrossRef]
10. Oh, Y.-K.; Park, T.G. siRNA delivery systems for cancer treatment. *Adv. Drug Deliv. Rev.* **2009**, *61*, 850–862. [CrossRef]
11. Whitehead, K.A.; Langer, R.; Anderson, D.G. Knocking down barriers: Advances in siRNA delivery. *Nat. Rev. Drug Discov.* **2009**, *8*, 129–138. [CrossRef] [PubMed]
12. Burnett, J.C.; Rossi, J.J.; Tiemann, K. Current progress of siRNA/shRNA therapeutics in clinical trials. *Biotechnol. J.* **2011**, *6*, 1130–1146. [CrossRef] [PubMed]
13. Lee, J.-M.; Yoon, T.-J.; Cho, Y.-S. Recent developments in nanoparticle-based siRNA delivery for cancer therapy. *BioMed Res. Int.* **2013**, *2013*, 782041. [CrossRef] [PubMed]
14. Zhou, J.; Shum, K.-T.; Burnett, J.C.; Rossi, J.J. Nanoparticle-based delivery of RNAi therapeutics: Progress and challenges. *Pharmaceuticals* **2013**, *6*, 85–107. [CrossRef] [PubMed]
15. Kanasty, R.; Dorkin, J.R.; Vegas, A.; Anderson, D. Delivery materials for siRNA therapeutics. *Nat. Mater.* **2013**, *12*, 967–977. [CrossRef] [PubMed]
16. Yin, H.; Kanasty, R.L.; Eltoukhy, A.A.; Vegas, A.J.; Dorkin, J.R.; Anderson, D.G. Non-viral vectors for gene-based therapy. *Nat. Rev. Genet.* **2014**, *15*, 541–555. [CrossRef]
17. Sercombe, L.; Veerati, T.; Moheimani, F.; Wu, S.Y.; Sood, A.K.; Hua, S. Advances and challenges of liposome assisted drug delivery. *Front. Pharmacol.* **2015**, *6*, 286. [CrossRef]
18. Barba, A.A.; Lamberti, G.; Sardo, C.; Dapas, B.; Abrami, M.; Grassi, M.; Farra, R.; Tonon, F.; Forte, G.; Musiani, F. Novel Lipid and Polymeric Materials as Delivery Systems for Nucleic Acid Based Drugs. *Curr. Drug Metabol.* **2015**, *16*, 427–452. [CrossRef]
19. Bochicchio, S.; Dalmoro, A.; Barba, A.A.; Grassi, G.; Lamberti, G. Liposomes as siRNA Delivery Vectors. *Curr. Drug Metabol.* **2014**, *15*, 882–892. [CrossRef]
20. Lechanteur, A.; Sanna, V.; Duchemin, A.; Evrard, B.; Mottet, D.; Piel, G. Cationic liposomes carrying siRNA: Impact of lipid composition on physicochemical properties, cytotoxicity and endosomal escape. *Nanomaterials* **2018**, *8*, 270. [CrossRef]
21. Ozpolat, B.; Sood, A.K.; Lopez-Berestein, G. Liposomal siRNA nanocarriers for cancer therapy. *Adv. Drug Deliv. Rev.* **2014**, *66*, 110–116. [CrossRef] [PubMed]
22. Chakraborty, C.; Sharma, A.R.; Sharma, G.; Doss, C.G.P.; Lee, S.-S. Therapeutic miRNA and siRNA: Moving from bench to clinic as next generation medicine. *Mol. Ther. Nucleic Acids* **2017**, *8*, 132–143. [CrossRef] [PubMed]
23. Tai, W. Current Aspects of siRNA Bioconjugate for In Vitro and In Vivo Delivery. *Molecules* **2019**, *24*, 2211. [CrossRef] [PubMed]
24. Bozzuto, G.; Molinari, A. Liposomes as nanomedical devices. *Int. J. Nanomed.* **2015**, *10*, 975. [CrossRef] [PubMed]

25. Akbarzadeh, A.; Rezaei-Sadabady, R.; Davaran, S.; Joo, S.W.; Zarghami, N.; Hanifehpour, Y.; Samiei, M.; Kouhi, M.; Nejati-Koshki, K. Liposome: Classification, preparation, and applications. *Nanoscale Res. Lett.* **2013**, *8*, 102. [CrossRef] [PubMed]
26. Çağdaş, M.; Sezer, A.D.; Bucak, S. Liposomes as potential drug carrier systems for drug delivery. In *Application of Nanotechnology in Drug Delivery*; IntechOpen: Istanbul, Turkey, 2014.
27. Laouini, A.; Jaafar-Maalej, C.; Limayem-Blouza, I.; Sfar, S.; Charcosset, C.; Fessi, H. Preparation, characterization and applications of liposomes: State of the art. *J. Colloid Sci. Biotechnol.* **2012**, *1*, 147–168. [CrossRef]
28. Li, M.; Du, C.; Guo, N.; Teng, Y.; Meng, X.; Sun, H.; Li, S.; Yu, P.; Galons, H. Composition design and medical application of liposomes. *Eur. J. Med. Chem.* **2019**, *164*, 640–653. [CrossRef]
29. Torchilin, V.P. Recent advances with liposomes as pharmaceutical carriers. *Nat. Rev. Drug Discov.* **2005**, *4*, 145–160. [CrossRef]
30. Sawant, R.R.; Torchilin, V.P. Liposomes as 'smart' pharmaceutical nanocarriers. *Soft Matter* **2010**, *6*, 4026–4044. [CrossRef]
31. Ahmed, K.S.; Hussein, S.A.; Ali, A.H.; Korma, S.A.; Lipeng, Q.; Jinghua, C. Liposome: Composition, characterisation, preparation, and recent innovation in clinical applications. *J. Drug Target.* **2019**, *27*, 742–761. [CrossRef]
32. Briuglia, M.-L.; Rotella, C.; McFarlane, A.; Lamprou, D.A. Influence of cholesterol on liposome stability and on in vitro drug release. *Drug Deliv. Transl. Res.* **2015**, *5*, 231–242. [CrossRef] [PubMed]
33. Moh'd Atrouse, O. The effects of liposome composition and temperature on the stability of liposomes and the interaction of liposomes with human neutrophils. *Pak. J. Biol. Sci.* **2002**, *5*, 948–951.
34. Bitounis, D.; Fanciullino, R.; Iliadis, A.; Ciccolini, J. Optimizing druggability through liposomal formulations: New approaches to an old concept. *ISRN Pharm.* **2012**, *2012*, 738432. [CrossRef] [PubMed]
35. Deodhar, S.; Dash, A.K. Long circulating liposomes: Challenges and opportunities. *Ther. Deliv.* **2018**, *9*, 857–872. [CrossRef] [PubMed]
36. Lin, Q.; Chen, J.; Zhang, Z.; Zheng, G. Lipid-based nanoparticles in the systemic delivery of siRNA. *Nanomedicine* **2014**, *9*, 105–120. [CrossRef] [PubMed]
37. Santel, A.; Aleku, M.; Keil, O.; Endruschat, J.; Esche, V.; Fisch, G.; Dames, S.; Löffler, K.; Fechtner, M.; Arnold, W. A novel siRNA-lipoplex technology for RNA interference in the mouse vascular endothelium. *Gene Ther.* **2006**, *13*, 1222. [CrossRef] [PubMed]
38. Santel, A.; Aleku, M.; Röder, N.; Möpert, K.; Durieux, B.; Janke, O.; Keil, O.; Endruschat, J.; Dames, S.; Lange, C. Atu027 prevents pulmonary metastasis in experimental and spontaneous mouse metastasis models. *Clin. Cancer Res.* **2010**, *16*, 5469–5480. [CrossRef] [PubMed]
39. Whitehead, K.A.; Dorkin, J.R.; Vegas, A.J.; Chang, P.H.; Veiseh, O.; Matthews, J.; Fenton, O.S.; Zhang, Y.; Olejnik, K.T.; Yesilyurt, V. Degradable lipid nanoparticles with predictable in vivo siRNA delivery activity. *Nat. Commun.* **2014**, *5*, 4277. [CrossRef] [PubMed]
40. Dong, Y.; Love, K.T.; Dorkin, J.R.; Sirirungruang, S.; Zhang, Y.; Chen, D.; Bogorad, R.L.; Yin, H.; Chen, Y.; Vegas, A.J. Lipopeptide nanoparticles for potent and selective siRNA delivery in rodents and nonhuman primates. *Proc. Natl. Acad. Sci. USA* **2014**, *111*, 3955–3960. [CrossRef]
41. Love, K.T.; Mahon, K.P.; Levins, C.G.; Whitehead, K.A.; Querbes, W.; Dorkin, J.R.; Qin, J.; Cantley, W.; Qin, L.L.; Racie, T. Lipid-like materials for low-dose, in vivo gene silencing. *Proc. Natl. Acad. Sci. USA* **2010**, *107*, 1864–1869. [CrossRef]
42. Akinc, A.; Zumbuehl, A.; Goldberg, M.; Leshchiner, E.S.; Busini, V.; Hossain, N.; Bacallado, S.A.; Nguyen, D.N.; Fuller, J.; Alvarez, R. A combinatorial library of lipid-like materials for delivery of RNAi therapeutics. *Nat. Biotechnol.* **2008**, *26*, 561. [CrossRef] [PubMed]
43. Akinc, A.; Goldberg, M.; Qin, J.; Dorkin, J.R.; Gamba-Vitalo, C.; Maier, M.; Jayaprakash, K.N.; Jayaraman, M.; Rajeev, K.G.; Manoharan, M. Development of lipidoid–siRNA formulations for systemic delivery to the liver. *Mol. Ther.* **2009**, *17*, 872–879. [CrossRef] [PubMed]
44. Lila, A.S.A.; Ishida, T. Liposomal delivery systems: Design optimization and current applications. *Biol. Pharm. Bull.* **2017**, *40*, 1–10. [CrossRef] [PubMed]
45. Lv, H.; Zhang, S.; Wang, B.; Cui, S.; Yan, J. Toxicity of cationic lipids and cationic polymers in gene delivery. *J. Control. Release* **2006**, *114*, 100–109. [CrossRef] [PubMed]

46. Schroeder, A.; Levins, C.G.; Cortez, C.; Langer, R.; Anderson, D.G. Lipid-based nanotherapeutics for siRNA delivery. *J. Intern. Med.* **2010**, *267*, 9–21. [CrossRef]
47. Xia, Y.; Tian, J.; Chen, X. Effect of surface properties on liposomal siRNA delivery. *Biomaterials* **2016**, *79*, 56–68. [CrossRef]
48. Constantinescu, C.A.; Fuior, E.V.; Rebleanu, D.; Deleanu, M.; Simion, V.; Voicu, G.; Escriou, V.; Manduteanu, I.; Simionescu, M.; Calin, M. Targeted Transfection Using PEGylated Cationic Liposomes Directed Towards P-Selectin Increases siRNA Delivery into Activated Endothelial Cells. *Pharmaceutics* **2019**, *11*, 47. [CrossRef]
49. Riaz, M.; Zhang, X.; Lin, C.; Wong, K.; Chen, X.; Zhang, G.; Lu, A.; Yang, Z. Surface functionalization and targeting strategies of liposomes in solid tumor therapy: A review. *Int. J. Mol. Sci.* **2018**, *19*, 195. [CrossRef]
50. Uckun, F.; Yiv, S. Nanoscale Small Interfering RNA De-livery Systems For Personalized Cancer Therapy. *Int. J. Nano Stud. Technol.* **2012**, *1*, 6–11.
51. Kleshchanok, D.; Tuinier, R.; Lang, P.R. Direct measurements of polymer-induced forces. *J. Phys. Condens. Matter* **2008**, *20*, 073101. [CrossRef]
52. Bulbake, U.; Doppalapudi, S.; Kommineni, N.; Khan, W. Liposomal formulations in clinical use: An updated review. *Pharmaceutics* **2017**, *9*, 12. [CrossRef] [PubMed]
53. Xu, C.-f.; Wang, J. Delivery systems for siRNA drug development in cancer therapy. *Asian J. Pharm. Sci.* **2015**, *10*, 1–12. [CrossRef]
54. Fehring, V.; Schaeper, U.; Ahrens, K.; Santel, A.; Keil, O.; Eisermann, M.; Giese, K.; Kaufmann, J. Delivery of therapeutic siRNA to the lung endothelium via novel lipoplex formulation DACC. *Mol. Ther.* **2014**, *22*, 811–820. [CrossRef] [PubMed]
55. Saw, P.E.; Park, J.; Lee, E.; Ahn, S.; Lee, J.; Kim, H.; Kim, J.; Choi, M.; Farokhzad, O.C.; Jon, S. Effect of PEG pairing on the efficiency of cancer-targeting liposomes. *Theranostics* **2015**, *5*, 746. [CrossRef] [PubMed]
56. Kang, M.H.; Yoo, H.J.; Kwon, Y.H.; Yoon, H.Y.; Lee, S.G.; Kim, S.R.; Yeom, D.W.; Kang, M.J.; Choi, Y.W. Design of multifunctional liposomal nanocarriers for folate receptor-specific intracellular drug delivery. *Mol. Pharm.* **2015**, *12*, 4200–4213. [CrossRef] [PubMed]
57. Wang, S.; Lee, R.J.; Cauchon, G.; Gorenstein, D.G.; Low, P.S. Delivery of antisense oligodeoxyribonucleotides against the human epidermal growth factor receptor into cultured KB cells with liposomes conjugated to folate via polyethylene glycol. *Proc. Natl. Acad. Sci. USA* **1995**, *92*, 3318–3322. [CrossRef]
58. Mendonça, L.S.; Firmino, F.; Moreira, J.N.; Pedroso de Lima, M.C.; Simões, S. Transferrin receptor-targeted liposomes encapsulating anti-BCR-ABL siRNA or as ODN for chronic myeloid leukemia treatment. *Bioconj. Chem.* **2009**, *21*, 157–168. [CrossRef] [PubMed]
59. Zang, X.; Ding, H.; Zhao, X.; Li, X.; Du, Z.; Hu, H.; Qiao, M.; Chen, D.; Deng, Y.; Zhao, X. Anti-EphA10 antibody-conjugated pH-sensitive liposomes for specific intracellular delivery of siRNA. *Int. J. Nanomed.* **2016**, *11*, 3951. [CrossRef]
60. Mahanty, A.; Li, Y.; Yu, Y.; Banerjee, P.; Prasad, B.; Chaurasiya, B.; Tu, J.; Sun, C. Bubble liposome: A modern theranostic approach of new drug delivery. *World J. Pharm. Pharm. Sci.* **2017**, *6*, 1290–1314. [CrossRef]
61. Suzuki, R.; Takizawa, T.; Negishi, Y.; Utoguchi, N.; Maruyama, K. Effective gene delivery with novel liposomal bubbles and ultrasonic destruction technology. *Int. J. Pharm.* **2008**, *354*, 49–55. [CrossRef]
62. Zhang, Y.; Satterlee, A.; Huang, L. In vivo gene delivery by nonviral vectors: Overcoming hurdles? *Mol. Ther.* **2012**, *20*, 1298–1304. [CrossRef] [PubMed]
63. Suzuki, R.; Takizawa, T.; Negishi, Y.; Utoguchi, N.; Sawamura, K.; Tanaka, K.; Namai, E.; Oda, Y.; Matsumura, Y.; Maruyama, K. Tumor specific ultrasound enhanced gene transfer in vivo with novel liposomal bubbles. *J. Control. Release* **2008**, *125*, 137–144. [CrossRef] [PubMed]
64. Negishi, Y.; Omata, D.; Iijima, H.; Takabayashi, Y.; Suzuki, K.; Endo, Y.; Suzuki, R.; Maruyama, K.; Nomizu, M.; Aramaki, Y. Enhanced laminin-derived peptide AG73-mediated liposomal gene transfer by bubble liposomes and ultrasound. *Mol. Pharm.* **2010**, *7*, 217–226. [CrossRef] [PubMed]
65. Endo-Takahashi, Y.; Negishi, Y.; Nakamura, A.; Ukai, S.; Ooaku, K.; Oda, Y.; Sugimoto, K.; Moriyasu, F.; Takagi, N.; Suzuki, R. Systemic delivery of miR-126 by miRNA-loaded Bubble liposomes for the treatment of hindlimb ischemia. *Sci. Rep.* **2014**, *4*, 3883. [CrossRef] [PubMed]
66. Dua, J.; Rana, A.; Bhandari, A. Liposome: Methods of preparation and applications. *Int. J. Pharm. Stud. Res.* **2012**, *3*, 14–20.
67. Mozafari, M.R. Liposomes: An overview of manufacturing techniques. *Cell. Mol. Biol. Lett.* **2005**, *10*, 711. [PubMed]

68. Patil, Y.P.; Jadhav, S. Novel methods for liposome preparation. *Chem. Phys. Lipids* **2014**, *177*, 8–18. [CrossRef] [PubMed]
69. Bangham, A.D.; Horne, R. Negative staining of phospholipids and their structural modification by surface-active agents as observed in the electron microscope. *J. Mol. Biol.* **1964**, *8*, 660–668. [CrossRef]
70. Barba, A.A.; Bochicchio, S.; Lamberti, G.; Dalmoro, A. Ultrasonic energy in liposome production: Process modelling and size calculation. *Soft Matter* **2014**, *10*, 2574–2581. [CrossRef] [PubMed]
71. Wagner, A.; Vorauer-Uhl, K. Liposome technology for industrial purposes. *J. Drug Deliv.* **2010**, *2011*, 591325. [CrossRef]
72. MacLachlan, I. Liposomal formulations for nucleic acid delivery. In *Antisense Drug Technology: Principles, Strategies, and Applications*; Crooke, S.T., Ed.; Taylor & Francis: Boca Raton, FL, USA, 2008; pp. 237–270.
73. Piazza, O.; Russo, I.; Bocchicchio, S.; Barba, A.; Lamberti, G.; Zeppa, P.; Di Crescenzo, V.; Carrizzo, A.; Vecchione, C.; Ciacci, C. Cyclin D1 Gene Silencing by siRNA in ex vivo human tissue cultures. *Curr. Drug Deliv.* **2017**, *14*, 246–253. [CrossRef] [PubMed]
74. Bochicchio, S.; Dapas, B.; Russo, I.; Ciacci, C.; Piazza, O.; De Smedt, S.; Pottie, E.; Barba, A.A.; Grassi, G. In vitro and ex vivo delivery of tailored siRNA-nanoliposomes for E2F1 silencing as a potential therapy for colorectal cancer. *Int. J. Pharm.* **2017**, *525*, 377–387. [CrossRef] [PubMed]
75. Stuart, D.D.; Kao, G.Y.; Allen, T.M. A novel, long-circulating, and functional liposomal formulation of antisense oligodeoxynucleotides targeted against MDR1. *Cancer Gene Ther.* **2000**, *7*, 466. [CrossRef] [PubMed]
76. Dwivedi, C.; Verma, S. Review on Preparation and Characterization of Liposomes with Application. *Int. J. Sci. Innov. Res.* **2013**, *2*, 486–508.
77. Foged, C.; Nielsen, H.M.; Frokjaer, S. Liposomes for phospholipase A2 triggered siRNA release: Preparation and in vitro test. *Int. J. Pharm.* **2007**, *331*, 160–166. [CrossRef] [PubMed]
78. Szoka, F.; Papahadjopoulos, D. Procedure for preparation of liposomes with large internal aqueous space and high capture by reverse-phase evaporation. *Proc. Natl. Acad. Sci. USA* **1978**, *75*, 4194–4198. [CrossRef] [PubMed]
79. Stuart, D.; Allen, T. A new liposomal formulation for antisense oligodeoxynucleotides with small size, high incorporation efficiency and good stability. *Biochim. Biophys. Acta* **2000**, *1463*, 219–229. [CrossRef]
80. Mokhtarieh, A.A.; Cheong, S.; Kim, S.; Chung, B.H.; Lee, M.K. Asymmetric liposome particles with highly efficient encapsulation of siRNA and without nonspecific cell penetration suitable for target-specific delivery. *Biochim. Biophys. Acta* **2012**, *1818*, 1633–1641. [CrossRef] [PubMed]
81. Yu, B.; Lee, R.J.; Lee, L.J. Microfluidic methods for production of liposomes. *Methods Enzymol.* **2009**, *465*, 129–141.
82. Jahn, A.; Vreeland, W.N.; Gaitan, M.; Locascio, L.E. Controlled vesicle self-assembly in microfluidic channels with hydrodynamic focusing. *J. Am. Chem. Soc.* **2004**, *126*, 2674–2675. [CrossRef] [PubMed]
83. Massing, U.; Cicko, S.; Ziroli, V. Dual asymmetric centrifugation (DAC)—A new technique for liposome preparation. *J. Control. Release* **2008**, *125*, 16–24. [CrossRef] [PubMed]
84. Hirsch, M.; Ziroli, V.; Helm, M.; Massing, U. Preparation of small amounts of sterile siRNA-liposomes with high entrapping efficiency by dual asymmetric centrifugation (DAC). *J. Control. Release* **2009**, *135*, 80–88. [CrossRef] [PubMed]
85. Adrian, J.E.; Wolf, A.; Steinbach, A.; Rössler, J.; Süss, R. Targeted delivery to neuroblastoma of novel siRNA-anti-GD2-liposomes prepared by dual asymmetric centrifugation and sterol-based post-insertion method. *Pharm. Res.* **2011**, *28*, 2261–2272. [CrossRef] [PubMed]
86. Massing, U.; Ingebrigtsen, S.G.; Škalko-Basnet, N.; Holsæter, A.M. Dual Centrifugation-A Novel "in-vial" Liposome Processing Technique. In *Liposomes*; IntechOpen: London, UK, 2017.
87. Semple, S.C.; Akinc, A.; Chen, J.; Sandhu, A.P.; Mui, B.L.; Cho, C.K.; Sah, D.W.; Stebbing, D.; Crosley, E.J.; Yaworski, E. Rational design of cationic lipids for siRNA delivery. *Nat. Biotechnol.* **2010**, *28*, 172–176. [CrossRef] [PubMed]
88. Saffari, M.; Shirazi, F.H.; Oghabian, M.A.; Moghimi, H.R. Preparation and in-vitro evaluation of an antisense-containing cationic liposome against non-small cell lung cancer: A comparative preparation study. *Iran. J. Pharm. Res.* **2013**, *12*, 3. [PubMed]
89. Saad, M.; Garbuzenko, O.B.; Minko, T. Co-delivery of siRNA and an anticancer drug for treatment of multidrug-resistant cancer. *Nanomedicine* **2008**, *3*, 761–776. [CrossRef] [PubMed]

90. Fenske, D.B.; Cullis, P.R. Entrapment of small molecules and nucleic acid–based drugs in liposomes. In *Methods in Enzymology*; Elsevier: Vancouver, BC, Canada, 2005; Volume 391, pp. 7–40.
91. Yeo, Y. *Nanoparticulate Drug Delivery Systems: Strategies, Technologies, and Applications*; John Wiley & Sons: San Francisco, CA, USA, 2013.
92. Jeffs, L.B.; Palmer, L.R.; Ambegia, E.G.; Giesbrecht, C.; Ewanick, S.; MacLachlan, I. A scalable, extrusion-free method for efficient liposomal encapsulation of plasmid DNA. *Pharm. Res.* **2005**, *22*, 362–372. [CrossRef]
93. Estanqueiro, M.; Vasconcelos, H.; Lobo, J.M.S.; Amaral, H. Delivering miRNA modulators for cancer treatment. In *Drug Targeting and Stimuli Sensitive Drug Delivery Systems*; Elsevier: Bucharest, Romania, 2018; pp. 517–565.
94. Gujrati, M.; Lu, Z.-R. Targeted systemic delivery of therapeutic siRNA. In *Gene Therapy of Cancer*; Elsevier: Amsterdam, The Netherlands, 2014; pp. 47–65.
95. Judge, A.D.; Bola, G.; Lee, A.C.; MacLachlan, I. Design of noninflammatory synthetic siRNA mediating potent gene silencing in vivo. *Mol. Ther.* **2006**, *13*, 494–505. [CrossRef]
96. Wilner, S.E.; Levy, M. Synthesis and characterization of aptamer-targeted SNALPs for the delivery of siRNA. In *Nucleic Acid Aptamers*; Springer: New York, NY, USA, 2016; pp. 211–224.
97. Morrissey, D.V.; Lockridge, J.A.; Shaw, L.; Blanchard, K.; Jensen, K.; Breen, W.; Hartsough, K.; Machemer, L.; Radka, S.; Jadhav, V.; et al. Potent and persistent in vivo anti-HBV activity of chemically modified siRNAs. *Nat. Biotechnol.* **2005**, *23*, 1002–1007. [CrossRef]
98. Geisbert, T.W.; Hensley, L.E.; Kagan, E.; Yu, E.Z.; Geisbert, J.B.; Daddario-DiCaprio, K.; Fritz, E.A.; Jahrling, P.B.; McClintock, K.; Phelps, J.R. Postexposure protection of guinea pigs against a lethal ebola virus challenge is conferred by RNA interference. *J. Infect. Dis.* **2006**, *193*, 1650–1657. [CrossRef]
99. Geisbert, T.W.; Lee, A.C.; Robbins, M.; Geisbert, J.B.; Honko, A.N.; Sood, V.; Johnson, J.C.; de Jong, S.; Tavakoli, I.; Judge, A. Postexposure protection of non-human primates against a lethal Ebola virus challenge with RNA interference: A proof-of-concept study. *Lancet* **2010**, *375*, 1896–1905. [CrossRef]
100. Zimmermann, T.S.; Lee, A.C.; Akinc, A.; Bramlage, B.; Bumcrot, D.; Fedoruk, M.N.; Harborth, J.; Heyes, J.A.; Jeffs, L.B.; John, M. RNAi-mediated gene silencing in non-human primates. *Nature* **2006**, *441*, 111–114. [CrossRef] [PubMed]
101. Elouahabi, A.; Ruysschaert, J.-M. Formation and intracellular trafficking of lipoplexes and polyplexes. *Mol. Ther.* **2005**, *11*, 336–347. [CrossRef] [PubMed]
102. Buyens, K.; De Smedt, S.C.; Braeckmans, K.; Demeester, J.; Peeters, L.; van Grunsven, L.A.; de Mollerat du Jeu, X.; Sawant, R.; Torchilin, V.; Farkasova, K. Liposome based systems for systemic siRNA delivery: Stability in blood sets the requirements for optimal carrier design. *J. Control. Release* **2012**, *158*, 362–370. [CrossRef] [PubMed]
103. Elbashir, S.M.; Harborth, J.; Lendeckel, W.; Yalcin, A.; Weber, K.; Tuschl, T. Duplexes of 21-nucleotide RNAs mediate RNA interference in cultured mammalian cells. *Nature* **2001**, *411*, 494. [CrossRef] [PubMed]
104. Available online: https://www.clinicaltrials.gov/ (accessed on 20 June 2019).
105. Tabernero, J.; Shapiro, G.I.; LoRusso, P.M.; Cervantes, A.; Schwartz, G.K.; Weiss, G.J.; Paz-Ares, L.; Cho, D.C.; Infante, J.R.; Alsina, M. First-in-Humans Trial of an RNA Interference Therapeutic Targeting VEGF and KSP in Cancer Patients with Liver Involvement. *Cancer Discov.* **2013**, *3*, 406–417. [CrossRef]
106. Jayaraman, M.; Ansell, S.M.; Mui, B.L.; Tam, Y.K.; Chen, J.; Du, X.; Butler, D.; Eltepu, L.; Matsuda, S.; Narayanannair, J.K. Maximizing the potency of siRNA lipid nanoparticles for hepatic gene silencing in vivo. *Angew. Chem. Int. Ed.* **2012**, *51*, 8529–8533. [CrossRef]
107. Frank-Kamenetsky, M.; Grefhorst, A.; Anderson, N.N.; Racie, T.S.; Bramlage, B.; Akinc, A.; Butler, D.; Charisse, K.; Dorkin, R.; Fan, Y. Therapeutic RNAi targeting PCSK9 acutely lowers plasma cholesterol in rodents and LDL cholesterol in nonhuman primates. *Proc. Natl. Acad. Sci. USA* **2008**, *105*, 11915–11920. [CrossRef]
108. Fitzgerald, K.; Frank-Kamenetsky, M.; Shulga-Morskaya, S.; Liebow, A.; Bettencourt, B.R.; Sutherland, J.E.; Hutabarat, R.M.; Clausen, V.A.; Karsten, V.; Cehelsky, J. Effect of an RNA interference drug on the synthesis of proprotein convertase subtilisin/kexin type 9 (PCSK9) and the concentration of serum LDL cholesterol in healthy volunteers: A randomised, single-blind, placebo-controlled, phase 1 trial. *Lancet* **2014**, *383*, 60–68. [CrossRef]

109. Coelho, T.; Adams, D.; Silva, A.; Lozeron, P.; Hawkins, P.N.; Mant, T.; Perez, J.; Chiesa, J.; Warrington, S.; Tranter, E.; et al. Safety and Efficacy of RNAi Therapy for Transthyretin Amyloidosis. *N. Engl. J. Med.* **2013**, *369*, 819–829. [CrossRef]
110. Solomon, S.D.; Adams, D.; Kristen, A.; Grogan, M.; González-Duarte, A.; Maurer, M.S.; Merlini, G.; Damy, T.; Slama, M.S.; Brannagan, T.H., III. Effects of Patisiran, an RNA Interference Therapeutic, on Cardiac Parameters in Patients with Hereditary Transthyretin-Mediated Amyloidosis: Analysis of the APOLLO Study. *Circulation* **2019**, *139*, 431–443. [CrossRef] [PubMed]
111. Minamisawa, M.; Claggett, B.; Adams, D.; Kristen, A.V.; Merlini, G.; Slama, M.S.; Dispenzieri, A.; Shah, A.M.; Falk, R.H.; Karsten, V. Association of Patisiran, an RNA Interference Therapeutic, With Regional Left Ventricular Myocardial Strain in Hereditary Transthyretin Amyloidosis: The APOLLO Study. *JAMA Cardiol.* **2019**, *4*, 466–472. [CrossRef] [PubMed]
112. Adams, D.; Suhr, O.B.; Dyck, P.J.; Litchy, W.J.; Leahy, R.G.; Chen, J.; Gollob, J.; Coelho, T. Trial design and rationale for APOLLO, a Phase 3, placebo-controlled study of patisiran in patients with hereditary ATTR amyloidosis with polyneuropathy. *BMC Neurol.* **2017**, *17*, 181. [CrossRef] [PubMed]
113. Landen, C.N.; Chavez-Reyes, A.; Bucana, C.; Schmandt, R.; Deavers, M.T.; Lopez-Berestein, G.; Sood, A.K. Therapeutic EphA2 gene targeting in vivo using neutral liposomal small interfering RNA delivery. *Cancer Res.* **2005**, *65*, 6910–6918. [CrossRef] [PubMed]
114. Halder, J.; Kamat, A.A.; Landen, C.N.; Han, L.Y.; Lutgendorf, S.K.; Lin, Y.G.; Merritt, W.M.; Jennings, N.B.; Chavez-Reyes, A.; Coleman, R.L. Focal adhesion kinase targeting using in vivo short interfering RNA delivery in neutral liposomes for ovarian carcinoma therapy. *Clin. Cancer Res.* **2006**, *12*, 4916–4924. [CrossRef]
115. Gray, M.J.; Van Buren, G.; Dallas, N.A.; Xia, L.; Wang, X.; Yang, A.D.; Somcio, R.J.; Lin, Y.G.; Lim, S.; Fan, F. Therapeutic targeting of neuropilin-2 on colorectal carcinoma cells implanted in the murine liver. *J. Natl. Cancer Inst.* **2008**, *100*, 109–120. [CrossRef] [PubMed]
116. Wagner, M.J.; Mitra, R.; McArthur, M.J.; Baze, W.; Barnhart, K.; Wu, S.Y.; Rodriguez-Aguayo, C.; Zhang, X.; Coleman, R.L.; Lopez-Berestein, G. Preclinical mammalian safety studies of EPHARNA (DOPC nanoliposomal EphA2-targeted siRNA). *Mol. Cancer Ther.* **2017**, *16*, 1114–1123. [CrossRef]
117. Aleku, M.; Schulz, P.; Keil, O.; Santel, A.; Schaeper, U.; Dieckhoff, B.; Janke, O.; Endruschat, J.; Durieux, B.; Röder, N. Atu027, a liposomal small interfering RNA formulation targeting protein kinase N3, inhibits cancer progression. *Cancer Res.* **2008**, *68*, 9788–9798. [CrossRef]
118. Santel, A.; Aleku, M.; Keil, O.; Endruschat, J.; Esche, V.; Durieux, B.; Löffler, K.; Fechtner, M.; Röhl, T.; Fisch, G. RNA interference in the mouse vascular endothelium by systemic administration of siRNA-lipoplexes for cancer therapy. *Gene Ther.* **2006**, *13*, 1360–1370. [CrossRef]
119. Strumberg, D.; Schultheis, B.; Traugott, U.; Vank, C.; Santel, A.; Keil, O.; Giese, K.; Kaufmann, J.; Drevs, J. Phase I clinical development of Atu027, a siRNA formulation targeting PKN3 in patients with advanced solid tumors. *Int. J. Clin. Pharmacol. Ther.* **2012**, *50*, 76. [CrossRef]
120. Schultheis, B.; Strumberg, D.; Santel, A.; Vank, C.; Gebhardt, F.; Keil, O.; Lange, C.; Giese, K.; Kaufmann, J.; Khan, M. First-in-human phase I study of the liposomal RNA interference therapeutic Atu027 in patients with advanced solid tumors. *J. Clin. Oncol.* **2014**, *32*, 4141–4148. [CrossRef] [PubMed]
121. Schultheis, B.; Strumberg, D.; Kuhlmann, J.; Wolf, M.; Link, K.; Seufferlein, T.; Kaufmann, J.; Gebhardt, F.; Bruyniks, N.; Pelzer, U. A phase Ib/IIa study of combination therapy with gemcitabine and Atu027 in patients with locally advanced or metastatic pancreatic adenocarcinoma. *J. Clin. Oncol.* **2016**, *34*, 385. [CrossRef]
122. Zabludoff, S.; Liu, Y.; Liu, J.; Zhang, J.; Xia, F.; Quimbo, A.; Yao, J.; Clamme, J.-P.; Siegmund, A.P.; Maruyama, K. Late Breaking Abstract-ND-L02-s0201 treatment leads to efficacy in preclinical IPF models. *Eur. Respir. Soc.* **2017**, *50*. [CrossRef]
123. Soule, B.; Tirucherai, G.; Kavita, U.; Kundu, S.; Christian, R. Safety, tolerability, and pharmacokinetics of BMS-986263/ND-L02-s0201, a novel targeted lipid nanoparticle delivering HSP47 siRNA, in healthy participants: A randomised, placebo-controlled, double-blind, phase 1 study. *J. Hepatol.* **2018**, *68*, S112. [CrossRef]
124. Sato, Y.; Murase, K.; Kato, J.; Kobune, M.; Sato, T.; Kawano, Y.; Takimoto, R.; Takada, K.; Miyanishi, K.; Matsunaga, T.; et al. Resolution of liver cirrhosis using vitamin A-coupled liposomes to deliver siRNA against a collagen-specific chaperone. *Nat. Biotechnol.* **2008**, *26*, 431–442. [CrossRef] [PubMed]
125. Meyers, R.A. *Translational Medicine: Molecular Pharmacology and Drug Discovery*; John Wiley & Sons: San Francisco, CA, USA, 2018.

126. Steegmaier, M.; Hoffmann, M.; Baum, A.; Lénárt, P.; Petronczki, M.; Krššák, M.; Gürtler, U.; Garin-Chesa, P.; Lieb, S.; Quant, J. BI 2536, a potent and selective inhibitor of polo-like kinase 1, inhibits tumor growth in vivo. *Curr. Biol.* **2007**, *17*, 316–322. [CrossRef] [PubMed]
127. Reagan-Shaw, S.; Ahmad, N. Silencing of polo-like kinase (Plk) 1 via siRNA causes induction of apoptosis and impairment of mitosis machinery in human prostate cancer cells: Implications for the treatment of prostate cancer. *FASEB J.* **2005**, *19*, 611–613. [CrossRef]
128. Bu, Y.; Yang, Z.; Li, Q.; Song, F. Silencing of polo-like kinase (Plk) 1 via siRNA causes inhibition of growth and induction of apoptosis in human esophageal cancer cells. *Oncology* **2008**, *74*, 198–206. [CrossRef]
129. Ramanathan, R.K.; Hamburg, S.I.; Borad, M.J.; Seetharam, M.; Kundranda, M.N.; Lee, P.; Fredlund, P.; Gilbert, M.; Mast, C.; Semple, S.C.; et al. Abstract LB-289: A phase I dose escalation study of TKM-080301, a RNAi therapeutic directed against PLK1, in patients with advanced solid tumors. In Proceedings of the AACR 104th Annual Meeting 2013, Washington, DC, USA, 6–10 April 2013.
130. Ramanathan, R.K.; Hamburg, S.I.; Halfdanarson, T.R.; Borad, M. A Phase I/II Dose Escalation Study of TKM-080301, a RNAi therapeutic directed against PLK1, in patients with advanced solid tumors, with an expansion cohort of patients with NET or ACC. *Pancreas* **2014**, *43*, 502–1502.
131. El Dika, I.; Lim, H.Y.; Yong, W.P.; Lin, C.C.; Yoon, J.H.; Modiano, M.; Freilich, B.; Choi, H.J.; Chao, T.Y.; Kelley, R.K.; et al. An Open-Label, Multicenter, Phase I, Dose Escalation Study with Phase II Expansion Cohort to Determine the Safety, Pharmacokinetics, and Preliminary Antitumor Activity of Intravenous TKM-080301 in Subjects with Advanced Hepatocellular Carcinoma. *Oncologist* **2019**, *24*, 747-e218. [CrossRef]
132. Feldmann, H.; Geisbert, T.W. Ebola haemorrhagic fever. *Lancet* **2011**, *377*, 849–862. [CrossRef]
133. Bixler, S.L.; Duplantier, A.J.; Bavari, S. Discovering drugs for the treatment of Ebola virus. *Curr. Treat. Opt. Infect. Dis.* **2017**, *9*, 299–317. [CrossRef] [PubMed]
134. Haque, A.; Hober, D.; Blondiaux, J. Addressing therapeutic options for Ebola virus infection in current and future outbreaks. *Antimicrob. Agents Chemother.* **2015**, *59*, 5892–5902. [CrossRef] [PubMed]
135. MacLachlan, I. Progress in the Development of Lipid Nanoparticle-RNA Therapeutics. In Proceedings of the ASGCT 17th Annual Meeting, Washington, DC, USA, 21 May 2014.
136. Kraft, C.S.; Hewlett, A.L.; Koepsell, S.; Winkler, A.M.; Kratochvil, C.J.; Larson, L.; Varkey, J.B.; Mehta, A.K.; Lyon III, G.M.; Friedman-Moraco, R.J. The use of TKM-100802 and convalescent plasma in 2 patients with Ebola virus disease in the United States. *Clin. Infect. Dis.* **2015**, *61*, 496–502. [CrossRef] [PubMed]
137. Dunning, J.; Sahr, F.; Rojek, A.; Gannon, F.; Carson, G.; Idriss, B.; Massaquoi, T.; Gandi, R.; Joseph, S.; Osman, H.K. Experimental treatment of Ebola virus disease with TKM-130803: A single-arm phase 2 clinical trial. *PLoS Med.* **2016**, *13*, e1001997. [CrossRef] [PubMed]
138. Mühlberger, E.; Hensley, L.L.; Towner, J.S. *Marburg-and Ebolaviruses: From Ecosystems to Molecules*; Springer: Cham, Switzerland, 2017; Volume 411.
139. Martin, S.; Chiramel, A.I.; Schmidt, M.L.; Chen, Y.-C.; Whitt, N.; Watt, A.; Dunham, E.C.; Shifflett, K.; Traeger, S.; Leske, A. A genome-wide siRNA screen identifies a druggable host pathway essential for the Ebola virus life cycle. *Genome Med.* **2018**, *10*, 58. [CrossRef] [PubMed]
140. Dudek, H.; Wong, D.H.; Arvan, R.; Shah, A.; Wortham, K.; Ying, B.; Diwanji, R.; Zhou, W.; Holmes, B.; Yang, H.; et al. Knockdown of [beta]-catenin with Dicer-Substrate siRNAs Reduces Liver Tumor Burden In vivo. *Mol. Ther.* **2014**, *22*, 92–101. [CrossRef] [PubMed]
141. Tolcher, A.W.; Papadopoulos, K.P.; Patnaik, A.; Rasco, D.W.; Martinez, D.; Wood, D.L.; Fielman, B.; Sharma, M.; Janisch, L.A.; Brown, B.D. Safety and activity of DCR-MYC, a first-in-class Dicer-substrate small interfering RNA (DsiRNA) targeting MYC, in a phase I study in patients with advanced solid tumors. *J. Clin. Oncol.* **2015**, *33*, 11006. [CrossRef]
142. Anselmo, A.C.; Mitragotri, S. Nanoparticles in the clinic. *Bioeng. Transl. Med.* **2016**, *1*, 10–29. [CrossRef] [PubMed]
143. Flisiak, R.; Jaroszewicz, J.; Łucejko, M. siRNA drug development against hepatitis B virus infection. *Expert Opin. Biol. Ther.* **2018**, *18*, 609–617. [CrossRef] [PubMed]
144. Agarwal, K.; Gane, E.; Cheng, W.; Sievert, W.; Roberts, S.; Ahn, S.H.; Kim, Y.J.; Streinu-Cercel, A.; Denning, J.; Symonds, W.; et al. HBcrAg, HBV-RNA declines in a phase 2a study evaluating the multi-dose activity of ARB-1467 in HBeAg-positive and negative virally suppressed subjects with hepatitis B. *Hepatology* **2017**, *66*.

145. Ghasemiyeh, P.; Mohammadi-Samani, S. Solid lipid nanoparticles and nanostructured lipid carriers as novel drug delivery systems: Applications, advantages and disadvantages. *Res. Pharm. Sci.* **2018**, *13*, 288. [PubMed]
146. Golan, T.; Hubert, A.; Shemi, A.; Segal, A.; Dancour, A.; Khvalevsky, E.Z.; Ben-David, E.; Raskin, S.; Goldes, Y.; Inbar, Y. A phase I trial of a local delivery of siRNA against k-ras in combination with chemotherapy for locally advanced pancreatic adenocarcinoma. *J. Clin. Oncol.* **2013**, *31*, 4037.
147. Rohovie, M.J.; Nagasawa, M.; Swartz, J.R. Virus-like particles: Next-generation nanoparticles for targeted therapeutic delivery. *Bioeng. Transl. Med.* **2017**, *2*, 43–57. [CrossRef] [PubMed]
148. Springer, A.D.; Dowdy, S.F. GalNAc-siRNA conjugates: Leading the way for delivery of RNAi therapeutics. *Nucleic Acid Ther.* **2018**, *28*, 109–118. [CrossRef] [PubMed]
149. Zatsepin, T.S.; Kotelevtsev, Y.V.; Koteliansky, V. Lipid nanoparticles for targeted siRNA delivery–going from bench to bedside. *Int. J. Nanomed.* **2016**, *11*, 3077.

Review

Strategies for Delivery of siRNAs to Ovarian Cancer Cells

Rossella Farra [1,†], Matea Maruna [1,†], Francesca Perrone [1], Mario Grassi [2], Fabio Benedetti [3], Marianna Maddaloni [1], Maguie El Boustani [4,5], Salvo Parisi [4,5], Flavio Rizzolio [4,6], Giancarlo Forte [7], Fabrizio Zanconati [8], Maja Cemazar [9,10], Urska Kamensek [10], Barbara Dapas [1,*] and Gabriele Grassi [1]

1 Department of Life Sciences, Cattinara University Hospital, Trieste University, Strada di Fiume 447, I-34149 Trieste, Italy; rfarra@units.it (R.F.); mateamaruna93@gmail.com (M.M.); francesca.perrone@phd.units.it (F.P.); marianna.maddaloni@libero.it (M.M.); ggrassi@units.it (G.G.)
2 Department of Engineering and Architecture, University of Trieste, Via Valerio 6/A, I-34127 Trieste, Italy; mario.grassi@dia.units.it
3 Dipartimento di Scienze Chimiche e Farmaceutiche, Università degli Studi di Trieste, I-34127 Trieste, Italy; benedett@units.it
4 Pathology Unit, IRCCS CRO Aviano-National Cancer Institute, I-33081 Aviano, Italy; elboustanymaguie@gmail.com (M.E.B.); kika.grassi.0671@gmail.com (S.P.); frizzolio1@gmail.com (F.R.)
5 Doctoral School in Molecular Biomedicine, University of Trieste, I-34127 Trieste, Italy
6 Department of Molecular Sciences and Nanosystems, Ca' Foscari University of Venice, I-30123 Venezia-Mestre, Italy
7 International Clinical Research Center (ICRC), St Anne's University Hospital, CZ-65691 Brno, Czech Republic; giancarlo.forte@fnusa.cz
8 Department of Medical, Surgical and Health Sciences, University of Trieste, Cattinara Hospital, Strada di Fiume, 447, I-34149 Trieste, Italy; fabrizio.zanconati@asuits.sanita.fvg.it
9 Department of Experimental Oncology, Institute of Oncology Ljubljana, Zaloska 2, SI-1000 Ljubljana, Slovenia; MCemazar@onko-i.si
10 Faculty of Health Sciences, University of Primorska, Polje 42, SI-6310 Izola, Slovenia; UKamensek@onko-i.si
* Correspondence: bdapas@units.it; Tel.: +39-040-399-6228
† These authors contributed equally to this work.

Received: 20 September 2019; Accepted: 18 October 2019; Published: 22 October 2019

Abstract: The unmet need for novel therapeutic options for ovarian cancer (OC) deserves further investigation. Among the different novel drugs, small interfering RNAs (siRNAs) are particularly attractive because of their specificity of action and efficacy, as documented in many experimental setups. However, the fragility of these molecules in the biological environment necessitates the use of delivery materials able to protect them and possibly target them to the cancer cells. Among the different delivery materials, those based on polymers and lipids are considered very interesting because of their biocompatibility and ability to carry/deliver siRNAs. Despite these features, polymers and lipids need to be engineered to optimize their delivery properties for OC. In this review, we concentrated on the description of the therapeutic potential of siRNAs and polymer-/lipid-based delivery systems for OC. After a brief description of OC and siRNA features, we summarized the strategies employed to minimize siRNA delivery problems, the targeting strategies to OC, and the preclinical models available. Finally, we discussed the most interesting works published in the last three years about polymer-/lipid-based materials for siRNA delivery.

Keywords: ovarian cancer; siRNA; polymer; lipid; delivery

1. Introduction

Effective therapeutic approaches are lacking for ovarian carcinoma (OC), the most lethal gynecological neoplasm. Thus, the development of novel strategies is of the utmost urgency. In this review, we focused our attention on the use of small interfering RNAs (siRNAs), short double stranded RNA molecules with the ability to downregulate the expression of virtually any gene causing disease in humans [1,2]. However, due to their fragile nature in the biological environment, siRNAs, as many other nucleic acid based molecules [3–5], need to be complexed with adequate delivery materials. Among these, we focused herein on polymers and lipids, which show high biocompatibility and the ability to deliver siRNAs. Other reviews have covered additional delivery systems [6–9].

For the development of effective delivery strategies and novel therapeutic drugs for OC, different aspects should be considered. The first is the biological/anatomical features of OC. For this reason, we have dedicated specific sections to this topic (Sections 1.1–1.3). The second aspect is related to the specific nature of siRNAs, as well as the problems of their delivery in general and in particular to OC cells (Sections 2.1 and 2.2). The third element involves the features of polymer- and lipid-based delivery materials, the general characteristics of which have been described (Sections 3.1 and 3.2). A fourth aspect relates to the possibility of generating siRNA delivery complexes able to target OC cells (Section 4.1) and/or OC-specific gene products (Section 4.2) to minimize side effects and improve effectiveness. The fifth aspect relates to the employment of models adequate to produce data with potential value in humans (Sections 5.1–5.3). Finally, we have presented the most recent works (from the last three years, Section 6) related to the delivery of siRNAs, also making methodological considerations (Section 7).

1.1. Ovarian Carcinoma

Ovarian carcinoma (OC), the seventh most common cause of death among women in industrialized countries, is the most lethal gynecological neoplasm [10]. In women under 40 years of age, OC is rare; however, the risk increases above age 40, peaking in the late 70s [10]. Annually, 240,000 new diagnoses are registered worldwide, with a higher incidence in the Caucasian population in the countries of northwestern Europe and in the USA compared to Asian, African, and South American countries [11]. As OC does not have a specific symptomatology in the initial phases, about 70% of OCs are diagnosed in advanced stages when there are already abdominal metastases [12]. For the advanced forms, the combined association of surgical removal of cancer and chemotherapy has a five-year survival rate of 25% [13]. Moreover, disease recurrence within two years of diagnosis is frequent [14], and when this occurs, OC very frequently displays a high resistance to chemotherapy. Together, these factors explain why the five-year survival in OC has remained almost stable (38% 1990–1994; 41% 2000–2004) in recent years.

1.2. Classification and Biology of OCs

From a histological point of view, OCs can be divided into three subtypes based on the cells of origin. Epithelial tumors, representing more than 90% of OCs, originate from the epithelial cells that line the surface of the ovaries or the fimbriae. Germ cell tumors, accounting for just 5% of OCs, originate from germ cells and are almost exclusively found in juvenile populations. Finally, stromal tumors, constituting about 4% of OCs, originate from the gonadal stroma, which supports ovary tissue. OCs of epithelial origin have been further subdivided into (Figure 1): type I tumors, which contain well-differentiated cells (such as low grade serous carcinoma and endometrioid carcinoma) and are characterized by low malignity; type II tumors, which are rather aggressive and arise directly from the epithelial tissue of the organ without going through a precancerous phase (typically represented by the high-grade serous ovarian carcinomas—HGSOCs) [15]. Gene expression profiling has revealed the existence of five molecular subtypes of HGSOCs: mesenchymal; proliferative, with the worst overall survival; differentiated, with an intermediate prognosis; anti-mesenchymal; and immunoreactive, with a better prognosis [16–18].

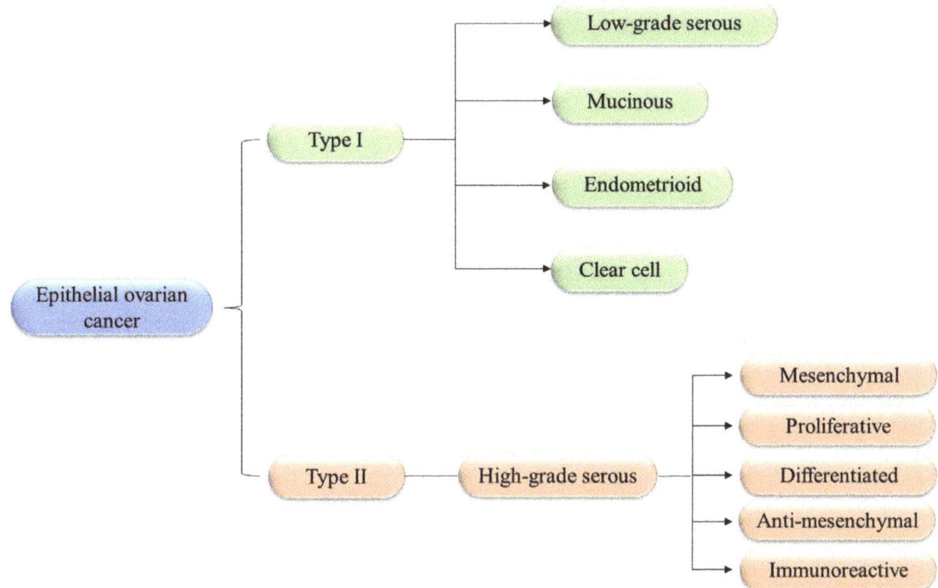

Figure 1. Histological classification of epithelial ovarian carcinoma.

HGSOC is the most common (70%) and most aggressive OC histotype [19]. The survival for early-stage and advanced-stage patients 10 years from diagnosis is 55%, and 15%, respectively [20]. In contrast to many other tumors, HGSOC does not require the blood or lymph to metastasize, as it can spread by direct contact to the neighboring organs within the peritoneal cavity or through cell exfoliation from the primary tumor [19]. Although, in principle, every organ within the peritoneal cavity may be involved in tumor dissemination, the involvement of a part of the neighboring peritoneum named omentum is the most frequent. The peritoneum is a large serous membrane extending from the stomach and covering the intestines (Figure 2), which contains energy-dense adipocytes. It has been proposed that HGSOC predilection for the omentum is due to cancer cell metabolism, which requires fatty-acid β-oxidation [21].

1.3. Current Therapies

The primary intervention for HGSOC is based on the use of surgery with the aim to remove the macroscopic lesions including metastasis [22], which are, as mentioned above, predominantly localized in the omentum. Notably, because of the specific biology of HGSOC, metastasis removal can improve overall survival. After surgery, the vast majority of patients undergo chemotherapy, which is the same for any type of OC [23]. Current chemotherapy is based on the use of the combined systemic administration of cisplatin–paclitaxel (PTX) [24]. More recently, intra-peritoneal (IP) administration of drugs has been proposed. As HGSOC is mostly confined to the peritoneal cavity, the IP route enables a higher local concentration [24]. Notably, a recent meta-analysis showed that IP administration significantly improved progression-free survival in advanced HGSOC patients compared to systemic administration [25]. However, significant toxicity is still an issue to be solved.

About 70% of patients treated for HGSOC respond to the first-line, PTX-based chemotherapy [26]. However, about 80% of these relapse soon or later [22]. Within relapsing patients, about 50% are still responsive to PTX, even though persistent side effects from previous treatment may become problematic. For PTX-resistant patients, alternative options are limited and poorly effective [22]. Novel therapeutic approaches include the use of poly (ADP-Ribose) polymerase (PARP) inhibitors. PARP mediates base excision repair of single-strand breaks to resolve spontaneous DNA damage [27]. Thus, PARP inhibition can expose the cancer cell to increased DNA damage, eventually leading to cell death.

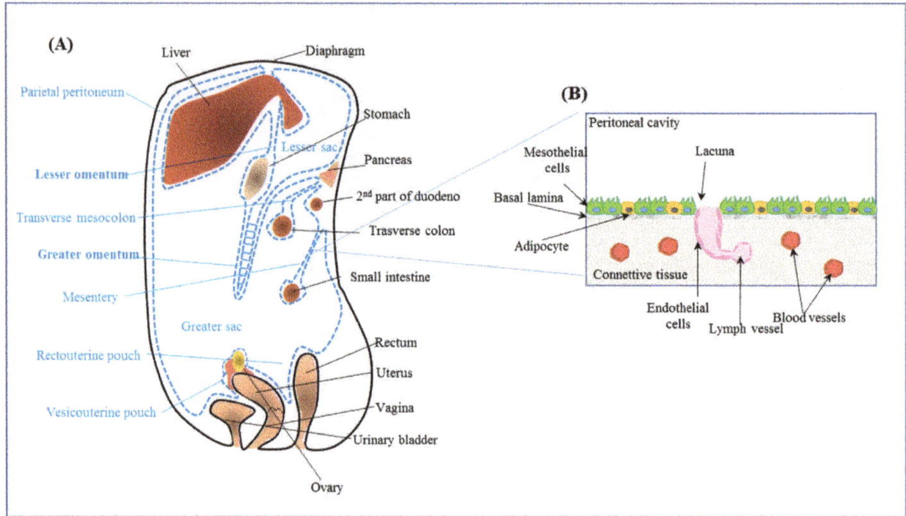

Figure 2. (**A**) Anatomical distribution of the peritoneum; the blue dotted line indicates the peritoneum. (**B**) Cellular structure of the peritoneum.

2. Small Interfering RNAs

Small interfering RNAs (siRNAs) were first described about three decades ago in plants [28], and then in *Caenorhabtidis elegans* [29]. Their therapeutic potential was first reported in 2001 [30], when it was shown that a short, double-stranded RNA duplex (about 21 nucleotide long) can induce target mRNA degradation, thus suppressing gene expression in human cells. The mechanism of action (Figure 3) begins with the uptake of one of the two RNA filaments (antisense strand) of the siRNA by a cellular protein complex termed RISC (RNA-induced silencing complex). While the other filament (sense strand) is discarded, the antisense strand drives RISC to a target RNA via a perfect sequence complementarity. This, in turn, allows RISC to induce the degradation of the target RNA, leading to gene expression inhibition. It is possible to design siRNAs targeted against virtually any deleterious (m)RNA, including viral RNA [31,32]. Moreover, their chemical synthesis is neither complex not particularly expensive. The above characteristics make the therapeutic applicability of siRNA potentially very broad [2,33–36].

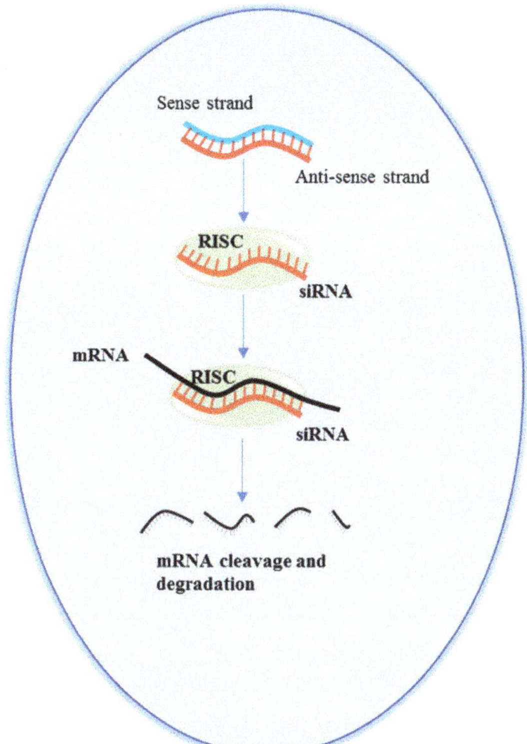

Figure 3. Short interfering RNA (siRNA) mechanism of action.

2.1. Delivery Barriers for Systemic Administration

Made of RNA, siRNAs are characterized by a negative electric charge and have poor stability in the biological environment. Thus, siRNAs need to be embedded into specific carriers for protection and to efficiently reach the target cells. It should be considered that for systemic administration, as could be the case in treating ovarian cancer, siRNAs encounter a number of obstacles that can dramatically reduce the possibility of their reaching the target cancer cells. Once in the blood stream, siRNA can: (1) be degraded by blood nucleases, (2) be eliminated by the phagocytic system, (3) be cleared from blood via kidney filtration and/or sequestered by the liver [37] and activate the innate immune response [38] (Figure 4). Once the siRNAs reach the target tissue, additional obstacles remain, like (4) crossing the vessel wall (extravasation), (5) the migration through the extracellular matrix (ECM) and then (6) crossing the cellular membrane (Figure 4). This last step is particularly inefficient for naked siRNA as their global negative charge, derived from the phosphate groups of their backbone, induces the repulsion of siRNA from the negatively charged molecules present on the outer side of the cell membrane. Moreover, the hydrophilic nature of siRNA substantially prevents its crossing through the hydrophobic inner layer of the cell membrane. Only a reduced amount of siRNA can be internalized via endocytosis. Finally, once into the target cell, siRNAs can be entrapped in endosomes (7), an event that can further reduce the amount of siRNA able to reach the target.

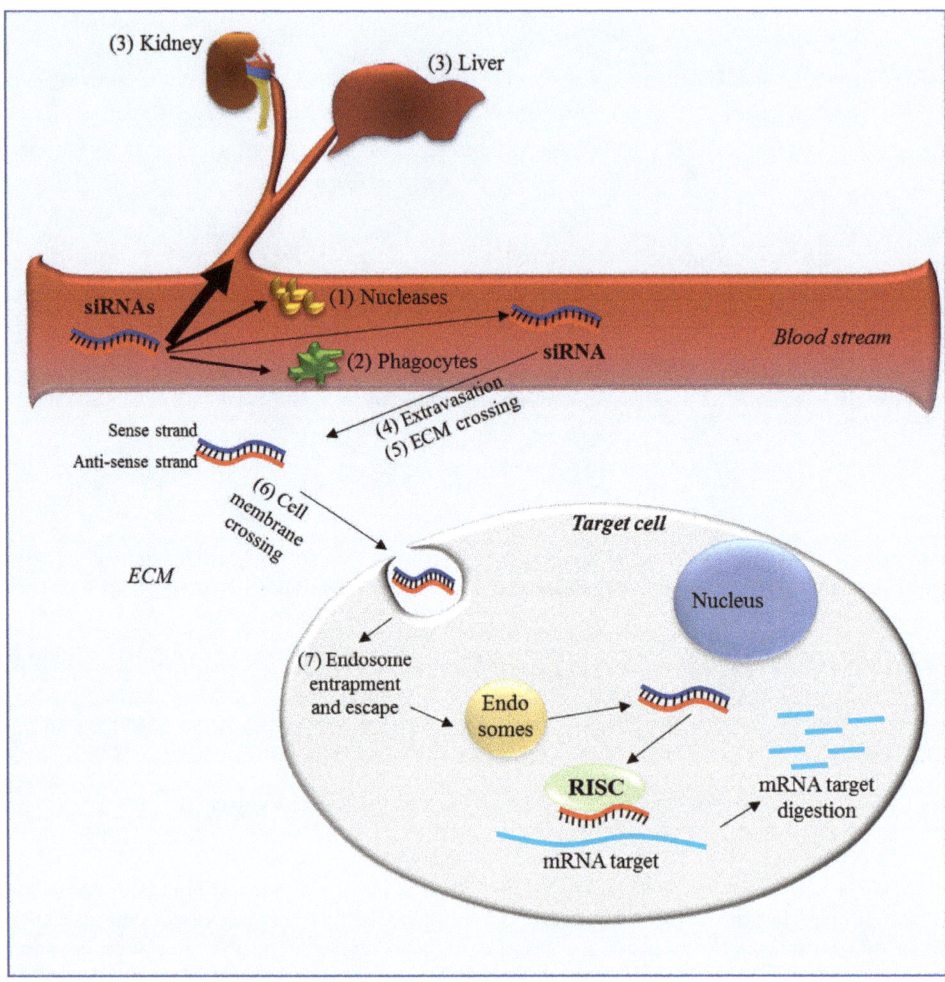

Figure 4. Obstacles for systemic siRNA delivery.

2.2. *Delivery Barriers for IP Administration*

An emerging drug administration strategy for ovarian cancer is the IP route. Thus, it might feasible to develop an IP delivery system for siRNAs as well. Considering that ovarian cancer metastasis very frequently targets the peritoneum/omentum, knowledge of the biology of this serous membrane is crucial to designing optimal siRNA delivery systems. The surface of the peritoneum contains a single layer of mesothelial cells (Figure 2) with a cuboidal shape and adipocytes (reviewed as reported by Sarfarazi et al [39]). Tight junctions and plasmalemma interdigitations connect mesothelial cells. Underneath the mesothelial cells there is basal lamina, which separates the cells from the connective tissue layer. This last layer, composed of collagen fibers, elastin, and fibroblasts, is characterized by the presence of irregular oval openings with diameters ranging from 3 to 15 µm, distributed irregularly in the layer. Dispersed on the peritoneum surface are lymphatic lacunae, lined by endothelial cells. The lacunae are in connection with lymphatic capillaries. In lacunae, MS and endothelial cells are close to each other and share a common basal lamina. The diameter of lacuna aperture is 1.8–6 µm in most animal species, and 3.2–3.6 μm in humans. Blood vessels are abundantly scattered in the connective tissue layer [39].

Following IP administration, delivery systems with a size similar to the diameter of the lymphatic lacuna are preferentially drained from the peritoneum via lacunae. Smaller delivery systems (<10 nm diameter) are mostly cleared via the blood, as they can more efficiently penetrate blood capillary walls. Moreover, the higher flow rate (100–500-fold) in blood vessels compared to lymphatic capillaries favors the diffusion of small molecules into the blood [40]. Thus, the retention of delivery systems in the peritoneal cavity can be modulated by their size. Delivery systems larger than the diameter of the aperture of the lymphatic lacunae will persist longer in the peritoneal cavity; those inferior in size will be predominantly cleared via the lacunar route, while the very small ones (nm range) will be mostly drained by blood vessels. Peritoneal persistence can be also improved by using positively charged delivery systems that can easily interact with the negatively charged membrane of MS [41]. Prolonged peritoneal persistence can present an advantage, as ovarian cancer metastasis typically colonizes the mesothelial cell layer without affecting the basal lamina [42]. Obviously, tumor cell membrane crossing, the presence of cytosolic nucleases, and endosome entrapment still remain obstacles for siRNA action in the case of IP administration.

3. Strategies to Optimize siRNA Delivery

The fragile nature of siRNAs in the cellular and extracellular environment prevents their use as naked molecules. To improve their stability and thus effectiveness, it is possible to chemically modify their structure to make them more resistant to degradation [2,33]. An additional and complementary strategy consists of the combination of siRNAs with delivery vectors able to preserve their integrity and drive them to the target cell [2,33]. Additionally, electroporation can be also used [43]. Often, but not always, the two strategies are combined together. However, as chemical modifications can impair siRNA effectiveness, their employment should be carefully considered. Here, we focus on a brief description of the most commonly used delivery materials, i.e., those based on polymers and lipids (Figure 5).

3.1. *Polymer-Based Delivery Vectors*

Polymers have been frequently used as siRNA delivery systems due to their versatility, low cost of production/isolation, and biocompatibility. The most commonly used polymers for siRNA delivery are polyethylenimine (PEI), polyethylenglycole (PEG), polycaprolactone (PCL), poly(lactic-*co*-glycolic acid) (PLGA), chitosan (CH), and hyaluronic acid (HA). These polymers are frequently combined with each other and/or with other molecules to create delivery particles with optimized delivery features. Regardless of the chemical nature of the polymer, they need to have positively charged resides to bind the negatively charged siRNAs via electrostatic interactions. Moreover, the resulting polymer–siRNA complex should have an appropriate surface charge: excessively strong cationic complexes can lead to nonspecific cellular uptake [44–46] and nonspecific membrane disruption. On the other hand, anionic particles that are too strong can have problems interacting with the negatively charged cellular membrane, thus resulting in inefficient intracellular delivery of siRNA.

Repeated units composed of amine groups and two carbon aliphatic CH_2–CH_2 spacers (Figure 5A) form the structure of PEI. It can exist in both linear and branched forms: in the first form, it contains all secondary amines, while in the second form it contains primary, secondary, and tertiary amino groups. The amino groups confer a cationic nature to PEI, suitable for binding siRNAs. High-molecular-weight PEI displays a superior ability to carry and deliver siRNAs compared to low-molecular-weight [47]. However, high-molecular-weight PEI is more cytotoxic than low-molecular-weight PEI, a problem that can be attenuated by the addition of hydrophilic and hydrophobic moieties or cell-/tissue-specific ligands [48]. Finally, due to its ability to promote the "proton sponge effect", PEI favors siRNA endosomal escape [49].

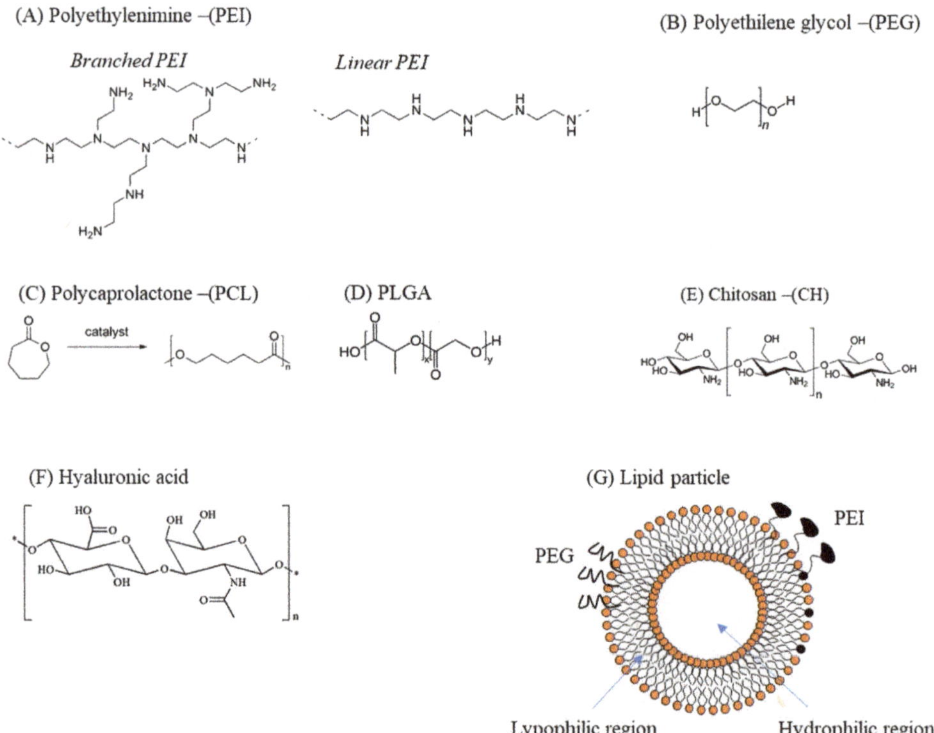

Figure 5. Structure of polymer- and lipid-based delivery materials, (**A**) polyethylenimine (PEI), (**B**) polyethylene glycol (PEG), (**C**) polycaprolactone (PCL), (**D**) poly(lactic-*co*-glycolic acid) (PLGA), (**E**) chitosan (CH), (**F**) hyaluronic acid (HA), and (**G**) lipid particles.

The structure of PEG is represented by the formula $H–(O–CH_2–CH_2)_n–OH$ (Figure 5B). Being substantially non-toxic [50] and soluble in water, PEG can be used in many biological applications. PEG can reduce the toxicity of other delivery materials and, because of this property, it is often added (PEGylation) to various delivery particles [51]. Moreover, it can be also used as a linker for specific ligands on the surface of delivery particles [52].

PCL is obtained by the ring opening polymerization of the cyclic monomer ε-caprolactone (Figure 5C) [53]. It is biodegradable and, in the field of drug delivery, it is typically used to tune the physical, chemical, and mechanical properties of different materials when co-polymerized with polymers such as PEG. Its main limitation in siRNA delivery is the lack of cationic groups able to interact with the negative siRNA backbone [54]. Thus, PCL is often combined with PEI and/or PEG.

PLGA is a copolymer of poly(D, L-lactic acid; LA) and poly(glycolic acid; GA) linked together through ester linkages [54] (Figure 5D). PLGA, biodegradable into its original monomers LA and GA, has been approved for human use by the US Food and Drug administration. PLGA gained attention as an siRNA delivery material due to its favorable safety profile and sustained-release characteristics. However, one of the weaknesses of PLGA is the modest siRNA loading capacity due to the repulsion between the anionic acid groups in PLGA and the negative phosphate backbone of siRNA. To minimize this problem, PLGA is often combined with cationic molecules such as dioleyltrimethylammoniumpropane (DOTAP) or PEI.

It is possible to commercially obtain CH by deacetylation of chitin, a substance contained in the exoskeletons of crustaceans and the cell walls of fungi, [55] (Figure 5E). Chitosan is a linear polysaccharide with a carbohydrate backbone with two types of repeating residues, 2-amino-2-deoxy-glucose (glucosamine) and 2-*N*-acetyl-2-deoxy-glucose (*N*-glucosamine), linked by a (1-4)-β-glycosidic linkage. The amino groups confer to CH a positive charge, which allows binding with the negatively charged phosphate groups of siRNAs. CH conjugation with PEI and PEG can bypass the problems of low solubility and delivery effectiveness. Alternatively, it is possible to tune the degree of acetylation and modify the molecular weight [56].

HA, a polymer of natural origin, is a linear polysaccharide formed by glucuronic acid and N-acetylglucosamine units linked via alternating –1,4 and –1,3 glycosidic bonds (Figure 5F). HA is biocompatible, biodegradable, and characterized by low toxicity. In the field of siRNA delivery, it is often used in combination with other cationic polymers. Indeed, its anionic nature attenuates positive charges, thus resulting in reduced toxicity and improved stability of the complex [57]. Finally, the presence of hyaluronic acid receptors (CD44, see Section 4) in many healthy and diseased human (cancer) tissues makes HA particularly suited as a targeting agent [58].

3.2. Lipid-Based Delivery Vectors

Liposomes are spherical vesicles formed by concentric lipid bilayers (Figure 5G) with a hydrophilic region localized in the core [59,60]. Because of this structure, liposomes can entrap hydrophilic molecules in the water compartment and hydrophobic molecules in the lipid layers. Frequently, liposomes are made up of neutral (cholesterol; Chol) or zwitterionic (dioleoylphosphatidylethanolamine; DOPE) lipids, giving origin at the physiological pH to neutral liposomes [60]. While neutral liposomes have good biocompatibility and favorable pharmacokinetics, they display poor interactions with negatively charged siRNA. An alternative to the use of neutral lipids is the employment of positively charged lipids, such as 3β[*N*-(*N*′,*N*′-dimethylaminoethane)-carbamoyl]DC-cholesterol, 1,2-dioleoyl-3-trimethylammonium propane (DOTMA) [60]. Cationic liposomes are composed by positively charged lipids within a matrix of neutral lipids, required for cellular internalization. Cationic liposomes allow easy and effective loading of negative molecules such as siRNA. Moreover, they facilitate the endosomal escape of siRNA, promoting the fusion between liposomal and endosomal bi-layers. Despite being suitable for siRNA delivery to the cells, cationic liposomes have some drawbacks [59]. For example, they are more toxic than neutral liposomes. They also have a low "fusogenicity" attitude, i.e., they are less effective in interacting with cell membranes to allow cell uptake. To attenuate this limitation, cationic lipids are often combined with non-cationic lipids such as DOPE, which display enhanced fusogenic activity. Another limitation of cationic liposomes is represented by the fact that, when applied to the blood stream, they are cleared by the reticulo-endothelial system. Thus, their effectiveness in siRNA delivery can be significantly diminished. To minimize this problem, cationic liposomes are often equipped with shielding molecules on their surface such as a hydrophilic polymer like PEG. However, once they reach the target cell, the PEG has to be removed to allow efficient siRNA delivery and to permit efficient endosome escape.

4. Targeting OC: Delivery Systems and siRNA Targets

The ideal therapeutic approach to OC should be able to selectively deliver the drug (siRNA in our case) to tumor cells, leaving the normal tissues as untouched as possible. This goal can be approached with two complementary strategies: the first is to equip the siRNA delivery system with molecules able to specifically target tumor cells, and the second is to use siRNAs against OC-specific oncogenes. The combination of these two strategies should give the highest specificity level possible.

4.1. OC-Targeting Molecules

So far, different antigens present on the surface of OC cells have been considered suitable for OC targeting (Table 1).

Table 1. Specific surface antigens on OC cells.

Extended Name	Abbreviation	References
Epidermal growth factor receptor	EGFR	[61–64]
Erythropoietin-producing hepatocellular receptor A2	EphA2	[65,66]
Folic acid receptor	FR	[67,68]
CD44 surface transmembrane glycoprotein	CD44	[69–71]
CD133 glycosylated transmembrane protein	CD133	[72]
CD117 (also known as c-kit) receptor tyrosine kinase	CD117	[73–75]
CD24	CD24	[76]

Epidermal growth factor receptor (EGFR) is highly expressed in OC [61], as well as in other tumor cells. An extracellular ligand-binding domain, a transmembrane domain, and a cytoplasmic domain containing the tyrosine kinase region [62] characterize this surface antigen. Following ligand binding to EGFR, receptor auto-transphosphorylation in the tyrosine kinase region occurs, triggering a series of signaling events. These, in turn, promote cell proliferation, apoptosis inhibition, cell invasion, and stimulation of neovascularization [63]. Because of its pro-proliferative effects, EGFR has also been considered as a drug target [64].

Erythropoietin-producing hepatocellular receptor A2 (EphA2) belongs to a large family of receptor tyrosine kinases mainly involved in the regulation of cell proliferation and migration [65]. The extracellular domain contains a ligand-binding domain, a Sushi domain, an epidermal growth factor (EGF)-like domain, and two fibronectin type-III repeats. A transmembrane domain and a tyrosine kinase domain follow the extracellular domain. EphA2 is overexpressed in ovarian cancer [66] as well as in various other solid tumors, such as breast, prostate, pancreas, glioblastoma, neck, renal, lung, melanoma, bladder, gastric esophageal, colorectal, and cervical cancers. In OC, the N-terminals of EphA2 are processed by membrane-type 1 matrix metalloproteinase, triggering ligand-independent signal activation, which in turn promotes cell motility, invasion, and metastasis [66].

The folic acid (FA) receptor (FR) is a 38 kDa glycosyl-phosphatidylinositol membrane-anchored glycoprotein [67]. Under normal conditions, FR is present only in some polarized epithelia and it is localized to the apical/luminal cell surface. In tumor cells of epithelial origin, it is often overexpressed and it loses its polarized nature, covering the entire cell surface. FR acts as a transporter of folate into the cells. The fact that folates are essential for the biosynthesis of purines and thymidine, necessary for DNA synthesis, methylation, and repair, explains why actively growing tumor cells need higher FR compared to normal cells. Notably, FR overexpression is present in more than 80% of HGSOCs [68]. Moreover, increased expression level of FR correlates with OC stage, poor response to chemotherapy, and worse survival.

In addition to the above-mentioned OC-targeting antigens, other antigens associated with ovarian cancer stem cells (OCSCs) are gaining great interest. The discovery of quiescent OCSCs with the ability to escape chemotherapy and to regenerate OCs has directed attention towards the possibility of specifically targeting OCSCs [69]. CD44 is a surface transmembrane glycoprotein that acts as a receptor for different molecules, including hyaluronic acid (HA). CD44 influences the expression levels of genes related to cellular differentiation and cell–matrix adhesion. It is one of the most common OCSC surface markers, used to identify this kind of cell. Despite this, CD44 does not seem to have relevant prognostic value in OC [70]. Monoclonal antibodies and siRNAs [71] directed against CD44 have already been tested. Another OCSC-related antigen is CD133, a glycosylated transmembrane protein with prognostic value in OC. CD133 modulates several pathways involved in the control of cancer stemness and metastasis. Notably, even though CD133-positive cells represent a minority of OC cells, its toxin-mediated targeting effectively downregulated OC growth in in vitro and in vivo models [72]. Further potential targeting antigens for OC are CD117 and CD24. CD117, also known as c-kit, is a receptor tyrosine kinase devoted to the promotion of cell survival, metabolism, and differentiation. In OC, poor disease-free survival is directly associated with CD117 levels [73]. While CD117 targeting in vitro successfully downregulated OC cell survival [74], only modest effects were observed in a phase II clinical trial [75]. Finally, another potential targeting antigen is represented by CD24, highly expressed in many cancers with a percent positivity in OC of about 70% [76].

4.2. siRNA Targets in OC

Due to their mechanism of action, siRNAs are particularly suited to targeting gene products of which the exuberant expression is responsible for increased cell growth, cell survival, migration, angiogenesis, and drug resistance (Tables 2–4). Notably, a single gene product often controls many of the above cellular phenotypes.

Table 2. Molecular targets implicated in OC cell growth/migration.

Extended Name	Abbreviation	References
Epidermal growth factor receptor	*EGFR*	[64]
Metastasis associated in colon cancer 1	*MACC1*	[77]
Metastasis-associated gene 1	*MTA1*	[78]
Wilms tumor gene	*WT1*	[79]
Rac1 Rho family small GTPases	*Rac1*	[80,81]
Polo-like kinase 1	*Plk1*	[82,83]
Notch1 *Claudin3* *Nin one binding protein* *Cyclooxigenase-2*	*Notch1* *CLDN3* *NOB1p* *COX-2*	[61]
E2 promoter binding factor 1	*E2F1*	[84–87]
Peptidylprolyl cis–trans isomerase, NIMA-interacting 1	*PIN1*	[88,89]

Table 3. Molecular targets implicated in angiogenesis.

Extended Name	Abbreviation	References
Vascular endothelial growth factors	VEGFs	[61]
VEGFs tyrosine kinase receptors	VEGFR-1/Ftl-1, VEGFR-2, VEGFR-3/Ftl-4	[61]
Plexin domain containing 1	PLXDC1	[90]

Table 4. Molecular targets implicated in drug resistance.

Extended Name	Abbreviation	References
Multidrug resistance gene 1	*MDR1*	[91]
Survivin	*SVV*	[92,93]
Focal adhesion kinase	*FAK*	[94]
B-cell lymphoma 2	*BCL2*	[95]

4.2.1. Molecular Targets Implicated in Cell Growth/Migration

One of the most targeted gene products in OC is EGFR, as it promotes cell growth, apoptosis inhibition, cell invasion, and neovascularization (Table 2). This gene product is also particularly attractive as it represents a targetable surface antigen for an siRNA delivery system [64] (see Section 4.1). The expression of the *Metastasis associated in colon cancer 1* (*MACC1*) is often upregulated in many cancers. Its targeting by short hairpin shRNA in OC cells resulted in the downregulation of cell migration and proliferation [77]. *Metastasis-associated gene 1* (*MTA1*) is another gene product relevant for invasion and metastasis in OC. Notably, its siRNA-mediated down-regulation not only reduced cell migration, it also promoted cell anoikis [78]. This is a form of apoptosis occurring in cells detached from the extracellular matrix. Anoikis resistance is relevant for the survival of OC cells in ascites.

The *Wilms tumor gene* (*WT1*) is overexpressed in OC but not in normal ovarian tissue; thus, it is of particular interest in terms of the specificity of siRNA effects (limited to OC cells). *WT1* targeting resulted in OC cell arrest in the G0–G1 phase of the cell cycle and promoted cell apoptosis [79]. Rac1 is a well characterized member of the Rho family that interacts with effector molecules and regulates cytoskeleton organization and membrane trafficking [80]. Rac-1 is implicated in OC cell survival, tumor angiogenesis, resistance to therapeutics, and contribution to extraperitoneal dissemination [81]. Deregulation of *Polo-like kinase 1 (Plk1)* was shown to be responsible for mitotic defects by affecting cell cycle checkpoints, thus resulting in aneuploidy and tumorigenesis [82]. Overexpression of *Plk1* has been observed in many cancerous tissues, including OC [83], and was shown to correlate with tumor stage and grade and poor patient prognosis. *Notch1, Claudin3 (CLDN3), Nin one binding protein (NOB1p)*, and *Cyclooxigenase-2 (COX-2)* (reviewed in [61]), represent other suitable targets related to OC growth and migration.

In addition to the above-discussed siRNA targets, there are others, such as *E2 promoter binding factor 1* (*E2F1*) and *peptidylprolyl cis–trans isomerase, NIMA-interacting 1* (*PIN1*), which we believe might merit the attention of future experimentation. *E2F1* belongs to a family of transcription factors with the ability to activate or repress the expression of genes promoting cell proliferation. *E2F1* in particular mostly triggers the transcription of pro-proliferative genes, thus favoring cell growth. In OC, its level directly correlates to unfavorable disease-free and overall survival, particularly in HGSOC (reviewed in Reference [84]). Moreover, recent evidence connects *E2F1* over-expression with reduced drug sensitivity [85]. Together, these findings make *E2F1* an attractive novel molecular target in OC. However, as *E2F1* also plays a relevant role in non-tumor cells, the use of an OC-targeted delivery system would be desirable. Among *E2F1* transcription targets, there is *PIN1* [86,87]. It accelerates the conversion of cis and trans isomers, thus inducing conformational changes able to regulate the functions of its substrates, such as *p53*, *p73*, $p27^{Kip1}$, $p21^{waf1/cip1}$, and *cyclin D1*, all proteins involved in the control of cell proliferation. The net result is the activation of oncogenes and inactivation of tumor suppressor genes in cancer cells. In OC (HGSOC), *PIN1* is overexpressed and when knocked down or chemically inhibited, OC cell death is induced [88,89]. The fact that *E2F1* regulates *PIN1* transcription and that both genes are involved in OC make this molecular circuit an interesting target for novel therapeutic approaches in OC.

4.2.2. *Molecular Targets Implicated in Angiogenesis*

Cancer neo-angiogenesis is necessary to support the increased demand for oxygen and nutrients of the growing tumor cells. Thus, the targeting of one or more of the elements of the angiogenic pathway has the potential to negatively regulate tumor cell growth. In several tumors, including OC, different components of the angiogenic pathway have been targeted with different therapeutic molecules [61]. Among these, the vascular endothelial growth factors (VEGFs) and their tyrosine kinase receptors (VEGFR-1/Ftl-1, VEGFR-2, VEGFR-3/Ftl-4) have often been considered (Table 3). Some studies have also considered the targeting of *plexin domain containing 1 (PLXDC1)*, previously known as TEM7, which is overexpressed in endothelial cells of all four tumor types [90]. *PLXDC1* promotes endothelial cell migration and invasion.

4.2.3. Molecular Targets Implicated in Drug Resistance

About 80% of OC patients, who respond to the first-line PTX chemotherapy relapse sooner or later [22]. Thus, the development of drug resistance in OC is a major problem. To try to circumvent this aspect, gene products responsible for multidrug resistance have been considered as possible targets for novel therapeutic approaches (Table 4). *Multidrug resistance gene 1 (MDR1)* is a membrane-bound P-glycoprotein overexpressed in chemo-resistant OC cells [91]. MDR1 favors drug efflux from the cells, thus resulting in a shortening of drug permanence into the cell. This, in turn, leads to the reduced therapeutic effects of the drug. Another gene product related to drug resistance is *survivin (SVV)*. As member of the *Inhibitor of apoptosis protein (IAP)* family [92], *SVV* induces chemotherapy and radiotherapy resistance in OC [93] through the inhibition of apoptosis. As this protein is overexpressed in many tumors compared to their normal tissue counterparts, it represents an attractive target for siRNA tumor-specific effects. Another target able to induce chemo-resistance is the *Focal adhesion kinase (FAK)*. This is a non-receptor tyrosine kinase of which the overexpression is present in 68% of epithelial OCs [94]. Moreover, *FAK* levels significantly correlate with shorter overall patient survival. Not only is *FAK* involved in the regulation of OC cell migration, invasion, adhesion, proliferation, and survival, it has recently been discovered that it promotes PTX resistance in OC as well as in other cancers [94]. Drug resistance is triggered via the promotion of *MDR-1* activity. Finally, *B-cell lymphoma 2 (BCL-2)* is the founding member of the *BCL-2* family of apoptosis regulatory proteins. In particular, *BCL-2* has anti-apoptotic function, and, recently, its involvement in the induction of chemo-resistance in OC has been described. Indeed, its targeting resulted in enhanced cisplatin-induced apoptosis in OC spheroids [95].

5. Experimental Models of OC

The development of effective anti OC siRNA and delivery strategies depends on the employment of adequate models of the disease. Available models include cellular and animal models. Recently, conglomerates of cells cultured in 3D and defined spheroids have been proposed as a useful model. Finally, mathematical modeling is gaining interest as a way to better understand the biological phenomena in siRNA effects and delivery. Herein, we have briefly summarized the features of the available models, highlighting their advantages and disadvantages.

5.1. Cellular Model

The ideal cellular model should be able to replicate the main phenotypic and genetic characteristics of OC cells. As HGSOC is the most problematic form of OC, the cellular models should ideally resemble the features of this form of OC. In HGSOC, copy number alterations (CNAs) are rather common; more frequent CNAs involve genes such as *MYC*, *MECOM*, *CCNE1*, and *KRAS* [96]. Additionally, some gene mutations are also common: among these, *TP53* mutations are present in 95% of the cells isolated from patients [97] and *BRCA1/BRCA2* have a frequency of about 10%. Notably, low-grade serous carcinoma has almost normal gene copy numbers and wild-type *TP53* [98]. Based on these and other molecular features, a classification of the cellular models that best resemble HGSOC has been proposed [97]. The most commonly used model cell lines, represented by SKOV3, A2780, OVCAR3, CAOV3, and IGROV1 [97], do not properly resemble HGSOC. OVCAR3 and CAOV3 possess *TP53* mutations like HGSOC, but have low CNAs compared to HGSOC. SKOV3 and A2780 do not have *TP53* mutations, but instead do have mutations frequently found in other histological subtypes, such as *ARID1A*, *BRAF*, *PIK3CA*, and *PTEN* mutations. IGROV1, characterized by few CNAs and many mutations not present in HGSOC, resembles the molecular features of the endometrioid carcinoma. The HeyA8 cell line has fewer *CNAs* than HGSOC and do not have mutated *TP53*, *BRCA1*, and *BRCA2*. In contrast, a cellular model that reasonably well resembles the molecular features of HGSOC is the KURAMOCHI cells. They have similar CNAs and mutation frequency in key oncogenes and tumor suppressors (*TP53*, *BRCA1*, *BRCA2*). Moreover, in contrast to the above-mentioned cell lines, the mRNA expression profiles of KURAMOCHI are similar to those of HGSOC [97]. A drawback of KURAMOCHI, however, is linked to the difficulties of its grafting into xenograft mouse models.

An emerging OC model linking the cellular and the animal models is the so-called "spheroids". Spheroids are defined as 3D structure grown from stem cells and consisting of organ-specific cell types [99]. While spheroids can be originated from normal stem cells, they can be also originated from cancer cells [100]. OC spheroids, which recently became available [101,102], replicate the genomic and histological characteristics of the lesion from which they are derived. Being organized into a 3D structure, OC spheroids better resemble the in vivo structure of OC compared to 2D culture models. Thus, they are in principle more informative then 2D cultures when used to study the biology of OC and also drug effectiveness. Moreover, OC spheroids can be xenografted, thus allowing drug-sensitivity tests in vivo.

5.2. Animal Models

The only two non-human animals known to naturally develop OC are the egg-laying hen and the jaguar [103]. The egg-laying hen model has been underutilized due to the scarcity of chicken-specific experimental reagents and due to the logistical challenges of animal husbandry. Jaguar use has been largely impossible, mainly because these animals are endangered. Thus, the mouse represents the main animal model utilized so far to study anti-OC drugs. In mouse, three models are commonly used: the xenograft model, the syngenic model, and the genetically engineered mouse models (GEMM).

In xenograft mouse models, cells of different genetic backgrounds can be injected into immunocompromised animals via three different routes: subcutaneous (SC), intraperitoneally (IP), or intrabursally (IB, i.e., into the bursa that surrounds the mouse ovary) [104]. As in the SC approach, cells do not metastasize, this model is not suited to the study of ovarian metastasis. Moreover, the anatomical location/microenvironment of the injected cells is different from that in the ovary glands. However, the SC model is suited for investigation with imaging modalities; moreover, it is suitable for pilot studies to test the effectiveness of newly developed drugs. IP and IB better mimic the metastatic dissemination of OC. However, while IP cannot effectively reproduce the initial steps in metastasis (cells exit the bursa to spread throughout the peritoneal cavity), IB can. OC cell lines are most often employed to generate xenograft models. However, it is also possible to use cells freshly isolated from OC patients, thus resulting in models more adherent to reality. The problems with patients' cells are often linked to the modest tumor engraftment percentages and the time taken to develop tumors [105]. Regardless of the cell type used, all xenograft mouse models suffer from the fact that the animals are immunocompromised; therefore, the role of the immune system is completely ignored.

A syngenic mouse model has been generated using murine ovarian surface epithelial cells from C57BL/6 mice, which have been subsequently transformed (ID8 cells) and eventually re-injected into C57BL/6 mice [106]. An alternative, more recent model, again derived from the ovarian surface epithelium, is the STOSE model generated in the FVB/N mice strain [107]. While it is possible to consider the effects of immune systems in these models, the cells used are different from human OC cells.

GEMMs of OC are useful for improving our understanding of the origin(s) of OCs. Indeed, it is possible to generate GEMMs with ovarian or oviductal epithelia carrying specific mutations. It is then possible to study how the specific mutation drives the development of OC. In this regards, GEMMs carrying the most common OC mutations, such as those of *TP53* and *BRCA1/2*, have been developed [104]. A problem liked to GEMMs is represented by the frequent infertility of GEMM animals. The other drawback stems from the fact that OCs tend to arise over an extended course of time, and thus, there are difficulties in controlling tumor onset and size.

5.3. Mathematical Models

An additional tool useful for the investigation and description of the effects and delivery of siRNA is based on the use of mathematical models. This approach, not yet very common in the field, has the possibility of predicting siRNA behavior once the main biological parameters have been established via an experimental approach. Thus, the system has the great advantage of significantly reducing the experimental load, as it can easily simulate/predict the effects of many variables of the system via mathematical simulation [108]. Interestingly, the mathematical approach is in line with Leonardo da Vinci's conviction that "*niuna umana investigazione si può dimandare vera scientia s'essa non passa per le matematiche dimostrazioni*" (no human investigation can be defined true science if it cannot be mathematically demonstrated). [109]. In our opinion, the use of mathematical modeling in the siRNA field became possible after the publication of the seminal work of Bartlett and Davis [110]. Curiously, the mathematical modeling of siRNA delivery was not the only goal of that paper, but the mathematical model was also described in such a concise manner that it was very difficult to understand. The graphical translation of this model into two schemes [108] revealed its powerful theoretical basis. Indeed, this model, relaying on 12 ordinary differential equations, is organized into four modules that can be changed independently to modify model complexity as desired. The first module deals with siRNA "circulation/extracellular transport" (three equations), the second is about siRNA "cellular uptake and intracellular trafficking" (four equations), the third describes "the fate of free and bound activated RISC" (three equations) according to the Michaelis–Menten multiple turnover mechanism, and the fourth focuses on "cell growth and target protein production" (two equations). According to this modular organization, the model individuates two distinct phases: the external phase, involving the first module, and the cellular phase, involving the other three modules. Consequently, the description of in vivo experiments requires both phases, i.e., all four modules, while in vitro experiments imply the use of just the cellular phase, i.e., the last three modules. On the basis of this approach, Barlett and Davis [110] were able to fit and predict the silencing of a protein (luciferase) by means of siRNA. This clearly indicates the theoretical and practical strength of the model, which can help in optimizing (interpreting and planning) experimental activities.

6. siRNA Delivery to OC Cells

In the light of the above information/considerations (Sections 1–5), in this Section we describe and comment on the most recent works (from the last three years) related to the delivery of siRNAs based on the use of polymer-/lipid-based delivery systems. We concentrated on polymers and lipids due to their general biocompatibility, their ability to deliver siRNAs, and the possibility of their undergoing significant modifications of structure, a feature useful in generating optimized delivery materials. Polymeric approaches have been divided into systems: those which do not have an OC targeting strategy and those that do have one.

6.1. Polymeric Delivery Systems

6.1.1. Non-Targeted Delivery Systems

Polyak et al. [111] developed a polymer comprising a poly(α)glutamate (PGA) backbone conjugated with different amine moieties (PGAamine) (Figure 6 and Table 5). Due to their positive electrostatic charge, amine groups allow interaction with the negatively charged siRNA. Moreover, the positive charge of the PGAamine polymer allows interaction with the negatively charged cell membrane. Finally, the multiple amino groups in acidic compartments (lysosomes, endosomes) may favor the proton sponge phenomenon, allowing the rupture of endocellular vesicle (lysosomes, endosomes) membranes, thus allowing siRNA release into the cytoplasm. siRNA release to the cytoplasm is also favored by the fact that the PGAamine polymer is cut by cathepsin B, an enzyme present in lysosomes/endosomes that it is upregulated in many human tumors. Upon cathepsin B cleavage of PGAamine, siRNAs are released from the delivery complex. Finally, PGA is water-soluble and has low immunogenicity and low toxicity; thus, it is an attractive candidate for siRNA release. As siRNA targets, the authors chose Rac1 (siRac1) and Plk1 (siPlk1), both involved in tumorigenesis via the control of cell migration and proliferation, respectively. A PGAamine:siRac1 polyplex demonstrated plasma stability and immune and hemo-compatibility in the ex vivo blood compartment. The functionality of PGAamine:siRac1 was proven in vitro in SKOV3 cells, showing the inhibition of cellular migration. In vivo, the authors used an orthotopic SKOV3 human OC model in athymic nude female mice. Following three sequential IP injections (8 mg/kg siRNA), a 2.75-fold increase in siRac1 tumor accumulation was observed compared to treatment with siRNA alone.

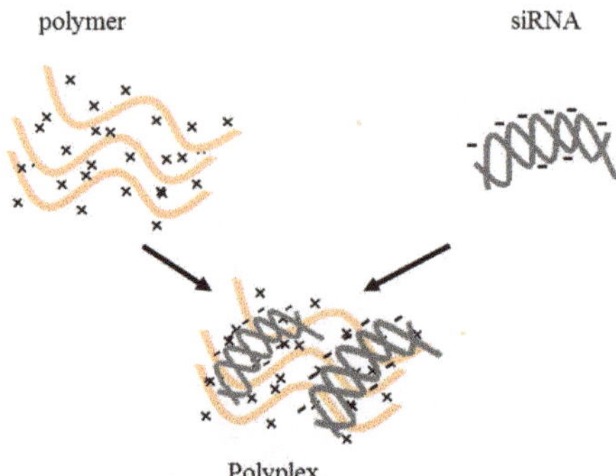

Figure 6. Polymeric delivery material developed in Reference [111]. Polymer: PGA conjugated with amine moieties.

This was paralleled by a 44% decrease of Rac1 expression in tumor tissue compared to siRac1-alone-treated mice. The effect on tumor growth was studied using PGAamine:siPlk1 particles in nu/nu mice bearing orthotopic IP tumors of mCherry-labeled SKOV3 cells. Following nine daily IP injections of PGAamine:siPlk1 (8 mg/kg siRNA), tumor growth inhibition was 73% compared to control siRNA (siCtrl)-treated mice. Moreover, PGAamine:siPlk1 improved animal survival. Given the particle size (160 ± 20 nm) it is likely that most particles were cleared via the lymphatic lacunae (see Section 2.2). Thus, it is feasible to assume that the therapeutic effects were exerted by the fraction of particles which escaped lymphatic lacunae drainage and made direct contact with tumor cells in the peritoneum.

One of the major problems with HGSOC is the occurrence of drug resistance. Thus, the possibility of attenuating this eventuality has been explored using siRNA against MDR1 and BCL2, both involved in the induction of drug resistance (see Section 4.2). Risnayanti et al. [112] developed a PLGA-based delivery system loaded with siRNA against MDR1 (siMDR1) and BCL2 (siBCL2) (Table 5). To obtain an efficient siRNA encapsulation, siRNAs were first complexed with the cationic poly-L-lysine (PLL), which can bind to the negative charged siRNAs. PLL was used in place of the more commonly used PEI to minimize the known toxicity associated with PEI. siMDR1/siBCL2–PLL particles were then encapsulated with the negatively charged PLGA. The PLGA–siMDR1/siBCL2–PLL particles, with a dimeter of 197.8 ± 5.2 nm, did not elicit any nonspecific cytotoxicity in the drug-resistant SKOV3-TR and A2780-CP20 cells. In this regard, it is somewhat strange that particles containing the siBCL2 did not trigger any cell death, considering that *BCL2* is an anti-apoptotic gene. By in vitro test, the authors also showed that the particles did not stimulate the innate immune system. The efficiency of particle uptake was studied by the use of cyanine5.5 fluorophore (Cy5.5) labeled siRNAs, and was 100% in both SKOV3-TR and A2780-CP20 cell lines. This was paralleled by a significant reduction in the mRNA and protein levels of both MDR1 and BCL2. From the functional point of view, PLGA–siMDR1/siBCL2–PLL particles were able to increase the permanence and concentration of PTX in the drug resistant SKOV3-TR and A2780-CP20. This resulted in an improved cell death compared to PTX alone. Notably, the use of particle carrying only one of the two siRNAs + PTX was less effective than the particle loaded with the two siRNAs +PTX. This suggests the importance of downregulating both inducers of drug resistance. Comparable results were obtained using cisplatin in place of PTX. While no in vivo data were provided, the data support the concept that it is, in principle, possible to improve drug sensitivity by targeting specific genes.

Classical PLGA–PEI particles were used by Hazekawa et al. [113] to deliver a siRNA (siGPC3) against Glypican-3 (GPC3), a member of the heparan sulfate proteoglycans, which has important functions in cellular signaling pathways, including cell growth (Table 5). While GPC3 is over expressed in chemo-resistant OC cells, its overexpression seems to occur mostly in clear OC cell adenocarcinoma rather than in HSGOC [114]. In vitro in the mouse HM-1 cells, PLGA–PEI–siGPC3 particles efficiently reduced GPC3 protein levels and reduced cell growth compared to controls. In a murine HM-1 peritoneal dissemination model, the IP injection of PLGA–PEI–siGPC3 resulted in a significant reduction in the tumor nodules compared to controls. Moreover, the GPC3 levels in the cell lysates of peritoneal cells from animals treated with PLGA–PEI–siGPC3 were also reduced compared to control. While the role of GOC3 in HSGOC is not completely clear, a merit of this work was the use of the HM-1 cell line in the syngeneic B6C3F1 mice strain, which allowed consideration of the contribution of the immune system.

Table 5. Polymer based delivery systems without an OC-targeting moiety.

First Author	Target	Delivery Material	Cell Model	siRNA Delivery Route	Animal Model	References
Polyak	Rac1; Plk1	Poly(α)glutamate; no specific targeting; 160 ± 20 nm (Particle size)	SKOV3	Intra-peritoneal (nine every other day intraperitoneal injections 8 mg/kg siRNA)	Orthotopic SKOV3 cells in athymic nude female mice (intra-peritoneal tumor cell injection)	[111]
Risnayanti	MDR1; BCL2	PLGA–PLL; 197.8 ± 5.2 nm (Particle size)	SKOV3-TR and A2780-CP20	-	-	[112]
Hazekawa	Gpc3	PLGA–PEI; 108.5 ± 2.5 nm (Particle size)	HM-1	Intra-peritoneal (100 pmol of siRNA on Day 1)	Syngeneic orthotopic intra-peritoneal HM-1 injection in syngeneic B6C3F1 mouse strain	[113]
Lou	Luciferase	POEGMA/PVTC; PVTC; 8–25 nm (particle sizes)	SKOV-3-luciferase	-	-	[115]
Leung	LPP	CHITOSAN	Luciferase-labeled OVCA432	Intravenous (twice-weekly tail-vein 5 μg siRNA in combination with weekly i.p. injections of Paclitaxel, 3.5 mg/kg for 6 weeks)	Orthotopic luciferase-labeled OVCA432 in female nude mice (intra-peritoneal tumor cell injection)	[116]

A complex polymer-based delivery system based on poly(vinyl benzyl trimethylammonium chloride) (PVTC) and its block copolymer with poly(oligo(ethyleneglycol) methacrylate) (POEGMA) was developed by Luo et al. [115] (Table 5). PVTC was used as it contains positively charged amino groups, which allow siRNA binding. The authors studied particles of PVTC or PVTC–POEGMA. The generated particles had a diameter range of 8–25 nm. In SKOV3-luc expressing the luciferase gene, PVTC–POEGMA and PVTC at the concentration of 15 μg/mL were significantly less toxic than PEI, used as a control polymer. Loading the particles with an anti-luciferase siRNA (luc-siRNA), the authors observed that at the concentration of 200 nM luc-siRNA and at a nitrogen/phosphate (N/P, polymer/siRNA) of 16, PVTC/luc-siRNA particles were the most effective in reducing luciferase expression (60% reduction) compared to control. Notably, PVTC/luc-siRNA particles were more effective than POEGMA/PVTC/luc-siRNA particles (35% inhibition). This difference is most probably due to the reduced cellular uptake of POEGMA/PVTC/luc-siRNA compared to PVTC/luc-siRNA, visualized by using a fluorescently labeled siRNA. No in vivo data were provided. However, the size of the particles (8–25 nm) suggests that in vivo these particles, when IP injected, could have been preferentially drained by the blood vessel (see Section 2.2). It remains to be determined whether this could be an appropriate route by which to reach metastatic tumor cells in the peritoneum.

Leung et al. [116] proposed an original strategy to combat OC. The authors found that cancer-associated fibroblasts upregulate the lipoma-preferred partner gene (LPP) in tumor endothelial cells. LPP expression levels in intra-tumor endothelial cells correlate with survival and chemo-resistance in OC patients. Moreover, LPP increases endothelial cell motility and tumor vessel leakiness. The author reasoned that the targeting of LPP by a siRNA (siLLP) could reduce tumor vessel amount and leakiness, thus increasing PTX permanence in the tumor site. In an orthotopic model of OC generated with luciferase-labeled OVCA432 cells, the authors delivered intravenous (IV) siLLP embedded into chitosan particles (CH–siLLP) together with PTX. The levels of LPP in tumor mass of animals treated with CH–siLLP were significantly lower than in controls. Moreover, the same animals had reduced tumor burdens and microvessel densities. This last observation suggests that LPP silencing reduced tumor angiogenesis. Finally, the authors demonstrated that tumor vessel leakiness was also reduced by CH–siLLP, and that this was paralleled by an increased PTX bioavailability. Notably, the combination of CH–siLLP with PTX gave a more pronounced reduction in tumor mass than the single treatment with either PTX or CH–siLLP, thus suggesting an additive effect of the combined treatment.

6.1.2. Targeted Delivery Systems

The last example of the above section shows that it is reasonable to attack OCs via the targeting of tumor endothelial cells. An additional strategy to target tumor endothelial cells was described by Kim et al. [117] (Figure 7 and Table 6). The authors generated chitosan particles (CH–siRNA) coated with HA (HA–CH–siRNA) to target the CD44 receptor present on tumor endothelial cells. Particles were loaded with siRNA against PLXDC1 (siPLXDC1), which is involved in the promotion of cell migration and invasion of tumor endothelial cells (see also Section 4.2). PLXDC1 and CD44 are overexpressed in OC-associated endothelial cells [117]. In vitro, the authors showed that in two models of endothelial cells (HUVEC and MOEC), *PLXDC1* silencing by HA–CH–siPLXDC1 resulted in a significant inhibition of cell migration, invasion, and tube formation compared to particles carrying a control inactive siRNA. Using a Cy5-labeled siRNA, it was also possible to show that HA–CH–Cy5 siRNA efficiently entered the MOEC (CD44-positive) but not the A2780 OC model cells (CD44-negative). In an IP mouse model of OC generated with A2780 cells, HA–CH–Cy5 siRNA delivered IV (150 mg/kg) had improved tumor localization compared to CH–Cy5 siRNA, indicating the targeting role of HA. Using the HA–CH–siPLXDC1 particles (150 mg/kg, IV twice per week), the authors showed a significant decrease in the level of *PLXDC1* mRNA in the tumor mass compared to CH–control siRNA and CH–siPLXDC1. Unfortunately, no data about the effects on animal survival were reported. Despite this, as A2780 are CD44-negative and do not over express PLXDC1, the data presented suggest that the therapeutic effect was mostly due to an anti-angiogenic action. In a variation of the above in vivo test, the authors used an IP mouse model of OC generated with HeyA8 cells, which overexpress CD44 and PLXDC1. The results were quantitatively comparable to those obtained using the A2780. In this second test, it is reasonable to assume that the HA–CH–siPLXDC1 particles could have targeted both tumor endothelial cells and OC HeyA8 cells. Thus, an additive/synergistic effect could have been expected. However, this was neither visible from the data provided, nor did the authors discuss this aspect. Finally, it would have been interesting to test the developed vector in a drug-resistance model of OC, as endothelial cell targeting could be an alternative strategy to bypass drug resistance.

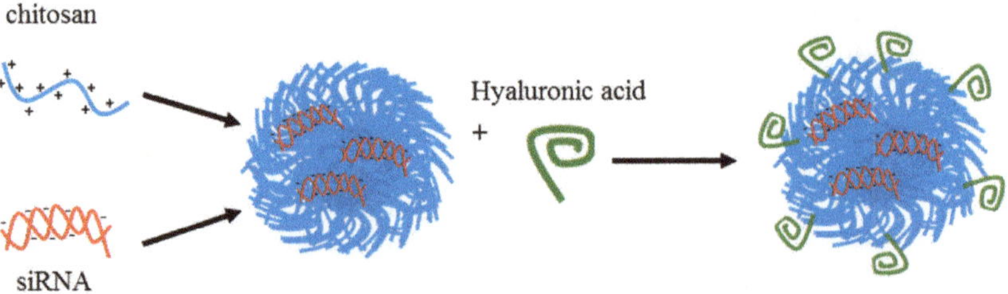

Figure 7. Polymeric delivery material developed in Reference [117].

Byeon et al. [118] described the use of PLGA conjugated with HA (PLGA–HA) to target OC cells overexpressing CD44 (see Section 4.1) (Table 6). The authors used a siRNA directed against FAK (siFAK), which is overexpressed in OC, correlates with shorter overall patient survival, and promotes drug resistance (see Section 4.2). In addition to the siFAK, PLGA–HA also contained PTX, commonly used in OC therapy (see Section 1.3). The authors proved the effective uptake of fluorescently labeled PLGA–HA particles in the CD44 overexpressing HeyA8 and SKOV3 cells. In the HeyA8 and SKOV3 variants which are PTX resistant (HeyA8-MDR1 and SKOV3 TR), the authors showed that the presence of siFAK improved the effects of PTX. Effective tumor cell uptake in vivo was proven following a single IV injection of FITC-labeled HA–PLGA–NPs into HeyA8-bearing female BALB/c nude mice. Effectiveness studies were conducted in HeyA8-MDR and SKOV3-TR orthotopic models of OC, treating the animals twice weekly via IV injection of 200 mg/kg siFAK and 1.4 mg/kg PTX. In both OC animal models, the PLGA–HA–siFAK–PTX showed reduction of tumor weight, number of tumor nodules, and levels of FAK mRNA compared to controls. Notably, particles lacking HA invariably showed a reduced effect compared to those bearing HA, thus proving the targeting relevance of HA. Unfortunately, the authors did not compare the effects of PLGA–HA–siFAK–PTX with PLGA–HA–PTX; this test could have provided information about the role of siFAK alone in the therapeutic effect. This comparison was limited to the effects on animal survival, which indicated extended animal survival for PLGA–HA–siFAK–PTX compared to PLGA–HA–PTX. The authors also confirmed the effectiveness of their approach by loading the particles with a different siRNA against survivin (siSVV), a gene product known to induce chemo-resistance and to protect cells from apoptosis (see Section 4.2). The authors further confirmed their findings by testing the PLGA–HA–siFAK–PTX particles in a mouse model generated using OC cells derived from a patient that had developed PTX–carboplatin resistance. Altogether, these findings support the robustness of the approach undertaken.

A rather novel approach to particle targeting to OC was developed by Hong et al. [119]. The strategy was based on the equipment of PEG–PEI particles with a short peptide mimicking the sequence of the follicle-stimulating hormone (FSH) (Table 6). FSH is secreted by the pituitary gland and exerts its action on ovaries via the FSH receptor (FSHR). As FSHR is expressed at high levels in reproductive tissues and at very low levels in other tissues, FSH-equipped particles can almost be directed to the ovarian tissue. Notably, approximately 70% of ovarian cancers express FSHR [120]. The authors loaded the PEG–PEI–FSH particles with a siRNA directed against the growth-regulated oncogene α (gro-α), (sigro). Gro-α induces malignant transformation, tumor growth, and metastatic spread in OC cells, where it is overexpressed compared to healthy cells [121]. As the cellular model, Hey cells, which express both FSHR and gro-α, were employed. As controls, SKOV3 cells, which have low FSHR expression but elevated gro-α, were used. In vitro, the PEG–PEI–FSH–sigro reduced the target mRNA level down to 47.3% of that of the control. Notably, particles lacking the FSH peptide (PEG–PEI–sigro) were less effective (66.3% reduction). In the FSHR-negative SKOV3, PEG–PEI–FSH–sigro reduced target level down to only 79%, indicating the effectiveness of the targeting strategy. In Hey cells, the PEG–PEI–FSH–sigro particles more effectively reduced cell proliferation, migration, and invasion compared to control. In an in vivo xenograft subcutaneous mouse model of OC (Hey cell grafted), the IV administration of PEG–PEI–FSH–sigro particles resulted in a significant reduction in tumor mass compared to saline and empty-particle-treated animals. Unfortunately, no control siRNA was included in tests; thus, it is not possible to evaluate the specificity of the sigro chosen.

A commonly used targeting strategy in siRNA delivery system is based on the addition of FA, which can interact with FR, to the delivery material (see also Section 4.1). Jones et al. [122] prepared triblock copolymers consisting of hyperbranched polyethylenimine-graft-polycaprolactone-block-poly(-ethylene glycol) (hyPEI-*g*-PCL-*b*-PEG) with and without a FA ligand for FR targeting (Table 6). In vitro in SKOV3 cells, the authors observed that the polymer with or without FA, and loaded with indium-111 labeled siRNA, had comparable results in term of cellular uptake. This was mostly because uptake was evaluated just 4 h after treatment. Despite this, the cellular distribution in the presence or absence of FA influenced siRNA localization.

Internalization via FR resulted in an even distribution of the siRNA within the cells, most likely due to the receptor-mediated internalization mechanisms. In the absence of FA, the distribution was dotted, probably reflecting the accumulation in endosomes, typical of an adsorptive endocytosis. In vivo, in an orthotropic model of OC, obtained using SKOV3 expressing the luciferase gene (SKOV3-luc), it was observed that when injected IV (35 μg of siRNA), the particles had short circulation half-lives. This was true for both the FR-targeted and -non-targeted particles, and was probably the consequence of the fast extravasation with predominant accumulation in the liver. FR-targeted particles displayed only a somewhat reduced liver localization compared to FR-non-targeted particles (38% and 53% of normalized injected dose per gram, respectively). Consequently, tumor accumulation was low and was characterized by an only slightly higher value for FR-targeted particles vs. -non-targeted (3.4 and 2.4%, respectively). When injected IP (35 μg of siRNA), both FR-targeted and -non-targeted particles accumulated predominantly in the kidney (7.36% and 7.78%, respectively) and in the tumor mass (5.63 and 5.28%, respectively). These data suggest that the effect of FR targeting was minimal, and tumor accumulation was most likely driven by passive targeting. From a functional point of view, the authors observed that the injection of an anti-luciferase siRNA delivered via the FR-targeted particles resulted in a significant reduction of luciferase expression 24 and 48 h post IP administration (35 μg siRNA-luc) compared to a control siRNA. From one point of view, this paper attenuated the enthusiasm for targeted delivery systems. On the other hand, the fact that just a small fraction of the siRNA reaching the tumor was sufficient to exert a clear anti-tumor effect is encouraging.

Table 6. Polymer based delivery systems with an OC-targeting moiety.

First Author	Target	Delivery Material	Cell Model	siRNA Delivery Route	Animal Model	References
Kim	PLXDC1	HA–CHITOSAN (targeting to CD44) 200 ± 10 nm (particle size)	HUVEC; MOEC; A2780; HeyA8	Intravenous (150 mg/kg twice per week)	Orthotopic A2780, HeyA8- Female BALB/c nude mice (intra-peritoneal tumor cell injection)	[117]
Byeon	FAK, surviving	HA–PLGA (targeting to CD44) 200–220 nm (particle size)	HeyA8; SKOV3; HeyA8-MDR; SKOV3-TR	Intravenous (200 mg/kg siFAK and 1.4 mg/kg PTX)	Orthotopic HeyA8-MDR, SKOV3-TR, patient-derived cells in female BALB/c nude mice (intra-peritoneal tumor cell injection)	[118]
Hong	Gro-α	PEG–PEI–FSH (targeting FSHR) 142.0 ± 3.8 nm	Hey	Intravenous (5 mg/kg)	Subcutaneous xenograft Hey BALB/c nude mice	[119]
Jones	Luciferase	hyPEI-g-PCL-b-PEG with and without FA 150 nm (particle size)	SKOV3	Intravenous (35 μg siRNA) Intra-peritoneal (35 μg siRNA)	Orthotopic SKOV3female nude mice (intra-peritoneal tumor cell injection)	[122]

6.2. Lipid-Based Delivery Systems

Lee et al. [123] used particles comprising DC–Chol cationic lipids and DOPE neutral lipids for siRNA delivery (Table 7). To overcome the problems of short blood circulation time and potential aggregation of cationic lipids, the authors added PEG molecules to the lipid components (lipid–PEG). A siRNA against the kinesin spindle protein (siKSP) was used. KSP regulates microtubule organization and spindle assembly during eukaryotic cell division [124]; thus, its targeting can abrogate cell proliferation. Using a Cy5.5-labeled siRNA, it was possible to show that the lipid–PEG–Cy5.5–siRNA efficiently entered the cytoplasm of SKOV3 cells, but not the nucleus. Moreover, the lipid–PEG–Cy5.5–siKSP was able to efficiently reduce the mRNA and protein level of the target, as well as the proliferation of SKOV3. Following the IV injection of lipid–PEG–Cy5.5–siKSP in a subcutaneous xenograft model of OC, the authors observed an accumulation of the particles in the kidney and liver, with a detectable amount also found in the tumor mass. Notably, the particles did not elicit immunological activation, at least with regard to the levels of the cytokines TNF-α and

IFN-α 1, 1 h post injection. Finally, lipid–PEG–siKSP reduced tumor mass growth compared to control. Unfortunately, no data about the effects on animal survival were reported. While the approach is, in principle, interesting, it was not determined whether KSP can be considered an appropriate target in OC treatment.

An interesting molecule to target in OC is thymidylate synthase (TS). This is an important rate-limiting enzyme in both normal and tumor DNA biosynthesis. Moreover, TS is highly expressed in both original and metastatic (peritoneal) OC lesions [125]. Its expression level seems to also be a key determinant for the efficacy of TS-targeting drugs. Iizuka et al. [126] developed delivery particles comprising dioleoylphosphatidylcholine (DOPC), DOPE, and DC, which were loaded with an anti-TS siRNA, giving origin to the preparation DFP-10825 (Figure 8 and Table 7). The effects of DFP-10825 alone or in combination with PTX were tested in a xenograft orthotopic model of OC generated by SKOV3-luc implantation in the peritoneum. Up to 24 h from IP administration of DFP-10825 (1 mg/kg of TS siRNA), the anti-TS siRNA could be detected in the ascitic fluids, with a pick of 4.26 ± 1.00 nM at 2 h. Notably, at the same time point, TS siRNA in the blood only reached the concentration of 0.005 ± 0.002 nM. Given the average size of DFP-10825 (395 ± 32 nm), this result is not surprising. Indeed, particles with diameters greater than 10 nm are scarcely drained by the blood vessels (see Section 2.2). Verified the favorable retention of DFP-10825 in the ascitic fluids, the authors tested the efficacy and found that the IP injection of DFP-10825 (0.5–2 mg/kg every third day) resulted in a time- and dose-dependent reduction of tumor mass. Unfortunately, no control siRNA was used, and thus the real specificity of the siRNA remains undetermined. No variation in animal weight was observed over the 28 days of the experiment, thus suggesting mild toxicity of DFP-10825. In an additional experiment, PTX at a dose of 15 mg/kg was IP administered alone or in combination with DFP-10825. Due to the potent antitumor activity of PTX alone, addition of DFP-10825 only resulted in a modest augmentation of antitumor activity. The antitumor effects resulted in an increased animal survival for both DFP-10825- or PTX-treated animals. Notably, the combination of DFP-10825 and PTX produced an additive effect on animal survival. Finally, the authors proved that DFP-10825, but not PTX, reduced TS mRNA levels in tumor cells collected from ascites.

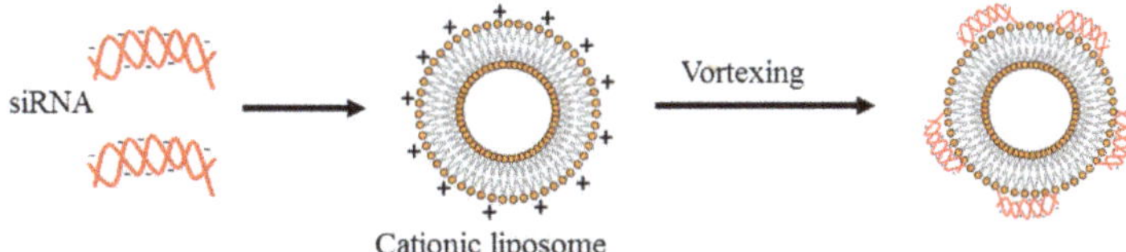

Figure 8. DOPC, DOPE, and DC delivery material developed in Reference [126].

While lipid particles most often contain lipids only, non-lipid molecules are sometimes added to improve the delivery properties. This was the case in the work presented by Mendes et al. [127] (Table 7). The authors developed lipid particles containing phosphatidylcholine (PC), Chol, and 1,2-dioleoyl-*sn*-glycero-3-phosphoethanolamine-*N*-(glutaryl) (NGPE) covered by low-molecular-weight branched PEI (PEIPOS). Three kinds of particles were prepared: PEIPOS alone, and PEIPOS with 0.1% and 0.5% by moles of PEI on the surface. PEI was added to improve siRNA delivery (see Section 3.1). Compared to PEIPOS and 0.1%PEIPOS, 0.5%PEIPOS particles had the highest uptake in A2780-ADR and SKOV3-TR cells, resistant to the drugs adriamycin and PTX, respectively. Most likely, the higher PEI amount in 0.5%PEIPOS was better able to exteriorize PEI delivery properties. In 3D tumor spheroids generated from cervical adenocarcinoma HeLa cells (not related to OC), 0.5%PEIPOS association to the tumor cells was higher (88%) than PEIPOS and 0.1%PEIPOS. Moreover, the authors showed that 0.5%PEIPOS could penetrate deeper into a spheroid structure by inter- and intra-cellular routes. Although HeLa spheroids are not directly pertinent to OC study, these data are encouraging for further tests in OC spheroids. In cultured A2780-ADR cells PEIPOS, 0.1%PEIPOS, and 0.5%PEIPOS loaded with PTX improved cytotoxicity. However, no significant differences were noted between

PEIPOS and 0.5%PEIPOS, leaving open the possibility that PEI might not have been essential for siRNA delivery in the specific experimental model. In the same cellular model, 100 nM siMDR1 (N/P of 13) coupled with 0.5%PEIPOS reduced the target protein level by about 20%. An in vivo model, a xenograft subcutaneous mice model generated using A2780-ADR, was employed. Animals were treated continuously with 0.5%PEIPOS/PTX/siMDR1 on alternate days until 12 injections has been administered, reaching the total doses of 66 mg/kg and 9.6 mg/kg of PTX and siMDR1, respectively. It is unclear why the authors used A2780-ADR in combination with PTX and not adriamycin, for which these cells show resistance. Only the 0.5%PEIPOS/PTX/siMDR1 treatment resulted in a reduction of tumor growth compared to free PTX and 0.5%PEIPOS/PTX. The authors suggested that the reason for the lack of effectiveness of the 0.5%PEIPOS/PTX may have been in the low dose used. Unfortunately, 0.5%PEIPOS/PTX/sicontrol was not tested; thus, the specificity of the siRNA remains undetermined. Moreover, 0.5%PEIPOS/siMDR1 was not tested, so it is not possible to evaluate the effects of the siRNA per se. Finally, the 0.5%PEIPOS/PTX/siMDR1 treatment had no significant side effects as evaluated by animal weight and by the levels of transaminases.

Recently, pressurized IP aerosol chemotherapy (PIPAC) has been studied as a novel drug delivery method in the treatment of OC peritoneal metastasis. Therapeutic drugs are nebulized throughout the peritoneal cavity via the use of high pressure (≈20 bars). This is thought to guarantee a homogenous distribution of the aerosolized particles containing the drug in the peritoneal cavity [128]. Minnaert et al. [129] studied the effects of the nebulization on a lipid (commercial) complex carrying siRNA against the luciferase gene, siLuc (Table 7). The authors observed that, from a physical point of view, the nebulization process only resulted in a small increase in particle size (about 25%, from 141 ± 1 to 193 ± 8). However, using particles loaded with up to 2 pmol of siLuc, the ability to reduce the fluorescence of SKOV3-luc in vitro was decreased. This phenomenon was not observed using higher siLuc amounts and, in general, the authors observed a dose-dependent reduction in the luminescence of SKOV3-luc. To mimic the real pathological environment, particles were incubated with ascitic liquid. The authors observed that the incubation reduced siLuc effectiveness by about 20%. The authors suggested that this could have been due to the binding of negatively charged proteins of ascites to the particles. Alternatively, the ascites' protein content might have influenced the intracellular pathway of siRNA. While no data in an animal model were provided, the work suggests that for lipid–siRNA complexes, delivery via nebulization in the peritoneal cavity might be feasible.

Table 7. Lipid-based delivery systems.

First Author	Target	Delivery Material	Cell Model	siRNA Delivery Route	Animal Model	References
Lee	KSP	DC–Chol–DOPE–PEG 90–110 nm (Particle size)	SKOV3	Intravenous (1 mg/kg every other day for a total of eight injections)	Subcutaneous xenograft mice model Balb/c nude mice	[123]
Iizuka	TS	DOPC–DOPE–DC 395 ± 32 nm (Particle size)	-	Intra-peritoneal (0.5–2 mg/kg every third day)	Orthotopic SKOV3-luc cell in male SOD/SCID mice (intra-peritoneal tumor cell injection)	[126]
Mendes	MDR1	PC–Chol–NGPE–PEI 161 ± 9.4 nm (Particle size)	A2780-ADR and SKOV3-TR	Intravenous (total doses in multiple administration: 66 mg/kg and 9.6 mg/kg of PTX and siMDR1, respectively)	Subcutaneous xenograft mice (athymic nude mice) model using A2780-ADR	[127]
Minnaert	luciferase	Commercial lipid 193 ± 8 nm (Particle size) Following nebulization	SKOV3-luc	-	-	[129]

7. Final Considerations

The use of naked siRNA in vivo results in negligible effects due to their instability in biological fluids and sub-optimal pharmacokinetics [130–132]. The use of polymer-/lipid-based materials has the potential to bypass the above problems, allowing the development of siRNA delivery systems suitable for therapeutic applications. Despite this, almost all the works published so far (1930 publications found in PubMed using the key words: "ovarian cancer siRNA") tested the delivery systems in orthotopic mouse models of OC at best. We could find only one clinical trial (NCT01437007) mentioning the use of siRNA in OC [133]. In this case, the aim was to evaluate the effect of a lipid particle formulation containing siRNA against PLK1 (TKM-080301) in patients with liver metastases from different tumors, including OC. Following hepatic arterial administration (4 mg/m^2 every 2 weeks for up to 12 doses) the study aimed to establish the maximum tolerable dose and dose-limiting toxicities. Although the clinical trial was posted in 2011, no conclusive results have been reported so far.

Based on the above consideration, it is evident that improvements in the delivery techniques available are necessary. In this regard, different factors can be considered work towards improved delivery effectiveness.

The first deals with the use of appropriate models for OC. Most of the work published used cellular models, which do not share the phenotypic/molecular features of HGSOC. Thus, the results obtained can only barely be extrapolated to HGSOC. It is therefore advisable to use more appropriate cell models in the future [97], possibly in the form of spheroids that better resemble in vivo situations. The problem of suboptimal animal models is harder to circumvent. However, the recommendation is to use orthotopic mouse models, which better resemble the peritoneal metastasis observed in humans even if most cannot take into account the effects of the immune system. The usefulness of subcutaneous mouse models, where the effects on tumor mass are easily measurable, might be limited to pilot studies for the preliminary testing of completely novel drugs and/or delivery systems. To further improve the usefulness of cellular/animal models, we believe that the use of mathematical models should be more thoroughly considered. Indeed, through mathematical models, it is possible to reduce experimental testing, with the consequent reduction in experimental burden and costs.

A second factor involves the possibility of targeting siRNA to OC cells. This approach may, in principle, allow normal cells to be left untouched, thus minimizing side effects. However, targeting is not so easy to achieve. One reason for this is that the targeted antigen is often present (albeit at lower levels) also on non-tumor cells, and it remains unclear whether this level of difference is sufficient for effective discrimination. To improve the specificity, one possibility is to combine a targeted delivery system with an siRNA directed against the mRNA of genes expressed only or predominantly in OC cells. In this way, even if the siRNAs reach healthy cells, they cannot exert major detrimental effects, as the target gene is expressed at low or negligible levels. Despite the above considerations, OC targeting may be more difficult to achieve than we think. For example, the work of Jones et al. [122], discussed above, showed that targeted/non-targeted particles behave very similarly with regard to the body distribution when delivered via either IV or IP administration. Additionally, tumor accumulation is modest for both methods, most likely due to passive targeting. It is comforting, however, that even a modest tumor accumulation seems to be sufficient to exert a functional effect. Moreover, evidence of the possibility of a benefit from target delivery systems has been reported in References [117,118], commented on in this review.

A third factor to be considered is the optimal size of the vector delivery particles. In the case of IV administration, it is known that particles of 1–3 µm diameter tend to localize closer to the endothelial layer (margination effect) of the vessel, whereas smaller particles localize to the middle of the vessel [134]. Thus, 1–3 µm particles, being closer to the vessel wall, have a better chance to extravasate. However, particles in the nm range are more suited to crossing cell membranes compared to µm particles. Thus, a possible compromise might involve the preparation of µm particles that, upon extravasation, undergo fragmentation to generate nm-sized particles (chimeric systems). Notably, in all the works described here where IV administration was chosen, the particle size was 150–200 nm

(Tables 5–7), i.e., not optimized for the margination effect and thus extravasation. In the case of IP administration, the retention of delivery systems in the peritoneal cavity is a function of the particle size. Delivery systems larger than the diameter of the aperture of the lymphatic lacunae (3.2–3.6 µm in humans) will persist longer in the peritoneal cavity. Particles with smaller size will be predominantly cleared via the lacunar route, and the very small particles (nm range) will be mostly drained by blood vessels. In the works described where the IP route was chosen (Tables 5–7), particle size was 100–395 nm, and thus not optimal for long peritoneum persistence. Additionally, a chimeric system composed of µm particles which can dissociate into nm particles route may allow proper peritoneum persistence and subsequently the crossing of tumor cell membrane with IP administration.

A final factor, which should be considered to improve the effectiveness of delivered siRNAs, deals with biological aspects, i.e., the targeting of OC stem cells (OCSCs). Almost all the published studies to date have focused their attention on general OC cells. However, targeting OCSCs would represent an attractive strategy, as these cells have the ability to perpetuate the tumor, thus favoring its persistence and spread. Moreover, OCSCs seem to have the ability to escape chemotherapy [69]. To develop such a strategy, the precise recognition of OC stem cells would be necessary, a problem which does not seem too complex, as different antigens for OC stem cells are known (see Section 4.1).

In conclusion, although the above issues may dampen enthusiasm for future research in the OC area with regard to siRNAs, the development of effective delivery systems should be further pursued. In particular, we believe that by taking into account the physiology and molecular biology of peritoneum/vessel/OC cells, it is possible to properly optimize siRNA delivery systems for OC.

Author Contributions: R.F. took care about Sections 1.1–1.3 and contributed to the final writing of Section 6; M.M. (Matea Maruna) took care of the first selection from the literature of the works mentioned in Section 6; F.P. prepared the figures and tables; M.G. wrote Section 5.3; F.B. took care about Section 3; M.M. (Marianna Maddaloni) and M.E.B. took care about Sections 2.1 and 2.2; S.P. organized Section 4.1; F.R. and G.F. wrote Sections 5.1 and 5.2; F.Z. organized Section 4.2.1; M.C. and U.K. took care about Sections 4.2.2 and 4.2.3, B.D. wrote the final draft of Section 6, G.G. wrote the abstract, the conclusions, proposed the subject of the review to all the other authors and took care about the general planning of the work. Finally, all the authors contributed equally to the general organization of the manuscript and its revision, with helpful suggestions about the content and the style of the text.

Funding: This work was supported in part by the "Fondazione Benefica Kathleen Foreman Casali of Trieste", the "Beneficentia Stiftung" of Vaduz Liechtenstein and by FRA 2015, University of Trieste. R.F. and B.D. have been supported by POR FESR 2014-2020 FRIULI VENEZIA GIULIA, "ATeNA" 2017 J96G17000120005. G.F. was supported by the European Social Fund and European Regional Development Fund-Project MAGNET (Number CZ.02.1.01/0.0/0.0/15_003/0000492).

Conflicts of Interest: The authors declare no conflict of interest.

References

1. Farra, R.; Dapas, B.; Pozzato, G.; Scaggiante, B.; Agostini, F.; Zennaro, C.; Grassi, M.; Rosso, N.; Giansante, C.; Fiotti, N.; et al. Effects of E2F1–cyclin E1–E2 circuit down regulation in hepatocellular carcinoma cells. *Dig. Liver Dis.* **2011**, *43*, 1006–1014. [CrossRef]
2. Grassi, M.; Cavallaro, G.; Scirè, S.; Scaggiante, B.; Daps, B.; Farra, R.; Baiz, D.; Giansante, C.; Guarnieri, G.; Perin, D.; et al. Current Strategies to Improve the Efficacy and the Delivery of Nucleic Acid Based Drugs. *Curr. Signal Transduct. Ther.* **2010**, *5*, 92–120. [CrossRef]
3. Grassi, G.; Dawson, P.; Guarnieri, G.; Kandolf, R.; Grassi, M. Therapeutic potential of hammerhead ribozymes in the treatment of hyper-proliferative diseases. *Curr. Pharm. Biotechnol.* **2004**, *5*, 369–386. [CrossRef]
4. Grassi, G.; Marini, J.C. Ribozymes: Structure, function, and potential therapy for dominant genetic disorders. *Ann. Med.* **1996**, *28*, 499–510. [CrossRef]
5. Grassi, G.; Schneider, A.; Engel, S.; Racchi, G.; Kandolf, R.; Kuhn, A. Hammerhead ribozymes targeted against cyclin E and E2F1 cooperate to down-regulate coronary smooth muscle cell proliferation. *J. Gene Med.* **2005**, *7*, 1223–1234. [CrossRef] [PubMed]
6. Goldberg, M.S. siRNA delivery for the treatment of ovarian cancer. *Methods* **2013**, *63*, 95–100. [CrossRef] [PubMed]

7. Van den Brand, D.; Mertens, V.; Massuger, L.F.A.G.; Brock, R. siRNA in ovarian cancer—Delivery strategies and targets for therapy. *J. Control. Release* **2018**, *283*, 45–58. [CrossRef] [PubMed]
8. Aghamiri, S.; Mehrjardi, K.F.; Shabani, S.; Keshavarz-Fathi, M.; Kargar, S.; Rezaei, N. Nanoparticle-siRNA: A potential strategy for ovarian cancer therapy? *Nanomedicine* **2019**, *14*, 2083–2100. [CrossRef] [PubMed]
9. Halbur, C.; Choudhury, N.; Chen, M.; Kim, J.H.; Chung, E.J. siRNA-Conjugated Nanoparticles to Treat Ovarian Cancer. *SLAS Technol.* **2019**, *24*, 137–150. [CrossRef]
10. Webb, P.M.; Jordan, S.J. Epidemiology of epithelial ovarian cancer. *Best Pract. Res. Clin. Obstet. Gynaecol.* **2017**, *41*, 3–14. [CrossRef]
11. Reid, B.M.; Permuth, J.B.; Sellers, T.A. Epidemiology of ovarian cancer: A review. *Cancer Biol. Med.* **2017**, *14*, 9–32. [PubMed]
12. Singh, A.; Gupta, S.; Sachan, M. Epigenetic Biomarkers in the Management of Ovarian Cancer: Current Prospectives. *Front. Cell Dev. Biol.* **2019**, *7*, 182. [CrossRef] [PubMed]
13. Siegel, R.L.; Miller, K.D.; Jemal, A. Cancer statistics, 2016. *CA Cancer J. Clin.* **2016**, *66*, 7–30. [CrossRef] [PubMed]
14. Chang, S.J.; Hodeib, M.; Chang, J.; Bristow, R.E. Survival impact of complete cytoreduction to no gross residual disease for advanced-stage ovarian cancer: A meta-analysis. *Gynecol. Oncol.* **2013**, *130*, 493–498. [CrossRef] [PubMed]
15. Shih, I.; Kurman, R.J. Ovarian tumorigenesis: A proposed model based on morphological and molecular genetic analysis. *Am. J. Pathol.* **2004**, *164*, 1511–1518. [CrossRef]
16. Verhaak, R.G.; Tamayo, P.; Yang, J.Y.; Hubbard, D.; Zhang, H.; Creighton, C.J.; Fereday, S.; Lawrence, M.; Carter, S.L.; Mermel, C.H.; et al. Prognostically relevant gene signatures of high-grade serous ovarian carcinoma. *J. Clin. Investig.* **2013**, *123*, 517–525. [CrossRef]
17. Konecny, G.E.; Wang, C.; Hamidi, H.; Winterhoff, B.; Kalli, K.R.; Dering, J.; Ginther, C.; Chen, H.W.; Dowdy, S.; Cliby, W.; et al. Prognostic and therapeutic relevance of molecular subtypes in high-grade serous ovarian cancer. *J. Natl. Cancer Inst.* **2014**, *106*. [CrossRef]
18. Wang, C.; Armasu, S.M.; Kalli, K.R.; Maurer, M.J.; Heinzen, E.P.; Keeney, G.L.; Cliby, W.A.; Oberg, A.L.; Kaufmann, S.H.; Goode, E.L. Pooled Clustering of High-Grade Serous Ovarian Cancer Gene Expression Leads to Novel Consensus Subtypes Associated with Survival and Surgical Outcomes. *Clin. Cancer Res.* **2017**, *23*, 4077–4085. [CrossRef]
19. Lisio, M.A.; Fu, L.; Goyeneche, A.; Gao, Z.H.; Telleria, C. High-Grade Serous Ovarian Cancer: Basic Sciences, Clinical and Therapeutic Standpoints. *Int. J. Mol. Sci.* **2019**, *20*, 952. [CrossRef]
20. Narod, S. Can advanced-stage ovarian cancer be cured? *Nat. Rev. Clin. Oncol.* **2016**, *13*, 255–261. [CrossRef]
21. Nieman, K.M.; Kenny, H.A.; Penicka, C.V.; Ladanyi, A.; Buell-Gutbrod, R.; Zillhardt, M.R.; Romero, I.L.; Carey, M.S.; Mills, G.B.; Hotamisligil, G.S.; et al. Adipocytes promote ovarian cancer metastasis and provide energy for rapid tumor growth. *Nat. Med.* **2011**, *17*, 1498–1503. [CrossRef] [PubMed]
22. Matulonis, U.A.; Sood, A.K.; Fallowfield, L.; Howitt, B.E.; Sehouli, J.; Karlan, B.Y. Ovarian cancer. *Nat. Rev. Dis. Primers* **2016**, *2*, 16061. [CrossRef]
23. Gonzalez-Martin, A.; Sanchez-Lorenzo, L.; Bratos, R.; Marquez, R.; Chiva, L. First-line and maintenance therapy for ovarian cancer: Current status and future directions. *Drugs* **2014**, *74*, 879–889. [CrossRef] [PubMed]
24. Markman, M. Optimizing primary chemotherapy in ovarian cancer. *Hematol. Oncol. Clin. N. Am.* **2003**, *17*, 957–968. [CrossRef]
25. Marchetti, C.; De Felice, F.; Perniola, G.; Palaia, I.; Musella, A.; Di Donato, V.; Cascialli, G.; Muzii, L.; Tombolini, V.; Cascialli, G.; et al. Role of intraperitoneal chemotherapy in ovarian cancer in the platinum-taxane-based era: A meta-analysis. *Crit. Rev. Oncol. Hematol.* **2019**, *136*, 64–69. [CrossRef]
26. Bast, R.C., Jr.; Hennessy, B.; Mills, G.B. The biology of ovarian cancer: New opportunities for translation. *Nat. Rev. Cancer* **2009**, *9*, 415–428. [CrossRef]
27. Papa, A.; Caruso, D.; Strudel, M.; Tomao, S.; Tomao, F. Update on Poly-ADP-ribose polymerase inhibition for ovarian cancer treatment. *J. Transl. Med.* **2016**, *14*, 267. [CrossRef]
28. Napoli, C.; Lemieux, C.; Jorgensen, R. Introduction of a Chimeric Chalcone Synthase Gene into Petunia Results in Reversible Co-Suppression of Homologous Genes in trans. *Plant Cell* **1990**, *2*, 279–289. [CrossRef]
29. Fire, A.; Xu, S.; Montgomery, M.K.; Kostas, S.A.; Driver, S.E.; Mello, C.C. Potent and specific genetic interference by double-stranded RNA in Caenorhabditis elegans. *Nature* **1998**, *391*, 806–811. [CrossRef]

30. Elbashir, S.M.; Harborth, J.; Lendeckel, W.; Yalcin, A.; Weber, K.; Tuschl, T. Duplexes of 21-nucleotide RNAs mediate RNA interference in cultured mammalian cells. *Nature* **2001**, *411*, 494–498. [CrossRef]
31. Lang, C.; Sauter, M.; Szalay, G.; Racchi, G.; Grassi, G.; Rainaldi, G.; Mercatanti, A.; Lang, F.; Kandolf, R.; Klingel, K. Connective tissue growth factor: A crucial cytokine-mediating cardiac fibrosis in ongoing enterovirus myocarditis. *J. Mol. Med.* **2008**, *86*, 49–60. [CrossRef] [PubMed]
32. Grassi, G.; Pozzato, G.; Moretti, M.; Giacca, M. Quantitative analysis of hepatitis C virus RNA in liver biopsies by competitive reverse transcription and polymerase chain reaction. *J. Hepatol.* **1995**, *23*, 403–411. [CrossRef]
33. Scaggiante, B.; Dapas, B.; Farra, R.; Grassi, M.; Pozzato, G.; Giansante, C.; Fiotti, N.; Grassi, G. Improving siRNA bio-distribution and minimizing side effects. *Curr. Drug Metab.* **2011**, *12*, 11–23. [CrossRef] [PubMed]
34. Farra, R.; Grassi, M.; Grassi, G.; Dapas, B. Therapeutic potential of small interfering RNAs/micro interfering RNA in hepatocellular carcinoma. *World J. Gastroenterol.* **2015**, *21*, 8994–9001. [CrossRef] [PubMed]
35. Grassi, G.; Scaggiante, B.; Dapas, B.; Farra, R.; Tonon, F.; Lamberti, G.; Barba, A.; Fiorentino, S.; Fiotti, N.; Zanconati, F.; et al. Therapeutic potential of nucleic acid-based drugs in coronary hyper- proliferative vascular diseases. *Curr. Med. Chem.* **2013**, *20*, 3515–3538. [CrossRef] [PubMed]
36. Agostini, F.; Dapas, B.; Farra, R.; Grassi, M.; Racchi, G.; Klingel, K.; Kandolf, R.; Heidenreich, O.; Mercatahnti, A.; Rainaldi, G.; et al. Potential applications of small interfering RNAs in the cardiovascular field. *Drug Future* **2006**, *31*, 513–525. [CrossRef]
37. Huang, Y.; Hong, J.; Zheng, S.; Ding, Y.; Guo, S.; Zhang, H.; Zhang, X.; Du, Q.; Liang, Z. Elimination pathways of systemically delivered siRNA. *Mol. Ther.* **2011**, *19*, 381–385. [CrossRef]
38. Jackson, A.L.; Linsley, P.S. Recognizing and avoiding siRNA off-target effects for target identification and therapeutic application. *Nat. Rev. Drug Discov.* **2010**, *9*, 57–67. [CrossRef]
39. Sarfarazi, A.; Lee, G.; Mirjalili, S.A.; Phillips, A.R.J.; Windsor, J.A.; Trevaskis, N.L. Therapeutic delivery to the peritoneal lymphatics: Current understanding, potential treatment benefits and future prospects. *Int. J. Pharm.* **2019**, *567*, 118456. [CrossRef]
40. Trevaskis, N.L.; Kaminskas, L.M.; Porter, C.J. From sewer to saviour—Targeting the lymphatic system to promote drug exposure and activity. *Nat. Rev. Drug Discov.* **2015**, *14*, 781–803. [CrossRef]
41. Mirahmadi, N.; Babaei, M.H.; Vali, A.M.; Dadashzadeh, S. Effect of liposome size on peritoneal retention and organ distribution after intraperitoneal injection in mice. *Int. J. Pharm.* **2010**, *383*, 7–13. [CrossRef] [PubMed]
42. Lengyel, E. Ovarian cancer development and metastasis. *Am. J. Pathol.* **2010**, *177*, 1053–1064. [CrossRef] [PubMed]
43. Luft, C.; Ketteler, R. Electroporation Knows No Boundaries: The Use of Electrostimulation for siRNA Delivery in Cells and Tissues. *J. Biomol. Screen.* **2015**, *20*, 932–942. [CrossRef] [PubMed]
44. Leonetti, J.P.; Degols, G.; Lebleu, B. Biological activity of oligonucleotide-poly(L-lysine) conjugates: Mechanism of cell uptake. *Bioconjug. Chem.* **1990**, *1*, 149–153. [CrossRef] [PubMed]
45. Sardo, C.; Farra, R.; Licciardi, M.; Dapas, B.; Scialabba, C.; Giammona, G.; Grassi, M.; Grassi, G.; Cavallaro, G. Development of a simple, biocompatible and cost-effective Inulin-Diethylenetriamine based siRNA delivery system. *Eur. J. Pharm. Sci.* **2015**, *75*, 60–71. [CrossRef] [PubMed]
46. Cavallaro, G.; Licciardi, M.; Amato, G.; Sardo, C.; Giammona, G.; Farra, R.; Dapas, B.; Grassi, M.; Grassi, G. Synthesis and characterization of polyaspartamide copolymers obtained by ATRP for nucleic acid delivery. *Int. J. Pharm.* **2014**, *466*, 246–257. [CrossRef] [PubMed]
47. Xu, C.; Wang, J. Delivery systems for siRNA drug development in cancer therapy. *Asian J. Pharm. Sci.* **2015**, *10*, 1–12. [CrossRef]
48. Hobel, S.; Aigner, A. Polyethylenimines for siRNA and miRNA delivery in vivo. *Wiley Interdiscip. Rev. Nanomed. Nanobiotechnol.* **2013**, *5*, 484–501. [CrossRef]
49. Liu, L.; Zheng, M.; Librizzi, D.; Renette, T.; Merkel, O.M.; Kissel, T. Efficient and Tumor Targeted siRNA Delivery by Polyethylenimine-graft-polycaprolactone-block-poly(ethylene glycol)-folate (PEI-PCL-PEG-Fol). *Mol. Pharm.* **2016**, *13*, 134–143. [CrossRef]
50. Roberts, M.J.; Bentley, M.D.; Harris, J.M. Chemistry for peptide and protein PEGylation. *Adv. Drug Deliv. Rev.* **2002**, *54*, 459–476. [CrossRef]
51. Bao, Y.; Jin, Y.; Chivukula, P.; Zhang, J.; Liu, Y.; Liu, J.; Clamme, J.P.; Mahato, R.I.; Ng, D.; Ying, W.; et al. Effect of PEGylation on biodistribution and gene silencing of siRNA/lipid nanoparticle complexes. *Pharm. Res.* **2013**, *30*, 342–351. [CrossRef] [PubMed]

52. Muralidharan, P.; Mallory, E.; Malapit, M.; Hayes, D., Jr.; Mansour, H.M. Inhalable PEGylated Phospholipid Nanocarriers and PEGylated Therapeutics for Respiratory Delivery as Aerosolized Colloidal Dispersions and Dry Powder Inhalers. *Pharmaceutics* **2014**, *6*, 333–353. [CrossRef] [PubMed]
53. Azimi, B.; Nourpanak, P.; Rabiee, M.; Arab, S. Poly (ε-caprolactone) Fiber: An Overview. *J. Eng. Fibers Fabr.* **2014**, *9*, 74–90.
54. Xu, Z.; Wang, D.; Cheng, Y.; Yang, M.; Wu, L.P. Polyester based nanovehicles for siRNA delivery. *Mater. Sci. Eng. C Mater. Biol. Appl.* **2018**, *92*, 1006–1015. [CrossRef] [PubMed]
55. Posocco, B.; Dreussi, E.; de Santa, J.; Toffoli, G.; Abrami, M.; Musiani, F.; Grassi, M.; Farra, R.; Tonon, F.; Grassi, G.; et al. Polysaccharides for the Delivery of Antitumor Drugs. *Materials* **2015**, *8*, 2569–2615. [CrossRef]
56. Ahmed, T.; Aljaeid, B. Preparation characterization and potential application of chitosan, chitosan derivates, and chitosan metal nanoparticles in pharmaceutical drug delivery. *Drug Des. Dev. Ther.* **2016**, *10*, 483–507. [CrossRef] [PubMed]
57. Khan, W.; Hosseinkhani, H.; Ickowicz, D.; Hong, P.D.; Yu, D.S.; Domb, A.J. Polysaccharide gene transfection agents. *Acta Biomater.* **2012**, *8*, 4224–4232. [CrossRef]
58. Oh, E.J.; Park, K.; Kim, K.S.; Kim, J.; Yang, J.A.; Kong, J.H.; Lee, M.Y.; Hoffman, A.S.; Hahn, S.K. Target specific and long-acting delivery of protein, peptide, and nucleotide therapeutics using hyaluronic acid derivatives. *J. Control. Release* **2010**, *141*, 2–12. [CrossRef]
59. Barba, A.A.; Lamberti, G.; Sardo, C.; Dapas, B.; Abrami, M.; Grassi, M.; Farra, R.; Tonon, F.; Forte, G.; Musiani, F.; et al. Novel Lipid and Polymeric Materials as Delivery Systems for Nucleic Acid Based Drugs. *Curr. Drug Metab.* **2015**, *16*, 427–452. [CrossRef]
60. Bochicchio, S.; Dalmoro, A.; Barba, A.A.; Grassi, G.; Lamberti, G. Liposomes as siRNA delivery vectors. *Curr. Drug Metab.* **2014**, *15*, 882–892. [CrossRef]
61. Ayen, A.; Jimenez, M.Y.; Marchal, J.A.; Boulaiz, H. Recent Progress in Gene Therapy for Ovarian Cancer. *Int. J. Mol. Sci.* **2018**, *19*, 1930. [CrossRef] [PubMed]
62. Hsu, P.; Jablons, D.; Yang, C.; You, L. Epidermal Growth Factor Receptor (EGFR) Pathway, Yes-Associated Protein (YAP) and the Regulation of Programmed Death-Ligand 1 (PD-L1) in Non-Small Cell Lung Cancer (NSCLC). *Int. J. Mol. Sci.* **2019**, *20*, 3821. [CrossRef] [PubMed]
63. Shepard, H.M.; Brdlik, C.M.; Schreiber, H. Signal integration: A framework for understanding the efficacy of therapeutics targeting the human EGFR family. *J. Clin. Investig.* **2008**, *118*, 3574–3581. [CrossRef] [PubMed]
64. Dickerson, E.B.; Blackburn, W.H.; Smith, M.H.; Kapa, L.B.; Lyon, L.A.; McDonald, J.F. Chemosensitization of cancer cells by siRNA using targeted nanogel delivery. *BMC Cancer* **2010**, *10*, 10. [CrossRef]
65. Zhou, Y.; Sakurai, H. Emerging and Diverse Functions of the EphA2 Noncanonical Pathway in Cancer Progression. *Biol. Pharm. Bull.* **2017**, *40*, 1616–1624. [CrossRef]
66. Takahashi, Y.; Hamasaki, M.; Aoki, M.; Koga, K.; Koshikawa, N.; Miyamoto, S.; Nabeshima, K. Activated EphA2 Processing by MT1-MMP Is Involved in Malignant Transformation of Ovarian Tumours In Vivo. *Anticancer Res.* **2018**, *38*, 4257–4266. [CrossRef]
67. Cortez, A.J.; Tudrej, P.; Kujawa, K.A.; Lisowska, K.M. Advances in ovarian cancer therapy. *Cancer Chemother. Pharmacol.* **2018**, *81*, 17–38. [CrossRef]
68. Cheung, A.; Bax, H.J.; Josephs, D.H.; Ilieva, K.M.; Pellizzari, G.; Opzoomer, J.; Bloomfield, J.; Fittall, M.; Grigoriadis, A.; Figini, M.; et al. Targeting folate receptor alpha for cancer treatment. *Oncotarget* **2016**, *7*, 52553–52574. [CrossRef]
69. Zong, X.; Nephew, K.P. Ovarian Cancer Stem Cells: Role in Metastasis and Opportunity for Therapeutic Targeting. *Cancers* **2019**, *11*, 934. [CrossRef]
70. Bartakova, A.; Michalova, K.; Presl, J.; Vlasak, P.; Kostun, J.; Bouda, J. CD44 as a cancer stem cell marker and its prognostic value in patients with ovarian carcinoma. *J. Obstet. Gynaecol.* **2018**, *38*, 110–114. [CrossRef]
71. Shah, V.; Taratula, O.; Garbuzenko, O.B.; Taratula, O.R.; Rodriguez-Rodriguez, L.; Minko, T. Targeted nanomedicine for suppression of CD44 and simultaneous cell death induction in ovarian cancer: An optimal delivery of siRNA and anticancer drug. *Clin. Cancer Res.* **2013**, *19*, 6193–6204. [CrossRef] [PubMed]
72. Skubitz, A.P.; Taras, E.P.; Boylan, K.L.; Waldron, N.N.; Oh, S.; Panoskaltsis-Mortari, A.; Vallera, D.A. Targeting CD133 in an in vivo ovarian cancer model reduces ovarian cancer progression. *Gynecol. Oncol.* **2013**, *130*, 579–587. [CrossRef] [PubMed]

73. Yang, B.; Yan, X.; Liu, L.; Jiang, C.; Hou, S. Overexpression of the cancer stem cell marker CD117 predicts poor prognosis in epithelial ovarian cancer patients: Evidence from meta-analysis. *Onco Targets Ther.* **2017**, *10*, 2951–2961. [CrossRef] [PubMed]
74. Shaw, T.J.; Vanderhyden, B.C. AKT mediates the pro-survival effects of KIT in ovarian cancer cells and is a determinant of sensitivity to imatinib mesylate. *Gynecol. Oncol.* **2007**, *105*, 122–131. [CrossRef] [PubMed]
75. Schilder, R.J.; Sill, M.W.; Lee, R.B.; Shaw, T.J.; Senterman, M.K.; Klein-Szanto, A.J.; Miner, Z.; Vanderhyden, B.C. Phase II evaluation of imatinib mesylate in the treatment of recurrent or persistent epithelial ovarian or primary peritoneal carcinoma: A Gynecologic Oncology Group Study. *J. Clin. Oncol.* **2008**, *26*, 3418–3425. [CrossRef] [PubMed]
76. Nakamura, K.; Terai, Y.; Tanabe, A.; Ono, Y.J.; Hayashi, M.; Maeda, K.; Fujiwara, S.; Ashihara, K.; Nakamura, M.; Tanaka, Y.; et al. CD24 expression is a marker for predicting clinical outcome and regulates the epithelial-mesenchymal transition in ovarian cancer via both the Akt and ERK pathways. *Oncol. Rep.* **2017**, *37*, 3189–3200. [CrossRef] [PubMed]
77. Zhang, R.; Shi, H.; Chen, Z.; Wu, Q.; Ren, F.; Huang, H. Effects of metastasis-associated in colon cancer 1 inhibition by small hairpin RNA on ovarian carcinoma OVCAR-3 cells. *J. Exp. Clin. Cancer Res.* **2011**, *30*, 83. [CrossRef]
78. Rao, Y.; Ji, M.; Chen, C.; Shi, H. Effect of siRNA targeting MTA1 on metastasis malignant phenotype of ovarian cancer A2780 cells. *J. Huazhong Univ. Sci. Technol. Med. Sci.* **2013**, *33*, 266–271. [CrossRef]
79. Huo, X.; Ren, L.; Shang, L.; Wang, X.; Wang, J. Effect of WT1 antisense mRNA on the induction of apoptosis in ovarian carcinoma SKOV3 cells. *Eur. J. Gynaecol. Oncol.* **2011**, *32*, 651–656.
80. De, P.; Aske, J.; Dey, N. RAC1 Takes the Lead in Solid Tumors. *Cells* **2019**, *8*, 382. [CrossRef]
81. Hudson, L.; Gillette, J.; Kang, H.; Rivera, M.; Wandinger-Ness, A. Ovarian Tumor Microenvironment Signaling:Convergence on the Rac1 GTPase. *Cancers* **2018**, *10*, 358. [CrossRef] [PubMed]
82. Lu, L.; Yu, X. The balance of Polo-like kinase 1 in tumorigenesis. *Cell Div.* **2009**, *4*, 4. [CrossRef] [PubMed]
83. Zhang, R.; Shi, H.; Ren, F.; Liu, H.; Zhang, M.; Deng, Y.; Li, X. Misregulation of polo-like protein kinase 1, P53 and P21WAF1 in epithelial ovarian cancer suggests poor prognosis. *Oncol. Rep.* **2015**, *33*, 1235–1242. [CrossRef] [PubMed]
84. Farra, R.; Dapas, B.; Grassi, M.; Benedetti, F.; Grassi, G. E2F1 as a molecular drug target in ovarian cancer. *Expert Opin. Ther. Targets* **2019**, *23*, 161–164. [CrossRef] [PubMed]
85. Oku, Y.; Nishiya, N.; Tazawa, T.; Kobayashi, T.; Umezawa, N.; Sugawara, Y.; Uehara, Y. Augmentation of the therapeutic efficacy of WEE1 kinase inhibitor AZD1775 by inhibiting the YAP-E2F1-DNA damage response pathway axis. *FEBS Open Bio* **2018**, *8*, 1001–1012. [CrossRef] [PubMed]
86. Ryo, A.; Liou, Y.C.; Wulf, G.; Nakamura, M.; Lee, S.W.; Lu, K.P. PIN1 is an E2F target gene essential for Neu/Ras-induced transformation of mammary epithelial cells. *Mol. Cell. Biol.* **2002**, *22*, 5281–5295. [CrossRef] [PubMed]
87. Liou, Y.C.; Zhou, X.Z.; Lu, K.P. Prolyl isomerase Pin1 as a molecular switch to determine the fate of phosphoproteins. *Trends Biochem. Sci.* **2011**, *36*, 501–514. [CrossRef]
88. Russo, S.C.; De, S.L.; Palazzolo, S.; Salis, B.; Granchi, C.; Minutolo, F.; Tuccinardi, T.; Fratamico, R.; Crotti, S.; D'Aronco, S.; et al. Liposomal delivery of a Pin1 inhibitor complexed with cyclodextrins as new therapy for high-grade serous ovarian cancer. *J. Control. Release* **2018**, *281*, 1–10. [CrossRef]
89. Russo, S.C.; De, S.L.; Poli, G.; Granchi, C.; El, B.M.; Ecca, F.; Grassi, G.; Grassi, M.; Canzonieri, V.; Giordano, A.; et al. Virtual screening identifies a PIN1 inhibitor with possible antiovarian cancer effects. *J. Cell. Physiol.* **2019**. [CrossRef]
90. Van Beijnum, J.R.; Petersen, K.; Griffioen, A.W. Tumor endothelium is characterized by a matrix remodeling signature. *Front. Biosci. (Sch. Ed.)* **2009**, *1*, 216–225. [CrossRef]
91. Ren, F.; Shen, J.; Shi, H.; Hornicek, F.J.; Kan, Q.; Duan, Z. Novel mechanisms and approaches to overcome multidrug resistance in the treatment of ovarian cancer. *Biochim. Biophys. Acta* **2016**, *1866*, 266–275. [CrossRef] [PubMed]
92. Trnski, D.; Gregoric, M.; Levanat, S.; Ozretic, P.; Rincic, N.; Vidakovic, T.M.; Kalafatic, D.; Maurac, I.; Oreskovic, S.; Sabol, M.; et al. Regulation of Survivin Isoform Expression by GLI Proteins in Ovarian Cancer. *Cells* **2019**, *8*, 128. [CrossRef] [PubMed]

93. Togashi, K.; Okada, M.; Yamamoto, M.; Suzuki, S.; Sanomachi, T.; Seino, S.; Yamashita, H.; Kitanaka, C. A Small-molecule Kinase Inhibitor, CEP-1347, Inhibits Survivin Expression and Sensitizes Ovarian Cancer Stem Cells to Paclitaxel. *Anticancer Res.* **2018**, *38*, 4535–4542. [CrossRef]
94. Levy, A.; Alhazzani, K.; Dondapati, P.; Alaseem, A.; Cheema, K.; Thallapureddy, K.; Kaur, P.; Alobid, S.; Rathinavelu, A. Focal Adhesion Kinase in Ovarian Cancer: A Potential Therapeutic Target for Platinum and Taxane-Resistant Tumors. *Curr. Cancer Drug Targets* **2019**, *19*, 179–188. [CrossRef] [PubMed]
95. Yang, Y.; Li, S.; Sun, Y.; Zhang, D.; Zhao, Z.; Liu, L. Reversing platinum resistance in ovarian cancer multicellular spheroids by targeting Bcl-2. *Onco Targets Ther.* **2019**, *12*, 897–906. [CrossRef] [PubMed]
96. Integrated genomic analyses of ovarian carcinoma. *Nature* **2011**, *474*, 609–615. [CrossRef] [PubMed]
97. Domcke, S.; Sinha, R.; Levine, D.A.; Sander, C.; Schultz, N. Evaluating cell lines as tumour models by comparison of genomic profiles. *Nat. Commun.* **2013**, *4*, 2126. [CrossRef]
98. Cho, K.R.; Shih, I. Ovarian cancer. *Annu. Rev. Pathol.* **2009**, *4*, 287–313. [CrossRef]
99. Clevers, H. Modeling Development and Disease with Organoids. *Cell* **2016**, *165*, 1586–1597. [CrossRef]
100. Jabs, J.; Zickgraf, F.M.; Park, J.; Wagner, S.; Jiang, X.; Jechow, K.; Kleinheinz, K.; Toprak, U.H.; Schneider, M.A.; Meister, M.; et al. Screening drug effects in patient-derived cancer cells links organoid responses to genome alterations. *Mol. Syst. Biol.* **2017**, *13*, 955. [CrossRef]
101. Kopper, O.; de Witte, C.J.; Lohmussaar, K.; Valle-Inc, J.E.; Hami, N.; Kester, L.; Balgobind, A.V.; Korving, J.; Proost, N.; Begthel, H.; et al. An organoid platform for ovarian cancer captures intra- and interpatient heterogeneity. *Nat. Med.* **2019**, *25*, 838–849. [CrossRef] [PubMed]
102. Maru, Y.; Tanaka, N.; Itami, M.; Hippo, Y. Efficient use of patient-derived organoids as a preclinical model for gynecologic tumors. *Gynecol. Oncol.* **2019**, *154*, 189–198. [CrossRef] [PubMed]
103. McCloskey, C.W.; Rodriguez, G.M.; Galpin, K.J.C.; Vanderhyden, B.C. Ovarian Cancer Immunotherapy: Preclinical Models and Emerging Therapeutics. *Cancers* **2018**, *10*, 244. [CrossRef] [PubMed]
104. Bobbs, A.S.; Cole, J.M.; Cowden Dahl, K.D. Emerging and Evolving Ovarian Cancer Animal Models. *Cancer Growth Metastasis* **2015**, *8*, 29–36. [CrossRef]
105. Scott, C.L.; Becker, M.A.; Haluska, P.; Samimi, G. Patient-derived xenograft models to improve targeted therapy in epithelial ovarian cancer treatment. *Front. Oncol.* **2013**, *3*, 295. [CrossRef]
106. Roby, K.F.; Taylor, C.C.; Sweetwood, J.P.; Cheng, Y.; Pace, J.L.; Tawfik, O.; Persons, D.L.; Smith, P.G.; Terranova, P.F. Development of a syngeneic mouse model for events related to ovarian cancer. *Carcinogenesis* **2000**, *21*, 585–591. [CrossRef]
107. McCloskey, C.W.; Goldberg, R.L.; Carter, L.E.; Gamwell, L.F.; Al-Hujaily, E.M.; Collins, O.; Macdonald, E.A.; Garson, K.; Daneshmand, M.; Carmona, E.; et al. A new spontaneously transformed syngeneic model of high-grade serous ovarian cancer with a tumor-initiating cell population. *Front. Oncol.* **2014**, *4*, 53. [CrossRef]
108. Barba, A.A.; Cascone, S.; Caccavo, D.; Lamberti, G.; Chiarappa, G.; Abrami, M.; Grassi, G.; Grassi, M.; Tomaiuolo, G.; Guido, S.; et al. Engineering approaches in siRNA delivery. *Int. J. Pharm.* **2017**, *525*, 343–358. [CrossRef]
109. Chiarappa, G.; Abrami, M.; Dapas, B.; Farra, R.; Trebez, F.; Musiani, F.; Musiani, F.; Grassi, G.; Grassi, M. Mathematical Modeling of Drug Release from Natural Polysaccharides Based Matrices. *Nat. Prod. Commun.* **2017**, *12*, 873–880. [CrossRef]
110. Bartlett, D.W.; Davis, M.E. Insights into the kinetics of siRNA-mediated gene silencing from live-cell and live-animal bioluminescent imaging. *Nucleic Acids Res.* **2006**, *34*, 322–333. [CrossRef]
111. Polyak, D.; Krivitsky, A.; Scomparin, A.; Eliyahu, S.; Kalinski, H.; Avkin-Nachum, S.; Satchi-Fainaro, R. Systemic delivery of siRNA by aminated poly(α)glutamate for the treatment of solid tumors. *J. Control. Release* **2017**, *257*, 132–143. [CrossRef] [PubMed]
112. Risnayanti, C.; Jang, Y.S.; Lee, J.; Ahn, H.J. PLGA nanoparticles co-delivering MDR1 and BCL2 siRNA for overcoming resistance of paclitaxel and cisplatin in recurrent or advanced ovarian cancer. *Sci. Rep.* **2018**, *8*, 7498. [CrossRef] [PubMed]
113. Hazekawa, M.; Nishinakagawa, T.; Kawakubo-Yasukochi, T.; Nakashima, M. Glypican-3 gene silencing for ovarian cancer using siRNA-PLGA hybrid micelles in a murine peritoneal dissemination model. *J. Pharmacol. Sci.* **2019**, *139*, 231–239. [CrossRef] [PubMed]
114. Stadlmann, S.; Gueth, U.; Baumhoer, D.; Moch, H.; Terracciano, L.; Singer, G. Glypican-3 expression in primary and recurrent ovarian carcinomas. *Int. J. Gynecol. Pathol.* **2007**, *26*, 341–344. [CrossRef] [PubMed]

115. Lou, B.; Beztsinna, N.; Mountrichas, G.; van den Dikkenberg, J.B.; Pispas, S.; Hennink, W.E. Small nanosized poly(vinyl benzyl trimethylammonium chloride) based polyplexes for siRNA delivery. *Int. J. Pharm.* **2017**, *525*, 388–396. [CrossRef] [PubMed]
116. Leung, C.S.; Yeung, T.L.; Yip, K.P.; Wong, K.K.; Ho, S.Y.; Mangala, L.S.; Sood, A.K.; Lopez-Berestein, G.; Sheng, J.; Wong, S.T.; et al. Cancer-associated fibroblasts regulate endothelial adhesion protein LPP to promote ovarian cancer chemoresistance. *J. Clin. Investig.* **2018**, *128*, 589–606. [CrossRef]
117. Kim, G.H.; Won, J.E.; Byeon, Y.; Kim, M.G.; Wi, T.I.; Lee, J.M.; Park, Y.Y.; Lee, J.W.; Kang, T.H.; Jung, I.D.; et al. Selective delivery of PLXDC1 small interfering RNA to endothelial cells for anti-angiogenesis tumor therapy using CD44-targeted chitosan nanoparticles for epithelial ovarian cancer. *Drug Deliv.* **2018**, *25*, 1394–1402. [CrossRef]
118. Byeon, Y.; Lee, J.W.; Choi, W.S.; Won, J.E.; Kim, G.H.; Kim, M.G.; Wi, T.I.; Lee, J.M.; Kang, T.H.; Jung, I.D.; et al. CD44-Targeting PLGA Nanoparticles Incorporating Paclitaxel and FAK siRNA Overcome Chemoresistance in Epithelial Ovarian Cancer. *Cancer Res.* **2018**, *78*, 6247–6256.
119. Hong, S.S.; Zhang, M.X.; Zhang, M.; Yu, Y.; Chen, J.; Zhang, X.Y.; Xu, C.J. Follicle-stimulating hormone peptide-conjugated nanoparticles for targeted shRNA delivery lead to effective gro-alpha silencing and antitumor activity against ovarian cancer. *Drug Deliv.* **2018**, *25*, 576–584. [CrossRef]
120. Zhang, X.Y.; Chen, J.; Zheng, Y.F.; Gao, X.L.; Kang, Y.; Liu, J.C.; Cheng, M.J.; Sun, H.; Xu, C.J. Follicle-stimulating hormone peptide can facilitate paclitaxel nanoparticles to target ovarian carcinoma in vivo. *Cancer Res.* **2009**, *69*, 6506–6514. [CrossRef]
121. Yang, G.; Rosen, D.G.; Zhang, Z.; Bast, R.C., Jr.; Mills, G.B.; Colacino, J.A.; Mercado-Uribe, I.; Liu, J. The chemokine growth-regulated oncogene 1 (Gro-1) links RAS signaling to the senescence of stromal fibroblasts and ovarian tumorigenesis. *Proc. Natl. Acad. Sci. USA* **2006**, *103*, 16472–16477. [CrossRef]
122. Jones, S.K.; Douglas, K.; Shields, A.F.; Merkel, O.M. Correlating quantitative tumor accumulation and gene knockdown using SPECT/CT and bioluminescence imaging within an orthotopic ovarian cancer model. *Biomaterials* **2018**, *178*, 183–192. [CrossRef] [PubMed]
123. Lee, J.; Ahn, H.J. PEGylated DC-Chol/DOPE cationic liposomes containing KSP siRNA as a systemic siRNA delivery Carrier for ovarian cancer therapy. *Biochem. Biophys. Res. Commun.* **2018**, *503*, 1716–1722. [CrossRef] [PubMed]
124. She, Z.Y.; Yang, W.X. Molecular mechanisms of kinesin-14 motors in spindle assembly and chromosome segregation. *J. Cell Sci.* **2017**, *130*, 2097–2110. [CrossRef] [PubMed]
125. Fujiwaki, R.; Hata, K.; Nakayama, K.; Fukumoto, M.; Miyazaki, K. Thymidylate synthase expression in epithelial ovarian cancer: Relationship with thymidine phosphorylase expression and prognosis. *Oncology* **2000**, *59*, 152–157. [CrossRef] [PubMed]
126. Iizuka, K.; Jin, C.; Eshima, K.; Hong, M.H.; Eshima, K.; Fukushima, M. Anticancer activity of the intraperitoneal-delivered DFP-10825, the cationic liposome-conjugated RNAi molecule targeting thymidylate synthase, on peritoneal disseminated ovarian cancer xenograft model. *Drug Des. Dev. Ther.* **2018**, *12*, 673–683. [CrossRef] [PubMed]
127. Mendes, L.P.; Sarisozen, C.; Luther, E.; Pan, J.; Torchilin, V.P. Surface-engineered polyethyleneimine-modified liposomes as novel carrier of siRNA and chemotherapeutics for combination treatment of drug-resistant cancers. *Drug Deliv.* **2019**, *26*, 443–458. [CrossRef]
128. Blanco, A.; Giger-Pabst, U.; Solass, W.; Zieren, J.; Reymond, M.A. Renal and hepatic toxicities after pressurized intraperitoneal aerosol chemotherapy (PIPAC). *Ann. Surg. Oncol.* **2013**, *20*, 2311–2316. [CrossRef]
129. Minnaert, A.K.; Dakwar, G.R.; Benito, J.M.; Garcia Fernandez, J.M.; Ceelen, W.; De Smedt, S.C.; Remaut, K. High-Pressure Nebulization as Application Route for the Peritoneal Administration of siRNA Complexes. *Macromol. Biosci.* **2017**, *17*, 1700024. [CrossRef]
130. Dykxhoorn, D.M.; Palliser, D.; Lieberman, J. The silent treatment: siRNAs as small molecule drugs. *Gene Ther.* **2006**, *13*, 541–552. [CrossRef]
131. Cejka, D.; Losert, D.; Wacheck, V. Short interfering RNA (siRNA): Tool or therapeutic? *Clin. Sci.* **2006**, *110*, 47–58. [CrossRef] [PubMed]
132. Bochicchio, S.; Dapas, B.; Russo, I.; Ciacci, C.; Piazza, O.; De Smedt, S.; Pottie, E.; Barba, A.A.; Grassi, G. In vitro and ex vivo delivery of tailored siRNA-nanoliposomes for E2F1 silencing as a potential therapy for colorectal cancer. *Int. J. Pharm.* **2017**, *2*, 377–387. [CrossRef] [PubMed]

133. TKM 080301 for Primary or Secondary Liver Cancer. Available online: https://clinicaltrials.gov/ct2/show/record/NCT01437007?term=siRNA&cond=Ovarian+Cancer&rank=1 (accessed on 3 August 2018).
134. D'Apolito, R.; Tomaiuolo, G.; Taraballi, F.; Minardi, S.; Kirui, D.; Liu, X.; Cevenini, A.; Palomba, R.; Ferrari, M.; Salvatore, F.; et al. Red blood cells affect the margination of microparticles in synthetic microcapillaries and intravital microcirculation as a function of their size and shape. *J. Control* **2015**, *217*, 263–272. [CrossRef] [PubMed]

pharmaceutics

Review

Aptamers as Delivery Agents of siRNA and Chimeric Formulations for the Treatment of Cancer

Ana Paula Dinis Ano Bom [1], Patrícia Cristina da Costa Neves [1], Carlos Eduardo Bonacossa de Almeida [2], Dilson Silva [1] and Sotiris Missailidis [1,*]

1 Instituto de Tecnologia em Imunobiológicos (Bio-Manguinhos), Fundação Oswaldo Cruz. Av. Brasil, 4365-Manguinhos, Rio de Janeiro/RJ CEP 21040-900, Brazil; adinis@bio.fiocruz.br (A.P.D.A.B.); Pcristina@bio.fiocruz.br (P.C.d.C.N.); dilson.silva@bio.fiocruz.br (D.S.)

2 Laboratório de Radiobiologia, Divisão de Física Médica, Instituto de Radioproteção e Dosimetria, Comissão Nacional de Energia Nuclear. Av. Salvador Allende S/N., Rio de Janeiro/RJ CEP 22783-127, Brazil; ce.bonacossa@gmail.com

* Correspondence: sotiris.missailidis@bio.fiocruz.br

Received: 30 September 2019; Accepted: 30 October 2019; Published: 16 December 2019

Abstract: Both aptamers and siRNA technologies have now reached maturity, and both have been validated with a product in the market. However, although pegaptanib reached the market some time ago, there has been a slow process for new aptamers to follow. Today, some 40 aptamers are in the market, but many in combination with siRNAs, in the form of specific delivery agents. This combination offers the potential to explore the high affinity and specificity of aptamers, the silencing power of siRNA, and, at times, the cytotoxicity of chemotherapy molecules in powerful combinations that promise to delivery new and potent therapies. In this review, we report new developments in the field, following up from our previous work, more specifically on the use of aptamers as delivery agents of siRNA in nanoparticle formulations, alone or in combination with chemotherapy, for the treatment of cancer.

Keywords: siRNA; aptamers; cancer; nanoparticles

1. Introduction

Small interfering RNA siRNA suppress expression of genes by targeting the mRNA expression. Targeted delivery of siRNA to specific cells is highly desirable for safe and efficient RNAi-based therapeutics [1]. However, the half-life of nucleic acids in the bloodstream is short due to the degradation by endo or exonucleases and rapid clearance [2]. One strategy to solve this challenge is developing siRNA delivery systems. Nanoparticles can be defined as particles less than 100 nm in diameter, these systems can be composed by different materials and are employed according to their purpose [3,4]. For this area, the most widely used systems are polymeric particles, nanoemulsions, nanocrystals, solid lipid nanoparticles, and liposomes [5]. The organic particles used for drug delivery application are micelles, liposomes, polymers, dendrimers, and nanogels. They have versatile surface building blocks for efficient endocytosis and loading [6]. There are numerous advantages to using nanoparticles: (I) Increased bioavailability, (II) dose proportionality, (III) decreased toxicity, (IV) smaller dosage form, (V) stability of drugs dosage forms, and (VI) increased active agent surface area resulting in a faster dissolution [7]. Ideally, nanoparticles should be stable in circulation to protect and deliver their therapeutic load (drug) into recipient tissue; have good penetration and retention in the target tissue so that drug release occurs within the therapeutic window; and ultimately be organically excreted to avoid long term accumulation toxicity [8]. Approaches to drug targeting and delivery may be facilitated by the enhanced permeability and retention (EPR) effect. This effect occurs due to the large endothelial tissue fenestrations which are characteristic of the rapid growth of tumor blood vessels. Therefore, the nanoparticles passively diffuse through the microenvironment targeting the tumor tissues [8].

Although nanocarrier technology has improved, its lack of target specificity limits its widespread use, to overcome this issue and address the lack of specificity is the generation of functionalized nanoparticles, i.e., second generation nanoparticles [8]. Nanoparticle surface functionalization occurs through the fixation of a ligand that interacts with specific tissue-specific receptors, to optimize the administration of the target, selectively transporting it to the binding site [9]. One of the advantages of taking drugs directly to specific tissues is the ability to use relatively more toxic and efficient drugs with less risk of collateral damage to other body tissues. In the case of cancer, drugs could be targeted at tumors, avoiding the systemic side effects of traditional therapies. The functionalization includes surface conjugation of chemicals or bio molecules, like folic acid, biotin molecules, peptides, antibodies, aptamers, short, single stranded RNA or DNA oligonucleotides, proteins, and oligosaccharides, to enhance the properties and hit the target with high precision [10]. In order to provide targetability, aptamers have been widely used due to (I) their capacity of binding to target proteins with a high affinity and specificity, (II) having already been shown to have antibody-like characteristics, and (III) the fact that they are relatively smaller and less immunogenic. All of these useful properties make aptamers attractive in therapeutic and diagnostic fields [11].

2. Aptamer in the Delivery of Therapeutic Nanoparticles Containing siRNA, shRNA, and miRNA

The origin of siRNA nanoparticles targeted delivery through aptamers dates to 1998. Guo et al. treated T cells with an RNA nanoparticle consisting of a dimer of the packaging RNA (pRNA) derived from the DNA-packaging motor of bacteriophage phi29 loaded with a siRNA for survivin mRNA and conjugated with a CD4 specific aptamer [12]. Hu et al. used the same platform to create a nanoparticle containing an siRNA for ICAM 1 conjugated with aptamer FB4 directed against the mouse transferrin receptor. The in vitro results showed a decreased of ICAM-1 expression and blocked the adhesion of monocytes [13].

From this date, many types of platforms and supports were developed so that siRNA could be targeted through aptamers. To target cells using aptamers, Afonin et al. designed multifunctional siRNA nanoparticles aiming at the silence of multiple HIV-1 genes [14]. To show the feasibility of siRNA delivery using nanorings, Li et al. developed a nanorings construct functionalized with J18 RNA aptamers specific for the human epidermal growth factor receptor (EGFR) [15]. The nanoring design can achieve cell-targeting properties, but the internalization and functional effects of siRNAs bound to aptamer-nanorings were not studied [16]. Zhao et al. developed a nanoparticle to treat anaplastic large cell lymphoma (ALCL). The nanocarrier incorporated both anaplastic lymphoma kinase (ALK) siRNA and a CD30 aptamer onto nano-polyethyleneimine-citrate leading to growth arrest and apoptosis in vitro [17]. Powell et al. tested the possibility of conjugating an aptamer A6 which can bind to human epidermal growth factor receptor 2 (HER-2) receptors on breast cancer cells to P-gp siRNA siRNA-containing nanoparticles. The knockdown of P-gp was increased significantly, decreasing a resistance to chemotherapeutics in the cells [18]. To overcome multi-drug resistance (MDR), an MD- specific aptamer-conjugated grapefruit-derived nanovectors (GNVs) was used for co-delivery of siRNA and doxorubicin (DOX), resulting in a potent anti-tumor activity [19].

Stable nucleic acid lipid particles (SNALPs) also used to deliver siRNA through aptamers. In this study, the authors coupled the transferrin receptor aptamer in SNALP loaded with siRNA. The aptamer-targeting enhanced siRNA uptake and target gene knockdown in cells [20]. Another aptamer–liposome–siRNA delivery system was developed by Alshaer et al. In this model, siRNA complexed with protamine and aptamer anti-CD44, a cell surface biomarker overexpressed in many tumors, was used as targeting delivery for the first time [21]. They confirmed the silencing disease-related genes in tumors [18].

To target prostate cancer cells, Kim et al. used shRNAs against the gene BCL-xL and the PSMA (prostate-specific membrane antigens) aptamer (A10–3) conjugated to a polyethyleneimine (PEI)-PEG construction. In order to amplify the therapeutic response and use different drugs in synergy, the authors also included DOX in aptamer-functionalized nanoparticle, and this approach led to the death of the cancer cells, demonstrating the effectiveness of the combination therapy [22]. Likewise, the PSMA aptamer was used to target delivery of miRNA (miR-15a and miR-16–1) identified as tumor suppressor genes in prostate cancer. The nanoparticle based on polyamidoamine and PEG allowed selective killing of prostate cancer cells [23]. Xu et al. used a multifunctional envelope-type nanoparticle platform constituted of two polymers that co-incorporate siRNA. For the nanoparticle functionalization was used *S,S*-2-(3-(5-amino-1-carboxypentyl)ureido)pentanedioic acid (ACUPA) ligand that can specifically bind to PSMA. This study reached an efficient gene silencing in prostate cancer tumor cells in vivo [24].

Lv et al. used Polyamidoamine (PAMAM) dendrimers functionalized with EGFR aptamers to co-deliver shRNA to survivin and erlotinib. Survivin shRNA is used to activate apoptosis while erlotinib is a tyrosine kinase inhibitor. This therapeutic approach in association with Chloroquine acted synergistically to reverse erlotinib resistance in EGFR mutation-positive of non-small-cell-lung cancer (NSCLC) [25].

Another strategy is use of siRNA-aptamer chimera to compose the nanoparticle, and the siRNA-aptamer chimera has emerged as a promising approach for efficient delivery of siRNA to specific cell types, owing to its low immunogenicity, ease of chemical synthesis and modification, and the outstanding targeting specificity of the aptamer [26]. However, in some cases, endosome escape has been reported. To avoid this problem, it has been used siRNA-aptamer chimeras in nanocarriers. Bagalkot and Gao developed a new technology for linking siRNA-aptamer chimeras to carrier nanoparticles [27]. In this new technology, the siRNA block should become the anchor point for interaction with cationic nanoparticles for reduced enzymatic degradation and nonspecific interaction with cells and tissues, whereas the aptamer block should stay on the outside with minimized interaction with the nanoparticle surface. This approach opened new opportunities in targeted delivery based on siRNA-aptamer chimeras.

In recent study, to evaluate cell viability and genes expression in metastatic breast cancer cells, Jafari et al. designed chitosan nanoparticles for co-delivery of Docetaxel and insulin-like growth factor receptor 1 (IGF-1R) siRNA. Docetaxel is a cytotoxic anti-cancer drug approved for the treatment of metastatic breast cancer. IGF-1R signaling is important in tumor growth, development, and its metastasis. The MUC1 aptamer was conjugated to a chitosan based nanocarrier. MUC1-aptamers have been used to targeted drug delivery to several types of cancers, including breast cancer cells [28]. Another study in the cancer field shows an efficiency of EGFR aptamer-conjugated liposome loaded with special AT-rich sequence binding protein 1 (SATB1) siRNA. The genome organizer (SATB1) expression decreased in vitro and in vivo studies with a consequent decrease in choriocarcinoma [29]. A novel targeted delivery platform to mediated gene-silencing in cancer cells was described by Ayatollahi et al.; the 10-bromodecanoic acid (10C) and 10C-PEG was loaded with shRNA plasmid for specific knockdown of BCL-xL protein and was conjugated to AS1411 aptamer for targeting to nucleolin on cancer cells [30]. A synergistic cancer cell death was achieved using a (AS1411) nucleolin aptamer-carbon nanotubes (CNTs) as biological carriers containing BCL-xL-specific shRNA and DOX [31].

Recently, a new types of siRNA delivery by aptamers have been demonstrated. Xu et al. used the three-way-junction (3WJ) RNA to couple epithelial cell adhesion molecule (EpCAM) aptamers aiming to deliver Delta-5-Desaturase (D5D) siRNA. This enzyme acts on dihomo-γ-linolenic acid DGLA, preventing the production of 8-hydroxyloctanoic acid (8-HOA), which is important for inhibiting histone deacetylation in cancer [32]. Li et al. combined pRNA dimers with siRNA and aptamer, and they demonstrated that pRNA, which is a component of a phi29 bacteriophage, can be useful for efficient siRNA delivery. The pRNA-siRNA-aptamer was encapsulated in folate and PEG functionalized chitosan nanoparticles. In that study, the aptamer used was FB4 for a transferrin domain, and the siRNA was c-myc [33]. Wu et al. demonstrated that functionalized cationic nanobubbles decorated with PSMA aptamers are able to carry and deliver FoxM1 siRNA to prostate cancer positive cells and xenographic tumors [34]. One recent study developed a programmable DNA nanostructure that is considered a potential system for drug delivery for non-small cell lung cancer. DNA Prism nanostructure conjugated with MUC-1 aptamers to deliver Rab26 siRNA presented a great cellular uptake efficiency and anti-tumoral activity [35].

In the field of theranostics, this approach can be useful. Theranostics is defined as nanomedicine that can deliver therapeutics to an intended region, diagnose disease, and follow the progression of disease [36]. Recently, interdisciplinary research in this field, which applies theranostics to diagnose and treat cancer, has explosively progressed. Kim et al. 2017 developed a theranostic liposomal system for carrying diagnostic Quantum Dots (QD) and therapeutic siRNA that were coupled to aptamer molecules against EGFR. The theranostic liposomes were evaluated in terms of cancer-targeted siRNA delivery and QD imaging in vitro and in vivo [11]. The anti-EGFR aptamer was used recently as a theranostic delivery system. The authors used two siRNA to provide a combinatorial anti-cancer treatment using BCL-2 siRNA, which interfere with tumor cell proliferation, and siRNA PKC-ι for inhibition of tumor cell migration. Moreover, these vehicles were able to simultaneously delivered QD provided fluorescence signals in internal organs and tumors. This study shows an efficiently delivered targeting of anti-cancer therapeutic siRNAs, as well as fluorescent QDs to tumor tissues, enabling the use of theranostic delivery systems in the treatment and fluorescence imaging of cancers [37].

In this review, we compiled several studies with different types of nanoparticles; despite different types of nanoparticles, aptamer functionalized nanoparticle loading therapeutic siRNA design can generally be done in two ways: (1) incorporated in the nanoparticle itself or (2) conjugated to the nanoparticle surface.

Over the past few years, the investment in therapeutic oligonucleotides (e.g., aptamers, siRNAs, and antisense technology) is constantly increasing. It is projected that clinical testing of aptamer-targeted drugs will materialize over the next 2–5 years However, there are several challenges in the use of aptamers as nanoparticle drivers, including the aptamer structural design, in vivo stability, immunogenicity, cellular uptake, and endosomal escape, that should be considered [16]. There is the possibility of target specificity, RNAi efficacy, and stability of siRNA-aptamers chimeras being affected by chemical modifications. Therefore, it is crucial to find an optimal alternative for each siRNA aptamers chimera to maximize its therapeutic efficacy [38].

In order to increase extracellular stability, incorporation of protective groups, such as thiol-phosphate, 20-fluoro, 20-amino, and 2′-*O*-methyl, in phosphate or sugar of nucleotides improves nuclease resistance of aptamer-siRNA chimeras [39,40]. Another possibility for improving stability is the aptamers PEG conjugation. Furthermore, it has been shown that PEG formulation reduces immune responses in vivo [41]. The thermal stability of siRNA is another parameter to be considered in gene silencing activity when it is conjugated to aptamers [38]. In this context, the computational modeling has been used to guide RNA aptamer truncations, maintaining their functionality, and preventing and minimizing therapy challenges using siRNA.

Several approaches have been developed to facilitate siRNA uptake and nanoparticle entry into endosomes. One possibility is to perform the conjugation of chimeras to short cell penetration cationic peptides (CPPs) that are important for the efficiency in intracellular transport of a variety of macromolecules [42]. Another possibility is the conjugation of the chimeras with peptides from the transactivator of transcription (TAT)region; Diao et al. demonstrated that the survivin siRNA was delivering specifically and efficiently and suppressed tumor growth in prostate cancer in vivo [43]. Inclusion of endosome-disrupting molecules, including PEI and poly (histidine), as well as fusogenic lipids and pH-sensitive lipids in aptamer-siRNA chimeras, have been useful for enhancing siRNA endosomal evasion [24,44,45]. Table 1 summarizes carriers aptamers cited in this section.

Table 1. Aptamers for delivery of siRNA.

siRNA/miRNA/shRNA Target Gene	Carrier Aptamer	Efficacy Tests	Ref.
Survivin	CD4	In vitro knock down antiapoptosis factor survivin	[12]
ICAM 1	FB4 (bind to transferrin receptors)	In vitro results showed a decreased of ICAM-1 expression and blocked the adhesion of monocytes	[13]
Human immunodeficiency virus (HIV)-1	epidermal growth factor receptor (EGFR)	In vitro and in vivo gene silencing	[14]
Anaplastic lymphoma kinase (ALK)	CD30	In vitro results of growth arrest and apoptosis	[17]
P-gp	A6 (bind to human epidermal growth factor receptor 2 (HER-2) receptors)	In vitro results about decreasing of resistance to chemotherapeutics	[18]
P-gp	LA1 (specific to multi-drug resistance (MDR))	In vitro and in vivo anti-tumor activity	[19]
GFP	Transferrin	In vitro enhanced siRNA uptake and target gene knockdown	[20]
Luciferase	CD44	In vitro and in vivo results of targeted gene silencing in CD44-positive breast cancer	[21]
P-gp	HER2	In vitro confirmation of silencing disease-related genes in tumors	[18]
BCL-xL	Prostate-specific membrane antigen (PSMA)	In vitro anti-tumor activity	[22]
Genes in prostate cancer	PSMA	In vitro anticancer effect	[23]
Prohibitin 1	PSMA	In vivo gene silencing in prostate cancer tumor cells	[24]
Survivin	EGFR	In vitro and in vivo antitumor effects	[25]
GFP	PSMA	In vitro GFP silencing/Fluorescence imaging (quantum dots)	[27]
Insulin-like growth factor receptor 1 (IGF-1R)	MUC-1	In vitro augment the targeting of pathways involved in tumorigenesis and metastasis	[28]
Special AT-rich sequence binding protein 1 (SATB1)	EGFR	In vitro and in vivo studies of gene expression decreased	[29]
BCL-xL	Nucleolin	In vitro gene silencing and apoptosis	[30]
BCL-xL	Nucelolin	In vitro gene silencing and tumoricidal efficacy	[31]
D5D	EpCAM	In vitro and in vivo inhibition of D5D expression	[32]
c-myc	Transferrin	In vitro and in vivo anti-tumor activity	[33]
FoxM1	PSMA	In vitro and in vivo anti-tumor activity in prostate cancer cells and xenografts in mice	[34]
Rab26	MUC-1	In vitro anti-tumor activity	[35]
bcl-2	EGFR	In vitro interference of tumor cell proliferation	[11]
bcl-2 and PKC-ι	EGFR	In vivo delivery and therapeutic efficacy	[37]

3. Cancer Immunotherapy Using Aptamers-siRNA

Immunotherapy has emerged as the best clinical alternative for cancer treatment during the last years. It is mostly based in monoclonal antibodies targeting immunological checkpoint pathways, as well as in genetically-engineered cells that directly attack the tumor, like CAR-T cells. Despite the great success of immunotherapy, it works for 20–40% of the patients, depending on tumor type and disease progression stage [46]. So, other classes of molecules and treatments, such as those based in RNA and aptamers, appeared as good alternatives for cancer immunological treatment.

RNA-based immunotherapy is, in concept, more advantageous than those based in recombinant protein expression platforms. It has lower cost, lower immunogenicity, and better penetration in the tumor tissue, which makes it worth developing. In addition, in the last years, a number of pre-clinical and clinical data proved its efficacy, using both aptamer alone and RNAi conjugates.

Antagonistic and agonistic aptamers, acting similar to monoclonal antibodies, were developed to bind to immunomodulating molecules, such as CTLA-4, PD-1, IL-10 R, and LAG-3. The tetrameric antagonistic aptamer selected for CTLA-4, the first one to be developed, was very effective during melanoma murine model studies in vivo [47]. In the same way, multimeric structures of aptamers were shown to have a better performance for agonistic ligands, bringing them together during the activation. It is the case of OX-40 agonistic aptamer, Pratico et al. constructed a multimerized anti OX-40 aptamer using biotinylation and demonstrated a much better biologic effect than the single aptamer, with increased IFN-© production and proliferation, using peripheral blood mononuclear cells as an in vitro model of T cells' activation [48].

Some receptors of the immune system have an interesting characteristic: They get internalized after T cell activation as a feedback mechanism of controlling over-stimulation. Taking advantage of this feature, some aptamers have been developed as delivery systems to siRNA. The first one was 4-1BB aptamer used to deliver siRNA to T cells, aiming to disrupt mTOR pathway and attenuate IL-2R signaling in $CD8^+$ T cells, which, in turn, leads to $CD8^+$ T cells persistence inside the tumor site and its elimination [49]. This same aptamer was used to address siRNA for Smad-4, a protein from TGF-β signaling cascade. TGF-β is a key mediator of immune system suppression during progression. This approach was successful in suppressing the negative effect of TGF-βon infiltrating $CD8^+$ cells elicited by vaccination with tumor antigens [50].

Cluster of Differentiation (CD) aptamers also have been shown as good alternatives for RNAi delivery. CD molecules are used traditionally for identification of subpopulations of immune cells during flow cytometry studies and, as well as 4-1BB, are frequently internalized after cells activation. For example, using CD4 aptamers complexed with short hairpin RNAs (shRNA), it was possible to silence ROR©T pathway, the key for T_H17 cells development, opening a possibility of treatment for a number of conditions, mainly related to autoimmune diseases [51]. In the cancer field, Soldevilla et al. used the previously developed agonistic CD40 aptamer to construct shRNA-aptamer chimeras to inhibit SMG1, a kinase that is essential for nonsense-mediated mRNA decay (NMD) initiation. Mice treated with the CD40 aptamer-shRNA chimera showed higher tumor infiltration of lymphocytes [52,53]. Table 2 summarizes aptamers-siRNAs used for immunotherapy cited in this section.

Table 2. Aptamers and siRNA for immunotherapy.

siRNA Target Gene	Carrier Aptamer	Efficacy Tests	Ref.
IL-2 Rα (CD25)	41-BB	In vitro using CHO cells and murine $CD8^+$ cells and in vivo using C57BL/6 mice adoptively transferred with OT-I cells and breast tumor 4T1 model	[49]
Smad-4	41-BB	In vitro using murine $CD8^+$ cells and in vivo using breast tumor 4T1 model	[50]
RORγT	CD4	In vitro using CD4+ Karpas 299 cells and primary human CD4+ T cells	[51]
SMG-1	CD40 agonist	In vitro using human B-lymphocytes and in vivo using B-cell lymphoma-bearing mice	[52]

4. Tumor-Targeted Aptamers Complexed Directly to siRNAs and Other Agents for Cancer Therapy

Thanks to the unique features of aptamers, such as stability and tumor penetration, its use as tools for delivery of therapeutic agents, including siRNA, has gained great attention from researchers during the last few years.

The idea of a tailored drug for the treatment of a diseased cell avoiding affecting the surrounding healthy cells is not recent, as recalled by Kruspe and Giangrande [54]. It was initially proposed by Paul Ehrlich about one century ago, when he first defined the "magic bullet" and its inherent concept of cell-surface receptor [55]. The route scientists have threaded to reach the current status has grooved a persistent tentative of diverse therapy strategies, some among them based on the conjugation of aptamer with siRNA.

Aptamers, however, due to their anionic characteristics, do not easily transit across the cell lipid bilayer, requiring adequate strategies to overcome the action of those charges, using diverse adjuvants in different structures based on lipids/liposomes, nanoparticles (as seen above), cationic polymers, and peptides.

Aptamer-siRNA conjugates, or AsiCs as they are frequently referred to, comprise an ever-increasing number of structures since 2006 when the delivery of siRNA by an aptamer was described simultaneously by two groups working with prostate membrane antigen (PSMA).

The first aptamer to be conjugated to an siRNA was to treat prostate cancer. The strategy consisted in addressing the tumor trough PSMA, a protein overexpressed in the surface of prostate cancer cells and tumor vascular endothelium, with a PSMA aptamer linked to PLK-1 and BCL-2 siRNAs. The siRNAs were responsible for disrupting the protection against apoptosis of the tumor cells. The tool was successful in eliminating a xenograft prostate cancer in a murine model [56]. After that, many other groups developed similar strategies. Interesting results were obtained for glioblastoma, the most frequent and aggressive primary brain tumor in adults, with a very poor prognosis. Expression and activation of the signal transducer and activator of transcription-3 (STAT3) has been reported as a key regulator of this kind of tumor, so a chimera aptamer antagonistic for PDGFRbr (a receptor overexpressed in glioblastoma, as well as other tumor types) carrying siRNA for STAT 3 was designed. This chimera was able to reduce the anti-apoptotic factors Poly (ADP-ribose) polymerase PARP and BCL-xL, inducing cell death in two glioblastoma lineages in vitro. Moreover, Gint4.T-STAT3 chimera treatment induced significant reduction of tumor growth rate in comparison to control group, as well reduced pro-tumoral factors in a xenograft mouse model of glioblastoma [57]. Similar results were obtained for HER+ breast cancer using a HER-2 sense and antisense bivalent aptamer conjugated to EGFR siRNA. It was able to be delivered to breast cancer cells overexpressing HER-2 and the

EGFR gene was efficiently downregulated in vitro, whilst a xenografted tumor growth was supressed in vivo [58]. Another study from the same group showed the performance of a bivalent HER-2-HER-3 aptamer with an EGFR siRNA between them. It was even better in reducing the expression of all three receptors, in inducing cell cycle arrest and apoptosis in vitro and inhibiting tumor growth with a little of target effect in vivo [59]. Aptamer-siRNA chimeras were also used to overcome multidrug resistance by Jeong et al. They engineered a mucin multiaptamer (E18 units) conjugated to RNAsi for BCL-2, with the intercalation of doxorubicin (DOX) between the double helix of the aptamers. Promising results were obtained using this construct to treat MUC-1 overexpressing MCF-7 breast cancer cells. The complex was efficient in delivering doxorubicin to multi drug resistant MCF-7 cells because of increased endocytosis efficiency due to the cluster effect. BCL-2 RNAsi acted synergistically with DOX, increasing the sensitivity of cells for apoptosis and, in turn, decreasing cell viability [60].

An interesting work has used DOX to inactivate cancer stem cells (CSC) by conjugating this drug to engineered RNA aptamers raised against CSC surface marker epithelial cellular adhesion molecule (EpCAM). A modified DNA-RNA hybrid EpCAM aptamer (10-bp GC at stem region) was loaded with 2,5 molecules of DOX per aptamer and this conjugate was capable to release approximated 89.2% of the intercalated DOX just after endocytosis. The authors proved the release of DOX was dependent on low pH that in turn limited the availability of the drug in systemic circulation and tissue interstitium (physiological pH) to affect sensitive organs. Additionally, the complex Apt-DOX was able to target and deliver DOX to EpCAM positive HT29 colorectal cancer cells in dose and time dependent manner and lead to the accumulation and retention of the drug in the cell nucleus. All these characteristics made this construct very efficient in controlling tumor growth overcoming chemoresistance in both tumorsphere formation assay and immunodeficient mice with LDA xenotransplanted with CSC. The aptamer deliver approach developed by Xiang et al. could in fact improve therapeutic index delivering a sufficient drug dose to the critical subcellular target for enough time to eliminate precisely CSCs [61]. The same group extended the investigation to new approaches to tackle cancer stem cells using EpCAM aptamer. They use the aptamer as a driver to survivin RNAi in colorectal cancer cells, both in vitro as well as in colorectal tumor bearing mice. The combination of the surivin downregulation with the action of 5-fluorouacil cytotoxic chemotherapeutic could increase the lethality of colorectal cancer stem cells, as well as tumor control, in the colorectal cancer xenograph model and, consequently, improvement of animal survival [62].

A proof of concept of an approach using an aptamer-liposomal-DOX to target HER3ECD (epitope to trastuzumab binding epitope antigen) with minimum undesirable toxicity is described by Dou Xi et al. The strategy was to conjugate a single-stranded DNA aptamer against HER3 to liposome to improve targeted delivery with less unspecific action of DOX. The aptamer raised against HER3CD was able to bind to HER3 in MCF-7 $^{HER3+}$, BT 474^{HER3+} but not in 293T $^{HER3-}$ cells. This aptamer originates the apt-Lip-DOX conjugate that in fact increased the sustained drug release and was able to increase the growth inhibition of MCF-7 $^{HER3+}$ and BT 474^{HER3+} cells when compared to liposome-DOX and DOX alone challenge. The Apt-Lip-DOX formulation had many advantages when compared to Lip-DOX treatment. There was a higher uptake and retention by MCF-7 tumors in mice. The survival rate of the animals group treated with the apt-Lip-DOX 60 days after chemotherapy was 40% against 0% of the DOX group attained by 36th day after treatment. It is interesting to notice there was any lethality in the Apt-Lip-DOX treated control group that means a very low toxicity of this construct. Additionally, liver and cardio toxicity in the Apt-Lip-DOX treated group was compared to the control NaCl treated group. Finally, the author have demonstrated that the use of aptamers as a driver to target cells unequivocally reduced the availability of DOX in tissues (reduced biodistribution) other than tumor, reducing significantly the cardio toxicity after chemotherapy when compared to the formulation without aptamer [63].

Prusty et al. (2018) developed a very elegant photo-switchable hybrid-aptameric nanoconstruct that efficiently release DOX at high concentration to HGFR (hepatocyte growth factor receptor–cMet) expressing cells. The HyApNc-DOX nanoconstruct was obtained by the combination of two lipid self-assembly single motifs. The first motif was constituted by a lipid-functionalized aptamer trCLN3 with four lipid-modified dU-phosphoramidite 1 at the 5′end of the aptamer. The so-called trCLN3-L4 retained nanomolar affinity to the cMet receptor in tumor cells despite the modification. The second motif was a lipid functionalized 5′-CG rich hairpin ODN designed to carry DOX by intercalation but with a 2′,6′-dimethylzobenzene (DMAB) photoswitch incorporated in the hairpin nanoscaffold. After UV irradiation, DMAB is responsible for the controlled release of DOX inside cell. A lipid-mediated self-assembly approach to build a multi component supramolecular structure resulted in an improved survival to serum nucleases attack and in a better efficiency in the uptake by the target cell and also a precise subcellular distribution of DOX to the cell nucleus, after endocytosis, when compared to the native aptamer [64].

Another approach using aptamers to deliver DOX was developed to target Glioblastoma multiforme tumor cells. A GMT-3 ssDNA aptamer raised against glioblastoma multiforme A-172 cells demonstrates a high affinity to different glioblastoma cell lines (Kd, ~75 nM). Aptamer GMT-3 charged with DOX (molar ratio 1.2 DOX per aptamer) could selectively bind to and inactivate A-172 glioblastoma cells in comparison to the effect noticed in MCF-7 control cells [65].

A great challenge in the treatment of Glioblastoma using chemotherapeutics is to overcome the blood brain barrier (BBB) and to avoid chemotherapeutic drug deliver to the non-tumor tissues. Luo et al. (2017) developed a nanoconjugate delivery system using AS1411 aptamer as a driver to target nucleolin receptor that is overexpressed in glioblastoma U87 MG cells and neo-vascular endothelial cells in way to deliver chemotherapeutics with more precision. AS 1411-Nucleolin complex undergoes endocytosis inhibiting DNA synthesis inducing apoptosis and also inhibiting angiogenesis, that means it has dual targets. The authors incorporated a more efficient cytotoxic drug to AS1411 synthetizing a poly (L-*c*-glutamyl-glutamine)-paclitaxel nanoconjugate (PGGPTX) to achieve an improvement of aqueous solubility and permanence in plasma. All the results obtained by the group demonstrated that the presence of AS1441 aptamer in the complex AS1411-PGG-PTX improved significantly the U87 MG, as well as neo-vascular endothelial cells uptake and internalization, U87 MG spheroid tumor penetration, and U87 MG cell growth inhibition. The same effect could be demonstrated in vivo with an enhanced accumulation of AS1411-PGG-PTX nanoconjugates in glioblastoma cells, its extended retention time in circulation, and its pronounced penetration in glioblastoma tumors. The authors observed that AS1411-PGG-PTX was able to promote an increase in the survival of the tumor bearing mice and an efficient PCX delivery to glioblastoma tissue with a consequently strong cytotoxic effect [66].

One interesting strategy was used by Balasubramanyam and collaborators in a recently published work, where they designed aptamer-siRNA chimeras with great specificity to cancer expressing EpCAM that can deliver siRNAs against chosen oncogenes. In the case, PLK1, BCL2, and STAT3, three important proteins with high relevance in tumor growth, were selected. They tested various chimeras with positive results in cancer cell lines of breast, lung, head, neck, liver, and retinoblastoma [67].

Li et al. (2017) demonstrated an interesting feature of the aptamers paving the way to the action of neoadjuvant cytotoxic drugs, not necessarily carrying drugs molecules in their own structure. They selected a new ssDNA named HL-1 with high affinity and specificity to Maver-I lymphoma cells (Kd = 70 pmol/L). The aptamer presents a G-quadruplex structure and, once internalized by endocytosis, promoted cell cycle arrest in S-phase in around half of cell population after three days of lymphoma cells treatment. They exploit this effect in a combined treatment of lymphoma cells with a cytotoxic cytarabine. The lethal action of cytarabine is predominant over lymphoma cells in S-phase, and they show that several rounds of aptamer treatment was able to synergistically sensitize these cells to the lethal effect of cytarabine [68].

Pancreatic ductal adenocarcinoma (PDAC) could receive a great benefit from an aptamer target drug deliver approach because it is one with the worst prognosis among all other cancers. The limitation of therapeutic drug treatment dose imposed by the side effects resulted from the non-specificity of the available cancer chemotherapeutics contributes significantly to a poor prognosis. Levy and Kratchmer

(2018) proposed a new treatment approach using the Waz aptamer (anti-transferrin receptor–TfR) and E07 aptamer (anti-epidermal growth factor—EGFR), two receptors overexpressed in PDAC that are internalized by clathrin mediated endocytosis. The product of the conjugation of these aptamers with thiol-reactive membrane-permeable MMAE (MC-VC-PAB-MMAE) bearing a valine-citrulline linker or the membrane-impermeable auristatin derivative MMAF (MC-MMAF) was tested in Panc-1, MIA PaCa-2, and BxPC3 PDAC cell lines, with promising results [69].

Finally, the MUC1 aptamers were also used to deliver chitosan nanoparticles containing both docetaxel and siRNAs for a co-delivery/co-treatment approached. The conjugation of the aptamer to the nanoparticle increased specificity and cellular uptake of the nanoconjugate, whereas a significant impact of the combination therapy was confirmed in terms of cell viability. Furthermore, successful silencing was confirmed. Although this was a preliminary study conducted in vitro, it also demonstrates the potential of this type of approach [70]. Table 3 summarizes all works cited in this section.

Table 3. Aptamers for delivery SiRNA/chemotherapeutics for cancer therapy.

Carrier Aptamers	siRNA or Drug Delivered	Target Cell Lines	Main Effects	Ref.
PDGFRβ (Gint4.T)	STAT-3 siRNA	U87MG and T98G glioblastoma cells	Reduction of anti-apoptotic factors PARP and BCL-xL, inducing cell death in two glioblastoma lineages in vitro. Reduction of tumor growth rate and pro-tumoral factors in a xenograft mouse model of glioblastoma	[57]
HER-2 sense& antisense bivalent	EGFR siRNA	BT474 and SKBR3 breast cancer cells	Downregulation of EGFR in vitro and suppression of xenografted tumor growth in vivo	[58]
bivalent HER-2-HER-3	EGFR siRNA	BT474 and SKBR3 breast cancer cells	Reduction of HER receptors, induction of cell cycle arrest and apoptosis in vitro and inhibition of tumor growth in vivo	[59]
MUC1 multiaptamer	BCL-2 siRNA and Doxorubicin (DOX)	MCF-7 breast cancer cells	More efficiency in delivering doxorubicin. BCL-2 siRNA acted synergistically with DOX increasing the sensitivity of cells for apoptosis	[60]
epithelial cell adhesion molecule (EpCAM)	Doxorubicin (DOX)	EpCAM positive HT29 colorectal cancer cells; SCOV-3 ovarian cancer cell line and T47D breast cancer cell line	Tumor growth control overcoming chemoresistance in both tumorsphere formation assay and in xenotransplanted mice	[61]
EpCAM	Survivin RNAi	EpCAM positive HT29 colorectal cancer cells	The combination of the surivin downregulation with the action of 5-fluorouacil cytotoxic chemotherapeutic could increase the lethality of colorectal cancer stem cells, as well as tumor control, in colorectal cancer xenograph model	[62]
HER-3ECD	Doxorubicin (DOX)	MCF-7 HER3+ and BT 474HER3+ breast cancer cells	Increase in the survival rate in a xenograft mouse model. Reduction of liver and cardio toxicity after chemotherapy when compared to the formulation without aptamer	[63]
cMET	Doxorubicin (DOX)	NCI-H1838 lung cancer cells	Better efficiency in the uptake by the target cell and a precise subcellular distribution of DOX to the cell nucleus, after endocytosis	[64]

Table 3. *Cont.*

Carrier Aptamers	siRNA or Drug Delivered	Target Cell Lines	Main Effects	Ref.
Anti A-172 cells aptamer	Doxorubicin (DOX)	Glioblastoma multiforme tumor cells A-172	Selective binding and inactivation of A-172 glioblastoma cells	[65]
Nucleolin (AS1411)	Paclitaxel (PCX)	Glioblastoma U87 MG cells and neo-vascular endothelial cells	Improvement of cell uptake and internalization, spheroid tumor penetration and cell growth inhibition. Enhanced accumulation of AS1411-PGG-PTX nanoconjugates in glioblastoma cells, retention time in circulation penetration in glioblastoma tumors. Increase survival of the tumor bearing mice and an efficient PCX delivery to glioblastoma tissue with consequently strong cytotoxic effect	[66]
EpCAM	PLK1, BCL2, and STAT3 siRNA	NCC RbC 51 retinoblastoma cells; MCF-7 breast cancer cells; Müller Glial Mio M1 cells; PMI 2650 Head and Neck cancer cells and HepG2 Hepatocellular cells	Cell death induction and tumor reduction on RB xenografts tumor model	[67]
HL-1 (anti Maver-I cells)	Cytarabine	Maver-I lymphoma cells	Aptamer treatment was able to sensitize synergistically lymphoma cells in S-phase to lethal effect of cytarabine	[68]
Waz (anti-transferrin receptor—TfR)	Auristatin modified toxins	Panc-1, MIA PaCa-2, and BxPC3 Pancreas Ductal Adenocarcinoma cells	Aptamer conjugates demonstrated to be toxic to cell lines in different extends	[69]
EGFR	Auristatin modified toxins	Panc-1, MIA PaCa-2 and BxPC3 Pancreas Ductal Adenocarcinoma cells	Aptamer-toxin conjugates demonstrated to be toxic to cell lines in different extends	[69]
MUC1	cMET siRNA and Docetaxel	SKBR3 breast cancer cells	Nanoparticle chitosan increased specificity and cellular uptake of the nanoconjugate and successful silencing was confirmed	[70]

5. Concluding Remarks

Today, there are 42 studies in clinical trials using the term "aptamer", and 64 completed or ongoing studies using the term "siRNA" for treating medical conditions, including: cancer, HIV, and diseases of the immune system (www.clinicaltrial.gov), which indicates that these two therapeutic agents are still on the frontier of knowledge and are tendency in the global market. In a study of market analysis, siRNA and aptamers are the new drugs that biopharmaceutical industries are investing, and the cancer is the most studied disease using this strategy. The global aptamer market was valued at nearly $1.0 billion USD in 2016 and is expected to expand with a CAGR of 20.0% over the period from 2017 to 2025 to attain the value of $5.0 billion USD by the end of 2025. This growth is attributable to technology advances and the introduction of companies at commercial panel. The main players operating in the global aptamer market are AM Biotech (Houston, TX, USA), Aptamer Group (York, UK), Aptamer Sciences Inc. (Gyeongsangbuk-do, Korea), Aptagen LLC (Jacobus, PA, USA), and Base Pair Biotechnologies (Pearland, TX, USA). In regard to siRNA, the therapeutics market is expected to expand significantly until 2025, since the first product approved by FDA is currently available in the market. Major players operating in the global small interfering RNA (siRNA) therapeutics market are GE Dharmacon Lafayette, USA), OPKO Health, Inc. (Miami, FL, USA), Alnylam Pharmaceuticals (Cambridge, MA, USA), Arrowhead Research Corp (Pasadena, CA, USA), Sanofi Genzyme (Cambridge, MA, USA), Genecon Biotechnologies Co., Ltd. (Baesweiler, Germany), Arbutus Biopharma Corp (Burnaby, Canada), Silence Therapeutics AG (London, UK), and Sylentis S.A. (Madrid, Spain) (www.grandviewresearch.com).

Ever since our last revision of the field in 2017 [71], more than 50 new works have been published and revised in this manuscript, which show the potential of this approach. Furthermore, a number of new works have appeared related to nanoparticle formulations of siRNA and other RNA forms, using different types of nanoparticles, and carried successfully by aptamers, whilst some approaches have favored the use of siRNAs in conjunction with chemotherapy agents, with very promising results. A review of some of these technologies in a graphic form is presented in the figure below (Figure 1).

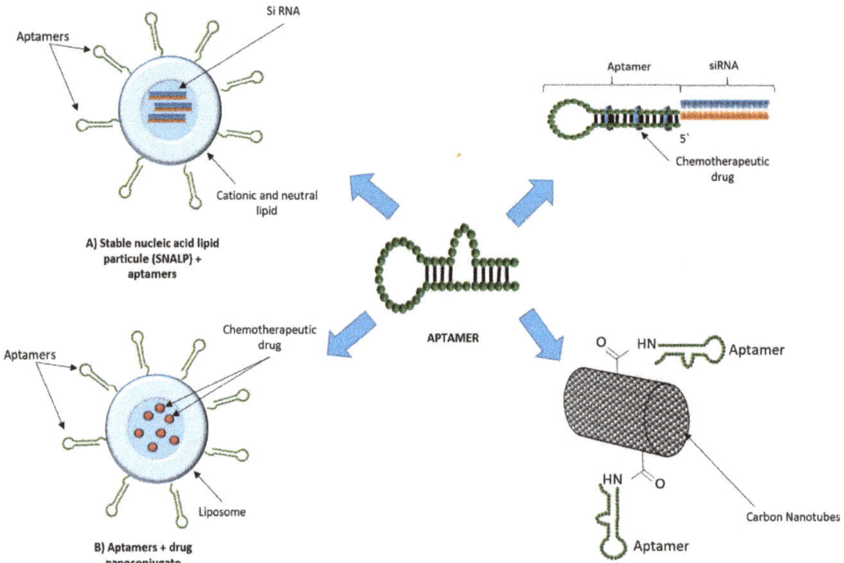

Figure 1. Uses of aptamers in therapeutic delivery.

Harnessing the targeting potential of aptamers, together with the silencing power of siRNAs and a long proved cytotoxic effect of chemotherapeutic agents, and bringing them together into a single therapeutic modality, has shown a great co-operative effect. It offers a therapy delivered directly and specifically to the tumor cell, with important proteins for tumor progression or survival silenced, whilst DNA acting cytotoxic agents act on tumor DNA, causing a specific therapy that has the potential to outmatch other approaches that focus on a single modalities.

Author Contributions: Conceptualization, S.M. and C.E.B.d.A.; writing—original draft preparation: Aptamer in the delivery of siRNA: A.P.D.A.B., Immunotherapy Using Aptamers-siRNA: P.C.d.C.N. and D.S., Aptamers-siRNA Chimeras: C.E.B.d.A., P.C.d.C.N. and D.S.; writing—review and editing, S.M.; supervision, S.M.

Funding: This work was supported by Instituto de Tecnologia em Imunobiológicos (Bio-Manguinhos).

Conflicts of Interest: The authors declare no conflict of interest.

References

1. Singh, A.; Trivedi, P.; Jain, N.K. Advances in siRNA delivery in cancer therapy. *Artif. Cells Nanomed. Biotechnol.* **2018**, *46*, 274–283. [CrossRef] [PubMed]
2. Mishra, D.K.; Balekar, N.; Mishra, P.K. Nanoengineered strategies for siRNA delivery: From target assessment to cancer therapeutic efficacy. *Drug Deliv. Transl. Res.* **2017**, *7*, 346–358. [CrossRef] [PubMed]
3. Goldberg, M.S. Improving cancer immunotherapy through nanotechnology. *Nat. Rev. Cancer* **2019**, *19*, 587–602. [CrossRef] [PubMed]
4. Nanjwade, B.K.; Sarkar, A.B.; Srichana, T. Design and Characterization of Nanoparticulate Drug Delivery. In *Characterization and Biology of Nanomaterials for Drug Delivery*; Elsevier: Amsterdam, The Netherlands, 2019.
5. Wang, X.X.; Li, Y.B.; Yao, H.J.; Ju, R.J.; Zhang, Y.; Li, R.J.; Yu, Y.; Zhang, L.; Lu, W.L. The use of mitochondrial targeting resveratrol liposomes modified with a dequalinium polyethylene glycol-distearoylphosphatidyl ethanolamine conjugate to induce apoptosis in resistant lung cancer cells. *Biomaterials* **2011**, *32*, 5673–5687. [CrossRef] [PubMed]
6. Faraji, A.H.; Wipf, P. Nanoparticles in cellular drug delivery. *Bioorg. Med. Chem.* **2009**, *17*, 2950–2962. [CrossRef]
7. Bhushan, B. *Springer Handbook of Nanotechnology*; Bhushan, B., Ed.; Springer: Berlin/Heidelberg, Germany, 2017; ISBN 978-3-662-54355-9.
8. Yao, V.J.; D'Angelo, S.; Butler, K.S.; Theron, C.; Smith, T.L.; Marchiò, S.; Gelovani, J.G.; Sidman, R.L.; Dobroff, A.S.; Brinker, C.J.; et al. Ligand-targeted theranostic nanomedicines against cancer. *J. Control. Release* **2016**, *240*, 267–286. [CrossRef] [PubMed]
9. Guaragna, A.; Chiaviello, A.; Paolella, C.; DAlonzo, D.; Palumbo, G.; Palumbo, G. Synthesis and evaluation of folate-based chlorambucil delivery systems for tumor-targeted chemotherapy. *Bioconjug. Chem.* **2012**, *23*, 84–96. [CrossRef]
10. Eloy, J.O.; Petrilli, R.; Raspantini, G.L.; Lee, R.J. Targeted Liposomes for siRNA Delivery to Cancer. *Curr. Pharm. Des.* **2018**, *24*, 2664–2672. [CrossRef]
11. Kim, M.W.; Jeong, H.Y.; Kang, S.J.; Choi, M.J.; You, Y.M.; Im, C.S.; Lee, T.S.; Song, I.H.; Lee, C.G.; Rhee, K.J.; et al. Cancer-targeted Nucleic Acid Delivery and Quantum Dot Imaging Using EGF Receptor Aptamer-conjugated Lipid Nanoparticles. *Sci. Rep.* **2017**, *7*, 9474. [CrossRef]
12. Guo, P.; Zhang, C.; Chen, C.; Garver, K.; Trottier, M. Inter-RNA interaction of phage θ29 pRNA to form a hexameric complex for viral DNA transportation. *Mol. Cell* **1998**, *2*, 149–155. [CrossRef]
13. Hu, J.; Xiao, F.; Hao, X.; Bai, S.; Hao, J. Inhibition of monocyte adhesion to brain-derived endothelial cells by dual functional RNA chimeras. *Mol. Ther. Nucleic Acids* **2014**, *3*, e209. [CrossRef] [PubMed]
14. Afonin, K.A.; Viard, M.; Koyfman, A.Y.; Martins, A.N.; Kasprzak, W.K.; Panigaj, M.; Desai, R.; Santhanam, A.; Grabow, W.W.; Jaeger, L.; et al. Multifunctional RNA Nanoparticles. *Nano Lett.* **2014**, *14*, 5662–5671. [CrossRef] [PubMed]
15. Li, N.; Ebright, J.N.; Stovall, G.M.; Chen, X.; Nguyen, H.H.; Singh, A.; Syrett, A.; Ellington, A.D. Technical and biological issues relevant to cell typing with aptamers. *J. Proteome Res.* **2009**, *8*, 2438–2448. [CrossRef] [PubMed]
16. Panigaj, M.; Reiser, J. Aptamer guided delivery of nucleic acid-based nanoparticles. *DNA RNA Nanotechnol.* **2016**, *2*, 42–52. [CrossRef]
17. Zhao, N.; Bagaria, H.G.; Wong, M.S.; Zu, Y. A nanocomplex that is both tumor cell-selective and cancer gene-specific for anaplastic large cell lymphoma. *J. Nanobiotechnol.* **2011**, *9*, 2. [CrossRef]
18. Powell, D.; Chandra, S.; Dodson, K.; Shaheen, F.; Wiltz, K.; Ireland, S.; Syed, M.; Dash, S.; Wiese, T.; Mandal, T.; et al. Aptamer-functionalized hybrid nanoparticle for the treatment of breast cancer. *Eur. J. Pharm. Biopharm.* **2017**, *114*, 108–118. [CrossRef]
19. Yan, W.; Tao, M.; Jiang, B.; Yao, M.; Jun, Y.; Dai, W.; Tang, Z.; Gao, Y.; Zhang, L.; Chen, X.; et al. Overcoming Drug Resistance in Colon Cancer by Aptamer-Mediated Targeted Co-Delivery of Drug and siRNA Using Grapefruit-Derived Nanovectors. *Cell. Physiol. Biochem.* **2018**, *50*, 79–91. [CrossRef]
20. Wilner, S.E.; Wengerter, B.; Maier, K.; Borba Magalhães, M.D.L.; Del Amo, D.S.; Pai, S.; Opazo, F.; Rizzoli, S.O.; Yan, A.; Levy, M. An RNA alternative to human transferrin: A new tool for targeting human cells. *Mol. Ther. Nucleic Acids* **2012**, *1*, e21. [CrossRef]

21. Alshaer, W.; Hillaireau, H.; Vergnaud, J.; Mura, S.; Deloménie, C.; Sauvage, F.; Ismail, S.; Fattal, E. Aptamer-guided siRNA-loaded nanomedicines for systemic gene silencing in CD-44 expressing murine triple-negative breast cancer model. *J. Control. Release* **2018**, *271*, 98–106. [CrossRef]
22. Kim, E.; Jung, Y.; Choi, H.; Yang, J.; Suh, J.S.; Huh, Y.M.; Kim, K.; Haam, S. Prostate cancer cell death produced by the co-delivery of Bcl-xL shRNA and doxorubicin using an aptamer-conjugated polyplex. *Biomaterials* **2010**, *31*, 4592–4599. [CrossRef]
23. Wu, X.; Ding, B.; Gao, J.; Wang, H.; Fan, W.; Wang, X.; Zhang, W.; Wang, X.; Ye, L.; Zhang, M.; et al. Second-generation aptamer-conjugated PSMA-targeted delivery system for prostate cancer therapy. *Int. J. Nanomed.* **2011**, *6*, 1747–1756.
24. Xu, X.; Wu, J.; Liu, Y.; Saw, P.E.; Tao, W.; Yu, M.; Zope, H.; Si, M.; Victorious, A.; Rasmussen, J.; et al. Multifunctional Envelope-Type siRNA Delivery Nanoparticle Platform for Prostate Cancer Therapy. *ACS Nano* **2017**, *11*, 2618–2627. [CrossRef]
25. Lv, T.; Li, Z.; Xu, L.; Zhang, Y.; Chen, H.; Gao, Y. Chloroquine in combination with aptamer-modified nanocomplexes for tumor vessel normalization and efficient erlotinib/Survivin shRNA co-delivery to overcome drug resistance in EGFR-mutated non-small cell lung cancer. *Acta Biomater.* **2018**, *76*, 257–274. [CrossRef] [PubMed]
26. Dassie, J.P.; Liu, X.Y.; Thomas, G.S.; Whitaker, R.M.; Thiel, K.W.; Stockdale, K.R.; Meyerholz, D.K.; McCaffrey, A.P.; McNamara, J.O.; Giangrande, P.H. Systemic administration of optimized aptamer-siRNA chimeras promotes regression of PSMA-expressing tumors. *Nat. Biotechnol.* **2009**, *27*, 839–846. [CrossRef]
27. Bagalkot, V.; Gao, X. SiRNA-aptamer chimeras on nanoparticles: Preserving targeting functionality for effective gene silencing. *ACS Nano* **2011**, *5*, 8131–8139. [CrossRef] [PubMed]
28. Jafari, R.; Majidi Zolbanin, N.; Majidi, J.; Atyabi, F.; Yousefi, M.; Jadidi-Niaragh, F.; Aghebati-Maleki, L.; Shanehbandi, D.; Soltani Zangbar, M.-S.; Rafatpanah, H. Anti-Mucin1 Aptamer-Conjugated Chitosan Nanoparticles for Targeted Co-Delivery of Docetaxel and IGF-1R siRNA to SKBR3 Metastatic Breast Cancer Cells. *Iran. Biomed. J.* **2019**, *23*, 21–33. [CrossRef]
29. Dong, J.; Cao, Y.; Shen, H.; Ma, Q.; Mao, S.; Li, S.; Sun, J. EGFR aptamer-conjugated liposome-polycation-DNA complex for targeted delivery of SATB1 small interfering RNA to choriocarcinoma cells. *Biomed. Pharmacother.* **2018**, *107*, 849–859. [CrossRef]
30. Ayatollahi, S.; Salmasi, Z.; Hashemi, M.; Askarian, S.; Oskuee, R.K.; Abnous, K.; Ramezani, M. Aptamer-targeted delivery of Bcl-xL shRNA using alkyl modified PAMAM dendrimers into lung cancer cells. *Int. J. Biochem. Cell Biol.* **2017**, *92*, 210–217. [CrossRef]
31. Taghavi, S.; Nia, A.H.; Abnous, K.; Ramezani, M. Polyethylenimine-functionalized carbon nanotubes tagged with AS1411 aptamer for combination gene and drug delivery into human gastric cancer cells. *Int. J. Pharm.* **2017**, *516*, 301–312. [CrossRef]
32. Xu, Y.; Pang, L.; Wang, H.; Xu, C.; Shah, H.; Guo, P.; Shu, D.; Qian, S.Y. Specific delivery of delta-5-desaturase siRNA via RNA nanoparticles supplemented with dihomo-γ-linolenic acid for colon cancer suppression. *Redox Biol.* **2019**, *21*, 101085. [CrossRef]
33. Li, L.; Hu, X.; Zhang, M.; Ma, S.; Yu, F.; Zhao, S.; Liu, N.; Wang, Z.; Wang, Y.; Guan, H.; et al. Dual Tumor-Targeting Nanocarrier System for siRNA Delivery Based on pRNA and Modified Chitosan. *Mol. Ther. Nucleic Acids* **2017**, *8*, 169–183. [CrossRef] [PubMed]
34. Wu, M.; Zhao, H.; Guo, L.; Wang, Y.; Song, J.; Zhao, X.; Li, C.; Hao, L.; Wang, D.; Tang, J. Ultrasound-mediated nanobubble destruction (UMND) facilitates the delivery of A10-3.2 aptamer targeted and siRNA-loaded cationic nanobubbles for therapy of prostate cancer. *Drug Deliv.* **2018**, *25*, 226–240. [CrossRef] [PubMed]
35. Liu, Q.; Wang, D.; Xu, Z.; Huang, C.; Zhang, C.; He, B.; Mao, C.; Wang, G.; Qian, H. Targeted Delivery of Rab26 siRNA with Precisely Tailored DNA Prism for Lung Cancer Therapy. *ChemBioChem* **2019**, *20*, 1139–1144. [CrossRef] [PubMed]
36. Jeelani, S.; Jagat Reddy, R.C.; Maheswaran, T.; Asokan, G.S.; Dany, A.; Anand, B. Theranostics: A treasured tailor for tomorrow. *J. Pharm. Bioallied Sci.* **2014**, *6* (Suppl. 1), S6. [CrossRef] [PubMed]
37. Kim, M.W.; Jeong, H.Y.; Kang, S.J.; Jeong, I.H.; Choi, M.J.; You, Y.M.; Im, C.S.; Song, I.H.; Lee, T.S.; Lee, J.S.; et al. Anti-EGF receptor aptamer-guided co-delivery of anti-cancer siRNAs and quantum dots for theranostics of triple-negative breast cancer. *Theranostics* **2019**, *9*, 837. [CrossRef] [PubMed]

38. Sivakumar, P.; Kim, S.; Kang, H.C.; Shim, M.S. Targeted siRNA delivery using aptamer-siRNA chimeras and aptamer-conjugated nanoparticles. *Wiley Interdiscip. Rev. Nanomed. Nanobiotechnol.* **2019**, *11*, e1543. [CrossRef]
39. Behlke, M.A. Chemical modification of siRNAs for in vivo use. *Oligonucleotides* **2008**, *18*, 305–320. [CrossRef]
40. Keefe, A.D.; Cload, S.T. SELEX with modified nucleotides. *Curr. Opin. Chem. Biol.* **2008**, *12*, 448–456. [CrossRef]
41. Mohamed, M.; Abu Lila, A.S.; Shimizu, T.; Alaaeldin, E.; Hussein, A.; Sarhan, H.A.; Szebeni, J.; Ishida, T. PEGylated liposomes: Immunological responses. *Sci. Technol. Adv. Mater.* **2019**, *20*, 710–724. [CrossRef]
42. Lächelt, U.; Wagner, E. Nucleic Acid Therapeutics Using Polyplexes: A Journey of 50 Years (and Beyond). *Chem. Rev.* **2015**, *115*, 11043–11078. [CrossRef]
43. Diao, Y.; Liu, J.; Ma, Y.; Su, M.; Zhang, H.; Hao, X. A specific aptamer-cell penetrating peptides complex delivered siRNA efficiently and suppressed prostate tumor growth in vivo. *Cancer Biol. Ther.* **2016**, *17*, 498–506. [CrossRef] [PubMed]
44. Subramanian, N.; Kanwar, J.R.; Athalya, P.K.; Janakiraman, N.; Khetan, V.; Kanwar, R.K.; Eluchuri, S.; Krishnakumar, S. EpCAM aptamer mediated cancer cell specific delivery of EpCAM siRNA using polymeric nanocomplex. *J. Biomed. Sci.* **2015**, *22*, 4. [CrossRef] [PubMed]
45. Tseng, Y.C.; Mozumdar, S.; Huang, L. Lipid-based systemic delivery of siRNA. *Adv. Drug Deliv. Rev.* **2009**, *61*, 721–731. [CrossRef] [PubMed]
46. Sharma, P.; Hu-Lieskovan, S.; Wargo, J.A.; Ribas, A. Primary, Adaptive, and Acquired Resistance to Cancer Immunotherapy. *Cell* **2017**, *168*, 707–723. [CrossRef]
47. Pastor, F.; Berraondo, P.; Etxeberria, I.; Frederick, J.; Sahin, U.; Gilboa, E.; Melero, I. An rna toolbox for cancer immunotherapy. *Nat. Rev. Drug Discov.* **2018**, *17*, 751–767. [CrossRef]
48. Pratico, E.D.; Sullenger, B.A.; Nair, S.K. Identification and characterization of an agonistic aptamer against the T cell costimulatory receptor, OX40. *Nucleic Acid Ther.* **2013**, *23*, 35–43. [CrossRef]
49. Rajagopalan, A.; Berezhnoy, A.; Schrand, B.; Puplampu-Dove, Y.; Gilboa, E. Aptamer-Targeted Attenuation of IL-2 Signaling in $CD8^+$ T Cells Enhances Antitumor Immunity. *Mol. Ther.* **2017**, *25*, 54–61. [CrossRef]
50. Puplampu-Dove, Y.; Gefen, T.; Rajagopalan, A.; Muheramagic, D.; Schrand, B.; Gilboa, E. Potentiating tumor immunity using aptamer-targeted RNAi to render $CD8^+$ T cells resistant to TGFβ inhibition. *Oncoimmunology* **2018**, *7*, e1349588. [CrossRef]
51. Song, P.; Chou, Y.K.; Zhang, X.; Meza-Romero, R.; Yomogida, K.; Benedek, G.; Chu, C.Q. CD4 aptamer-RORγt shRNA chimera inhibits IL-17 synthesis by human CD4 + T cells. *Biochem. Biophys. Res. Commun.* **2014**, *452*, 1040–1045. [CrossRef]
52. Soldevilla, M.M.; Villanueva, H.; Bendandi, M.; Inoges, S.; López-Díaz de Cerio, A.; Pastor, F. 2-fluoro-RNA oligonucleotide CD40 targeted aptamers for the control of B lymphoma and bone-marrow aplasia. *Biomaterials* **2015**, *67*, 274–285. [CrossRef]
53. Nozari, A.; Berezovski, M.V. Aptamers for CD Antigens: From Cell Profiling to Activity Modulation. *Mol. Ther. Nucleic Acids* **2017**, *6*, 29–44. [CrossRef] [PubMed]
54. Kruspe, S.; Giangrande, P.H. Aptamer-siRNA chimeras: Discovery, progress, and future prospects. *Biomedicines* **2017**, *5*, 45. [CrossRef] [PubMed]
55. Strebhardt, K.; Ullrich, A. Paul Ehrlich's magic bullet concept: 100 Years of progress. *Nat. Rev. Cancer* **2008**, *8*, 473–480. [CrossRef] [PubMed]
56. McNamara, J.O.; Andrechek, E.R.; Wang, Y.; Viles, K.D.; Rempel, R.E.; Gilboa, E.; Sullenger, B.A.; Giangrande, P.H. Cell type-specific delivery of siRNAs with aptamer-siRNA chimeras. *Nat. Biotechnol.* **2006**, *24*, 1005–1015. [CrossRef] [PubMed]
57. Esposito, C.L.; Nuzzo, S.; Catuogno, S.; Romano, S.; de Nigris, F.; de Franciscis, V. STAT3 Gene Silencing by Aptamer-siRNA Chimera as Selective Therapeutic for Glioblastoma. *Mol. Ther. Nucleic Acids* **2018**, *10*, 398–411. [CrossRef] [PubMed]
58. Xue, L.; Maihle, N.J.; Yu, X.; Tang, S.C.; Liu, H.Y. Synergistic Targeting HER2 and EGFR with Bivalent Aptamer-siRNA Chimera Efficiently Inhibits HER2-Positive Tumor Growth. *Mol. Pharm.* **2018**, *15*, 4801–4813. [CrossRef]
59. Yu, X.; Ghamande, S.; Liu, H.; Xue, L.; Zhao, S.; Tan, W.; Zhao, L.; Tang, S.C.; Wu, D.; Korkaya, H.; et al. Targeting EGFR/HER2/HER3 with a Three-in-One Aptamer-siRNA Chimera Confers Superior Activity against HER2+ Breast Cancer. *Mol. Ther. Nucleic Acids* **2018**, *10*, 317–330. [CrossRef]

60. Jeong, H.; Lee, S.H.; Hwang, Y.; Yoo, H.; Jung, H.; Kim, S.H.; Mok, H. Multivalent Aptamer–RNA Conjugates for Simple and Efficient Delivery of Doxorubicin/siRNA into Multidrug-Resistant Cells. *Macromol. Biosci.* **2017**, *17*, 1600343. [CrossRef]
61. Xiang, D.; Shigdar, S.; Bean, A.G.; Bruce, M.; Yang, W.; Mathesh, M.; Wang, T.; Yin, W.; Tran, P.H.L.; Shamaileh, H.A.; et al. Transforming doxorubicin into a cancer stem cell killer via EpCAM aptamer-mediated delivery. *Theranostics* **2017**, *7*, 4071. [CrossRef]
62. Alshamaileh, H.; Wang, T.; Xiang, D.; Yin, W.; Tran, P.H.L.; Barrero, R.A.; Zhang, P.Z.; Li, Y.; Kong, L.; Liu, K.; et al. Aptamer-mediated survivin RNAi enables 5-fluorouracil to eliminate colorectal cancer stem cells. *Sci. Rep.* **2017**, *7*, 5898. [CrossRef]
63. Dou, X.Q.; Wang, H.; Zhang, J.; Wang, F.; Xu, G.L.; Xu, C.C.; Xu, H.H.; Xiang, S.S.; Fu, J.; Song, H.F. Aptamer–drug conjugate: Targeted delivery of doxorubicin in a HER3 aptamer-functionalized liposomal delivery system reduces cardiotoxicity. *Int. J. Nanomed.* **2018**, *13*, 763–776. [CrossRef]
64. Prusty, D.K.; Adam, V.; Zadegan, R.M.; Irsen, S.; Famulok, M. Supramolecular aptamer nano-constructs for receptor-mediated targeting and light-triggered release of chemotherapeutics into cancer cells. *Nat. Commun.* **2018**, *9*, 535. [CrossRef] [PubMed]
65. Bayraç, A.T.; Akça, O.E.; Eyidoğan, F.İ.; Öktem, H.A. Target-specific delivery of doxorubicin to human glioblastoma cell line via ssDNA aptamer. *J. Biosci.* **2018**, *43*, 97–104. [CrossRef] [PubMed]
66. Luo, Z.; Yan, Z.; Jin, K.; Pang, Q.; Jiang, T.; Lu, H.; Liu, X.; Pang, Z.; Yu, L.; Jiang, X. Precise glioblastoma targeting by AS1411 aptamer-functionalized poly (L-γ-glutamylglutamine)–paclitaxel nanoconjugates. *J. Colloid Interface Sci.* **2017**, *490*, 783–796. [CrossRef] [PubMed]
67. Balasubramanyam, J.; Badrinarayanan, L.; Dhaka, B.; Gowda, H.; Pandey, A.; Subramanian, K.; Subadhra, L.B.; Elchuri, S. V EpCAM Aptamer siRNA chimeras: Therapeutic efficacy in epithelial cancer cells. *bioRxiv* **2019**. [CrossRef]
68. Li, H.; Yang, S.; Yu, G.; Shen, L.; Fan, J.; Xu, L.; Zhang, H.; Zhao, N.; Zeng, Z.; Hu, T.; et al. Aptamer internalization via endocytosis inducing s-phase arrest and priming maver-1 lymphoma cells for cytarabine chemotherapy. *Theranostics* **2017**, *7*, 1204–1213. [CrossRef] [PubMed]
69. Kratschmer, C.; Levy, M. Targeted Delivery of Auristatin-Modified Toxins to Pancreatic Cancer Using Aptamers. *Mol. Ther. Nucleic Acids* **2018**, *10*, 227–236. [CrossRef]
70. Zolbanin, N.M.; Jafari, R.; Majidi, J.; Atyabi, F.; Yousefi, M.; Jadidi-Niaragh, F.; Aghebati-Maleki, L.; Shanehbandi, D.; Zangbar, M.S.S.; Nayebi, A.M. Targeted co-delivery of docetaxel and cMET siRNA for treatment of mucin1 overexpressing breast cancer cells. *Adv. Pharm. Bull.* **2018**, *8*, 383–393. [CrossRef]
71. de Almeida, C.E.B.; Alves, L.N.; Rocha, H.F.; Cabral-Neto, J.B.; Missailidis, S. Aptamer delivery of siRNA, radiopharmaceutics and chemotherapy agents in cancer. *Int. J. Pharm.* **2017**, *525*, 334–342. [CrossRef]

Article

Light-Triggered Cellular Delivery of Oligonucleotides

Leena-Stiina Kontturi [1,2,3], Joep van den Dikkenberg [1], Arto Urtti [2,3,4], Wim E. Hennink [1] and Enrico Mastrobattista [1,*]

1. Department of Pharmaceutics, Utrecht Institute for Pharmaceutical Sciences (UIPS), Utrecht University, 3584 CG Utrecht, The Netherlands; leena.kontturi@helsinki.fi (L.-S.K.); J.B.vandenDikkenberg@uu.nl (J.v.d.D.); w.e.hennink@uu.nl (W.E.H.)
2. Drug Research Program, Division of Pharmaceutical Biosciences, Faculty of Pharmacy, University of Helsinki, 00790 Helsinki, Finland; arto.urtti@helsinki.fi
3. School of Pharmacy, Faculty of Health Sciences, University of Eastern Finland, 70211 Kuopio, Finland
4. Institute of Chemistry, St Petersburg State University, Petergoff, 198540 St Petersburg, Russia

* Correspondence: e.mastrobattista@uu.nl; Tel.: +31-6-2273-6567

Received: 9 January 2019; Accepted: 14 February 2019; Published: 21 February 2019

Abstract: The major challenge in the therapeutic applicability of oligonucleotide-based drugs is the development of efficient and safe delivery systems. The carriers should be non-toxic and stable in vivo, but interact with the target cells and release the loaded oligonucleotides intracellularly. We approached this challenge by developing a light-triggered liposomal delivery system for oligonucleotides based on a non-cationic and thermosensitive liposome with indocyanine green (ICG) as photosensitizer. The liposomes had efficient release properties, as 90% of the encapsulated oligonucleotides were released after 1-minute light exposure. Cell studies using an enhanced green fluorescent protein (EGFP)-based splicing assay with HeLa cells showed light-activated transfection with up to 70%–80% efficacy. Moreover, free ICG and oligonucleotides in solution transfected cells upon light induction with similar efficacy as the liposomal system. The light-triggered delivery induced moderate cytotoxicity (25%–35% reduction in cell viability) 1–2 days after transfection, but the cell growth returned to control levels in 4 days. In conclusion, the ICG-based light-triggered delivery is a promising method for oligonucleotides, and it can be used as a platform for further optimization and development.

Keywords: oligonucleotide delivery; light-activated release; intracellular release; liposome; indocyanine green

---

1. Introduction

Therapeutics based on oligonucleotides have significant potential for the treatment of a wide variety of diseases [1,2]. In principle, any disease with a known genetic origin can be treated by modifying genetic functions with oligonucleotide-based drugs. Compared to traditional pharmaceuticals, this approach has several advantages, including specificity, potency, and possibility for a rapid and rational drug design. The major limitation for the clinical translation of these therapeutics is the difficulty of in vivo delivery; as large, anionic macromolecules that are prone to degradation by nucleases, oligonucleotides require sophisticated carrier systems to enable delivery into the target cells [3,4]. An optimal carrier protects oligonucleotides from enzymatic degradation and clearance, and transfers them selectively into the cytoplasm of the target cells with minimal toxicity.

The most investigated synthetic vectors for oligonucleotides are lipid-based nanoparticles [5–7]. Commonly, these carriers contain cationic lipids that enable high loading capacity by complexing with negatively charged oligonucleotides and efficient intracellular delivery by interacting with the negatively charged cell membranes. However, the utility of cationic liposomes in vivo is limited, as excess positive charge results in toxicity, innate immune activation, and poor pharmacokinetic properties [8–10]. Neutral or anionic liposomes show less interaction with serum proteins and complement components and, consequently, are less toxic and have better pharmacokinetic profiles. Yet, as non-cationic lipids do not interact with cellular membranes as efficiently as cationic ones, cellular uptake and intracellular release of entrapped oligonucleotide cargo with neutral liposomes is usually poor. Currently, the most promising lipid formulations contain ionizable lipids [11–13]. These lipids switch charge pH-dependently, enabling neutral particles at physiological pH in the blood circulation and in the extracellular space of tissues, and positive charge in the acidic environment of endosomes after cellular internalization.

In general, liposomal drug delivery is associated with poorly controlled and insufficient cytosolic oligonucleotide release. To improve the control and effectiveness of cytosolic delivery and drug release at the target site, systems that are activated by external or internal signals, such as temperature, pH, ultrasound, specific enzymes, magnetic field and light, have been developed [14,15]. In the present study, we applied a previously developed light-triggered liposomal system for oligonucleotide delivery [16–18]. The system consists of thermosensitive liposomes with indocyanine green (ICG) as the photosensitizing agent. The light sensitivity is based on the photothermal ability of ICG to absorb light energy and convert it to heat [18,19]; when the temperature-sensitive liposomes containing ICG are exposed to light, the released heat creates a localized temperature increase, leading to fluidization of the thermosensitive lipid membranes and release of the encapsulated drug.

ICG injections have been approved by the US Food and Drug Administration (FDA) and European Medicines Agency (EMA) for fluorescence-based clinical imaging [20,21]. Compared to the most commonly used photothermal agents, gold and carbon nanomaterials, ICG has certain advantages: (1) ICG has absorption maximum at the near infrared (NIR) range, enabling excitation at a safe wavelength of 800 nm that penetrates into tissues [22,23]. (2) As an organic molecule, ICG can be conveniently incorporated into delivery systems. Also, processes for particle size control, such as extrusion and microfluidization, can be used, as the presence of ICG does not limit the size of the carrier [17]. (3) Since ICG is a fluorescent compound, it enables imaging-guided drug delivery in certain tissues [24–26]. (4) The safety profile of ICG is well-documented, while the long-term toxicity of non-biodegradable inorganic nanoparticles is unknown and they have not been approved for clinical use [27,28].

We have previously shown that the ICG-containing liposomes are functional in light-triggered release of small and large fluorescently labeled model compounds [16]. In the present work, we extended the concept to the delivery of oligonucleotides. Our aim was to investigate the effects of light induction and ICG on cellular delivery of oligonucleotides and liposomal oligonucleotides.

2. Materials and Methods

2.1. Materials

1,2-Dipalmitoyl-sn-glycero-3-phosphocholine (DPPC), 1,2-distearoyl-sn-glycero-3-phosphocholine (DSPC) and 1,2-distearoyl-sn-glycero-3-phosphoethanolamine-*N*-[methoxy(polyethylene glycol)-2000] (DSPE-PEG) were bought from Lipoid (Ludwigshafen, Germany). 1-stearoyl-2-hydroxy-sn-glycero-3-phosphocholine (Lyso PC) was from Avanti Polar Lipids, Inc. (Alabaster, AL, USA). The oligonucleotides used in this study were a splice switching antisense oligonucleotide (SSO) restoring correct splicing of EGFP [29] and an siRNA against luciferase. The sequences of the oligonucleotides are the following:

SSO: 5′-GCT ATT ACC TTA ACC CAG-3′
siRNA: sense 5′-CUUACGCUGAGUACUUCGAdTdT-3′
anti-sense 5′-UCGAAGUACUCAGCGUAAGdTdT-3′

Underlined bases indicate a 2′-*O*-methyl modification. dT indicates deoxyribonucleic acid bases with phosphorothioate (PS) bonds. The SSO consists completely of PS bonds. The SSO was purchased from Biosearch Technologies (Petaluma, CA, USA) and the siRNA from Integrated DNA technologies (Leuven, Belgium). Cell medium and supplements were from GibcoBRL, Thermo Fisher Scientific (Naarden, The Netherlands). Indocyanine green purchased from Sigma-Aldrich (St. Louis, MO, USA) was the United States Pharmacopeia (USP) Reference Standard (mw. 775 g/mol). All other compounds were bought from Sigma-Aldrich (St. Louis, MO, USA) unless otherwise mentioned. Fluorescence measurements of ICG and the Ribogreen assay were performed with a Jasco FP8300 Spectrofluorometer with micro-well plate reader (JASCO Benelux BV., De Meern, The Netherlands). An 808N10W laser system with a circular beam of 7 mm in diameter was used for the light triggering studies (Changchun Dragon Lasers Co., Ltd., Changchun, China). The output light intensities with different power settings were measured using a P-9710-1 optometer with RCH-102-2 custom-made detector head (Te Lintelo Systems BV, Zevenaar, The Netherlands). Light intensities (mW/cm^2) corresponding to the power settings of 1–10 W are shown in the supporting information (Table S1).

2.2. Liposome Preparation

Liposomes were prepared by a lipid film hydration method using a composition of DPPC/DSPC/Lyso PC/DSPE-PEG at a molar ratio of 75:15:10:4, respectively. The lipids dissolved in chloroform were mixed and dried to a thin lipid film by rotary evaporation. Residual chloroform was removed under a nitrogen flow for 30 min. The film was hydrated at 55 °C for 1 h with 2.5 mg/mL oligonucleotides dissolved in 20 mM HEPES buffer, pH 7.4. To promote encapsulation of oligonucleotides, a high lipid concentration of 100 mM (typically, 50 µmol of total lipids and 0.5 mL of hydration solution) was used. The formed liposomes were extruded 5 times through a track-etched polycarbonate membrane (Whatman Nuclepore, GE Healthcare, Chicago, IL, USA) with 100-nm pore size using a syringe extrusion device (Avanti Polar Lipids Inc., Alabaster, AL, USA). The free, unencapsulated oligonucleotides were separated by ultracentrifugation for 1 h at 55,000 rpm at 4 °C for 2–3 times, and the liposomes were dispersed in 20 mM HEPES, 140 mM NaCl, pH 7.4.

For ICG incorporation, the liposomes were incubated in ICG solution of 1 mg/mL (in 20 mM HEPES, 140 mM NaCl, pH 7.4) for 1 h in rotation at room temperature. The volumes of the incubated liposomes and ICG solution were adjusted to molar ratios in the range of 1/25–1/200 ICG to lipid. The amount of ICG in the liposomes was determined by separating the free ICG by ultracentrifugation (1 h at 55,000 rpm at 4 °C) after incubation. The amount of free ICG in the supernatant was determined based on ICG fluorescence that was measured at 770/810 nm (excitation/emission wavelengths), and the concentration was determined using a calibration curve. As the percentage of non-incorporated ICG was less than 10% even at the highest ICG concentration, the separation step by ultracentrifugation was not considered necessary and was not carried out in further experiments. The addition of ICG to the liposomes was done always immediately before the experiments.

2.3. Liposome Characterization

2.3.1. Size

The mean particle size and polydispersity index (PDI) were measured by dynamic light scattering (DLS) with a Malvern CGS-3 multiangle goniometer with He–Ne laser source (λ = 632.8 nm, 22 mW output power) using an angle of 90° (Malvern Instruments, Malvern, UK). For the measurement, liposome samples were diluted to 0.25 mM (lipid concentration) in 20 mM HEPES with 140 mM NaCl, pH 7.4.

2.3.2. Zeta-Potential

The zeta-potential of the liposomes was measured on a Zetasizer Nano-Z (Malvern Instruments) with samples diluted to 0.25 mM (lipid concentration) in 20 mM HEPES, pH 7.4.

2.3.3. Phase Transition Temperature (T_m)

Differential scanning calorimetry (DSC) (Discovery DSC, TA instruments, New Castle, DE, USA) was used to determine the T_m values of the liposomes. Liposome sample with a concentration of 50 mM and a reference sample (20 mM HEPES with 140 mM NaCl, pH 7.4) were placed in hermetically sealed aluminum pans, and heated from 20 to 60 °C at a rate of 0.5 °C/min. T_m represents the peak temperature of the endotherm recorded during the heating scan.

2.3.4. Encapsulation Efficiency

The amount of SSO or siRNA encapsulated inside the liposomes was determined using Quant-iT™ RiboGreen®RNA Assay Kit (Thermo Fischer Scientific, Waltham, MA, USA) according to the manufacturer's protocol. The measurement was performed for non-treated liposomes and for liposomes disrupted with 0.5% Triton-X 100. As the Ribogreen reagent does not penetrate liposomal membranes, the signal measured after treatment with Triton-X 100 represents the oligonucleotide concentration entrapped inside the liposomes. Concentrations of the samples were calculated based on calibration curves prepared with (1) oligonucleotides in the presence of empty liposomes and (2) oligonucleotides in the presence of empty liposomes and Triton-X 100. The encapsulation efficiency was calculated using the formula: % encapsulation = $(ON_t - ON_0)/ON_i \times 100$, where ON_t = oligonucleotide concentration of Triton-X 100 treated liposomes, ON_0 = oligonucleotide concentration of non-treated liposomes, and ON_i = initial oligonucleotide concentration. Measurements were done at wavelengths of 480/520 nm.

2.3.5. Light-Induced Oligonucleotide Release

The light-induced release of SSO or siRNA from the liposomes was determined by measuring the concentrations of non-treated liposomes, liposomes after light exposure, and liposomes treated with 0.5% Triton-X 100. The liposomes were diluted to 1 mM (lipid concentration) in buffer, heated to 37 °C on a thermomixer heating device and exposed to 808 nm light with the intensity of 370 mW/cm^2 for 1 min. Control samples were kept at similar conditions, but were shielded from the light. Non-treated samples at +4 °C represented the background signal and Triton-X 100 treated samples were set at 100% release. The oligonucleotide concentrations were measured immediately after the light exposure using Quant-iT RiboGreen RNA Assay Kit as described in the previous section, and the release percentage was calculated using the formula: % released = $(ON_l - ON_0)/(ON_t - ON_0) \times 100$, where ON_l = oligonucleotide concentration of light exposed liposomes, ON_0 = oligonucleotide concentration of non-treated liposomes, and ON_t = oligonucleotide concentration of Triton-X 100 treated liposomes.

Both the SSO- and siRNA-encapsulated ICG liposomes were prepared and characterized as described above, while cell studies were performed using only the SSO-encapsulated liposomes. The purpose of the siRNA-encapsulated liposomes was to investigate if the light-triggered release of siRNA ($M_W \sim 14$ kDa) differed from the release of SSO ($M_W \sim 7$ kDa).

2.4. Cell Studies

2.4.1. Cell Line

To study the ability of the light-activated liposomes to deliver oligonucleotides intracellularly, an EGFP-based splicing assay with HeLa S3 cells was used [29]. The assay was based on a construct where a C-to-T mutation at nucleotide 654 of the human β-globin intron-2 was inserted in the EGFP cDNA (IVS2-654), preventing correct translation of EGFP. Delivery of SSO (antisense oligonucleotide directed to position 654) blocked the aberrant splice site and restored the correct splicing of the EGFP precursor mRNA, generating properly translated EGFP. In this approach, antisense activity of SSO was directly proportional to up-regulation of EGFP in cells transfected with the IVS2-654 EGFP construct, thus providing a positive, quantitative readout.

2.4.2. Cell Culture

HeLa S3 IVS2-654 EGFP cells were cultured in Dulbecco's Modified Eagle's Medium with high glucose supplemented with 10% (*v*/*v*) fetal bovine serum and 400 μg/mL G418 at 37 °C under a humidified atmosphere containing 5% CO_2. Cells were routinely passaged twice a week and used for the experiments at passages 5–20. Cells were regularly tested for mycoplasma and found to be negative.

2.4.3. Transfection Studies

HeLa S3 IVS2-654 EGFP cells were seeded on white μView clear-bottom 96-well plates (Greiner Bio-One B.V., Alphen aan de Rijn, The Netherlands) at a density of 9000 cells/well. After attachment (5–6 h after seeding), liposomes diluted in growth medium were added to the cells for overnight incubation. Next day, the liposomes were removed, the cells were washed with phosphate buffered saline, and the light triggering was performed. In the light triggering set-up, the cell culture plate was placed on a thermomixer heater to keep the temperature at 37 °C and the cells were exposed to 808 nm light, while control samples on the same plate were shielded from the light. After light exposure, the cells were transferred back to the culture incubator (37 °C under a humidified atmosphere containing 5% CO_2). Transfection was typically measured either 24 or 48 h after light exposure (in some experiments, also the time points of 72 and 96 h were used). Detection was done by confocal imaging using a Cell Voyager CV7000s high-content confocal imager (Yokogawa, Tokyo, Japan). The nuclei were stained by incubating the cells in 2 μg/mL Hoechst 33342 (Life Technologies, Bleiswijk, The Netherlands) for 15 min at 37 °C. Imaging was done by acquiring 20–30 images with 3–4 z-stacks from each well using a 20× objective.

The effect of the following variables on transfection efficacy was investigated: (1) liposome concentration, (2) ICG concentration of the liposomes, (3) intensity of light, and (4) duration of light exposure. In addition, possible changes in transfection efficacy were followed for 48, 72, and 96 h after transfection and light triggering. Finally, the ability of free ICG or empty ICG liposomes in combination with free SSO to induce transfection upon light triggering was studied. Lipofectamine®3000 (Thermo Fisher Scientific, Waltham, MA, USA) was used as a positive control according to the manufacturer's protocol. As a negative control, an antisense oligonucleotide directed to position 705 of the cDNA was used [30]. This is an oligonucleotide of the same length as the active SSO, but complementary to a different position and will thus not result in altered splicing.

2.4.4. Cytotoxicity

Cytotoxicity of light exposure and/or the liposome formulation were measured by calculating Hoechs-labeled nuclei (2 μg/mL Hoechst 33342 for 15 min at 37 °C) from confocal images. Numbers of the treated cells were compared to numbers of cells without any treatment and reported as percentage values (cells without treatment set as 100%). Effect of light intensity (370–1500 mW/cm^2), liposome concentration (0.5–1.4 mM, overnight incubation), and the combination of these were studied, and the cell numbers were measured at 48, 72, and 96 h after treatment.

2.4.5. Image Analysis

Customized image analysis protocols were developed with Columbus Software (version 2.7.1; PerkinElmer Inc., Waltham, MA, USA). The determination of transfection efficacy was based on the intensity of EGFP fluorescence signal in cell cytoplasm. The percentage of transfected cells was obtained by setting a threshold value for the EGFP intensity and by separating the cell population using this value into transfected (intensity above the threshold) and non-transfected (intensity below the threshold) cells. The threshold was set to give less than 1 as the transfection percentage (<1% transfection efficacy) for the non-treated control cells. Details of the analysis can be found from supplementary information (Figure S1). Cell numbers were determined by calculating Hoechst-labeled nuclei.

3. Results

3.1. Characterization of the Oligonucleotide-Encapsulated ICG Liposomes

The mean diameter of the liposomes was ~150 nm with a PDI < 0.15. The incorporation of ICG did not significantly affect the liposome size. Zeta-potential values of the liposomes without ICG were negative (~ −8 mW), and inclusion of ICG further lowered the zeta-potentials. The T_m values were 42.6–42.7 °C, and the incorporation of ICG lowered these values by 0.4–0.8 °C. Encapsulation efficiency of oligonucleotides was 5%–8%. Light-induced release of oligonucleotides was effective; approximately 90% of the SSO and siRNA were released after exposure to 808 nm light with the intensity of 370 mW/cm^2 for 1 min. Results on the characterization of the liposomes are shown in Table 1; Table 2 and in Figure 1. In general, there were no significant differences in the characteristics between the liposomes encapsulated with the SSO or siRNA or empty liposomes.

Table 1. Size, PDI, and encapsulation efficiency of the liposomes. SD, standard deviation (n = 3) and PDI, polydispersity index.

Liposome Type	ICG Concentration	Diameter ± SD (nm)	PDI	Encapsulation % ± SD
SSO liposomes	without ICG	145 ± 3	0.123	6.7 ± 0.9
	with 1/25 ICG/lipid	150 ± 1	0.102	
siRNA liposomes	without ICG	153 ± 2	0.141	5.8 ± 0.7
	with 1/25 ICG/lipid	155 ± 2	0.116	

Table 2. Zeta-potential and transition temperature (T_m) of the liposomes. SD, standard deviation (n = 3).

Liposome Type	ICG Concentration	Zeta-Potential ± SD (mV)	T_m (°C)
SSO liposomes	without ICG	−8.1 ± 0.1	42.7
	with 1/50 ICG/lipid	−12.9 ± 0.3	42.1
	with 1/25 ICG/lipid	−16.4 ± 0.3	41.9
siRNA liposomes	without ICG	−8.1 ± 0.3	42.6
	with 1/50 ICG/lipid	−12.2 ± 0.3	42.0
	with 1/25 ICG/lipid	−16.5 ± 0.3	42.1
Empty liposomes	without ICG	−8.2 ± 0.1	42.7
	with 1/50 ICG/lipid	−11.9 ± 0.4	42.3
	with 1/25 ICG/lipid	−15.9 ± 0.2	42.2

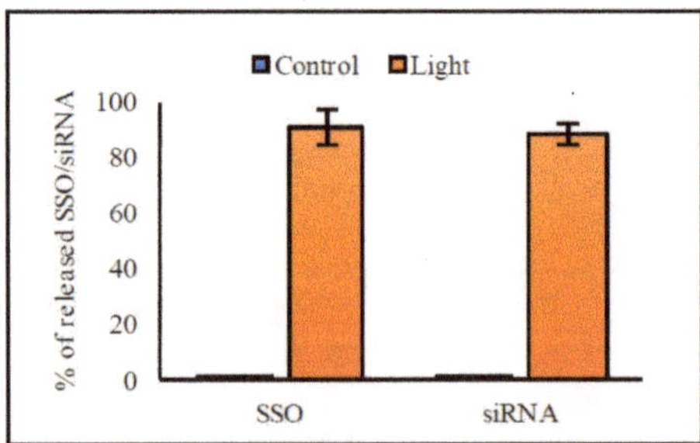

Figure 1. Light-triggered release of splice switching antisense oligonucleotide (SSO) and siRNA from the indocyanine green (ICG) liposomes. The liposomes were kept at 37 °C and exposed to 808 nm light with the intensity of 370 mW/cm^2 for 1 min (light-triggered samples, orange columns) or shielded from the light (control samples, blue columns). The columns represent average values of released SSO or siRNA (n = 3) with error bars as standard deviation.

3.2. *Transfection Experiments*

We studied the effects of liposomal formulation (ICG and lipid concentrations) and light exposure (intensity, duration) on transfection efficacy. Moreover, transfection efficacy at different time points was measured. Successful transfection was seen as increased EGFP fluorescence in the cell cytoplasm (Figure 2). Treatment with ICG liposomes encapsulated with the negative control oligonucleotide induced no detectable transfection (Figure S2). Light triggering was shown not to affect the activity of the SSO by comparing transfection efficacy of Lipofectamin-transfected cells with and without light exposure (Figure S3).

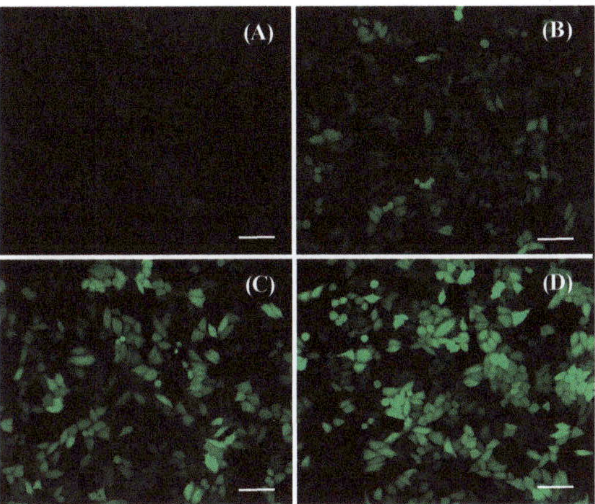

Figure 2. Confocal images of HeLa S3 IVS2-654 enhanced green fluorescent protein (EGFP) cells. Action of SSO restores EGFP expression. The cells were administered with **(A)** 0, **(B)** 0.5, **(C)** 0.7, and **(D)** 1.4 mM of SSO-encapsulated ICG liposomes and exposed to 808 nm light with the intensity of 370 mW/cm^2 for 2 min. The green color represents EGFP fluorescence. The concentrations refer to lipid concentrations of the liposome dispersions. The liposomes contained ICG at a molar ratio of 1/50 ICG to lipid. Scale bar = 100 μm.

3.2.1. Liposome Concentration

Liposome concentration had a clear effect on transfection efficacy (Figure 2). Increasing liposome concentrations led to higher transfection percentages: when the concentration was approximately tripled (from 0.5 mM to 1.4 mM), the transfection percentage increased roughly 3-fold (from 20% to 60% and from 30% to 80%) (Figure 3A,B). An increased transfection was detected in the cells treated with liposomes and exposed to light, as well as in the cells that were incubated with liposomes without light exposure. This was especially evident for liposomes having a higher concentration of ICG; with 1.4 mM liposomes of 1/25 ICG-to-lipid ratio, 14% of the cells were transfected without light exposure (Figure 3B). Yet, the transfection efficacy was considerably higher (78%) with light exposure than without light. In general, the transfection efficacies obtained with the light-triggered liposomes were in the same range as with Lipofectamine®3000 that was used as a positive control. However, compared to Lipofectamine, the results with the light-triggered delivery were much more consistent (Figure S4).

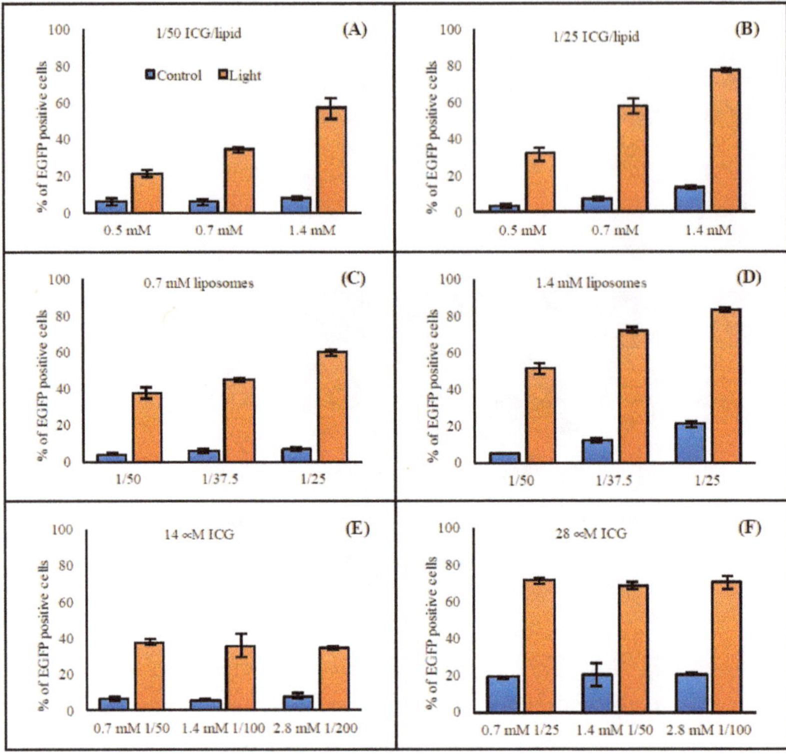

Figure 3. Effect of (**A**,**B**) liposome concentration and (**C**–**F**) ICG concentration of the liposomes on the transfection efficacy. The HeLa S3 IVS2-654 EGFP cells were incubated with liposomes having the following lipid concentrations and ICG-to-lipid molar ratios: (**A**) 0.5 mM–1.4 mM, 1/50 ICG/lipid, (**B**) 0.5 mM–1.4 mM, 1/25 ICG/lipid, (**C**) 0.7 mM, 1/25–1/50 ICG/lipid, (**D**) 1.4 mM, 1/25–1/50 ICG/lipid, (**E**) 0.7 mM–2.8 mM, 1/50–1/200 ICG/lipid, and (**F**) 0.7 mM–2.8 mM, 1/25–1/100 ICG/lipid. In (**E**) and (**F**), all the treatments had the same ICG concentration, 14 μM in (**E**) and 28 μM in (**F**). After incubation, the cells were exposed to 808 nm light (370 mW/cm^2 for 2 min). The columns represent average values of EGFP-positive cells (n = 3) with error bars as standard deviation. Cells without treatment had less than 1% of EGFP-positive cells.

3.2.2. ICG Concentration

Increasing the ICG concentration that was added to the liposomes led to an increased transfection efficacy. The higher the ICG concentration of the liposomes, the higher the percentage of transfected cells (Figure 3C,D). Instead, cells treated with liposomes having different lipid concentrations but same ICG concentration had equal transfection percentages (Figure 3E,F). As a conclusion, efficacy of the transfection process is mainly dependent on the ICG concentration, and both lipid and SSO concentrations have a smaller impact. Moreover, the liposomes with a higher ICG concentration induced transfection in the absence of light as well. At the highest ICG concentration, about 20% of the cells were EGFP positive without light exposure (Figure 3D,F).

3.2.3. Illumination Time

In general, the transfection efficacy was not affected when the illumination time was increased to longer than 1 min (Figure 4A,B). Effects of illumination times shorter than 1 min were dependent on other variables, the major determinant being ICG concentration. At a higher ICG concentration, changes in transfection efficacy with illumination times of 15 s, 30 s, and 1 min were more prominent than those observed for lower ICG concentrations. In addition, at very high ICG concentrations, some increase in transfection efficacy could be seen between 1 min and 2 min light exposures (data not shown). Consequently, the illumination time of 2 min was chosen for further studies.

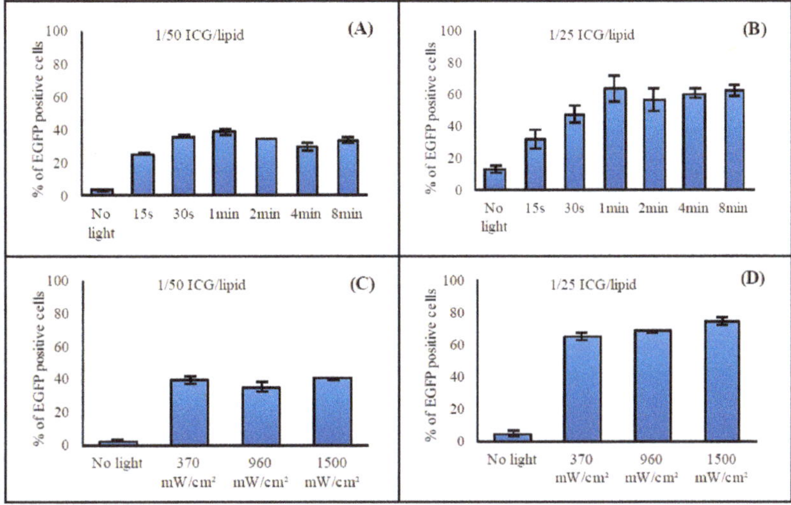

Figure 4. Effect of (**A**,**B**) illumination time and (**C**,**D**) light intensity on transfection efficacy. The HeLa S3 IVS2-654 EGFP cells were incubated with 0.7 mM liposomes having (**A**,**C**) 1/50 and (**B**,**D**) 1/25 ICG-to-lipid molar ratio. After incubation, the cells were exposed to 808 nm light with (**A**,**B**) the intensity of 370 mW/cm^2 for 15 s to 8 min or (**C**,**D**) the intensities of 370–1500 mW/cm^2 for 2 min. The columns represent average values of EGFP-positive cells (n = 3) with error bars as standard deviation. Cells without treatment had less than 1% of EGFP-positive cells.

3.2.4. Light Intensity

The intensity of 808 nm light used to trigger the liposomes did not affect the transfection efficacy in the range of 370–1500 mW/cm^2 (Figure 4C,D). The laser instrument used in this study limited the intensities to 370 mW/cm^2 and higher levels. Intensities higher than 1500 mW/cm^2 led to variability in results between repeats, and in general did not significantly improve the transfection efficacy (data not shown). As a conclusion, light intensities in the range of 370–1500 mW/cm^2 were suitable for induction of the oligonucleotide release.

3.2.5. Free ICG and SSO

During the studies, we observed that ICG and SSO can transfect cells upon light triggering without incorporation into liposomes. Therefore, experiments with free ICG or empty ICG liposomes and free SSO were performed. The transfection efficacies of free ICG and SSO, and SSO-encapsulated ICG liposomes were equal (Figure 5). In addition, a mixture of empty ICG liposomes and free-SSO transfected cells, but less effectively compared to free ICG and SSO or the SSO-encapsulated ICG liposomes (transfection efficacies of 19% versus 40% with 14 μM ICG; 45 versus 60% with 35 μM ICG, respectively). In the absence of ICG, treatment with SSO did not result in detectable levels of transfected cells.

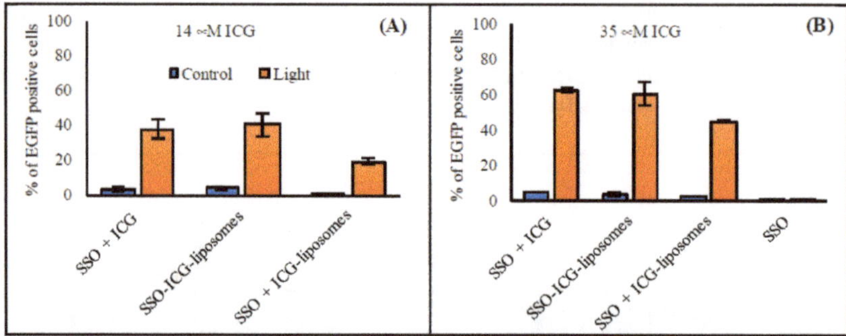

Figure 5. Free ICG- and SSO-induced transfection after light exposure. The HeLa S3 IVS2-654 EGFP cells were incubated with free ICG or empty ICG liposomes and SSO corresponding to the following concentrations of SSO-encapsulated ICG liposomes: (**A**) 0.7 mM, 1/50 ICG-to-lipid ratio (240 nM SSO, 14 µM ICG) and (**B**) 1.4 mM, 1/40 ICG-to-lipid ratio (480 nM SSO, 35 µM ICG). After incubation, the cells were exposed to 808 nm light (370 mW/cm^2 for 2 min). The columns represent average values of EGFP-positive cells (n = 3) with error bars as standard deviation. Cells without treatment had less than 1% of EGFP-positive cells.

3.2.6. Time after Transfection and Light Triggering

At higher liposome concentrations of 0.7 and 1.4 mM, transfection efficacies increased from 48 to 96 h after light triggering: transfection percentages increased from ~35% to 50% with 0.7 mM liposomes and from ~55% to 70% with 1.4 mM liposomes (Figure 6). At the lower liposome concentrations of 0.25 and 0.5 mM, differences between the time points were smaller and a slight increase in the transfection percentages could be seen only between 48 and 72 h (from 10% to 14% with 0.25 mM liposomes; from 22% to 27% with 0.5 mM liposomes). After 4 days, the HeLa cells grew over-confluent making quantitative analysis of the results impossible.

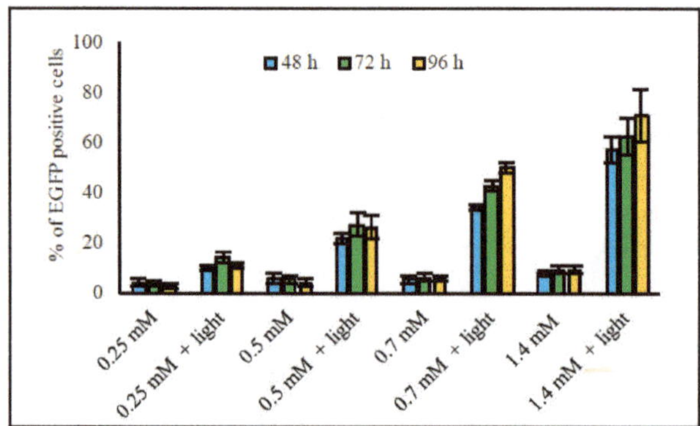

Figure 6. Effect of time after transfection and light triggering on transfection efficacy. The HeLa S3 IVS2-654 EGFP cells were incubated with 4 different liposome concentrations having 1/50 ICG-to-lipid molar ratio. Light exposure was performed at 808 nm (370 mW/cm^2 for 2 min), and transfection efficacy was measured at 48, 72, and 96 h after light triggering. The columns represent average values of EGFP-positive cells (n = 3) with error bars as standard deviation. Cells without treatment had less than 1% of EGFP-positive cells.

3.3. Cytotoxicity

The light exposure or liposome incubation separately showed no toxic effects (Figure 7A,B). Slightly lower cell numbers were seen after incubation with 1.0 and 1.4 mM liposomes: cell numbers were 92%–96% of the control cell numbers at the first time point, 48 h after liposome incubation. At 72 and 96 h, cell numbers were similar in all the groups. The combination of liposomes and light exposure resulted in moderate toxic effects (Figure 7C). Especially, 48 h after light triggering, the cell numbers were lower than in the control group: 75% in cells treated with 0.5 and 0.7 mM liposomes and 65% in cells treated with 1.0 and 1.4 mM liposomes. However, cell numbers in the treated groups had increased at 72 h after light triggering and at 96 h, the cell numbers did not differ from the non-treated control group.

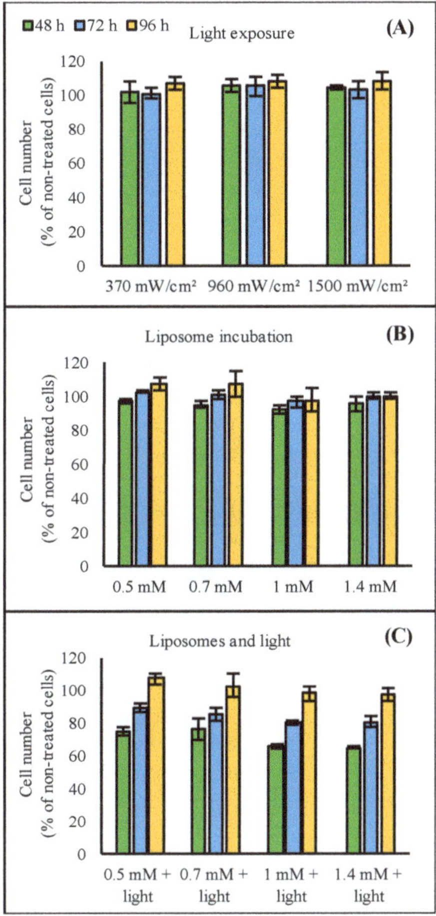

Figure 7. Effect of (**A**) light exposure, (**B**) the ICG liposome concentration, or (**C**) light exposure and the ICG liposome concentration on cell growth. Cells were (**A**) exposed to 808 nm light with the intensities of 370–1500 mW/cm^2 for 2 min or (**B**) incubated with 0.5–1.4 mM liposomes having 1/50 ICG-to-lipid molar ratio overnight. In (**C**), light exposure was performed after overnight liposome incubation. The columns represent average values of cell numbers (n = 3) calculated as percentage of non-treated control cells with error bars as standard deviation.

4. Discussion

The light-triggered liposome formulation used in this work is based on our previous studies on ICG liposomes [16,17], but ICG was now incorporated into the liposome dispersion. This approach avoided any interference of ICG on oligonucleotide encapsulation. As an amphiphilic compound,

ICG can associate with the hydrophilic surface layer of liposomes or penetrate the hydrophobic lipid bilayer. The post-inserted ICG may be partly dissolved in the lipid bilayer, because the T_m values of liposomes with ICG were slightly lower than the T_m values of liposomes without ICG (42.6–42.7 °C and 41.9–42.3 °C for liposomes without and with ICG, respectively, Table 2). Based on molecular modeling simulations [16,31], ICG molecules can also bind to the PEG chains on the surface of the lipid bilayers. Thus, ICG is likely to be partly solubilized in the lipid bilayer and interacts with its more polar regions with surface-grafted PEG.

Most studies on liposomal oligonucleotide delivery systems have utilized either cationic or ionizable lipids. We chose non-cationic lipids for two reasons. Firstly, we wanted to specifically investigate the light-activated delivery. Since cationic as well as ionizable lipids were known to interact with cell membranes and in this way induced intracellular delivery and release [32–34], we preferred to study the light-activated process with neutral lipids and exclude the possible effects of charged lipids. Secondly, compared to cationic liposomes, neutral liposomes were less toxic, more stable, and showed improved pharmacokinetics [35–37]. However, neutral liposomes were generally less efficient in intracellular delivery and transfection, and their endosomal escape and intracellular release properties were poor [38,39]. We aimed to improve the drawbacks of neutral liposomes with light-induced release of oligonucleotides.

We demonstrated effective light-induced release and cell transfection with the ICG liposomes. ICG was required for the light-induced contents' release from the liposomes (Figure S5), and the ICG dose was the major factor influencing the transfection efficacy (Figure 3A–D). The oligonucleotide and lipid concentrations (Figure 3E,F) and extent of light exposure (Figure 4) were less important. The liposomes efficiently released both the encapsulated SSO (M_W ~ 7 kDa) and siRNA (M_W ~ 14 kDa) upon light activation (Figure 1), suggesting that the system was functional with oligonucleotides of different sizes. Light exposure did not affect the activity of the SSO (Figure S3).

Interestingly, free ICG and SSO in solution also led to transfection upon light activation (Figure 5), indicating that the intracellular delivery of ICG and light was not dependent on the liposomal components. However, light triggering without ICG did not lead to any detectable transfection. Hence, ICG is a necessary component for the light-triggered delivery, while the liposomes are optional. The use of non-liposomal, free ICG and oligonucleotides can be applied as a simple delivery system to certain tissues. For example, in the treatment of eye diseases, ICG and oligonucleotides may be injected into the vitreous (to treat posterior eye diseases) or instilled topically (to treat corneal diseases). The small size of naked oligonucleotides and ICG compared to liposomes may be beneficial for penetration into certain targets. The combination of ICG and oligonucleotides to a liposomal formulation has, however, certain advantages. Firstly, this approach avoids problems related to differences in clearance kinetics of oligonucleotides versus ICG. Secondly, it ensures that ICG and the oligonucleotides are taken up by the same cell, which is a requirement for the light-induced transfection. Thirdly, liposomes stabilize ICG in vivo [40,41]. Fourthly, liposomes can be targeted to selected cells by attaching suitable ligands on their surface [42]. Dual targeting with ligands and light triggering could maximize the therapeutic effect in target cells and minimize the exposure of non-target cells.

In general, oligonucleotides should cross at least two cellular membranes to reach their sites of action inside cells in the cytosol or in the nucleus: firstly, they should permeate the cellular membrane surrounding the cell (cellular uptake) and secondly, the endosomal membrane after cellular uptake (endosomal escape). As large and anionic molecules, naked oligonucleotides are not easily taken up by negatively charged cells. Further, the endocytosed oligonucleotides are not released from endosomes, but transferred to lysosomes, where they are degraded enzymatically [43–45]. Cationic and ionizable lipids interact with negatively charged cellular membranes inducing cellular uptake and endosomal escape [32,46]. With ICG and light triggering, the mechanism must be different, as transfection takes place without cationic lipids or even without any delivered lipids (free ICG, Figure 5). A possible mechanism for the increased cellular uptake and endosomal escape induced by ICG and light is partial fluidization of the cellular membranes, allowing oligonucleotides to permeate and cross the cellular lipid bilayers. This hypothesis is supported by the ability of free ICG without lipids to induce transfection, while in the absence of ICG, no transfection takes place (Figure 5B).

We suggest that the increased fluidity of cell membranes and consequent increase in cellular permeability by this delivery system is dependent on two factors: Firstly, the ability of ICG to diffuse into cellular membranes, and this way, to create leakiness. Secondly, released heat of ICG after light exposure causes changes in the organization of lipid bilayers, leading to fluidization of the membranes. The first factor is supported by the ability of free ICG and ICG liposomes to induce some transfection in the absence of light (20% transfection efficacy without light exposure in Figure 3D,F). Yet, the major effect on cellular membrane permeability and successful transfection is most likely the effect of increased temperature after light triggering. We have previously shown using gold nanoparticle-containing liposomes that light activation increases endosomal escape and intracellular release [47]. It is, therefore, likely that a similar mechanism is involved in the endosomal release of oligonucleotides with ICG liposomes and light.

The light exposure or liposome administration separately did not affect cell viability (Figure 7A,B). However, the combination of these treatments reduced cell numbers indicating cytotoxicity (Figure 7C). In general, the observed toxic effects were mostly dependent on the ICG dose. Yet, the effects on cell viability were transient, as the cells reverted to normal state in 4 days (Figure 7C). Apparently, the transfection process causes stress for the cells, leading to slower growth and decreased viability for a certain period. Importantly, the percentage of transfected cells also increased over time (Figure 6). This proves that the observed increase in cell numbers was not solely dependent on growth of the cells that were non-transfected, but also on the transfected cells as they remained viable. A likely mechanism for the cytotoxicity is the local heat production by ICG after light exposure, causing toxic hyperthermia for the cells. However, no increased temperature in the medium surrounding the cells could be detected after treatment with ICG liposomes and light triggering (data not shown). We assume that the quick and transient increase in temperature produced by ICG [18] leads to highly localized temperature effects not detectable in the bulk surroundings.

As such, this light-activated delivery system is potential for the local delivery of oligonucleotides into tissues that can be exposed to light, including the eye, skin, lungs, the gastrointestinal tract, and tumours. The system is excited with NIR light of 800 nm that penetrates deep into tissues [28]. Moreover, utilization of fiberoptic technologies enables light exposure of tissues that cannot be reached by superficial illumination. For example, the posterior segment of the eye is an interesting target. Retinal disorders are the leading causes of impaired vision and blindness, but drug delivery to this tissue is challenging [48,49]. Since transparent ocular tissues allow straightforward light exposure of the posterior eye and neutral nanoparticles diffuse in the vitreous without aggregation, the light-activated liposomes might be suitable for the delivery of therapeutic oligonucleotides in the posterior eye tissues via intravitreal injection [17,50,51]. Another potential application is cancer therapy, where the site- and time-controlled delivery could be used to reduce the off-target effects of toxic anti-cancer drugs. The treatments with ICG-containing liposomes can be further optimized by utilizing the detection of ICG fluorescence in vivo, allowing imaging-guided drug delivery.

5. Conclusions

The present study describes a light-induced method for the cellular delivery of oligonucleotides. The light-triggered ICG liposomes released and delivered oligonucleotides into cultured cells. Interestingly, free ICG also facilitated oligonucleotide delivery to cells in the presence of low intensity NIR light. This method of oligonucleotide delivery avoids the need for cationic lipids that are often accompanied with the activation of innate immune pathways, and at high doses even hepatotoxicity. As such, the light-triggered liposomes might be a good system for local delivery of therapeutic oligonucleotides at places where light penetration is not limited.

Supplementary Materials: The following are available online at http://www.mdpi.com/1999-4923/11/2/90/s1, Table S1: Intensity outputs of the laser with different power settings, Figure S1: Analysis file exported from

Columbus software used to determine the transfection percentages of HeLa S3 IVS2-654 EGFP cells, Figure S2: Transfection efficacy of the negative control oligonucleotide in HeLa S3 IVS2-654 EGFP cells, Figure S3: Effect of light exposure on transfection efficacy of Lipofectamine-SSO complexes in HeLa S3 IVS2-654 EGFP cells, Figure S4: Transfection efficacy and cytotoxicity of Lipofectamine-SSO complexes in HeLa S3 IVS2-654 EGFP cells, Figure S5: Light-triggered release of SSO and siRNA from temperature-sensitive liposomes without ICG.

Author Contributions: Conceptualization, L.-S.K., A.U. and E.M.; methodology, L.-S.K. and E.M.; validation, L.-S.K.; formal analysis, L.-S.K.; investigation, L.-S.K. and J.v.d.D.; data curation, L.-S.K.; writing—original draft preparation, L.-S.K.; writing—review and editing, A.U., W.E.H., and E.M.; visualization, L.-S.K.; supervision, E.M. and W.E.H.; funding acquisition, L.-S.K., A.U., W.E.H., and E.M.

Funding: This research was funded by the Academy of Finland, projects 294315 and 307088.

Acknowledgments: Arto Urtti acknowledges the support by the Government of Russian Federation Mega-Grant 14.W03.031.0025 "Biohybrid technologies for modern biomedicine".

Conflicts of Interest: The authors declare no conflict of interest. The funders had no role in the design of the study; in the collection, analyses, or interpretation of data; in the writing of the manuscript, or in the decision to publish the results.

References

1. Miller, C.M.; Harris, E.N. Antisense oligonucleotides: Treatment strategies and cellular internalization. *RNA Dis.* **2016**, *3*, e1393. [CrossRef] [PubMed]
2. Bhandari, B.; Chopra, D.; Wardha, N. Antisense oligonucleotide: Basic concept and its therapeutic application. *J. Res. Pharm. Sci.* **2014**, *2*, 1–13.
3. Juliano, R.L. The delivery of therapeutic oligonucleotides. *Nucleic Acids Res.* **2016**, *44*, 6518–6548. [CrossRef] [PubMed]
4. Lächelt, U.; Wagner, E. Nucleic Acid Therapeutics Using Polyplexes: A Journey of 50 Years (and Beyond). *Chem. Rev.* **2015**, *115*, 11043–11078. [CrossRef] [PubMed]
5. Wang, Y.; Miao, L.; Satterlee, A.; Huang, L. Delivery of oligonucleotides with lipid nanoparticles. *Adv. Drug Deliv. Rev.* **2015**, *87*, 68–80. [CrossRef] [PubMed]
6. de Jesus, M.B.; Zuhorn, I.S. Solid lipid nanoparticles as nucleic acid delivery system: Properties and molecular mechanisms. *J. Control. Release* **2015**, *201*, 1–13. [CrossRef] [PubMed]
7. Petrilli, R.; Eloy, J.O.; Marchetti, J.M.; Lopez, R.F.; Lee, R.J. Targeted lipid nanoparticles for antisense oligonucleotide delivery. *Curr. Pharm. Biotechnol.* **2014**, *15*, 847–855. [CrossRef]
8. Audouy, S.A.; de Leij, L.F.; Hoekstra, D.; Molema, G. In vivo characteristics of cationic liposomes as delivery vectors for gene therapy. *Pharm. Res.* **2002**, *19*, 1599–1605. [CrossRef]
9. Knudsen, K.B.; Northeved, H.; Kumar, P.E.; Permin, A.; Gjetting, T.; Andresen, T.L.; Larsen, S.; Wegener, K.M.; Lykkesfeldt, J.; Jantzen, K.; et al. In vivo toxicity of cationic micelles and liposomes. *Nanomedicine* **2015**, *11*, 467–477. [CrossRef]
10. Zelphati, O.; Uyechi, L.S.; Barron, L.G.; Szoka, F.C., Jr. Effect of serum components on the physico-chemical properties of cationic lipid/oligonucleotide complexes and on their interactions with cells. *Biochim. Biophys. Acta* **1998**, *1390*, 119–133. [CrossRef]
11. Semple, S.C.; Akinc, A.; Chen, J.; Sandhu, A.P.; Mui, B.L.; Cho, C.K.; Sah, D.W.; Stebbing, D.; Crosley, E.J.; Yaworski, E.; et al. Rational design of cationic lipids for siRNA delivery. *Nat. Biotechnol.* **2010**, *28*, 172–176. [CrossRef] [PubMed]
12. Jayaraman, M.; Ansell, S.M.; Mui, B.L.; Tam, Y.K.; Chen, J.; Du, X.; Butler, D.; Eltepu, L.; Matsuda, S.; Narayanannair, J.K.; et al. Maximizing the potency of siRNA lipid nanoparticles for hepatic gene silencing in vivo. *Angew. Chem.* **2012**, *51*, 8529–8533. [CrossRef] [PubMed]
13. Evers, M.J.W.; Kulkarni, J.A.; van der Meel, R.; Cullis, P.R.; Vader, P.; Schiffelers, R.M. State-of-the-art design and rapid-mixing production techniques of lipid nanoparticles for nucleic acid delivery. *Small Methods* **2018**, *2*, 1700375. [CrossRef]
14. Mura, S.; Nicolas, J.; Couvreur, P. Stimuli-responsive nanocarriers for drug delivery. *Nat. Mater.* **2013**, *12*, 991–1003. [CrossRef] [PubMed]
15. Movahedi, F.; Hu, R.G.; Becker, D.L.; Xu, C. Stimuli-responsive liposomes for the delivery of nucleic acid therapeutics. *Nanomedicine* **2015**, *11*, 1575–1584. [CrossRef] [PubMed]

16. Lajunen, T.; Kontturi, L.S.; Viitala, L.; Manna, M.; Cramariuc, O.; Róg, T.; Bunker, A.; Laaksonen, T.; Viitala, T.; Murtomäki, L.; et al. Indocyanine Green-Loaded Liposomes for Light-Triggered Drug Release. *Mol. Pharm.* **2016**, *13*, 2095–2107. [CrossRef]
17. Lajunen, T.; Nurmi, R.; Kontturi, L.S.; Viitala, L.; Yliperttula, M.; Murtomäki, L.; Urtti, A. Light activated liposomes: Functionality and prospects in ocular drug delivery. *J. Control. Release* **2016**, *244*, 157–166. [CrossRef]
18. Viitala, L.; Pajari, S.; Lajunen, T.; Kontturi, L.S.; Laaksonen, T.; Kuosmanen, P.; Viitala, T.; Urtti, A.; Murtomäki, L. Photothermally triggered lipid bilayer phase transition and drug release from gold nanorod and indocyanine green encapsulated liposomes. *Langmuir* **2016**, *32*, 4554–4563. [CrossRef]
19. Desmettre, T.; Devoisselle, J.; Mordon, S. Fluorescence properties and metabolic features of indocyanine green (ICG) as related to angiography. *Surv. Ophthalmol.* **2000**, *45*, 15–27. [CrossRef]
20. Yannuzzi, L.A. Indocyanine green angiography: A perspective on use in the clinical setting. *Am. J. Ophthalmol.* **2011**, *151*, 745–751. [CrossRef]
21. Alander, J.T.; Kaartinen, I.; Laakso, A.; Pätilä, T.; Spillmann, T.; Tuchin, V.V.; Venermo, M.; Välisuo, P. A review of indocyanine green fluorescent imaging in surgery. *Int. J. Biomed. Imaging* **2012**, *2012*, 940585. [CrossRef] [PubMed]
22. Eichler, J.; Knof, J.; Lenz, H. Measurements on the depth of penetration of light (0.35–1.0 μm) in tissue. *Radiat. Environ. Biophys.* **1977**, *14*, 239–242. [CrossRef] [PubMed]
23. Kielbassa, C.; Roza, L.; Epe, B. Wavelength dependence of oxidative DNA damage induced by UV and visible light. *Carcinogenesis* **1997**, *18*, 811–816. [CrossRef] [PubMed]
24. Wang, H.; Li, X.; Tse, B.W.C.; Yang, H.; Thorling, C.A.; Liu, Y.; Touraud, M.; Chouane, J.B.; Liu, X.; Roberts, M.S.; et al. Indocyanine green-incorporating nanoparticles for cancer theranostics. *Theranostics* **2018**, *8*, 1227–1242. [CrossRef] [PubMed]
25. Lin, L.; Liang, X.; Xu, Y.; Yang, Y.; Li, X.; Dai, Z. Doxorubicin and indocyanine green loaded hybrid bicelles for fluorescence imaging guided synergetic chemo/photothermal therapy. *Bioconjug. Chem.* **2017**, *28*, 2410–2419. [CrossRef] [PubMed]
26. Zhu, Z.; Si, T.; Xu, R.X. Microencapsulation of indocyanine green for potential applications in image-guided drug delivery. *Lab Chip* **2015**, *15*, 646–649. [CrossRef]
27. Fratoddi, I.; Venditti, I.; Cametti, C.; Russo, M.V. How toxic are gold nanoparticles? The state-of-the-art. *Nano Res.* **2015**, *8*, 1771. [CrossRef]
28. Lee, K.; Jeong, J.; Choy, J.H. Toxicity evaluation of inorganic nanoparticles: Considerations and challenges. *Mol. Cell. Toxicol.* **2013**, *9*, 205–210.
29. Sazani, P.; Kang, S.H.; Maier, M.A.; Wei, C.; Dillman, J.; Summerton, J.; Manoharan, M.; Kole, R. Nuclear antisense effects of neutral, anionic and cationic oligonucleotide analogs. *Nucleic Acids Res.* **2001**, *29*, 3965–3974. [CrossRef]
30. Kang, S.H.; Cho, M.J.; Kole, R. Up-regulation of luciferase gene expression with antisense oligonucleotides: Implications and applications in functional assay development. *Biochemistry* **1998**, *37*, 6235–6239. [CrossRef]
31. Lajunen, T.; Nurmi, R.; Wilbie, D.; Ruoslahti, T.; Johansson, N.G.; Korhonen, O.; Rog, T.; Bunker, A.; Ruponen, M.; Urtti, A. The effect of light sensitizer localization on the stability of indocyanine green liposomes. *J. Control. Release* **2018**, *284*, 213–223. [CrossRef] [PubMed]
32. Zelphati, O.; Szoka, F.C. Mechanism of oligonucleotide release from cationic liposomes. *Proc. Natl. Acad. Sci. USA* **1996**, *93*, 11493–11498. [CrossRef] [PubMed]
33. Ruozi, B.; Battini, R.; Tosi, G.; Forni, F.; Vandelli, M.A. Liposome-oligonucleotides interaction for in vitro uptake by COS I and HaCaT cells. *J. Drug Target* **2005**, *13*, 295–304. [CrossRef] [PubMed]
34. Gordon, S.P.; Berezhna, S.; Scherfeld, D.; Kahya, N.; Schwille, P. Characterization of interaction between cationic lipid-oligonucleotide complexes and cellular membrane lipids using confocal imaging and fluorescence correlation spectroscopy. *Biophys. J.* **2005**, *88*, 305–316. [CrossRef] [PubMed]
35. Zhao, W.; Zhuang, S.; Qi, X.R. Comparative study of the in vitro and in vivo characteristics of cationic and neutral liposomes. *Int. J. Nanomed.* **2011**, *6*, 3087–3098. [CrossRef]
36. Fröhlich, E. The role of surface charge in cellular uptake and cytotoxicity of medical nanoparticles. *Int. J. Nanomed.* **2012**, *7*, 5577–5591. [CrossRef]
37. Ernsting, M.J.; Murakami, M.; Roy, A.; Li, S.D. Factors controlling the pharmacokinetics, biodistribution and intratumoral penetration of nanoparticles. *J. Control. Release* **2013**, *172*, 782–794. [CrossRef]

38. Miller, C.R.; Bondurant, B.; McLean, S.D.; McGovern, K.A.; O'Brien, D.F. Liposome-cell interactions in vitro: Effect of liposome surface charge on the binding and endocytosis of conventional and sterically stabilized liposomes. *Biochemistry* **1998**, *37*, 12875–12883. [CrossRef]
39. Xia, Y.; Tian, J.; Chen, X. Effect of Surface Properties on Liposomal siRNA Delivery. *Biomaterials* **2016**, *79*, 56–68. [CrossRef]
40. Kraft, J.C.; Ho, R.J. Interactions of indocyanine green and lipid in enhancing near-infrared fluorescence properties: The basis for near-infrared imaging in vivo. *Biochemistry* **2014**, *53*, 1275–1283. [CrossRef]
41. Kim, T.H.; Chen, Y.; Mount, C.W.; Gombotz, W.R.; Li, X.; Pun, S.H. Evaluation of temperature-sensitive, indocyanine green-encapsulating micelles for noninvasive near-infrared tumor imaging. *Pharm. Res.* **2010**, *27*, 1900–1913. [CrossRef] [PubMed]
42. Riaz, M.K.; Riaz, M.A.; Zhang, X.; Lin, C.; Wong, K.H.; Chen, X.; Zhang, G.; Lu, A.; Yang, Z. Surface functionalization and targeting strategies of liposomes in solid tumor therapy: A review. *Int. J. Mol. Sci.* **2018**, *19*, 195. [CrossRef] [PubMed]
43. Wan, Y.; Moyle, P.M.; Toth, I. Endosome escape strategies for improving the efficacy of oligonucleotide delivery systems. *Curr. Med. Chem.* **2015**, *22*, 3326–3346. [CrossRef] [PubMed]
44. Varkouhi, A.K.; Scholte, M.; Storm, G.; Haisma, H.J. Endosomal escape pathways for delivery of biologicals. *J. Control. Release* **2011**, *151*, 220–228. [CrossRef] [PubMed]
45. Juliano, R.L.; Carver, K. Cellular uptake and intracellular trafficking of oligonucleotides. *Adv. Drug Deliv. Rev.* **2015**, *87*, 35–45. [CrossRef] [PubMed]
46. Zelphati, O.; Szoka, F.C., Jr. Intracellular distribution and mechanism of delivery of oligonucleotides mediated by cationic lipids. *Pharm. Res.* **1996**, *13*, 1367–1372. [CrossRef] [PubMed]
47. Paasonen, L.; Sipilä, T.; Subrizi, A.; Laurinmäki, P.; Butcher, S.J.; Rappolt, M.; Yaghmur, A.; Urtti, A.; Yliperttula, M. Gold-embedded photosensitive liposomes for drug delivery: Triggering mechanism and intracellular release. *J. Control. Release* **2010**, *147*, 136–143. [CrossRef]
48. Urtti, A. Challenges and obstacles of ocular pharmacokinetics and drug delivery. *Adv. Drug Deliv. Rev.* **2006**, *58*, 1131–1135. [CrossRef]
49. Del Amo, E.M.; Urtti, A. Current and future ophthalmic drug delivery systems: A shift to the posterior segment. *Drug Discov. Today* **2008**, *13*, 135–143. [CrossRef]
50. Christie, J.G.; Kompella, U.B. Ophthalmic light sensitive nanocarrier systems. *Drug Discov. Today* **2008**, *13*, 124–134. [CrossRef]
51. Martens, T.F.; Vercauteren, D.; Forier, K.; Deschout, H.; Remaut, K.; Paesen, R.; Ameloot, M.; Engbersen, J.F.; Demeester, J.; De Smedt, S.C.; et al. Measuring the intravitreal mobility of nanomedicines with single-particle tracking microscopy. *Nanomedicine* **2013**, *8*, 1955–1968. [CrossRef] [PubMed]

Article

Targeted Co-Delivery of siRNA and Methotrexate for Tumor Therapy via Mixed Micelles

Fei Hao [1], Robert J. Lee [1,2], Chunmiao Yang [1], Lihuang Zhong [1], Yating Sun [1], Shiyan Dong [1], Ziyuan Cheng [1], Lirong Teng [1], Qingfan Meng [1], Jiahui Lu [1], Jing Xie [1,*] and Lesheng Teng [1,*]

1 School of Life Sciences, Jilin University, Changchun 130012, China; haofei16@mails.jlu.edu.cn (F.H.); lee.1339@osu.edu (R.J.L.); yangcm425@163.com (C.Y.); zhonglh16@mails.jlu.edu.cn (L.Z.); 18744026372@sina.cn (Y.S.); sydong16@mails.jlu.edu.cn (S.D.); chengzy1316@mails.jlu.edu.cn (Z.C.); tenglirong@jlu.edu.cn (L.T.); mengqf@jlu.edu.cn (Q.M.); lujh@jlu.edu.cn (J.L.)

2 College of Pharmacy, The Ohio State University, Columbus, OH 43210, USA

* Correspondence: xiejing@jlu.edu.cn (J.X.); tenglesheng@jlu.edu.cn (L.T.); Tel.: +86-138-4300-4264 (J.X.); +86-138-4418-1693 (L.T.)

Received: 24 December 2018; Accepted: 14 February 2019; Published: 21 February 2019

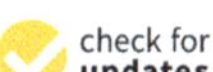

Abstract: A combination of chemotherapeutic drugs and siRNA is emerging as a new modality for cancer therapy. A safe and effective carrier platform is needed for combination drug delivery. Here, a functionalized mixed micelle-based delivery system was developed for targeted co-delivery of methotrexate (MTX) and survivin siRNA. Linolenic acid (LA) was separately conjugated to branched polyethlenimine (b-PEI) and methoxy-polyethyleneglycol (mPEG). MTX was then conjugated to LA-modified b-PEI (MTX-bPEI-LA) to form a functionalized polymer-drug conjugate. Functionalized mixed micelles (M-MTX) were obtained by the self-assembly of MTX-bPEI-LA and LA-modified mPEG (mPEG-LA). M-MTX had a narrow particle size distribution and could successfully condense siRNA at an N/P ratio of 16/1. M-MTX/siRNA was selectively taken up by HeLa cells overexpressing the folate receptor (FR) and facilitated the release of the siRNA into the cytoplasm. In vitro, M-MTX/siRNA produced a synergy between MTX and survivin siRNA and markedly suppressed survivin protein expression. In tumor-bearing mice, M-MTX/Cy5-siRNA showed an elevated tumor uptake. In addition, M-MTX/siRNA inhibited tumor growth. Immunohistochemistry and a western blot analysis showed a significant target gene downregulation. In conclusion, M-MTX/siRNA was highly effective as a delivery system and may serve as a model for the targeted co-delivery of therapeutic agents.

Keywords: drug co-delivery; methotrexate; siRNA; antitumor effect; mixed micelles; targeted delivery system

1. Introduction

The combination of chemotherapy drugs and siRNA is recently emerging as a strategy for cancer therapy [1–3]. Rational design for the combination therapy strategy is essential to obtain maximum efficacy with minimum dosage, side effects, and drug tolerance. Methotrexate (MTX) is a dihydrofolate reductase (DHFR) inhibitor and has shown anticancer activity [4,5]. However, MTX monotherapy at high dosage is often associated with systemic toxicity, drug resistance, and low efficacy [6,7]. Therefore, MTX needs to be combined with other agents to reduce side effects and enhance tumor efficacy [8,9]. Small interfering RNA (siRNA) with a length of 20–25 base pairs has drawn much attention from researchers and is regarded as a potential therapeutic modality for cancer [10,11]. A siRNA could specifically silence the expression of the targeted gene through RNA interference [12]. Survivin has recently been found to be a crucial protein to tumor growth and metastasis and is a promising therapeutic target for tumor [13]. Moreover, studies had indicated that the survivin could facilitate tumor drug resistance. The inhibition of survivin expression may boost the chemotherapeutic efficacy of cancer [14]. Thus, a co-delivery of siRNA targeting survivin expression and MTX may be a promising approach to overcome cancer drug resistance [1]. In addition, the distinct mechanisms of MTX and survivin siRNA suggest that their combination may produce synergy [15].

Nanocarriers such as liposomes, micelles, dendrimers, or supramolecular systems have been evaluated as vehicles for the co-delivery of chemotherapeutic drugs and gene agents [2,16]. Micelles for the co-delivery of chemotherapeutic drugs and nucleic acid have been shown to have good stability and to control drug release. They can increase the effectiveness of drug combination therapy and can reduce drug resistance [17–20]. Compared to the pristine micelles, mixed micelles self-assembled from two or more amphiphilic polymers provide greater flexibility. They have recently drawn much attention for use in combination cancer therapy [21,22]. The mixed micelles are easy to optimize, in terms of kinetic stability, drug loading capacity, size distribution, and the preparation of multifunctional carriers [21–24]. The mixed micelles also could be prepared with simplified procedures, could achieve a desirable antitumor efficacy, and could reduce variability when expanding to a large scale for clinical application [25,26]. However, the efficacy of the mixed micelles is still restricted due to rapid drug release especially for the co-delivery of two or more different types of therapeutic drugs in blood circulation. Polymer-drug conjugates have been studied as nanomedicine for the improvement of disease treatment efficacy recently [27]. Drug covalently bound to the polymers could avoid the drug dissociation and rapid clearance and could improve the drug stability during dilution or exposure to components in blood. Therefore, a system for the co-delivery of siRNA and MTX was developed based on mixed micelles consisted of functionalized polymer-drug conjugates [28].

MTX conjugated to dendrimers or polymers had previously been shown to retain good antitumor activity in vitro and in vivo [29,30]. Moreover, MTX has been suggested to act both as a targeting ligand and a therapeutic agent in recent studies [31]. When conjugated to dendrimers, MTX may directly target the folic acid (FA) receptor, be internalized into the cell, and then act on its target [32]. Polyethylenimine (PEI) has been used extensively for the delivery of nucleic acids such as siRNA, miRNA, and oligonucleotides because of its superior ability to electrostatically complex with nucleic acids and to facilitate the endosomal escape through its proton sponge effect. However, PEI has had limited clinical use due to its toxicity [33,34]. We previously demonstrated that fatty acid modified PEI showed reduced toxicity and enhanced the efficiency for oligonucleotides delivery [35,36]. Here, MTX was conjugated to linolenic acid modified branched PEI (MTX-bPEI-LA). The MTX-bPEI-LA has several advantages: (i) MTX conjugated to the polymers by amide bond is stable and the MTX conjugated micelles have a long circulation time; (ii) it reduced systemic toxicity compared to free MTX [37]; (iii) bPEI-LA with MTX conjugation may be less toxic than bPEI due to a reduction in charged amino groups [38], and (iv) polymer-MTX conjugates may act in dual roles of a targeting and a therapeutic agent [31,32]. In this study, we also synthesized linolenic acid (LA)-modified methoxy-polyethyleneglycol (mPEG-LA), which was combined with MTX-bPEI-LA to formed mixed micelles (M-MTX) (Scheme 1). In order to evaluate the co-delivery efficiency of the system, we studied their cellular uptake, cytotoxicity, siRNA target modulation, biodistribution and the therapeutic effect in vitro and in vivo.

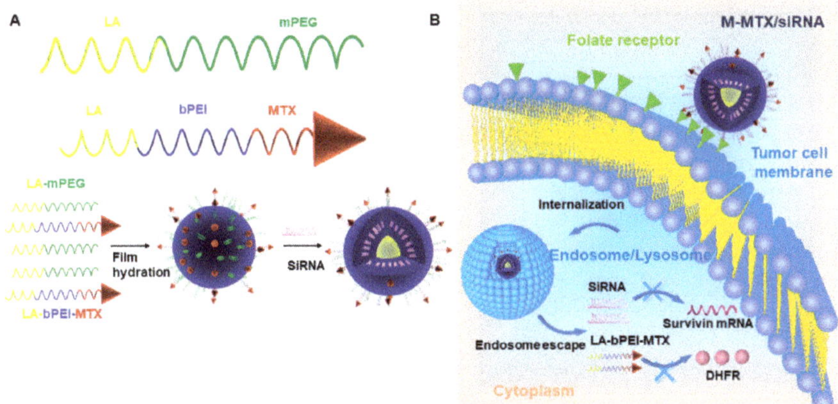

Scheme 1. The preparation of the methotrexate (MTX)-conjugated mixed micelles (M-MTX) and the co-delivery of MTX and siRNA to the cytoplasm. (**A**) The amphiphilic polymers of mPEG-LA (linolenic acid-modified methoxy-polyethyleneglycol) and MTX-bPEI-LA (MTX conjugated to linolenic acid-modified branch polyethylenimine) were first synthesized and then self-assembled in one step to form M-MTX and then incubated with siRNA to form the M-MTX/siRNA complexes (M-MTX/siRNA). (**B**) M-MTX/siRNA complexes could be efficiently taken up by the tumor cells through the folate receptor (FR)-mediated endocytosis, could successfully release the loadings to the cytoplasm and could produce a synergy between survivin siRNA and MTX-bPEI-LA with gene silencing and reduced enzyme activity.

2. Materials and Methods

2.1. Materials

Branched polyethylenimine (bPEI, 25 kDa) was purchased from Sigma-Aldrich (St. Louis, MO, USA). Methotrexate (MTX) and 3-(4,5-dimethyl-2-thiazolyl)-2,5-diphenyl-2-H-tetrazolium bromide (MTT) were purchased from Shanghai Yuanye Biological Technology (Shanghai, China). mPEG-NH_2 (2000 Da) was purchased from Yarebio (Shanghai, China). Linolyl chloride (LC) was obtained from Tokyo Chemical Industry Co., Ltd. (Shanghai, China). Survivin siRNA: Sense (5′–3′): mGCAGGUUCCUmUAUCUGUCAdTdT; Antisense (5′–3′): UGAmCAGAmUAAGGAACCUGmCdTdT; Survivin siRNA negative control: Sense (5′–3′): mUUCUCCGAACmGUGUCACGUdTdT; Antisense (5′–3′): ACGmUGACmACGUUCGGAGAmAdTdT, and Cy3, 5′-FAM and Cy5-labeled survivin siRNA for cellular uptake and biodistribution studies were synthesized by Ribo Biochemistry (Guangzhou, China). HeLa cells were purchased from ATCC (Rockefeller, MD, USA). 4′,6-Diamidino-2-phenylindole (DAPI) and Lyso Tracker™ Green DND-99 were purchased from Invitrogen Co. (Carlsbad, CA, USA). The Dihydrofolate Reductase Assay Kit was also purchased from BioVision (S. Milpitas Blvd., Milpitas, CA, USA). All chemical reagents used were of analytical grade.

2.2. Synthesis and Characterization of the Amphiphilic Polymers

LA was separately conjugated to the mPEG and b-PEI as shown in Figure S1 using a previously reported method [35,36]. Briefly, linolenic chloride (LC) dissolved in anhydrous dichloromethane (DCM) (Sinopharm Chemical Reagent Co., Ltd., Shanghai, China) was added dropwise to the mPEG2000-NH_2 and bPEI (25 kDa) anhydrous DCM solution, respectively. After 12 h, the reaction mixture was precipitated and washed three times by diethyl ether (Sinopharm Chemical Reagent Co., Ltd., Shanghai, China). The products mPEG-LA and bPEI-LA were obtained by removing organic solvent in a rotary evaporator (Shanghai Yukang Scientific Instrument Co., Ltd.,

Shanghai, China) and then vacuumed for 2 h. MTX modified bPEI-LA(MTX-bPEI-LA) was prepared through the reaction of the amino groups of bPEI-LA and carboxy groups of MTX (Figure S1B). The activated reagents of 1-hydroxybenzotriazole (HOBT) (Xiya Chemical Industry Co., Ltd., Linshu, China), *O*-benzotriazole-*N*,*N*,*N*′,*N*′-tetramethyl-uroniumhexafluorophosphate (HBTU) (Xiya Chemical Industry Co., Ltd., Linshu, China), and *N*,*N*-diisopropylethylamine (DIEA) (Sinopharm Chemical Reagent Co., Ltd., Shanghai, China) were first added to the MTX solution to activate the carboxyl groups of MTX for 2 h. Then, the bPEI-LA solution dissolved in anhydrous methanol (Sinopharm Chemical Reagent Co., Ltd., Shanghai, China) was added dropwise to the MTX solution. The mixture was incubated at room temperature for 24 h under nitrogen atmosphere. The reaction mixture was placed in a dialysis bag with a molecular weight cutoff (MWCO) of 8000 to 14,000 Da and dialyzed against deionized water. The dialysate was changed every 4 h. After 48 h, MTX-bPEI-LA was freeze-dried on a Christ epsilon 2-6D LSC (Osterode, Germany). The structures of mPEG-LA, bPEI-LA, and MTX-bPEI-LA were confirmed by ^{1}H NMR on a spectrometer from Bruker (Fällanden, Switzerland). mPEG-LA and bPEI-LA were dissolved in deuterated chloroform ($CDCl_3$, Cambridge Isotope Laboratories, Inc., Tewksbury, MA, USA). MTX-bPEI-LA was dissolved in deuterated water (D_2O, (Cambridge Isotope Laboratories, Inc., Tewksbury, MA, USA). The concentration of MTX in MTX-bPEI-LA was determined based on a calibration curve of MTX, and the drug reaction efficiency was calculated. The reaction efficiency of MTX was defined as the ratio of the weight of MTX which was conjugated to the bPEI-LA to the total weight of MTX added to the reaction. The drug loading efficiency was obtained by calculating the ratio of the weight of MTX conjugated to bPEI-LA to the total weight of MTX-bPEI-LA.

2.3. Preparation of MTX-Conjugated Mixed Micelles (M-MTX)

M-MTX was prepared by the self-assembly of MTX-bPEI-LA and mPEG-LA. A specified volume of MTX-bPEI-LA and mPEG-LA solution (with a molar ratio of 1:200) dissolved in chloroform (Sinopharm Chemical Reagent Co., Ltd., Shanghai, China) was mixed together and sonicated for 3 min. Then, the mixture solution was evaporated by a rotary evaporator to remove the chloroform at 37 °C and further under vacuum for 2 h to remove the residual organic solvent and to obtain a film on the flask. To prepare M-MTX, diethyl pyrocarbonate (DEPC)-treated water (Coolaber, Beijing, China) was added to the flask and sonicated for 2 min. The particle size, zeta potential, and polydispersity index (PDI) of M-MTX were measured on a Zeta-sizer Nano ZS90 from Malvern Instruments (Malvern, UK) at 25 °C.

2.4. Preparation and Characterization of M-MTX/siRNA Complexes

Gel retardation assays were performed to investigate the ability of M-MTX to complex siRNA using agarose gel electrophoresis. M-MTX with different concentrations and survivin siRNA solutions were first diluted to prepare the M-MTX/siRNA complexes with different N/P ratios. The desired amount of siRNA solution was then mixed with an equal volume of the M-MTX solution by gentle pipetting. The complexes were incubated for 10 min at room temperature before use. Then, 10 μL of the M-MTX/siRNA complexes with different N/P ratios were mixed with 2 μL of 6× loading dye and loaded into a 2% agarose gel. The voltage of electrophoresis (BIO-RAD Laboratories, Hercules, CA, USA) was set up at 100 V and run for 10 min in a Tris-acetate-EDTA (TAE) buffer (Beijing Dingguo Changsheng Biotechnology Co., Ltd., Beijing, China). After that, the gel was placed in a staining solution containing Molecular Probes SYBR®Gold nucleic acid (Invitrogen, Ltd., Willow Creek Road, Eugene, OR, USA) for 30 min. Free siRNA in the complexes could be detected as a band on the gel with a GelDoc-It Ts Imaging System (Analytik Jena US LLC., Upland, CA, USA). The particle size and the zeta potential of the M-MTX/siRNA complexes were measured by Nano ZS90 (Malvern, UK).

2.5. In Vitro siRNA Release

The in vitro siRNA release curve of FAM-siRNA loaded M-MTX (M-MTX/FAM-siRNA) in phosphate buffer saline (PBS) was studied. 1 mL M-MTX/FAM-siRNA complexes were transferred into a dialysis bag (MWCO 100 KDa, Shanghai Yuanye Biological Technology, Shanghai, China). The dialysis bag was immersed in 40 mL of PBS and stirred at 37 °C at a speed of 100 rpm. At fixed time intervals, 100 μL of the external solution was withdrawn and replaced with the same volume of fresh PBS. The fluorescence intensity of FAM-siRNA was measured by Bio Tek SYNERGY4 (Winooski, VT, USA) at λ_{ex} = 485 nm and λ_{em} = 535 nm, and the concentrations of FAM-siRNA were measured based on a calibration curve of FAM-labeled siRNA with known concentrations.

2.6. Cell Culture

HeLa cells with high folate receptor (FR) expression were used to evaluate the cellular uptake of the M-MTX/siRNA complexes. HeLa cells were cultured in DMEM (Carlsbad, CA, USA) which contained 10% fetal bovine serum (FBS) (Gemini, Woodland, CA, USA) and 1% penicillin-streptomycin (Carlsbad, CA, USA) at 37 °C in a humidified atmosphere of 5% CO_2.

2.7. Hemolytic Analysis of M-MTX and MTX-bPEI-LA on Murine Erythrocytes

Fresh blood samples from healthy mice were collected from the orbital sinus in heparin-coated tubes. Red blood cells (RBCs) were collected by centrifuging at 3000 rpm for 5 min and washed three times with physiological saline solution. Then, the RBCs were dispersed in the physiological saline solution to obtain a 2% erythrocyte standard dispersion (*v/v*). Then, 200 μL various concentrations of MTX and MTX-bPEI-LA (equivalent to MTX-bPEI-LA at concentrations of 0, 40, 80, 100, 150, and 200 μg/mL) were incubated with 1 mL erythrocyte standard dispersion for 3 h. The suspensions were centrifuged, and the absorbance of supernatant (100 μL) was measured at 450 nm.

2.8. The Viability of Cell Cultures Exposed to MTX-bPEI-LA and M-MTX

The carrier cytotoxicity of the MTX-bPEI-LA and M-MTX was studied. The HeLa cells were plated in 96-well microtiter plates (5000 cells per well) and cultured overnight. M-MTX and MTX-bPEI-LA at 3 concentrations (1, 5, and 20 μg/mL) were added. After another 24 h, 20 μL of the MTT solution (5 mg/mL) was added and incubated for 4 h to form formazan crystals. The medium was removed, and the formazan crystals were dissolved by adding 150 μL DMSO and were incubated for 15 min at 37 °C. Absorbance values at 490 nm were measured on Bio Tek SYNERGY4 (Winooski, VT, USA). Relative cell viability was determined and was presented as a viability percentage of the untreated cells.

2.9. Cellular Uptake of the M-MTX/Cy3-Labeled siRNA Complexes

Flow cytometry was first used to investigate the cellular uptake of M-MTX/siRNA complexes. The HeLa cells were seeded in 12-well cell culture plates at a concentration of 1×10^5 per well and cultured for 24 h to be attached to the plates. In order to study the FR targeting ability of the M-MTX/Cy3-labeled siRNA (Cy3-siRNA) complexes, the HeLa cells were first preincubated with free FA (0.1 mM and 1 mM) for 1 h to competitively bind to FR. Mixed micelles without an MTX conjugation but loaded with Cy3-siRNA (M/Cy3-siRNA) were set as a non-FR targeting control. Naked Cy3-siRNA, M/Cy3-siRNA, and M-MTX/Cy3-siRNA with an equivalent siRNA concentration of 100 nM were then added to the plates. After 4 h of incubation, the HeLa cells treated with different formulations were trypsinized, harvested by centrifugation, washed with cold PBS, and resuspended with 4% formaldehyde solution (*w/v*) (Beijing Dingguo Changsheng Biotechnology Co., Ltd., Beijing, China). The fluorescence intensity of the cells was measured on a Beckman Coulter EPICS XL flow cytometer (Brea, CA, USA). The cellular uptake of M-MTX/Cy3-siRNA was further visualized on a confocal laser scanning microscopy (CLSM). The HeLa cells were collected, counted, and then seeded at the bottom of glass flasks for 12 h. The medium was replaced with fresh opti-MEM (Thermo scientific, Rockford, IL, USA) and preincubated with free folic acid (1 mM) for 1 h. The cells were then treated with naked siRNA, M/Cy3-siRNA, and M-MTX/Cy3-siRNA complexes with an equivalent siRNA concentration of 100 nM at 37 °C. After 4 h of incubation, the medium was removed and the cells were washed gently three times with PBS (0.01 M, pH 7.4). Then, the cells were fixed with 4% (*w/v*) formaldehyde for 15 min at room temperature and washed repeatedly with PBS three times to remove the formaldehyde. Subsequently, the nuclei were stained with DAPI for 10 min, and the cells were collected after washing with PBS to remove the residual dye. The uptake of the M-MTX/Cy3-siRNA complexes in the HeLa cells was observed using an LSM710 microscope from Carl Zeiss (Oberkochen, Germany).

2.10. Internalization and Endosome Escape of M-MTX/FAM-siRNA Complexes in HeLa Cells

The HeLa cells were seeded at the bottom of glass flasks at a cell concentration of 1×10^5 and cultured for 24 h. The medium was replaced with fresh opti-MEM and incubated with M-MTX/ FAM-siRNA complexes (100 nM) for 1 h, 2 h, and 4 h, respectively. After washing with PBS, the cells were then incubated with Lyso Tracker™ Red DND-99 for 30 min. Then, the supernatant was removed, and the cells were gently washed. The cells were then sequentially fixed with 4% (*w/v*) formaldehyde and stained with DAPI. After washing away the residual dye, the internalization and endosome escape of M-MTX/ FAM-siRNA complexes was observed on CLSM.

2.11. Cell Cytotoxicity of the M-MTX/Survivin-siRNA

The cytotoxicity of the M-MTX/survivin-siRNA complexes was investigated. The HeLa cells were plated in 96-well microtiter plates (1×10^4 cells per well) and cultured overnight. The cells of the designed wells were pretreated with free FA (1 mM) for 1 h. MTX, M-MTX, M-MTX/survivin siRNA negative control, M, M/survivin-siRNA, and M-MTX/survivin-siRNA were added to the wells (survivin siRNA 50 nM, MTX 0.242 μg/mL). M-MTX loaded with a survivin siRNA negative control was defined as an M-MTX/siRNA negative control. After being cultured for 48 h, 20 μL of the MTT solution (5 mg/mL) was added and incubated for 4 h to form formazan crystals. The medium was removed, and the formazan crystals were dissolved by adding 150 μL dimethyl sulfoxide (DMSO, Sinopharm Chemical Reagent Co., Ltd., Shanghai, China). The absorbance values at 490 nm were measured on Bio Tek SYNERGY4 (Winooski, VT, USA). The relative cell viabilities were presented as a viability percentage of the cells treated with different formulations compared to the untreated cell samples which the viabilities were referred to be 100%.

2.12. Western Blot Test

The survivin expression was analyzed by western blot. The HeLa cells were seeded in 6-well cell culture plates at a concentration of 1.5×10^5 cells per well for 24 h at 37 °C in a 5% CO_2 humidified atmosphere. Free FA (1 mM) was preincubated with the cells in the desired wells. MTX, M-MTX, M-MTX/siRNA, M-MTX/siRNA negative control, and M, M/siRNA with an equivalent siRNA concentration to 50 nM were then added to the wells. After 4 h of incubation, the medium was replaced with a fresh medium. After 48 h, the cells of different groups were collected, lysed in radio immunoprecipitation assay (RIPA) lysis buffer containing 1% protease inhibitor cocktail (Sigma-Aldrich, St. Louis, MO, USA) and 2% phenylmethanesulfonyl fluoride (Sigma-Aldrich, St. Louis, MO, USA) and plated on ice for 15 min. Protein fractions were collected by centrifugation at 10,000 rpm at 4 °C for 10 min, quantified by a bicinchoninic acid (BCA) Protein Assay Kit (Thermo scientific, Rockford, IL, USA), subjected to polyacrylamide gel electrophoresis containing 10% sodium dodecyl sulfate (SDS-PAGE), and then transferred to polyvinylidene fluoride (PVDF) membranes (0.45 m, Merck Millipore, Billerica, MA). The membranes were blocked with a 5% bovine serum albumin (BSA) (Sigma-Aldrich, St. Louis, MO, USA) solution (*w/v*) for 4 h at room temperature and incubated with survivin rabbit mAb (71G4B7E, Cell Signaling Technology Inc, Danvers, MA, USA) and GAPDH (ab181602, Abcam, cambridgeshire, UK) antibodies at 4 °C overnight, respectively. Horseradish peroxidase (HPR)-conjugated secondary antibody (Beijing Dingguo Changsheng Biotechnology Co., Ltd., Beijing, China) was added and incubated with the membranes at 4 °C for 4 h. The corresponding protein expression was measured by an electrochemiluminescence (ECL) detection kit (Merck Millipore, Billerica, MA, USA) and visualized by an imaging system (BioSpectrum 600, Analytik Jena US LLC., Upland, CA, USA).

2.13. Dose-Dependent Inhibition Efficiency of MTX and M-MTX on Dihydrofolate Reductase (DHFR) Activity

The inhibition efficiency of MTX and M-MTX on DHFR activity was carried out in accordance to the protocol of the Dihydrofolate Reductase Activity Kit (Colorimetric) from BioVision (S. Milpitas Blvd., Milpitas, CA, USA). Briefly, 40 μL of NADPH (500 μM), 60 μL of the DHFR substrate (15-fold dilution), and 50 μL of a series of concentrations of MTX or MTX-bPEI-LA (an equivalent MTX concentration 62.5 nM to 1000 nM) were added to each well of a 96-well clear plate. Finally, 50 μL of DHFR (250-fold dilution) was added to the wells to initiate the reaction with a total volume of 200 μL per well. The absorbance values at 340 nm were measured for 10 min at room temperature. The wells without MTX and MTX-bPEI-LA were set as the positive controls, and the wells without the DHFR enzyme were set as the negative controls. The Inhibition efficiency was defined as follows:

$$Inhibition\ efficiency\ (\%) = \frac{(\Delta OD\ positive\ control - \Delta OD\ sample)}{(\Delta OD\ positive\ control - \Delta OD\ negative\ control)}$$

2.14. Establishment of Tumor Model

The animal experimental protocol was in compliance with the institutional guidelines and was approved by the Experimental Animal Ethics Committee of the School of Life Sciences, Jilin University. The number for the permit for the animal experiment was 201805003 from the Experimental Animal Ethics Committee of the School of Life Sciences, Jilin University. BALB/c nude mice (female, 6–8 weeks) were obtained from Beijing Vital River Laboratory Animal Technology Co., Ltd. (Beijing, China). Tumor-bearing mice were established through the subcutaneous injection of 5×10^6 HeLa cells into the right rear leg of nude mice after the mice were adapted to the new environment for a week.

2.15. Accumulation of M-MTX/Cy5-Labeled siRNA (Cy5-siRNA) Complexes in Tumor Tissue

Tumor-bearing mice were injected with Cy5-siRNA, M/Cy5-siRNA, and M-MTX/Cy5-siRNA via the tail vein with an equivalent amount of siRNA (1 nmol). The biodistribution of the complexes was visualized by an IVIS®spectrum system from Caliper Life Sciences (Hopkinton, MA, USA) at the second, fourth, and sixth hour after administration. The mice were anesthetized by administrating a 1% (*w/v*) pentobarbital sodium solution to the abdomen, and the optimized parameter (excitation, 640 nm; emission, 680 nm) was set up for image acquisition at various time points. At the sixth hour after administration, the internal organs (Heart, Liver, Spleen, Lung, and Kidney) and tumors were dissected and then visualized by the In Vivo Imaging System.

2.16. In Vivo Antitumor Efficacy of M-MTX/Survivin siRNA Complexes

When the average tumor volume of nude mice was grown to approximately 100–150 mm^3 (Day 0), the animals were randomized into four groups, each group containing 5 nude mice. Tumor-bearing mice were then injected with saline, free MTX, the M-MTX/siRNA negative control, and M-MTX/siRNA via the tail vein, respectively. The formulations (MTX 500 μg/kg, siRNA 2 nmol) were administered every 3 days. Simultaneously, the volume of the tumor and body weight were measured every 4 days using a vernier caliper and scale. On day 24, the mice were anesthetized with sodium pentobarbital and sacrificed. The tumors were dissected, weighed, and fixed for hematoxylin and eosin (H&E) and immunohistochemistry staining. The expression of survivin in tumor tissue was also analyzed by a western blot assay. The pixel density of the survivin bands compared to the GAPDH bands was quantified for each sample using Image J software (National Institutes of Health, Bethesda, MD, USA).

2.17. Histopathologic Analysis

All mice were euthanized with a 1% pentobarbital sodium solution on day 24. Vital organs (Heart, liver, Spleen, Lung, and Kidney) and tumors were dissected in each group and fixed with 4% paraformaldehyde for histopathologic analysis. Tissue sections were cut into 5 microns thick and stained with H&E. For the immunohistochemistry analysis, the tumor tissue sections were first incubated with survivin rabbit mAb.

2.18. Statistical Analysis

Data were expressed as mean ± SEM and graphed by Origin 8.0 (OriginLab Corp., Northampton, MA, USA). The statistical analysis of two group differences and correlations was determined using Student's *t*-test. * $p < 0.05$ was considered statistically significant. ** $p < 0.01$ and *** $p < 0.001$ were considered highly significant.

3. Results

3.1. Synthesis and Characterization of MTX-bPEI-LA and mPEG-LA

The chemical structure characterization of mPEG-LA, bPEI-LA, and MTX-bPEI-LA was confirmed by ^{1}H NMR (Figures S2–S4). The peak assignment and peak integration were marked with character. In Figure S2, the new characteristic peak of mPEG-LA was found at 3.37 ppm for 1α-NC=O–C (C). The peak assignment of the 1α-NC=O–C indicated that the LC was successfully conjugated to mPEG-NH_2. Respectively, the characteristic peaks for 1-ethylene, methyl, and methylene of LA were seen at 5.34 ppm (A), 0.88 ppm (F), and 1.62 ppm (E). The characteristic peaks for mPEG were marked with B at 3.65 ppm for ethyl (–H–CH–CH–H). The characteristic peak for amide near 1α-C(=0)–N and 2α-C=C of mPEG-LA was at 2.04 ppm (D). The purity of the mPEG-LA was then calculated to be 91% based on the peak integration of C (1α-NC=O–C, δ = 3.37ppm, 1.01) and F (–H–CH_2–CH–H, δ = 0.88 ppm, 1.67). As shown in Figure S3, the characteristic peak assignments and peak integrations of PEI and LA were both studied. The characteristic peaks of PEI were marked with C (about 2.3–3.00 ppm, 14.27), and the characteristic peak of LA was marked with A (δ = 5.34 ppm, 0.93) for 1-ethylene, E (1β-C–(C=O)–N, δ = 1.56 ppm, 0.40), and F (–H–CH_2–CH–H, δ = 0.88 ppm, 0.81). A new characteristic peak was found in B (1α-N-(C)-C and 1-N–C, δ = 3.63 ppm, 1.14) indicating a successful synthesis of bPEI-LA (Figure S3). The purity of bPEI-LA was calculated to be 94.7% based on the peak integration of B and F (Figure S3). Then, the degree of functionalization of bPEI with LA moieties was calculated to 12. The ^{1}H NMR spectra of MTX-bPEI-LA conjugates contained not only the characteristic peaks of bPEI (G, δ = 2.89 ppm) and LC (D, δ = 5.32 ppm; F, δ = 0.90 ppm) but also the characteristic signals for 1-benzene (B, δ = 7.71 ppm; C, δ = 7.34 ppm) and 2-pyrazine (A, δ = 8.47 ppm) of MTX (Figure S4). Furthermore, the new characteristic peak at 3.52 ppm (E) due to the generation of an amide bond between bPEI and MTX was also characterized. The results indicated that the MTX was successfully conjugated to the bPEI-LA-NH_2. The purity of MTX-bPEI-LA was calculated to be 95.6% based on the characteristic peak integration of B (2.08), C (2.08), and E (4.50). The absorption curve peak of the MTX-bPEI-LA was measured at 303 nm. The MTX reaction efficiency and the drug loading efficiency were calculated to be 59.6% and 12.1% respectively through the standard curve of MTX with a known concentration measured by a UV spectrophotometer. The degree of functionalization of bPEI-LA with MTX moieties was also calculated to 15.

3.2. Preparation and Characterization of M-MTX and M-MTX/siRNA Complexes

M-MTX was prepared by self-assembly of MTX-bPEI-LA and mPEG-LA and had a narrow particle size of 141.8 ± 2.4 nm (PDI = 0.212 ± 0.012) (Figure 1A). SiRNA was then complexed by M-MTX with different N/P ratios (Figure 1B). The result showed that siRNA could be successfully condensed via electrostatic adsorption at an N/P ratio of 16/1 (Figure 1B). M-MTX/siRNA complexes also had a narrow particle size of 124.7 ± 9.1 nm (PDI = 0.231 ± 0.011) (Figure 1C). SiRNA adsorption to the cationic polymer of the carrier may contribute to a decrease in particle size and zeta potential (12.23 ± 0.47 mv decreased to 1.63 ± 0.14 mv). The MTX and the survivin siRNA drug loading efficiencies in the M-MTX/siRNA complexes were then calculated to be 0.17% and 2% (*w/w*) respectively with an N/P ratio of 16/1.

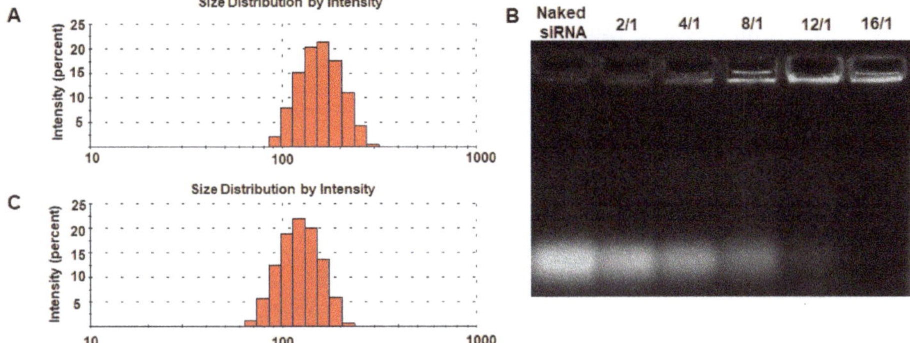

Figure 1. The characterization of the M-MTX and M-MTX /siRNA complexes. (**A**) The particle size of M-MTX was measured by Zeta-sizer Nano ZS90. (**B**) SiRNA was then condensed by M-MTX with a different N/P ratio through the electrostatic interaction. (**C**) The particle size of the M-MTX/siRNA complexes at an N/P ratio of 16/1 was further determined.

3.3. In Vitro FAM-siRNA Release

The in vitro siRNA release curve of FAM-siRNA loaded in M-MTX (M-MTX/FAM-siRNA) was shown in Figure S5. The cumulative release percentage of FAM-siRNA loaded in the M-MTX/siRNA complexes was less than 30% at 48 h. A strong interaction of siRNA with the cationic carrier may play an important role in the slow release of FAM-siRNA loaded in M-MTX. The results showed a controlled release of siRNA and high stability of the M-MTX/FAM-siRNA complexes in PBS (pH = 7.4). The release curve of free MTX in M-MTX in a dialysis bag with an MWCO of 100 KDa was also investigated in PBS (pH = 7.4). However, no free MTX was detected within 72 h. Since MTX was conjugated to the polymers (MTX-bPEI-LA) with a stable amide bond, MTX was hardly released in the release medium at pH = 7.4. The results were consistent with those of the previous study [39].

3.4. Carrier Toxicity of M-MTX and MTX-bPEI-LA

PEG has widely been applied to reduce the cytotoxicity, to prolong the circulation time in the blood, and to avoid uptake by the mononuclear phagocyte system of the cationic carriers due to its well-known good biocompatibility [23,24,40]. In order to investigate whether mPEG-LA had a good ability to reduce the toxicity of the mixed micelles, we performed a hemolytic activity assay and cytotoxicity analysis of M-MTX and MTX-bPEI-LA, as shown in Figures S6 and S7. As shown in Figure S6, MTX-bPEI-LA had a much stronger hemolysis effect when the concentrations of MTX-bPEI-LA were greater than 50 μg/mL compared with M-MTX at the same MTX-bPEI-LA concentration. In Figure S7, MTX-bPEI-LA and M-MTX had the same cytotoxicity when the concentration of MTX-bPEI-LA was low (1 μg/mL). When the concentration of MTX-bPEI-LA was 5 μg/mL, MTX-bPEI-LA showed significant carrier toxicity while M-MTX also had low cytotoxicity. When the concentration was 20 μg/mL, both MTX-bPEI-LA and M-MTX showed higher carrier toxicity. These results indicated that the mPEG-LA could reduce the toxicity of the cationic polymer (MTX-bPEI-LA) within a moderate concentration range which may interact with the cell membrane as the previous study suggested [23,24,40]. Thus, the mixed micelles were expected to have a greater potential in the clinical application compared with MTX-bPEI-LA.

3.5. Cellular Uptake of M-MTX/Cy3-siRNA in FR-Overexpressing HeLa Cells

The cellular uptake of M-MTX/Cy3-siRNA was investigated in the FR-overexpressing HeLa cells. Flow cytometry results showed that the M-MTX/Cy3-siRNA-treated groups had the highest mean fluorescence intensity and had significant improvement compared with the Cy3-siRNA- and M/Cy3-siRNA-treated groups (*** $p < 0.001$) (Figure 2A). Moreover, the fluorescence intensity of the M-MTX/Cy3-siRNA-treated group could be greatly reduced by adding a free FA (0.1 mM and 1 mM) to cells in advance. The similar results were further visualized by CLSM. The red signal in the cytoplasm treated with M-MTX/Cy3-siRNA was markedly stronger than that of the naked Cy3-siRNA and M/Cy3-siRNA and was also greatly reduced due to the free FA (1 mM) preincubated with HeLa cells (Figure 2B).

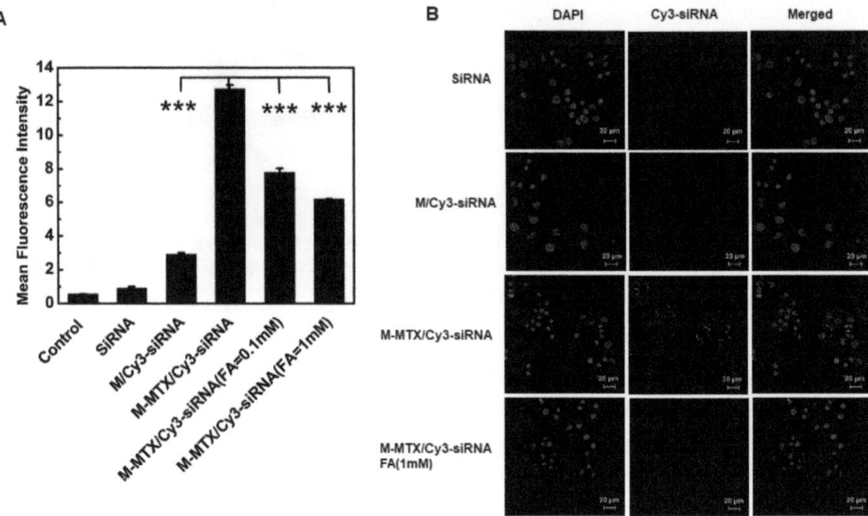

Figure 2. The cell uptake of M-MTX/Cy3-siRNA in the FR-expressing HeLa cells. (**A**) Flow cytometry was used to quantify the cellular uptake of different formulations. The HeLa cells were incubated with M-MTX/Cy3-siRNA, M/Cy3-siRNA, and Cy3-siRNA for 4 h at 37 °C. Two concentrations of free folic acid (FA) (0.1 mM and 1 mM) was added 1 h before the M-MTX/Cy3-siRNA (100 nM) was incubated. *** $p < 0.001$. (**B**) The confocal micrographs of HeLa cells were also obtained after incubation with M-MTX/Cy3-siRNA, M/Cy3-siRNA, and Cy3-siRNA for 4 h at 37 °C. Free FA (1 mM) was also added 1 h before the M-MTX/Cy3-siRNA (100 nM) addition. The green bar in the image was 20 μm. The values were the mean ± SEM (n = 3).

3.6. Internalization and Endosome Escape of M-MTX/ FAM-siRNA in FR-Overexpressing HeLa Cell

The M-MTX/FAM-siRNA complexes could be specifically taken up by HeLa cells. However, the process of internalization, cytoplasmic delivery, subcellular localization, and drug release of FAM-siRNA still remained unclear. Thus, we applied another fluorescent reagent, Lyso Tracker™ Red DND-99, to label the endosomes and to investigate whether FAM-siRNA could escape from the endosomes. As shown in Figure 3, M-MTX/ FAM-siRNA (green signal) started to be internalized to the cells, and the endosomes (red signal) were also found near the cell membrane at the first hour. At the second hour, the overlap of green and red signals gradually increased and the overlapping orange signal started to transport into the cytoplasm. At the fourth hour, the fluorescence signal was completely located in the cytoplasm and a rich green signal was observed to clearly separate from the red signal, indicating that FAM-siRNA were successfully released into the cytosol. The results demonstrated the endocytic process of M-MTX/FAM-siRNA, which finally achieved drug release in the HeLa cells. The proton sponge effect of the cationic polymer of bPEI may play the most important role in achieving endosome escape and drug release [34]. Although some of the amines of bPEI had been reacted with LA and MTX, M-MTX still had a proton sponge effect and achieved siRNA endosomal escape.

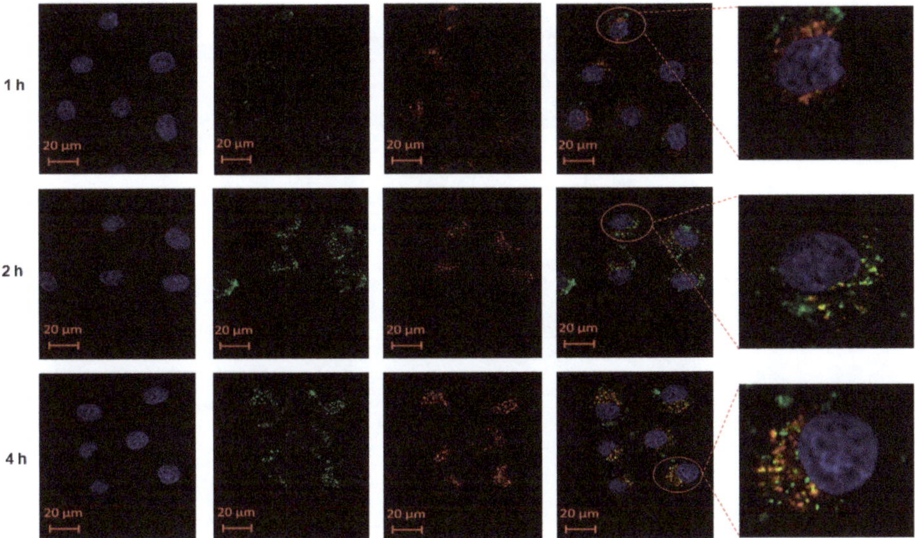

Figure 3. The internalization and endosome escape of M-MTX/siRNA in folate receptor (FR)-overexpressing HeLa cells: The internalization and endosome (red) escape of M-MTX loaded with FAM-siRNA (green) in the Hela cells was visualized by the confocal laser scanning microscopy (CLSM) at the first, second, and fourth hours. The red bar in the images was 20 µm.

3.7. In Vitro Biological Activities of M-MTX/Survivin-siRNA

The synergy of MTX and survivin-siRNA was first studied in HeLa cells by an MTT assay. M-MTX/survivin-siRNA showed enhanced cell cytotoxicity (*** $p < 0.001$) compared to MTX and siRNA alone in vitro (Figure 4A). Simultaneously, the cytotoxic efficiency of the M-MTX/survivin-siRNA-treated group was reversed by free FA(1 mM) while no significant cytotoxicity difference was found between the M/siRNA-treated group preincubated with free FA (1 mM) and the M/siRNA-treated group without FA (1 mM) addition (Figure 4A). The western blot results were also consistent with the cytotoxicity experiments (Figure 4B). M-MTX/survivin-siRNA had the lowest protein expression and could be reversed by the addition of FA (1 mM). MTX was conjugated to bPEI-LA through an amide bond which had previously been demonstrated to be a stable chemical bond and was difficult to degrade. Thus, we assumed that the inhibition of cell viability and protein expression was achieved by MTX-bPEI-LA. To verify this, we investigated the inhibitory activity of MTX and MTX-bPEI-LA on DHFR. As shown in Figure 4C, MTX-bPEI-LA with a larger structure also had a good inhibitory efficiency against DHFR, although slightly lower than MTX, and exhibited a concentration-dependent inhibitory effect. The large structural steric hindrance of MTX-bPEI-LA may result in the reduced binding ability to DHFR. Interestingly, M-MTX had a higher protein expression inhibitory effect in the western blot assay after 48 h compared with free MTX although MTX-bPEI-LA was shown to have a lower inhibitory activity against DHFR (Figure 4B). Survivin expression is closely related to the efficient delivery of survivin siRNA and MTX to the targeted site. This could be explained by a much higher uptake of M-MTX, which had a positive charge and was able to interact with the negatively charged cell membranes. Free MTX, M-MTX, M-MTX/siRNA negative control, and M-MTX/siRNA complexes were incubated with cells only for 4 h and replaced with fresh medium in the western blot assay. M-MTX had a significant decrease in survivin expression due to more M-MTX uptake compared to MTX and the M-MTX/siRNA negative control. The greatest cytotoxicity and protein inhibition caused by M-MTX/siRNA may depend primarily on the enhanced tumor cell uptake of MTX conjugates, while M-siRNA without MTX conjugates with a much smaller amount of cellular uptake showed lower cytotoxicity and more survivin expression. Thus, we proposed that M-MTX had enhanced the anticancer therapy compared to the free MTX and could serve as both a targeting agent and a chemotherapeutic agent for tumor therapy [31,32,41,42].

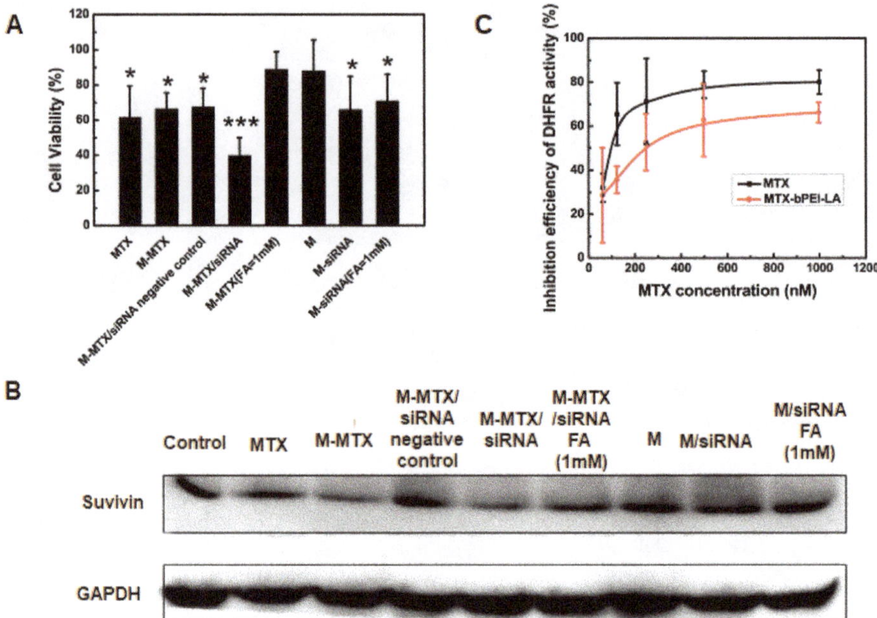

Figure 4. The cytotoxicity, protein expression, and dihydrofolate reductase (DHFR) inhibition activity of M-MTX/siRNA. (**A**) The cell viability was obtained by an MTT assay after incubation with MTX, M-MTX, M-MTX/siRNA negative control, M-MTX/siRNA, M, and M/siRNA for 48 h. * $p < 0.1$, *** $p < 0.001$. (**B**) Survivin expression was further measured by a western blot assay. The competitive inhibition was also investigated by preincubated with FA (1 mM) in advance. (**C**) The inhibition activity of MTX-bPEI-LA to DHFR was performed to evaluate the difference of inhibition activity of the conjugated MTX compared to free MTX.

3.8. Biodistribution of M-MTX/Cy5-siRNA in Tumor-Bearing Mice

The efficient delivery of siRNA and MTX to the targeting site was a prerequisite for a successful treatment. Therefore, we investigated the biodistribution of M-MTX/siRNA in tumor-bearing mice. Mice injected with M-MTX/Cy5-siRNA showed a higher accumulation (about 6-folds higher radiant efficiency) in solid tumors than the M/Cy5-siRNA-treated group without MTX conjugated (Figure 5). At the sixth hour, the red fluorescence signal in the tumor injected with Cy5-siRNA and M/ Cy5-siRNA was barely visible. The same results were also reflected in the anatomical organs (Heart, Liver, Spleen, and Lung). The fluorescence signal of the free Cy5-siRNA injection group was barely detectable in major organs and tumors other than the kidney. This indicated a fast clearance of free siRNA in vivo. However, M-MTX/Cy5-siRNA exhibited the lowest fluorescence intensity in the kidney. This suggested that the Cy5-siRNA loaded in M-MTX had a long circulation time. Compared to M-MTX/Cy5-siRNA, M/Cy5-siRNA had a higher fluorescence signal in the spleen, liver, lung, and kidney but had a lower tumor fluorescence signal. This may be due to the larger size, higher positive charge, and non-targeting ability of the M/Cy5-siRNA which was easily captured by the reticuloendothelial system. The high and specific accumulation of M-MTX/Cy5-siRNA in tumor tissues further indicated a potential and promising imaging application of the carrier in the clinic.

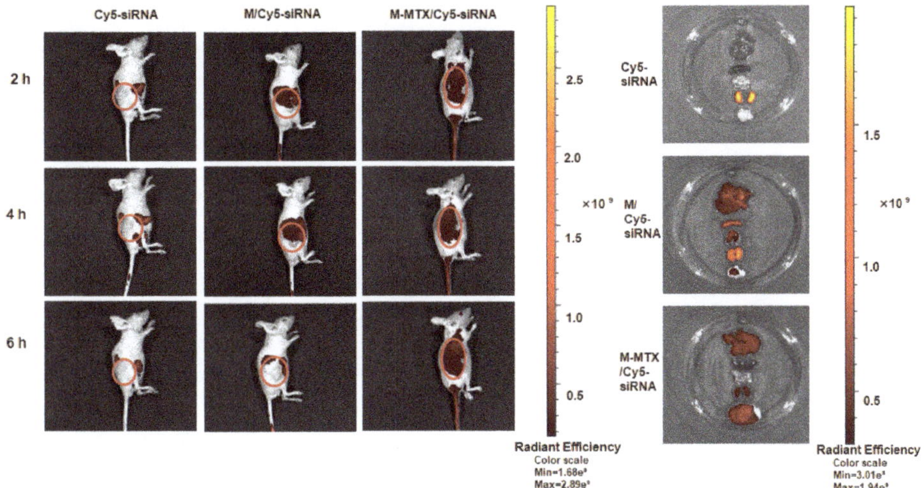

Figure 5. The biodistribution of M-MTX/Cy5-siRNA in tumor-bearing mice. (**Left**) The real-time fluorescence imaging of tumor-bearing mice injected with naked Cy5-siRNA, M/Cy5-siRNA, and M-MTX/Cy5-siRNA via tail vein were obtained at the second, fourth, and sixth hours after administration (Cy5-siRNA = 1 nmol). (**Right**) The ex vivo imaging of the tumors and organs excised from tumor-bearing mice at the sixth hour after injection. The images were obtained by IVIS®In Vivo Imaging System with an optimized parameter (excitation, 640 nm; emission, 680 nm).

3.9. Antitumor Efficacy of M-MTX/siRNA In Vivo

The synergistic antitumor efficacy of MTX and survivin siRNA was further assessed in tumor-bearing mice. Consistent with the in vitro results, tumor growth inhibition and disease remission were significantly achieved in the M-MTX/siRNA treated group with a low MTX and siRNA dose (** $p < 0.01$, $n = 5$) (Figure 6A–C). The M-MTX/siRNA-treated group had a large tumor volume and weight reduction (about 75% and 63%, respectively) compared to other groups. The HeLa cell lines are known for their high tumor growth rate. The Saline, MTX, and M-MTX/siRNA negative control groups showed a rapid increase in tumor size and weight (Figure 6A–C). Compared to MTX, M-MTX/siRNA negative control group had a lower tumor weight (Figure 6C). This suggested that the M-MTX/siRNA negative control had a more enhanced antitumor effect than free MTX, probably due to its extended circulation time (Figure 5) and an enhanced cellular uptake mediated by FR (Figure 2). The body weight of the mice was measured to evaluate the toxicity of the formulations. As shown in Figure 6D, the body weight of the different groups increased steadily, indicating that the formulations had no significant toxicity.

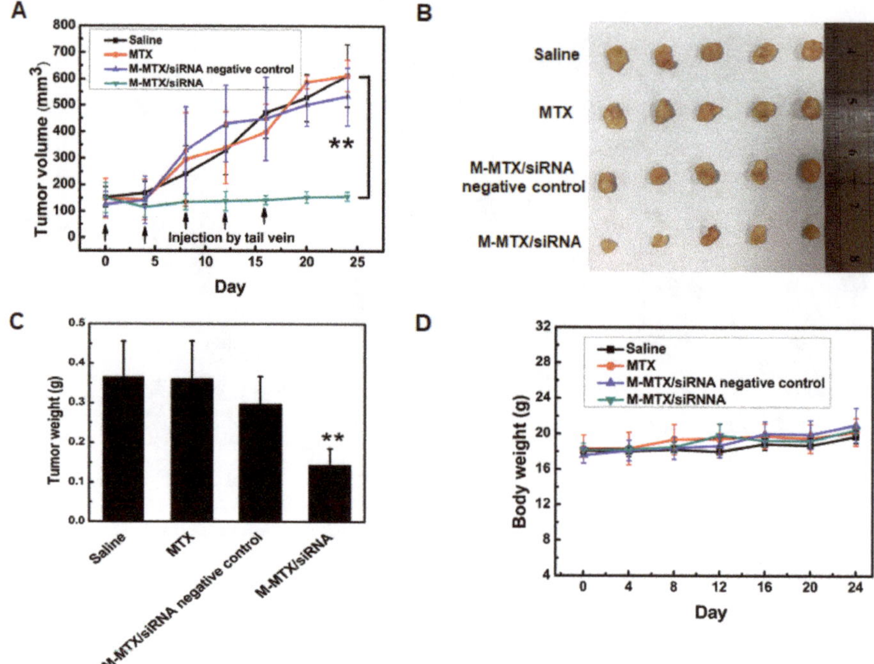

Figure 6. The therapeutic efficacy of M-MTX/siRNA in vivo. The tumor volume (**A**), photographs of the solid tumor (**B**), the tumor weight (**C**) of the animals treated with various drug formulations were obtained on day 24. ($n = 5$, ** $p < 0.01$, M-MTX/siRNA compared to saline). The body weight of the tumor-bearing mice treated with various drug formulations was also measured to assess the system toxicity (**D**). The values are the mean ± SEM, $n = 5$.

3.10. Protein Expression, Immunohistochemistry, and Histopathological Analysis

Survivin expression levels in tumors were also analyzed by a western blot and an immunohistochemical analysis (Figure 7A–C). The western blot results showed a 40% reduction in survivin expression in M-MTX/siRNA-treated tumors compared to the saline group (* $p < 0.05$, $n = 3$) (Figure 7A,B). The same results were also confirmed by the immunohistochemical analysis. The cells stained with brown and black nuclei were shown to be positive for survivin expression as indicated by the black arrow (Figure 7C). Furthermore, HE staining of the tumor tissues treated with M-MTX/siRNA showed a large amount of necrosis of the tumor cells as indicated by the black arrow (Figure 7D). The results of Figures 6 and 7 showed that M-MTX could efficiently deliver siRNA to the cells in tumor tissue, suppressing the related expression of proteins and finally leading to the death of a large number of tumor cells in tumor tissues. There was no significant toxicity in the major organs of mice treated with MTX, M-MTX/siRNA negative control, or M-MTX/siRNA compared to the saline group due to the reduced dose of MTX and siRNA (Figure 8). M-MTX exhibited good biocompatibility and low toxicity as a non-viral delivery for siRNA delivery.

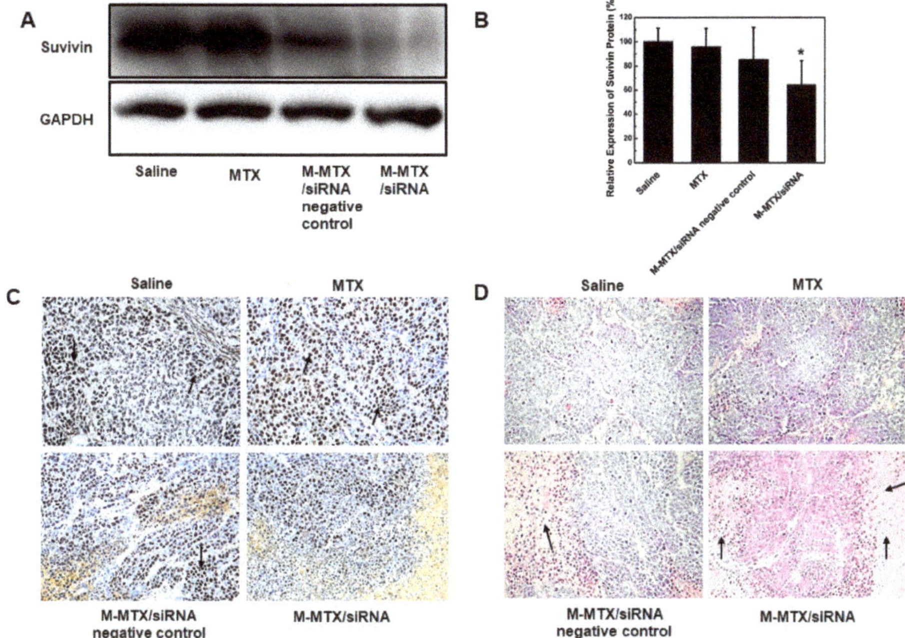

Figure 7. The protein expression, immunohistochemistry, and histopathological analysis of tumor tissues. Survivin expression in tumor tissue was determined by western blot assay (**A**) and then quantified by image J (**B**) (n = 3, * p < 0.05, M-MTX/siRNA compared to saline). The immunohistochemical analysis of protein expression (**C**) and the hematoxylin-eosin staining for tumor tissue necrosis (**D**) was also investigated.

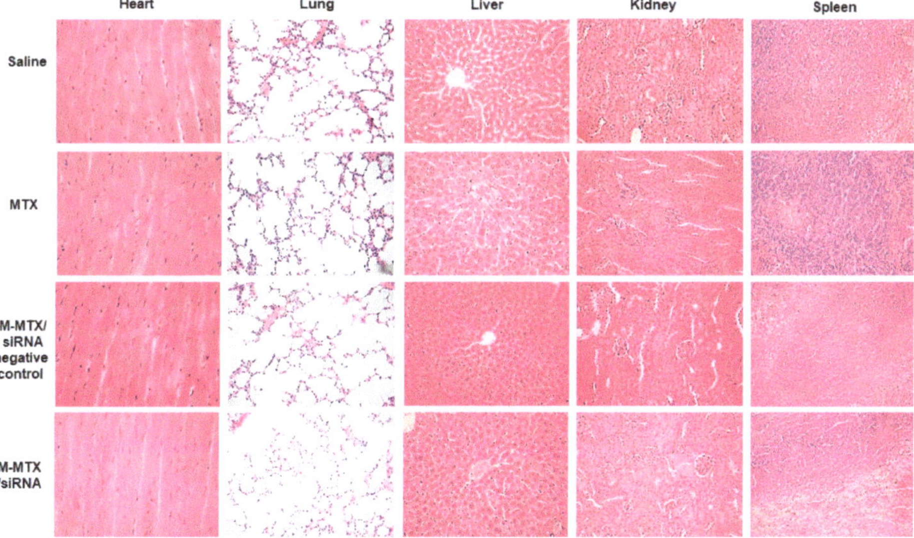

Figure 8. The organ toxicity analysis of different formulations: Major organs (Heart, Lung, Liver, Kidney, and Spleen) were dissected and stained with hematoxylin-eosin to assess the systemic toxicity of the drug formulations.

4. Discussion

MTX, a FA analog, has long been used for cancer therapy but is often associated with severe systemic toxicity, bone marrow suppression, and drug resistance [6,7,43]. In addition, MTX monotherapy has limited effectiveness. Therefore, the new combination strategy of MTX and siRNA as well as a novel drug delivery system that reduces MTX toxicity and improves cancer efficacy is desirable [8,9]. As one of the strongest tumor apoptosis inhibitors, survivin not only promotes tumor cell proliferation but also is closely related to the development of tumor resistance [14,44,45]. Thus, the combination therapy of MTX and survivin-siRNA demonstrated a new possibility for enhanced tumor efficacy. In order to efficiently deliver MTX and survivin-siRNA into the cells, modified cationic polymer-based mixed micelles having an MTX targeting ability were designed due to their simplicity in synthesis and preparation.

LA is a polyunsaturated fatty acid that is essential for the body and is nontoxic. Moreover, recent studies reported that LA also inhibited tumor cell growth and metastasis [46–48]. Therefore, LA was selected to be separately conjugated to the b-PEI and mPEG to form two amphiphilic polymers of bPEI-LA and mPEG-LA in this work. The hydrophobic molecule of LA conjugated to bPEI was designed to reduce the high toxicity of bPEI [38]. MTX was then conjugated to the bPEI-LA by an esterase-stable amide linkage (Figure S1). MTX was conjugated to bPEI-LA by an amide linker to prevent and avoid MTX release in the blood and to enhance the targeting ability of MTX to the FA receptor. Then, M-MTX self-assembled by MTX-bPEI-LA and mPEG-LA was prepared and applied to efficiently co-deliver MTX and siRNA. M-MTX was aimed to target the tumor cells overexpressing FR and successfully release loadings into the cytoplasm as shown in Scheme 1. Actually, some researchers had revealed that MTX conjugated to the dendrimers showed a potential FR targeting ability through the tight-binding of MTX to the folate binding protein [49]. From the reversed cellular uptake (Figure 2) and biological activities (Figure 4) of M-MTX/Cy5-siRNA-treated groups preincubated with free FA and the high cellular uptake (Figure 2), biological activities (Figure 4) in vitro, and accumulation of M-MTX/Cy5-siRNA in the solid tumor (Figure 5), we point out that M-MTX nanocarrier exhibits an FR targeting ability and a much higher internalization efficiency via FA receptor-mediated endocytosis than M/siRNA complexes [50]. In Figure 4C, when M-MTX and MTX were incubated with cells for 4 h, the M-MTX-treated group exhibited a lower protein expression compared to free MTX. This also indicated that M-MTX entered more into the cells and had a higher internalization efficiency compared to free methotrexate. The siRNA and MTX-bPEI-LA conjugates with stable chemical linkage achieved endosome escape (Figure 3) and were released probably by the proton sponge effect of cationic polymer of bPEI [33,34]. As expected, M-MTX/survivin siRNA achieved tumor growth inhibition and disease remission with a low MTX and siRNA dose in tumor-bearing mice. There was no significant difference in the antitumor effect among saline, MTX, and the M-MTX/siRNA negative control with a low dose of MTX, which was not sufficient to achieve disease remission. However, the M-MTX/siRNA complexes exhibited an unexpectedly potent effect compared to MTX and the M-MTX/siRNA negative control. This may largely depend on the specific targeting and efficient delivery of siRNA by the bifunctional vector of M-MTX. Moreover, previous studies have shown that survivin siRNA may have a positive effect on chemotherapeutic drug sensitivity and may reduce drug resistance [14]. Thus, we propose that the effect of mutual promotion and the sensitivity of the two drugs at a low dose also contribute to the desired efficacy. The M-MTX carrier exhibited good biocompatibility and low toxicity as a non-viral carrier for siRNA co-delivery in vivo. However, the specific and detailed mechanisms and synergy of MTX and siRNA warrant further investigation.

5. Conclusions

In this study, an M-MTX/siRNA co-delivery system was developed based on the mixed micelles composed of mPEG-LA and MTX-bPEI-LA. M-MTX/siRNA exhibited a good ability to target tumor cells overexpressing FR. M-MTX/siRNA was internalized by FR-mediated endocytosis and was able to release siRNA into the cytoplasm. Studies in tumor-bearing mice showed that M-MTX/siRNA could greatly target tumor tissue and inhibit tumor growth in vivo. All above, M-MTX could be applied as a promising and safe nanoplatform for siRNA or even other therapeutic agent deliveries and may find utility in anticancer therapy.

Supplementary Materials: The following are available online at http://www.mdpi.com/1999-4923/11/2/92/s1, Figure S1. The synthesis procedure of the amphiphilic polymers. mPEG-LA was synthesized by the reaction of oleic acid chloride and mPEG-NH2 (A). MTX-bPEI-LA was synthesized by two steps (B). BPEI-LA was first synthesized, and then, the MTX was conjugated to finally obtained MTX-bPEI-LA; Figure S2. The ^{1}H NMR of mPEG-LA. mPEG-LA was dissolved in $CDCl_3$ and measured in a nuclear magnetic resonance instrument (500 MH_Z). The characteristic peaks of mPEG and LA were marked with the colour characters A, B, C, D, E, and F; Figure S3. The ^{1}H NMR of bPEI-LA. bPEI-LA was dissolved in $CDCl_3$ and measured in a nuclear magnetic resonance instrument (500 MH_Z). The characteristic peaks of bPEI and LA were marked with the colour characters A, B, C, D, E, and F; Figure S4. The ^{1}H NMR of MTX-bPEI-LA. MTX-bPEI-LA was dissolved in D_2O and measured in a nuclear magnetic resonance instrument (500 MH_Z). The characteristic peaks of bPEI, MTX, and LA were marked with the colour characters A, B, C, D, E, F, and G; Figure S5. The in vitro siRNA release profile of FAM-siRNA-loaded mixed micelles in PBS (pH = 7.4); Figure S6. The hemolytic analysis of M-MTX and MTX-bPEI-LA on murine erythrocytes (n = 3, mean ± SEM). 200 μL various concentrations of M-MTX and MTX-bPEI-LA were incubated with 1 mL erythrocytes in a 2% phosphate buffer saline for 3 h. The suspensions were centrifuged, and the absorbance of supernatant (100 μL) was measured at 450 nm; Figure S7. The cytotoxicity of M-MTX and MTX-bPEI-LA (n = 3, mean ± SEM). The cells were plated in 96-well microtiter plates (5000 cells per well) and cultured overnight. M-MTX and MTX-bPEI-LA at 3 concentrations (1, 5, and 20 μg/mL) were added. After another 24 h, the relative cell viability was determined and was presented as a percentage of the viability of untreated cells.

Author Contributions: Conceptualization, L.T., J.X. and F.H.; methodology, F.H., C.Y., L.Z. and Y.S.; software, S.D.; validation, Z.C., Q.M. and J.L.; formal analysis, F.H.; investigation, F.H.; resources, L.T.; data curation, F.H.; writing—original draft preparation, F.H.; writing—review and editing, R.J.L.; visualization, F.H.; supervision, L.T. and Q.M.; project administration, L.T., L.T. and J.L.

Funding: This research received no external funding.

Acknowledgments: The biodistribution and histopathologic analysis in this article were offered help with the IVIS®spectrum system and technical support from the Key Laboratory of Pathology and Biology Teaching of Ministry of Education and Basic Medical Experimental Teaching Center of Jilin University.

Conflicts of Interest: The authors declare no conflict of interest.

References

1. Creixell, M.; Peppas, N.A. Co-delivery of siRNA and therapeutic agents using nanocarriers to overcome cancer resistance. *Nano Today* **2012**, *7*, 367–379. [CrossRef] [PubMed]
2. Kang, L.; Gao, Z.; Huang, W.; Jin, M.; Wang, Q. Nanocarrier-mediated co-delivery of chemotherapeutic drugs and gene agents for cancer treatment. *Acta Pharm. Sin. B* **2015**, *5*, 169–175. [CrossRef] [PubMed]
3. Saraswathy, M.; Gong, S. Recent developments in the co-delivery of siRNA and small molecule anticancer drugs for cancer treatment. *Mater. Today* **2014**, *17*, 298–306. [CrossRef]
4. Rahman, L.K.; Chhabra, S.R. The chemistry of methotrexate and its analogues. *Med. Res. Rev.* **1988**, *8*, 95–155. [CrossRef]
5. McGuire, J.J. Anticancer antiFAs: Current status and future directions. *Curr. Pharm. Des.* **2003**, *9*, 2593–2613. [CrossRef]
6. Moscow, J.A. Methotrexate transport and resistance. *Leuk. Lymphoma* **1998**, *30*, 215–224. [CrossRef]
7. Matherly, L.H.; Taub, J.W. Methotrexate pharmacology and resistance in childhood acute lymphoblastic leukemia. *Leuk. Lymphoma* **1996**, *21*, 359–368. [CrossRef]
8. Khan, Z.A.; Tripathi, R.; Mishra, B. Methotrexate: A detailed review on drug delivery and clinical aspects. *Expert Opin. Drug Deliv.* **2012**, *9*, 151–169. [CrossRef]
9. Abolmaali, S.S.; Tamaddon, A.M.; Dinarvand, R. A review of therapeutic challenges and achievements of methotrexate delivery systems for treatment of cancer and rheumatoid arthritis. *Cancer Chemother. Pharm.* **2013**, *71*, 1115–1130. [CrossRef]
10. Dorsett, Y.; Tuschl, T. siRNAs: Applications in functional genomics and potential as therapeutics. *Nat. Rev. Drug Discov.* **2004**, *3*, 318–329. [CrossRef]
11. Wittrup, A.; Lieberman, J. Knocking down disease: A progress report on siRNA therapeutics. *Nat. Rev. Genet.* **2015**, *16*, 543–552. [CrossRef]

12. Wilson, R.C.; Doudna, J.A. Molecular mechanisms of RNA interference. *Annu. Rev. Biophys.* **2013**, *42*, 217–239. [CrossRef] [PubMed]
13. Altieri, D.C. Targeting survivin in cancer. *Cancer Lett.* **2013**, *332*, 225–228. [CrossRef] [PubMed]
14. Singh, N.; Krishnakumar, S.; Kanwar, R.K.; Cheung, C.H.; Kanwar, J.R. Clinical aspects for survivin: A crucial molecule for targeting drug-resistant cancers. *Drug Discov. Today* **2015**, *20*, 578–587. [CrossRef] [PubMed]
15. Chi, X.; Gatti, P.; Papoian, T. Safety of antisense oligonucleotide and siRNA-based therapeutics. *Drug Discov. Today* **2017**, *22*, 823–833. [CrossRef] [PubMed]
16. Kim, H.J.; Kim, A.; Miyata, K.; Kataoka, K. Recent progress in development of siRNA delivery vehicles for cancer therapy. *Adv. Drug Deliv. Rev.* **2016**, *104*, 61–77. [CrossRef] [PubMed]
17. Wen, D.; Peng, Y.; Lin, F.; Singh, R.K.; Mahato, R.I. Micellar Delivery of miR-34a Modulator Rubone and Paclitaxel in Resistant Prostate Cancer. *Cancer Res.* **2017**, *77*, 3244–3254. [CrossRef]
18. Yao, C.; Liu, J.; Wu, X.; Tai, Z.; Gao, Y.; Zhu, Q.; Li, J.; Zhang, L.; Hu, C.; Gu, F.; et al. Reducible self-assembling cationic polypeptide-based micelles mediate co-delivery of doxorubicin and microRNA-34a for androgen-independent prostate cancer therapy. *J. Control. Release* **2016**, *232*, 203–214. [CrossRef]
19. Yoo, H.S.; Park, T.G. FA receptor targeted biodegradable polymeric doxorubicin micelles. *J. Control. Release* **2004**, *96*, 273–283. [CrossRef]
20. Kumar, V.; Mundra, V.; Peng, Y.; Wang, Y.; Tan, C.; Mahato, R.I. Pharmacokinetics and biodistribution of polymeric micelles containing miRNA and small-molecule drug in orthotopic pancreatic tumor-bearing mice. *Theranostics* **2018**, *8*, 4033–4049. [CrossRef]
21. Cagel, M.; Tesan, F.C.; Bernabeu, E.; Salgueiro, M.J.; Zubillaga, M.B.; Moretton, M.A.; Chiappetta, D.A. Polymeric mixed micelles as nanomedicines: Achievements and perspectives. *Eur. J. Pharm. Biopharm.* **2017**, *113*, 211–228. [CrossRef] [PubMed]
22. Bae, Y.; Diezi, T.A.; Zhao, A.; Kwon, G.S. Mixed polymeric micelles for combination cancer chemotherapy through the concurrent delivery of multiple chemotherapeutic agents. *J. Control. Release* **2007**, *122*, 324–330. [CrossRef] [PubMed]
23. Li, H.; Fu, Y.; Zhang, T.; Li, Y.; Hong, X.; Jiang, J.; Gong, T.; Zhang, Z.; Sun, X. Rational Design of Polymeric Hybrid Micelles with Highly Tunable Properties to Co-Deliver MicroRNA-34a and Vismodegib for Melanoma Therapy. *Adv. Funct. Mater.* **2015**, *25*, 7457–7469. [CrossRef]
24. Yang, L.; Wu, X.; Liu, F.; Duan, Y.; Li, S. Novel biodegradable polylactide/poly(ethylene glycol) micelles prepared by direct dissolution method for controlled delivery of anticancer drugs. *Pharm. Res.* **2009**, *26*, 2332–2342. [CrossRef] [PubMed]
25. Lo, C.L.; Lin, S.J.; Tsai, H.C.; Chan, W.H.; Tsai, C.H.; Cheng, C.H.; Hsiue, G.H. Mixed micelle systems formed from critical micelle concentration and temperature-sensitive diblock copolymers for doxorubicin delivery. *Biomaterials* **2009**, *30*, 3961–3970. [CrossRef] [PubMed]
26. Liu, T.; Qian, Y.; Hu, X.; Ge, Z.; Liu, S. Mixed polymeric micelles as multifunctional scaffold for combined magnetic resonance imaging contrast enhancement and targeted chemotherapeutic drug delivery. *J. Mater. Chem.* **2012**, *22*, 5020–5030. [CrossRef]
27. Greco, F.; Vicent, M.J. Combination therapy: Opportunities and challenges for polymer-drug conjugates as anticancer nanomedicines. *Adv. Drug Deliv. Rev.* **2009**, *61*, 1203–1213. [CrossRef]
28. Xiang, Y.; Oo, N.N.L.; Lee, J.P.; Li, Z.; Loh, X.J. Recent development of synthetic nonviral systems for sustained gene delivery. *Drug Discov. Today* **2017**, *22*, 1318–1335. [CrossRef]
29. Sanchis, J.; Canal, F.; Lucas, R.; Vicent, M.J. Polymer-drug conjugates for novel molecular targets. *Nanomedicine* **2010**, *5*, 915–935. [CrossRef]
30. Jiang, Y.Y.; Tang, G.T.; Zhang, L.H.; Kong, S.Y.; Zhu, S.J.; Pei, Y.Y. PEGylated PAMAM dendrimers as a potential drug delivery carrier: In vitro and in vivo comparative evaluation of covalently conjugated drug and noncovalent drug inclusion complex. *J. Drug Target* **2010**, *18*, 389–403. [CrossRef]
31. Thomas, T.P.; Huang, B.; Choi, S.K.; Silpe, J.E.; Kotlyar, A.; Desai, A.M.; Zong, H.; Gam, J.; Joice, M.; Baker, J.R., Jr. Polyvalent dendrimer-methotrexate as a FA receptor-targeted cancer therapeutic. *Mol. Pharm.* **2012**, *9*, 2669–2676. [CrossRef] [PubMed]
32. Wong, P.T.; Choi, S.K. Mechanisms and implications of dual-acting methotrexate in FA-targeted nanotherapeutic delivery. *Int. J. Mol. Sci.* **2015**, *16*, 1772–1790. [CrossRef]
33. von Harpe, A.; Petersen, H.; Li, Y.; Kissel, T. Characterization of commercially available and synthesized polyethylenimines for gene delivery. *J. Control. Release* **2000**, *69*, 309–322. [CrossRef]

34. Boussif, O.; Lezoualc'h, F.; Zanta, M.A.; Mergny, M.D.; Scherman, D.; Demeneix, B.; Behr, J.P. A versatile vector for gene and oligonucleotide transfer into cells in culture and in vivo: Polyethylenimine. *Proc. Natl. Acad. Sci. USA* **1995**, *92*, 7297–7301. [CrossRef]
35. Xie, J.; Teng, L.; Yang, Z.; Zhou, C.; Liu, Y.; Yung, B.C.; Lee, R.J. A polyethylenimine-linoleic acid conjugate for antisense oligonucleotide delivery. *Biomed. Res. Int.* **2013**, *2013*, 710502. [CrossRef]
36. Teng, L.S.; Xie, J.; Teng, L.R.; Lee, R.J. Enhanced siRNA Delivery Using Oleic Acid Derivative of Polyethylenimine. *Anticancer Res.* **2012**, *32*, 1267–1271. [PubMed]
37. Patri, A.K.; Kukowska-Latallo, J.F.; Baker, J.R., Jr. Targeted drug delivery with dendrimers: Comparison of the release kinetics of covalently conjugated drug and non-covalent drug inclusion complex. *Adv. Drug Deliv. Rev.* **2005**, *57*, 2203–2214. [CrossRef] [PubMed]
38. Kircheis, R.; Wightman, L.; Wagner, E. Design and gene delivery activity of modified polyethylenimines. *Adv. Drug Deliv. Rev.* **2001**, *53*, 341–358. [CrossRef]
39. Dang, W.; Colvin, O.M.; Brem, H.; Saltzman, W.M. Covalent coupling of methotrexate to dextran enhances the penetration of cytotoxicity into a tissue-like matrix. *Cancer Res.* **1994**, *54*, 1729–1735.
40. Rabanel, J.M.; Hildgen, P.; Banquy, X. Assessment of PEG on polymeric particles surface, a key step in drug carrier translation. *J. Control. Release* **2014**, *185*, 71–87. [CrossRef]
41. Kukowska-Latallo, J.F.; Candido, K.A.; Cao, Z.; Nigavekar, S.S.; Majoros, I.J.; Thomas, T.P.; Balogh, L.P.; Khan, M.K.; Baker, J.R., Jr. Nanoparticle targeting of anticancer drug improves therapeutic response in animal model of human epithelial cancer. *Cancer Res.* **2005**, *65*, 5317–5324. [CrossRef] [PubMed]
42. Majoros, I.J.; Thomas, T.P.; Mehta, C.B.; Baker, J.R., Jr. Poly(amidoamine) dendrimer-based multifunctional engineered nanodevice for cancer therapy. *J. Med. Chem.* **2005**, *48*, 5892–5899. [CrossRef] [PubMed]
43. Howard, S.C.; McCormick, J.; Pui, C.H.; Buddington, R.K.; Harvey, R.D. Preventing and Managing Toxicities of High-Dose Methotrexate. *Oncologist* **2016**, *21*, 1471–1482. [CrossRef] [PubMed]
44. Ryan, B.M.; O'Donovan, N.; Duffy, M.J. Survivin: A new target for anti-cancer therapy. *Cancer Treat. Rev.* **2009**, *35*, 553–562. [CrossRef]
45. Peery, R.C.; Liu, J.Y.; Zhang, J.T. Targeting survivin for therapeutic discovery: Past, present, and future promises. *Drug Discov. Today* **2017**, *22*, 1466–1477. [CrossRef] [PubMed]
46. Yang, L.; Yuan, J.; Liu, L.; Shi, C.; Wang, L.; Tian, F.; Liu, F.; Wang, H.; Shao, C.; Zhang, Q.; et al. Alpha-linolenic acid inhibits human renal cell carcinoma cell proliferation through PPAR-gamma activation and COX-2 inhibition. *Oncol. Lett.* **2013**, *6*, 197–202. [CrossRef] [PubMed]
47. Wiggins, A.K.A.; Kharotia, S.; Mason, J.K.; Thompson, L.U. alpha-Linolenic Acid Reduces Growth of Both Triple Negative and Luminal Breast Cancer Cells in High and Low Estrogen Environments. *Nutr. Cancer Int. J.* **2015**, *67*, 1001–1009. [CrossRef]
48. Kong, X.; Ge, H.; Chen, L.; Liu, Z.; Yin, Z.; Li, P.; Li, M. Gamma-linolenic acid modulates the response of multidrug-resistant K562 leukemic cells to anticancer drugs. *Toxicol. In Vitro* **2009**, *23*, 634–639. [CrossRef]
49. Van Dongen, M.A.; Rattan, R.; Silpe, J.; Dougherty, C.; Michmerhuizen, N.L.; Van Winkle, M.; Huang, B.; Choi, S.K.; Sinniah, K.; Orr, B.G.; et al. Poly(amidoamine) dendrimer-methotrexate conjugates: The mechanism of interaction with folate binding protein. *Mol. Pharm.* **2014**, *11*, 4049–4058. [CrossRef]
50. Huang, B.; Otis, J.; Joice, M.; Kotlyar, A.; Thomas, T.P. PSMA-targeted stably linked "dendrimer-glutamate urea-methotrexate" as a prostate cancer therapeutic. *Biomacromolecules* **2014**, *15*, 915–923. [CrossRef]

Article

Effect of Cationic Lipid Type in Folate-PEG-Modified Cationic Liposomes on Folate Receptor-Mediated siRNA Transfection in Tumor Cells

Yoshiyuki Hattori [1,*], Satono Shimizu [1], Kei-ichi Ozaki [2] and Hiraku Onishi [1]

[1] Department of Drug Delivery Research, Hoshi University, 2-4-41 Ebara, Shinagawa, Tokyo 142-8501, Japan; s.sato1911@gmail.com (S.S.); onishi@hoshi.ac.jp (H.O.)

[2] Education and Research Center for Pharmaceutical Sciences, Osaka University of Pharmaceutical Sciences, 4-20-1 Nasahara, Takatsuki, Osaka 569-1094, Japan; kozak@gly.oups.ac.jp

* Correspondence: yhattori@hoshi.ac.jp; Tel.: +81-3-5498-5766

Received: 12 March 2019; Accepted: 12 April 2019; Published: 15 April 2019

Abstract: In this study, we examined the effect of cationic lipid type in folate (FA)-polyethylene glycol (PEG)-modified cationic liposomes on gene-silencing effects in tumor cells using cationic liposomes/siRNA complexes (siRNA lipoplexes). We used three types of cationic cholesterol derivatives, cholesteryl (3-((2-hydroxyethyl)amino)propyl)carbamate hydroiodide (HAPC-Chol), *N*-(2-(2-hydroxyethylamino)ethyl)cholesteryl-3-carboxamide (OH-Chol), and cholesteryl (2-((2-hydroxyethyl)amino)ethyl)carbamate (OH-C-Chol), and we prepared three types of FA-PEG-modified siRNA lipoplexes. The modification of cationic liposomes with 1–2 mol % PEG-lipid abolished the gene-silencing effect in human nasopharyngeal tumor KB cells, which overexpress the FA receptor (FR). In contrast, FA-PEG-modification of cationic liposomes restored gene-silencing activity regardless of the cationic lipid type in cationic liposomes. However, the optimal amount of PEG-lipid and FA-PEG-lipid in cationic liposomes for selective gene silencing and cellular uptake were different among the three types of cationic liposomes. Furthermore, in vitro transfection of polo-like kinase 1 (PLK1) siRNA by FA-PEG-modified liposomes exhibited strong cytotoxicity in KB cells, compared with PEG-modified liposomes; however, in in vivo therapy, intratumoral injection of PEG-modified PLK1 siRNA lipoplexes inhibited tumor growth of KB xenografts, as well as that of FA-PEG-modified PLK1 siRNA lipoplexes. From these results, the optimal formulation of PEG- and FA-PEG-modified liposomes for FR-selective gene silencing might be different between in vitro and in vivo transfection.

Keywords: cationic liposome; folate; folate receptor; cationic cholesterol derivative; siRNA delivery; gene knockdown; tumor-targeting

---

1. Introduction

RNA interference (RNAi) is a powerful gene-silencing process that holds great promise in the field of cancer therapy. Small double-stranded RNAs, i.e., synthetic small interfering RNAs (siRNAs) suppress the expression of a target gene by triggering specific degradation of the complementary mRNA sequence [1], and therefore siRNA therapeutics have become an increasingly important strategy for anticancer therapy [2]. For siRNA therapy against tumors, siRNAs are designed for targeting mRNAs transcribed from a tumor-causing gene, and siRNAs are introduced into the cytoplasm of tumor cells to cause cleavage of the target mRNA. For example, polo-like kinase 1 (PLK1) is a key regulator for cell mitosis, and its expression is elevated in many types of human tumors [3,4]. The inhibition of PLK1 expression by PLK1 siRNA can lead to death of tumor cells with multiple stages of mitosis [5], and therefore PLK1 is expected to be one of potential targets for siRNA therapy against tumors. However, the success of siRNA therapy relies on the development of safe and efficacious delivery systems that can introduce siRNAs into target tumor cells [2,6,7].

For effective transfection of siRNAs into tumor cells, siRNA carriers such as cationic liposomes are currently the most widely validated means [8]. Many different cationic lipids have been synthesized for lipid-based gene delivery [9] and shown activity in delivering siRNA into cells [10]. For siRNA delivery by cationic liposomes, cationic cholesterol derivatives have often been used [11]. Recently, we reported that cationic liposomes composed of cholesteryl (3-((2-hydroxyethyl)amino)propyl)carbamate hydroiodide (HAPC-Chol), *N*-(2-(2-hydroxyethylamino)ethyl)cholesteryl-3-carboxamide (OH-Chol), or cholesteryl (2-((2-hydroxyethyl)amino)ethyl)carbamate (OH-C-Chol) could efficiently suppress the expression of target genes by siRNA in cells [12,13]. Regarding the liposomal formulation, we previously reported that cationic liposomes composed of OH-Chol/DOPE and OH-C-Chol/DOPE exhibited high gene silencing efficacies compared with cationic nanoparticles composed of OH-Chol/Tween80 and OH-C-Chol/Tween80, respectively [14]. DOPE is thought to improve transfection efficiency by destabilizing the endosomal membrane [15,16], thereby facilitating the release of siRNAs into the cytoplasm. Therefore, a combination of cationic cholesterol derivative with DOPE in cationic liposomes might be a suitable formulation in siRNA delivery. However, selective delivery of siRNAs into tumor cells by cationic liposomes must be achieved for clinical applications.

Folate receptors (FRs) have been found to be overexpressed in a wide range of tumors, including ovary, uterus, lung, kidney, breast, colon, prostate, and brain cancers [17–19]. The following four isoforms of FR have been identified: folate receptor alpha (FR-α), beta (FR-β), delta (FR-Δ), and gamma (FR-γ); and FR-α and FR-β are attached to the cell by a GPI-anchor [18]. Among these FR isoforms, FR-α is the most widely studied as a biomarker for tumors, because a few normal tissues have been found to express FR-α, although most express the protein at much lower levels than are detected in FR-α-positive carcinoma. Therefore, FR-mediated tumor targeting has emerged as an attractive method of active targeting of siRNAs into tumor cells by cationic liposomes. When folic acid (FA) or its conjugates bind to FRs, they are taken up into the cells via receptor-mediated endocytosis. Therefore, FA-polyethylene glycol (PEG)-modification has been employed in cationic liposomes to facilitate the uptake of siRNA lipoplexes into tumor cells. Generally, PEGylated cationic liposomes can significantly reduce nonspecific gene transfer. However, conjugation of folate to the PEG chain can restore the cellular association with FR-positive tumors. To the best of our knowledge, there have been no reports about the effect of cationic lipids in FA-PEG-modified cationic liposomes on FR-targeting, although several studies have investigated the application of FA-PEG-modified cationic liposomes for siRNA delivery into FR-expressing cells [20–23]. In this study, to examine the effect of cationic lipid type in FA-PEG-modified cationic liposomes on gene-silencing effects, we selected three kinds of cationic cholesterol derivatives, HAPC-Chol, OH-Chol, and OH-C-Chol, and prepared three types of FA-PEG-modified cationic liposomes composed of cationic cholesterol derivatives and DOPE for the evaluation of gene-silencing effects. Here, we found that in FR-selective siRNA delivery, the cationic lipid type in FA-PEG-modified cationic liposomes affected an optimal amount of FA-PEG$_{2000}$-DSPE in liposomal formulation, and the optimized formulation of FA-PEG-modified cationic liposomes by in vitro transfection was not necessarily correlated with formulation by in vivo transfection.

2. Materials and Methods

2.1. Materials

N-(2-(2-Hydroxyethylamino)ethyl)cholesteryl-3-carboxamide (OH-Chol) and cholesteryl (2-((2-hydroxyethyl)amino)ethyl)carbamate (OH-C-Chol) were synthesized as described previously [14,24,25]. Cholesteryl (3-((2-hydroxyethyl)amino)propyl)carbamate hydroiodide (HAPC-Chol) was synthesized as described previously [26]. Methoxy-poly(ethyleneglycol)-distearylphosphatidylethanolamine (PEG$_{2000}$-DSPE/SUNBRIGHT DSPE-020CN, and PEG$_{5000}$-DSPE/SUNBRIGHT DSPE-050CN, and PEG mean molecular weight, 2000 and 5000, respectively) and 1,2-dioleoyl-*sn*-glycero-3-phosphoethanolamine

(DOPE, COATSOME ME-8181) were obtained from NOF Co. Ltd. (Tokyo, Japan). All other chemicals were of the highest grade available.

2.2. Small Interfering RNAs

The siRNAs targeting nucleotides of firefly pGL4 luciferase (Luc), enhanced green fluorescent protein (EGFP), human polo-like kinase 1 (PLK1), and non-silencing siRNA control (Cont) as a negative control were synthesized by Sigma Genosys (Tokyo, Japan). Alexa Fluor®488-labeled AllStars Negative Control siRNA (AF-siRNA) was obtained from Qiagen (Valencia, CA, USA). The siRNA sequences for Luc siRNA: sense strand: 5′-GGACGAGGACGAGCACUUCUU-3′ and antisense strand: 5′-GAAGUGCUCGUCCUCGUCCUU-3. The siRNA sequences for EGFP siRNA: sense strand: 5′-GGCUACGUCCAGGAGCGCACCTT-3′ and antisense strand: 5′-GGUGCGCUCC UGGACGUAGCCTT-3. The siRNA sequences of the PLK1 siRNA were as reported previously [27]. The siRNA sequences of the Cont siRNA were as reported previously [28].

2.3. Preparation of Cationic Liposomes and siRNA Lipoplexes

FA-PEG-distearoylphosphatidylethanolamine (FA-PEG-DSPE) (mean MW of PEG: 2000 and 5000) was synthesized as described previously [29]. Cationic liposomes were prepared from HAPC-Chol/ DOPE (composition designated as LP-HAPC), OH-Chol/DOPE (composition designated as LP-OH), and OH-C-Chol/DOPE (composition designated as LP-OH-C), at a molar ratio of 3:2. PEG_{2000}- and FA-PEG_{2000}-modified cationic liposomes were incorporated at 1, 2, or 3 mol% PEG_{2000}-DSPE and FA-PEG_{2000}-DSPE, respectively, into the cationic liposomal formulation, as shown in Table 1. PEG_{5000}- and FA-PEG_{5000}-modified LP-HAPC (LP-HAPC-PEG_{5000} and LP-HAPC-FA-PEG_{5000}) were incorporated with 1 mol % PEG_{5000}-DSPE and FA-PEG_{5000}-DSPE, respectively, into the formulation of LP-HAPC.

For the preparation of cationic liposomes by a thin-film hydration method, cationic cholesterol derivative, DOPE, and PEG-DSPE or FA-PEG-DSPE were dissolved in chloroform, and then the chloroform was evaporated under vacuum on a rotary evaporator at 60 °C to obtain a thin film. The thin film was hydrated with water at 60 °C by vortex mixing. The liposomes were sonicated in a bath-type sonicator (Bransonic® 2510J-MTH, 100W, Branson UL Trasonics Co., Danbury, CT, USA) to produce liposomes with a final size of approximately 100 nm.

In in vitro transfection, to prepare cationic liposome/siRNA complexes (siRNA lipoplexes), each cationic liposome was added to 50 pmol siRNA at a charge ratio (+:−) of 7:1 with vortex mixing for 10 s and left at room temperature for 15 min. The charge ratio (+:−) of cationic liposome:siRNA is expressed as the molar ratio of cationic lipid to siRNA phosphate.

Table 1. Formulation of FA-modified cationic liposomes.

Liposomes	Formulation (mol%)							
	HAPC-Chol	OH-Chol	OH-C-Chol	DOPE	PEG_{2000}-DSPE	FA-PEG_{2000}-DSPE	PEG_{5000}-DSPE	FA-PEG_{5000}-DSPE
LP-HAPC	60	-	-	40	-	-	-	-
LP-HAPC-1mol%PEG_{2000}	59.4	-	-	39.6	1	-	-	-
LP-HAPC-1mol%PEG_{5000}	59.4	-	-	39.6	-	-	1	-
LP-HAPC-2mol%PEG_{2000}	58.8	-	-	39.2	2	-	-	-
LP-HAPC-3mol%PEG_{2000}	58.2	-	-	38.8	3	-	-	-
LP-HAPC-1mol%FA-PEG_{2000}	59.4	-	-	39.6	-	1	-	-
LP-HAPC-1mol%FA-PEG_{5000}	59.4	-	-	39.6	-	-	-	1
LP-HAPC-2mol%FA-PEG_{2000}	58.8	-	-	39.2	-	2	-	-
LP-HAPC-3mol%FA-PEG_{2000}	58.2	-	-	38.8	-	3	-	-
LP-OH	-	60	-	40	-	-	-	-
LP-OH-1mol%PEG_{2000}	-	59.4	-	39.6	1	-	-	-
LP-OH-2mol%PEG_{2000}	-	58.8	-	39.2	2	-	-	-
LP-OH-3mol%PEG_{2000}	-	58.2	-	38.8	3	-	-	-
LP-OH-1mol%FA-PEG_{2000}	-	59.4	-	39.6	-	1	-	-
LP-OH-2mol%FA-PEG_{2000}	-	58.8	-	39.2	-	2	-	-
LP-OH-3mol%FA-PEG_{2000}	-	58.2	-	38.8	-	3	-	-
LP-OH-C	-	-	60	40	-	-	-	-
LP-OH-C-1mol%PEG_{2000}	-	-	59.4	39.6	1	-	-	-
LP-OH-C-2mol%PEG_{2000}	-	-	58.8	39.2	2	-	-	-
LP-OH-C-3mol%PEG_{2000}	-	-	58.2	38.8	3	-	-	-
LP-OH-C-1mol%FA-PEG_{2000}	-	-	59.4	39.6	-	1	-	-
LP-OH-C-2mol%FA-PEG_{2000}	-	-	58.8	39.2	-	2	-	-
LP-OH-C-3mol%FA-PEG_{2000}	-	-	58.2	38.8	-	3	-	-

2.4. Size and ζ-Potential of Cationic Liposomes and siRNA Lipoplexes

The siRNA lipoplexes were formed by addition of cationic liposomes to 5 μg Cont siRNA at a charge ratio (+:−) of 7:1 with vortex mixing for 10 s. The particle size distributions of cationic liposomes and siRNA lipoplexes were measured by the cumulant method using a light-scattering photometer (ELS-Z2, Otsuka Electronics Co., Ltd., Osaka, Japan) at 25 °C after diluting the dispersion to an appropriate volume (~1.5 mL) with water. The ζ-potentials of cationic liposomes and siRNA lipoplexes were measured by electrophoresis light-scattering methods using ELS-Z2 at 25 °C after diluting the dispersion with an appropriate volume (~1.5 mL) with water.

2.5. Cell Culture

Human nasopharyngeal tumor KB cells (also known as a subline of human cervix adenocarcinoma HeLa) were obtained from the Cell Resource Center for Biomedical Research, Tohoku University (Miyagi, Japan). KB cells were cultured in RPMI-1640 medium with 10% heat-inactivated fetal bovine serum (FBS) and 100 μg/mL kanamycin in a humidified atmosphere containing 5% CO_2 at 37 °C.

For the preparation of KB cells stably expressing firefly pGL4 luciferase, KB cells were plated on 35-mm culture dishes. Twenty-four hours later, the cells were transfected with 2 μg of plasmid DNA encoding firefly pGL4 luciferase under the control of a cytomegalovirus (CMV) promoter (pGL4.51[luc2/CMV/Neo] Vector, Promega, Madison, WI, USA) using Lipofectamine 2000 transfection reagent (Invitrogen, Thermo Fisher Scientific, Inc., Carlsbad, CA, USA). The transfected cells were selected in medium with 1200 μg/mL G418 sulfate. G418-resistant colonies were subcultured and established as a permanent cell line expressing firefly pGL4 luciferase (KB-Luc).

For the preparation of KB cells stably expressing EGFP, KB cells were transfected with 2 μg of plasmid DNA encoding EGFP under the control of a CMV promoter (pEGFP-C1, Clontech, Palo Alto, CA, USA) using Lipofectamine 2000 transfection reagent. The transfected cells were selected in medium with 1 mg/mL G418 sulfate. G418-resistant colonies were subcultured and established as a permanent cell line expressing EGFP (KB-EGFP).

2.6. Gene Silencing Effect by FA-PEG-Modified siRNA Lipoplexes in KB-Luc Cells

The KB-Luc cells were seeded in 6-well culture plate at a density of 3×10^5 cells per well, 24 h prior to transfection. The siRNA lipoplexes were formed by the addition of cationic liposomes into 50 pmol Cont siRNA or Luc siRNA with vortex mixing for 10 s and left at room temperature for 15 min. For siRNA transfection, each siRNA lipoplex was diluted in 1 mL of folate-deficient RPMI-1640 medium (Invitrogen) supplemented with 10% FBS and then the mixture was added to the cells (final 50 nM siRNA concentration). Lipofectamine RNAiMAX lipoplexes (Invitrogen) were prepared in accordance with the manufacturer's protocol and then added to the cells. Forty-eight hours after the transfection, luciferase activity was measured as counts per sec (cps)/μg protein using the luciferase assay system (PicaGene MelioraStar-LT Luminescence Reagent, Toyo Ink Mfg. Co. Ltd., Tokyo, Japan) and BCA reagent (Pierce™ BCA Protein Assay Kit, Pierce, Rockford, IL, USA). Luciferase activity (%) was calculated as relative to the luciferase activity (cps/μg protein) of untransfected cells.

2.7. Gene Silencing Effect by FA-PEG-Modified siRNA Lipoplexes in KB-EGFP Cells

The KB-EGFP cells were seeded in 6-well culture plate at a density of 3×10^5 cells per well, 24 h prior to transfection. The siRNA lipoplexes were formed by addition of cationic liposomes into 50 pmol Cont siRNA or EGFP siRNA with vortex mixing for 10 s and left at room temperature for 15 min. For siRNA transfection, each siRNA lipoplex was diluted in 1 mL of folate-deficient RPMI-1640 medium supplemented with 10% FBS and then the mixture was added to the cells (final 50 nM siRNA concentration). Lipofectamine RNAiMAX lipoplexes were prepared in accordance with the manufacturer's protocol and then added to the cells. Forty-eight hours after transfection, EGFP expression levels in the cells were determined by examining fluorescence intensity using a

FACSVerse™ flow cytometer (Becton Dickinson, San Jose, CA, USA) equipped with a 488-nm argon ion laser. Data for 10,000 fluorescence events were obtained by recording forward scatter (FSC), side scatter (SSC), and green (530/30 nm) fluorescence. EGFP expression level (%) was calculated as relative to the fluorescence intensity of untransfected cells.

2.8. Cytotoxicity by FA-PEG-Modified siRNA Lipoplexes

The KB cells were seeded in 96-well culture plate, 24 h prior to transfection. Each siRNA lipoplex with 50 pmol Cont siRNA was diluted in 1 mL of folate-deficient RPMI-1640 medium supplemented with 10% FBS, and then the mixture (100 µL) was added to the cells at 50% confluency in the well (final 50 nM siRNA concentration). After a 24 h incubation period, cell numbers were determined using a WST-8 assay (Cell Counting Kit-8, Dojindo Laboratories, Kumamoto, Japan). Cell viability was expressed as relative to the absorbance at 450 nm of untransfected cells.

2.9. Cellular Association with FA-PEG-Modified siRNA Lipoplexes

The KB cells were seeded in 6-well culture plate at a density of 3×10^5 cells per well, 24 h prior to transfection. Each siRNA lipoplex with 50 pmol AF-siRNA was diluted in 1 mL of folate-deficient RPMI-1640 medium (50 nM siRNA) supplemented with 10% FBS and then added to the cells. After 3 h incubation, the cells were washed twice with 1 mL phosphate-buffered saline (PBS) to remove any unbound lipoplexes. The amount of AF-siRNA in the cells was determined by examining fluorescence intensity using a FACSVerse™ flow cytometer equipped with a 488-nm argon ion laser. Data for 10,000 fluorescence events were obtained by recording forward scatter (FSC), side scatter (SSC), and green (530/30 nm) fluorescence.

2.10. Gel Retardation Assay

One microgram of siRNAs was mixed with cationic liposomes at various charge ratios (+:−) from 1:1 to 4:1. After 15 min incubation of the siRNA lipoplexes, they were analyzed by 18% acrylamide gel electrophoresis for siRNA in Tris-Borate-EDTA buffer (pH 8.0) and detected by ethidium bromide staining, as reported previously [30].

2.11. Accessibility of siRNA in siRNA Lipoplexes

The siRNA association by cationic liposomes was analyzed by exclusion assay using an SYBR® Green I Nucleic Acid Gel Stain (Takara Bio Inc., Shiga, Japan). The siRNA lipoplexes were formed at charge ratios (+:−) of 1:1, 2:1, 3:1, and 4:1. The siRNA lipoplexes with 1 µg of siRNA in a volume of 100 µL of Tris-HCl buffer (pH 8.0) were mixed with 100 µL of 2500-fold diluted SYBR® Green I Nucleic Acid Gel Stain solution with Tris-HCl buffer, and then incubated for 30 min. The fluorescence was measured at an emission wavelength of 535 nm with an excitation wavelength of 485 nm using a fluorescence plate reader (ARVO X2, Perkin Elmer, Waltham, MA, USA). As a control, the value of fluorescence obtained upon addition of free siRNA solution was set as 100%. The amount of siRNA available to interact with the SYBR® Green I was expressed as a percentage of the control.

2.12. Antiproliferative Activity

The KB cells were seeded in 96-well culture plate, 24 h prior to transfection. Each siRNA lipoplex with 50 pmol Cont siRNA or PLK1 siRNA was diluted in 1 mL of folate-deficient RPMI-1640 medium supplemented with 10% FBS, and then the mixture (100 µL) was added to the cells at 20–30% confluency in the well (final 50 nM siRNA concentration). After the 48 h incubation period, cell viability was measured by a WST-8 assay as described above.

2.13. Measurement of Expression Level of PLK1 mRNA

For the knockdown of PLK1 mRNA by transfection with PLK1 siRNA, KB cells were plated into 6-well culture plate at a density of 3×10^5 cells/well. The siRNA lipoplexes with 50 pmol Cont siRNA or PLK1 siRNA were diluted in 1 mL of folate-deficient RPMI-1640 medium (50 nM siRNA) supplemented with 10% FBS and then added to the cells. At 24 h after transfection, total RNA was isolated using NucleoSpin RNA (Macherey-Nagel, GmbH, Düren, Germany). First strand cDNA was synthesized from 1 µg of total RNA using PrimeScript RTase (Takara Bio, Inc., Otsu, Japan). Quantitative (q)PCR was performed using a Roche Light Cycler 96 system (Roche Diagnostics, Basel, Switzerland) and TaqMan Gene expression assays (PLK1, Hs00983227_m1 and GAPDH, Hs02786624_g1, Applied Biosystems, Thermo Fisher Scientific Inc.). Samples were run in triplicate, and the expression levels of PLK1 mRNA were normalized by the amount of GAPDH mRNA in the same sample, and analyzed using the comparative Cq ($2^{-\Delta\Delta Cq}$) method [31].

2.14. In Vivo Anti-Tumor Effect

All animal experiments were conducted in accordance with the "Guide for the Care and Use of Laboratory Animals" published by the U.S. National Institutes of Health and the "Guide for the Care and Use of Laboratory Animals" adopted by the Institutional Animal Care and the Use Committee of Hoshi University (Tokyo, Japan) (which is accredited by the Ministry of Education, Culture, Sports, Science, and Technology, Japan). Ethical approval for this study was obtained from the Institutional Animal Care and the Use Committee of Hoshi University (permission number: 30-074).

To generate KB tumor xenografts, 1×10^7 cells suspended in 50 µL of PBS were inoculated subcutaneously in the flank region of female BALB/c nu/nu mice (8 weeks of age, CLEA Japan Inc., Tokyo, Japan). The tumor volume was calculated using the formula, tumor volume = $0.5 \times a \times b^2$, where a and b are the larger and smaller diameters, respectively. When the average volume of the xenograft tumors reached 80 mm^3 (day 0), LP-HAPC-2mol%PEG$_{2000}$ or LP-HAPC-2mol%FA-PEG$_{2000}$ lipoplexes of 10 µg of Cont siRNA or PLK1 siRNA per tumor were directly injected into xenografts on days 0, 2, and 4. Tumor volume was measured on days 0, 2, 4, 6, and 8. Tumor volume (%) was calculated as relative to each tumor volume at the day 0. At day 8, mice were sacrificed by cervical dislocation, and then the excised tumors were weighed.

2.15. Statistical Analysis

The statistical significance of differences between mean values was determined using Student's *t*-test. A *p*-value of 0.05 or less was considered significant.

3. Results

3.1. Characterization of FA-PEG-Modified Cationic Liposomes and siRNA Lipoplexes

Firstly, we examined whether the cationic lipid type in FA-PEG-modified cationic liposomes affected gene silencing after transfection of FA-PEG-modified siRNA lipoplexes into KB cells, which overexpressed FR-α [32]. Here, we used HAPC-Chol, OH-Chol, and OH-C-Chol (Figure 1) as cationic cholesterol derivatives for preparation of cationic liposomes. LP-HAPC, LP-OH, and LP-OH-C were prepared from HAPC-Chol/DOPE, OH-Chol/DOPE, and OH-C-Chol/DOPE, respectively, at a molar ratio of 3:2 (Table 1). For PEGylated cationic liposomes, 1, 2, or 3 mol% PEG$_{2000}$-DSPE was added to the formulation of LP-HAPC, LP-OH, and LP-OH-C (LP-HAPC-1–3mol%PEG$_{2000}$, LP-OH-1–3mol%PEG$_{2000}$, and LP-OH-C-1–3mol%PEG$_{2000}$, respectively). For FR-targeted cationic liposomes, 1, 2, or 3 mol% FA-PEG$_{2000}$-DSPE was added to the formulation of LP-HAPC, LP-OH, and LP-OH-C (LP-HAPC-1–3mol%FA-PEG$_{2000}$, LP-OH-1–3mol%FA-PEG$_{2000}$, and LP-OH-C-1–3mol%FA-PEG$_{2000}$, respectively). In addition, as PEG- or FA-PEG-modified cationic liposomes with a longer PEG chain, 1 mol% PEG$_{5000}$-DSPE and FA-PEG$_{5000}$-DSPE, respectively, were added to the formulation of LP-HAPC (LP-HAPC-1mol%PEG$_{5000}$, and LP-HAPC-1mol%FA-PEG$_{5000}$, respectively).

In HAPC-Chol-based liposomes, the sizes of LP-HAPC, LP-HAPC-1–3mol%PEG_{2000}, LP-HAPC-1–3mol%FA-PEG_{2000}, LP-HAPC-1mol%PEG_{5000}, and LP-HAPC-1mol%FA-PEG_{5000} were approximately 90–110 nm, polydispersity index (PDI) 0.21–0.26, and the ζ-potentials were approximately +35–47 mV (Table 2). In OH-Chol-based liposomes, the sizes of LP-OH, LP-OH-1– 3mol%PEG_{2000}, and LP-OH-1–3mol%FA-PEG_{2000} were approximately 84–110 nm (PDI 0.17–0.31), and the ζ-potentials were approximately +34–47 mV (Table 2). In OH-C-Chol-based liposomes, the sizes of LP-OH-C, LP-OH-C-1–3mol%PEG_{2000} and LP-OH-C-1–3mol%FA-PEG_{2000} were approximately 90–110 nm (PDI 0.12–0.26), and the ζ-potentials were approximately +43–52 mV (Table 2).

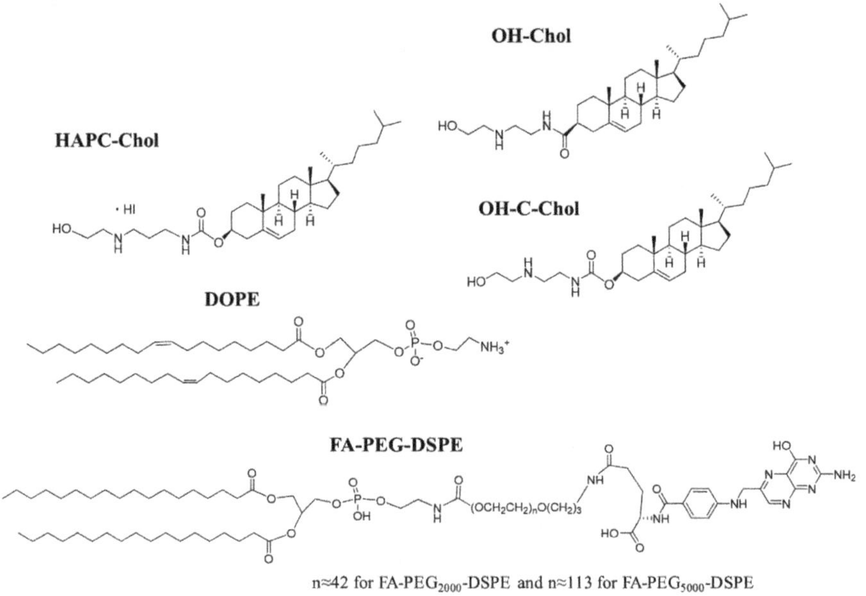

n≈42 for FA-PEG_{2000}-DSPE and n≈113 for FA-PEG_{5000}-DSPE

Figure 1. Structure of cationic cholesterol derivatives, neutral helper lipid, and FA-PEG-DSPE: HAPC-Chol; cholesteryl (3-((2-hydroxyethyl)amino)propyl)carbamate hydroiodide, OH-Chol; *N*-(2-(2-hydroxyethylamino)ethyl)cholesteryl-3-carboxamide, OH-C-Chol; cholesteryl (2-((2-hydroxyethyl)amino)ethyl)carbamate, DOPE; 1,2-dioleoyl-*sn*-glycero-3-phosphoethanolamine, FA-PEG- DSPE; FA-PEG-distearoylphosphatidylethanolamine.

Table 2. Particle size and ζ-potential of cationic liposomes and siRNA lipoplexes.

Liposome	Liposomes			Lipoplexes [(b)]		
	Size [(a)] (nm)	PDI	ζ-Potential [(a)] (mV)	Size [(a)] (nm)	PDI	ζ-Potential [(a)] (mV)
LP-HAPC	103.3 ± 0.3	0.21 ± 0.01	45.4 ± 2.2	183.6 ± 0.2	0.24 ± 0.01	39.5 ± 0.8
LP-HAPC-1mol%PEG_{2000}	98.2 ± 2.5	0.25 ± 0.01	46.8 ± 1.0	185.4 ± 3.0	0.28 ± 0.00	31.5 ± 0.1
LP-HAPC-1mol%PEG_{5000}	103.1 ± 1.0	0.24 ± 0.01	39.1 ± 0.5	168.8 ± 1.2	0.24 ± 0.01	23.8 ± 0.8
LP-HAPC-2mol%PEG_{2000}	101.3 ± 2.5	0.26 ± 0.00	40.7 ± 1.5	179.4 ± 4.1	0.27 ± 0.00	28.3 ± 0.5
LP-HAPC-3mol%PEG_{2000}	97.9 ± 2.5	0.22 ± 0.03	37.6 ± 1.1	172.9 ± 1.2	0.27 ± 0.02	31.0 ± 1.3
LP-HAPC-1mol%FA-PEG_{2000}	105.3 ± 0.8	0.23 ± 0.01	45.2 ± 0.5	208.1 ± 5.6	0.26 ± 0.02	32.9 ± 1.2
LP-HAPC-1mol%FA-PEG_{5000}	101.6 ± 0.7	0.22 ± 0.00	37.6 ± 1.0	204.5 ± 1.8	0.28 ± 0.01	21.4 ± 1.3
LP-HAPC-2mol%FA-PEG_{2000}	97.3 ± 0.1	0.23 ± 0.01	41.5 ± 1.1	172.9 ± 1.2	0.27 ± 0.02	28.3 ± 0.9
LP-HAPC-3mol%FA-PEG_{2000}	88.3 ± 0.6	0.23 ± 0.01	35.2 ± 0.4	175.7 ± 1.2	0.25 ± 0.01	29.0 ± 1.4
LP-OH	108.9 ± 1.2	0.21 ± 0.01	46.7 ± 1.7	173.5 ± 3.7	0.22 ± 0.02	45.3 ± 0.5
LP-OH-1mol%PEG_{2000}	104.1 ± 8.0	0.31 ± 0.02	39.9 ± 1.6	187.9 ± 0.6	0.27 ± 0.01	38.0 ± 2.2
LP-OH-2mol%PEG_{2000}	110.5 ± 4.7	0.25 ± 0.01	46.4 ± 3.2	292.1 ± 1.8	0.27 ± 0.01	29.0 ± 1.0
LP-OH-3mol%PEG_{2000}	104.4 ± 3.0	0.21 ± 0.01	44.1 ± 1.0	205.8 ± 5.6	0.28 ± 0.01	24.2 ± 0.8
LP-OH-1mol%FA-PEG_{2000}	92.6 ± 1.4	0.17 ± 0.01	46.4 ± 1.7	180.1 ± 2.8	0.24 ± 0.01	35.2 ± 1.4
LP-OH-2mol%FA-PEG_{2000}	106.4 ± 1.4	0.21 ± 0.01	39.5 ± 0.2	202.3 ± 9.7	0.24 ± 0.03	20.8 ± 0.3
LP-OH-3mol%FA-PEG_{2000}	83.5 ± 1.7	0.19 ± 0.01	34.2 ± 1.7	197.7 ± 1.5	0.28 ± 0.01	28.5 ± 0.6
LP-OH-C	91.4 ± 1.4	0.12 ± 0.02	49.1 ± 2.2	172.4 ± 1.9	0.24 ± 0.00	41.4 ± 2.9
LP-OH-C-1mol%PEG_{2000}	100.0 ± 0.6	0.24 ± 0.00	43.5 ± 0.7	220.8 ± 5.4	0.26 ± 0.01	38.9 ± 5.7
LP-OH-C-2mol%PEG_{2000}	108.5 ± 2.1	0.24 ± 0.01	47.1 ± 1.9	258.2 ± 9.3	0.26 ± 0.01	38.5 ± 1.0
LP-OH-C-3mol%PEG_{2000}	110.9 ± 0.9	0.25 ± 0.01	48.8 ± 0.7	201.5 ± 1.9	0.26 ± 0.00	29.8 ± 1.3
LP-OH-C-1mol%FA-PEG_{2000}	103.8 ± 2.4	0.23 ± 0.01	51.5 ± 1.6	245.5 ± 1.2	0.12 ± 0.00	31.6 ± 0.7
LP-OH-C-2mol%FA-PEG_{2000}	105.8 ± 2.7	0.22 ± 0.01	43.3 ± 1.6	188.8 ± 1.2	0.26 ± 0.01	29.1 ± 1.1
LP-OH-C-3mol%FA-PEG_{2000}	104.7 ± 1.9	0.26 ± 0.01	46.6 ± 1.8	415.4 ± 70.0	0.21 ± 0.03	22.0 ± 0.4

PDI: polydispersity index. [(a)] in water. [(b)] charge ratio (+:−) of cationic lipid to siRNA phosphate = 7:1. Each value represents the mean ± SD of three measurements per sample.

For the preparation of siRNA lipoplexes, their liposomes were mixed with siRNA at a charge ratio (+:−) of 7:1 as reported previously [13], because siRNA lipoplexes at this charge ratio (+:−) exhibited high gene silencing efficacy (Supplemental Figure S1). In HAPC-Chol-based lipoplexes, the sizes of LP-HAPC, LP-HAPC-1–3mol%PEG_{2000}, LP-HAPC-1–3mol%FA-PEG_{2000}, LP-HAPC-1mol%PEG_{5000}, and LP-HAPC-1mol%FA-PEG_{5000} lipoplexes were approximately 170–210 nm (PDI 0.24–0.28), and the ζ-potentials were approximately 21–40 mV (Table 2). In OH-Chol-based lipoplexes, the sizes of LP-OH, LP-OH-1–3mol%PEG_{2000}, and LP-OH-1–3mol%FA-PEG_{2000} lipoplexes were approximately 170–300 nm (PDI 0.22–0.28), and the ζ-potentials were approximately 21–45 mV (Table 2). In OH-C-Chol-based lipoplexes, the sizes of LP-OH-C, LP-OH-C-1–3mol%PEG_{2000}, and LP-OH-C-1–3mol%FA-PEG_{2000} lipoplexes were approximately 170–420 nm (PDI 0.12–0.26), and the ζ-potentials were approximately 22–39 mV (Table 2). Regardless the cationic lipid type in cationic liposomes, PEG-modification, or FA-PEG-modification of cationic liposomes trended to decrease the ζ-potentials of cationic liposomes and siRNA lipoplexes. However, the type of cationic cholesterol derivatives in cationic liposomes did not largely affect size and ζ-potential of cationic liposomes and siRNA lipoplexes.

3.2. *Effect of Cationic Lipid of FA-PEG-Modified Cationic Liposomes on In Vitro Gene Knockdown Efficacy*

To examine the gene silencing effect in FR-expressing cells by FA-PEG-modified siRNA lipoplexes, KB-Luc cells were incubated with siRNA lipoplexes modified with 1–3 mol% PEG_{2000}-DSPE or FA-PEG_{2000}-DSPE, and the gene silencing activity was assessed by assaying luciferase activity. LP-HAPC, LP-OH, and LP-OH-C lipoplexes with Luc siRNA strongly suppressed luciferase activity (>80% knockdown, compared with Cont siRNA) (Figure 2A,C,D). In HAPC-Chol- and OH-C-Chol-based liposomes, above 2 mol% PEG-modification of LP-HAPC and LP-OH-C lipoplexes with PEG_{2000}-DSPE completely abolished the gene silencing effect; however, LP-HAPC-2mol%FA-PEG_{2000} and LP-OH-C-2mol%FA-PEG_{2000} lipoplexes with Luc siRNA exhibited strong suppression of luciferase activity (Figure 2A,D). In contrast, in OH-Chol-based liposomes, above 1 mol% PEG-modification of LP-OH lipoplexes with PEG_{2000}-DSPE abolished the gene silencing effect; however, LP-OH-1mol%FA-PEG_{2000} lipoplexes with Luc siRNA exhibited strong suppression of luciferase activity (Figure 2C). From these results, in FR-mediated gene silencing, the optimal amount of PEG_{2000}-DSPE or FA-PEG_{2000}-DSPE in the liposomal formulation may be affected by the cationic lipid type in FA-PEG-modified liposomes. Regarding the length of the PEG chain between FA and lipid, both LP-HAPC-1mol%PEG_{5000} and LP-HAPC-1mol%FA-PEG_{5000} lipoplexes did not exhibit gene silencing activity (Figure 2B), indicating that a longer PEG chain inhibited FR-mediated uptake by cells. From these results, optimal modifications of PEG_{2000}-DSPE or FA-PEG_{2000}-DSPE in formulations of LP-HAPC, LP-OH, and LP-OH-C were 2 mol%, 1 mol%, and 2 mol%, respectively, for selective FR-mediated gene silencing in tumor cells.

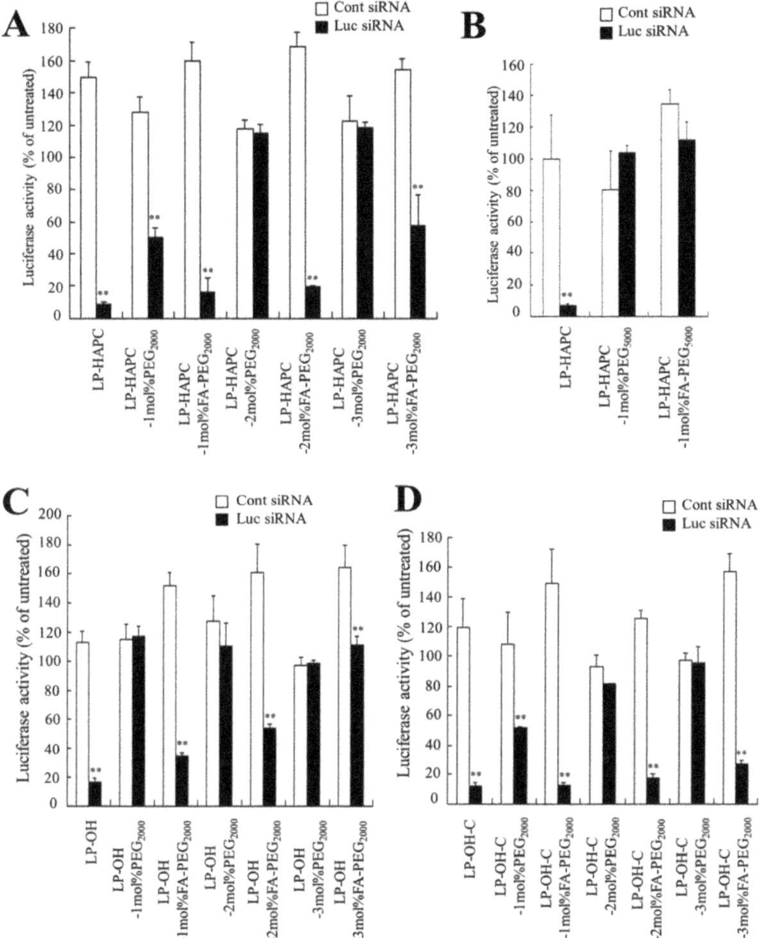

Figure 2. Effect of FA-PEG modification of cationic liposomes on suppression of luciferase expression in KB-Luc cells after transfection with FA-PEG-modified siRNA lipoplexes. (**A**) LP-HAPC-1–3mol%PEG_{2000} and LP-HAPC-1–3mol%FA-PEG_{2000}, (**B**) LP-HAPC-1mol%PEG_{5000} and LP-HAPC-1mol%FA-PEG_{5000}, (**C**) LP-OH-1–3mol%PEG_{2000} and LP-OH-1–3mol%FA-PEG_{2000}, (**D**) LP-OH-C-1–3mol%PEG_{2000} and LP-OH-C-1–3mol%FA-PEG_{2000} were used. The siRNA lipoplexes with Cont siRNA or Luc siRNA were added to KB-Luc cells at 50 nM siRNA, and the luciferase assay was carried out 48 h after incubation. Each column represents the mean + SD (n = 3). ** $p < 0.01$, compared with Cont siRNA.

3.3. *Cytotoxicity by FA-PEG-Modified siRNA Lipoplexes*

To examine whether FA-PEG-modification of cationic liposomes affected the cytotoxicity, we investigated cell viability at 24 h after transfection into KB cells with FA-PEG-modified siRNA lipoplexes. LP-HAPC, LP-OH, and LP-OH-C lipoplexes did not exhibit marked cytotoxicity (80–90% cell viability), and PEG-modification of their lipoplexes also did not affect cytotoxicity (approximately 90% cell viability) (Figure 3). However, FA-PEG-modification of LP-HAPC, LP-OH, and LP-OH-C slightly increased cytotoxicity (70–80% cell viability) with increasing the amounts of FA-PEG_{2000}-DSPE in the cationic liposomes, compared with the PEG-modification. These results indicated that

FA-PEG-modification of cationic liposomes might increase cytotoxicity by increasing the cellular uptake of siRNA lipoplexes.

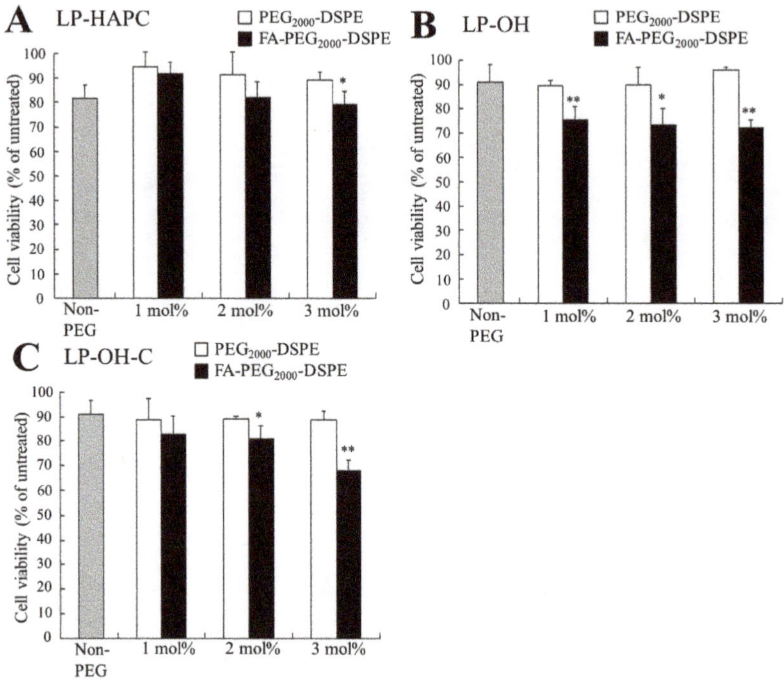

Figure 3. Effect of FA-PEG-modification of cationic liposomes on cell viability 24 h after transfection with FA-PEG-modified siRNA lipoplexes into KB-Luc cells. (**A**) LP-HAPC-1–3mol%PEG_{2000} and LP-HAPC-1–3mol%FA-PEG_{2000}, (**B**) LP-OH-1–3mol%PEG_{2000} and LP-OH-1–3mol%FA-PEG_{2000}, (**C**) LP-OH-C-1–3mol%PEG_{2000} and LP-OH-C-1–3mol%FA-PEG_{2000} were used. The siRNA lipoplexes were added to KB cells at 50 nM siRNA. Each column represents the mean + SD (n = 4). Each column represents the mean + SD (n = 3). * $p < 0.05$, ** $p < 0.01$, compared with PEG_{2000}-DSPE.

3.4. *Association of FA-PEG-Modified siRNA Lipoplexes with Cells*

To examine the effect of cationic lipid type on cellular association with FA-PEG-modified siRNA lipoplexes, we measured the siRNA amount taken up by KB cells at 3 h after transfection with FA-PEG-modified siRNA lipoplexes (Figure 4). In LP-HAPC, LP-OH, and LP-OH-C, the amount of siRNA in the cells decreased with increasing PEG-modification (Figure 4A–C). In contrast, 1–3 mol% FA-PEG-modification of LP-HAPC and LP-OH exhibited high cellular uptake of siRNA in the cells, compared with PEG-modified ones (Figure 4A,B). However, in LP-OH-C, with an increase of FA-PEG-modification, the amount of siRNA in the cells was decreased substantially, and LP-OH-C-3mol%FA-PEG_{2000} lipoplexes did not enhance cellular association compared with LP-OH-C-3mol%PEG_{2000} lipoplexes (Figure 4C). These results indicated that the cellular association of FA-PEG-modified siRNA lipoplexes might be strongly affected by cationic lipid type in cationic liposomes.

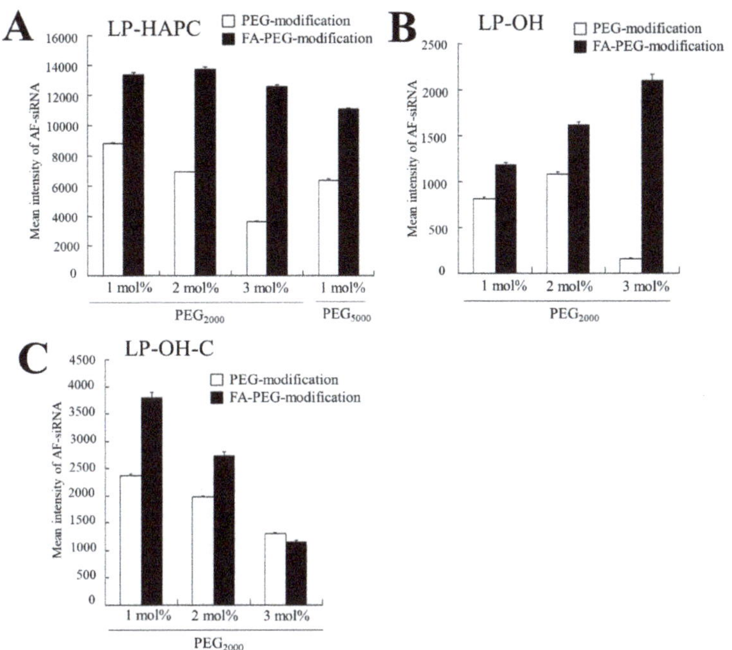

Figure 4. Effect of FA-PEG-modification of cationic liposomes on cellular association at 3 h after transfection of FA-PEG-modified siRNA lipoplexes. (**A**) LP-HAPC-1–3mol%PEG$_{2000}$ and LP-HAPC-1–3mol%FA-PEG$_{2000}$, (**B**) LP-OH-1–3mol%PEG$_{2000}$ and LP-OH-1–3mol%FA-PEG$_{2000}$, (**C**) LP-OH-C-1–3mol%PEG$_{2000}$ and LP-OH-C-1–3mol%FA-PEG$_{2000}$ were used. The siRNA lipoplexes were formed by mixing cationic liposomes with AF-siRNA, and they were added to KB cells at a final concentration of 50 nM siRNA. The association of siRNA lipoplexes with the cells was determined on the basis of Alexa Fluor®488-fluorescence by flow cytometry. Each column represents the mean fluorescent intensity + SD (n = 3).

3.5. Association of FA-PEG-Modified Cationic Liposomes with siRNA

Next, we evaluated the association of FA-PEG-modified cationic liposomes with siRNA. The association of siRNA with each cationic liposome was monitored by gel retardation electrophoresis (Figure 5). The migration pattern of siRNA in siRNA lipoplexes was changed when the siRNA was mixed with cationic liposomes at charge ratios (+:−) from 1:1 to 4:1, and the migration of siRNAs ceased gradually as the charge ratio (+:−) increased. In HAPC-Chol-based liposomes, no migration was observed beyond charge ratios (+:−) of 2:1 in LP-HAPC, LP-HAPC-1mol%PEG$_{2000}$, and LP-HAPC-1mol%FA-PEG$_{2000}$ lipoplexes, of 3:1 in LP-HAPC-2mol%PEG$_{2000}$, LP-HAPC-2mol%FA-PEG$_{2000}$, and LP-HAPC-3mol%PEG$_{2000}$, and of 4:1 in LP-HAPC-3mol%FA-PEG$_{2000}$, (Figure 5A).

This result suggested that the association of siRNA with the cationic liposomes was inhibited with increasing amounts of FA-PEG$_{2000}$-DSPE or PEG$_{2000}$-DSPE. In siRNA transfection, we used a charge ratio (+:−) of 7:1 for the preparation of siRNA lipoplexes; therefore, siRNAs were completely bound to LP-HAPC regardless the PEG- or FA-PEG-modification. In addition, in LP-HAPC-1mol%PEG$_{5000}$ and LP-HAPC-1mol%FA-PEG$_{5000}$ lipoplexes, no migration was observed beyond charge ratios (+:−) of 2:1 (Figure 5A), indicating that the decrease in gene silencing activity in LP-HAPC-1mol%FA-PEG$_{5000}$ lipoplexes (Figure 2B) was not caused by a decrease in the association of cationic liposomes with siRNA by the long PEG chain. Furthermore, LP-OH- and LP-OH-C-based liposomes, beyond charge ratios (+:−) of 4:1 in LP-OH, LP-OH-1mol%PEG$_{2000}$, LP-OH-1mol%FA-PEG$_{2000}$, LP-OH-C, LP-OH-C-2mol%PEG$_{2000}$, LP-OH-C-2mol%FA-PEG$_{2000}$ lipoplexes, no migration or decreased migration was observed (Figure 5B,C). These results indicated that OH-Chol- and OH-C-Chol-based liposomes might make a weaker association with siRNA than HAPC-Chol-based liposomes.

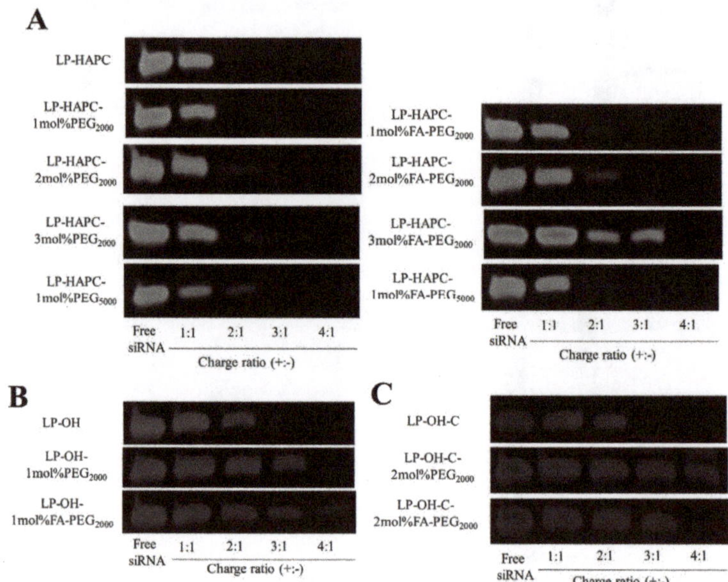

Figure 5. Effect of FA-PEG-modification of cationic liposomes on association of siRNA with FA-PEG-modified cationic liposomes. siRNA association by cationic liposomes was analyzed by gel retardation assay. (**A**) LP-HAPC-1–3mol%PEG_{2000}, LP-HAPC-1mol%PEG_{5000}, LP-HAPC-1–3mol%FA-PEG_{2000}, and LP-HAPC-1mol%FA-PEG_{5000}, (**B**) LP-OH-1–3mol%PEG_{2000} and LP-OH-1–3mol%FA-PEG_{2000}, (**C**) LP-OH-C-1–3mol%PEG_{2000} and LP-OH-C-1–3mol%FA-PEG_{2000} were used. Each liposome was formed with siRNA at various charge ratios (+:–) from 1:1 to 4:1, and were analyzed using 18% acrylamide gel electrophoresis.

Furthermore, we examined the association of siRNA with each cationic liposome using an exclusion assay with SYBR® Green I. SYBR® Green I is a DNA/RNA-intercalating agent whose fluorescence was dramatically enhanced upon binding to unbound siRNA in cationic lipoplexes. As a result, in all the cationic liposomes, the fluorescence of SYBR® Green I was markedly decreased by the addition of cationic liposomes into the siRNA solution beyond charge ratios (+:–) of 2:1 or 3:1, compared with that in siRNA solution (Figure 6A–C). This result suggested that siRNAs were completely bound to each cationic liposome regardless of PEG or FA-PEG modification of the cationic liposomes. Although a discrepancy between the results from the accessibility of SYBR® Green I (Figure 6) and gel retardation electrophoresis (Figure 5) was observed, siRNAs might be released from siRNA lipoplexes by electrophoresis due to the weak association between siRNA and the cationic liposomes. From the result shown in Figure 6, for all the cationic liposomes used in this study, siRNAs might be completely bound to the cationic liposomes when mixed beyond a charge ratio (+:–) of 3:1.

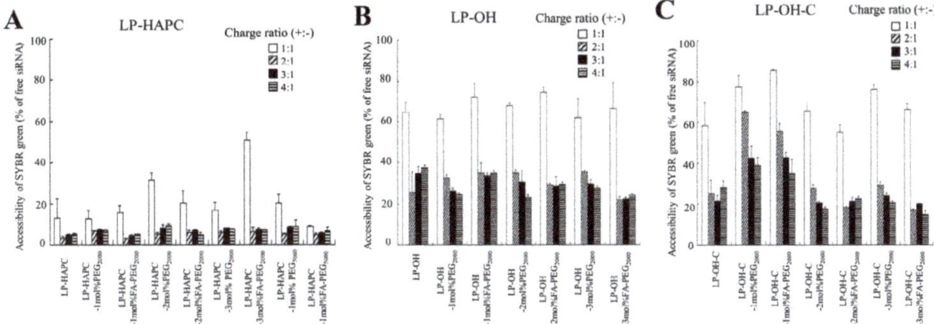

Figure 6. Effect of FA-PEG-modification of cationic liposomes on association of siRNA with FA-PEG-modified cationic liposome. siRNA association by cationic liposomes was analyzed by exclusion assay using an SYBR® Green I Nucleic Acid Gel Stain. (**A**) LP-HAPC-1–3mol%PEG_{2000}, LP-HAPC-1mol%PEG_{5000}, LP-HAPC-1–3mol%FA-PEG_{2000}, and LP-HAPC-1mol%FA-PEG_{5000}, (**B**) LP-OH-1–3mol%PEG_{2000} and LP-OH-1–3mol%FA-PEG_{2000}, (**C**) LP-OH-C-1–3mol%PEG_{2000} and LP-OH-C-1–3mol%FA-PEG_{2000} were used. The siRNA lipoplexes were formed at various charge ratios (+:−) from 1:1 to 4:1. As a control, the value of fluorescence obtained upon addition of free siRNA solution was set as 100%. The amount of siRNA available to interact with the SYBR® Green I was expressed as a percentage of the control. Each column represents the mean + SD (n = 3).

3.6. Suppression of EGFP Expression by FA-PEG-Modified siRNA Lipoplexes

To examine the effect of FA-PEG-modification in cationic liposomes on gene knockdown using FA-PEG-modified siRNA lipoplexes, KB-EGFP cells were incubated with siRNA lipoplexes, and then the gene silencing effect was assessed by assaying the fluorescence intensity in the cells. Here, we decided to use LP-HAPC-2mol%FA-PEG_{2000}, LP-OH-1mol%FA-PEG_{2000}, and LP-OH-C-2mol%FA-PEG_{2000} as optimal FA-PEG-modified liposomes. In addition, we used Lipofectamine RNAiMax as a commercially available in vitro transfection reagent for siRNAs.

LP-HAPC lipoplexes with EGFP siRNA strongly suppressed EGFP expression (~50% knockdown, compared with Cont siRNA); however, LP-OH and LP-OH-C lipoplexes with EGFP siRNA were suppressed moderately (20–30% knockdown, compared with Cont siRNA) (Figure 7). In contrast, LP-HAPC-2mol%PEG_{2000}, LP-OH-1mol%PEG_{2000}, and LP-OH-C-2mol%PEG_{2000} lipoplexes did not exhibit suppression of EGFP expression in the cells. However, LP-HAPC-2mol%FA-PEG_{2000}, LP-OH-1mol%FA-PEG_{2000}, and LP-OH-C-2mol%FA-PEG_{2000} lipoplexes restored the gene silencing effect by FA-PEG-modification. Among the FA-PEG modified siRNA lipoplexes, LP-HAPC-2mol%FA-PEG_{2000} lipoplexes strongly suppressed the expression of EGFP in the cells (~40% knockdown, compared with Cont siRNA), similar to Lipofectamine RNAiMax. Although LP-HAPC-2mol%FA-PEG_{2000}, LP-OH-1mol%FA-PEG_{2000}, and LP-HAPC-2mol%FA-PEG_{2000} lipoplexes exhibited strong gene silencing effects in KB-Luc cells (Figure 2), they showed moderate gene silencing efficacy in KB-EGFP cells (Figure 7). It has been reported that the $t_{1/2}$ of firefly luciferase protein is about 2–3 h [33,34], but those of GFP and EGFP are more than 24 h [35], indicating that EGFP expression could not be suppressed completely by siRNA due to the long $t_{1/2}$ of EGFP. Among the FA-PEG-modified cationic liposomes, LP-HAPC-2mol%FA-PEG_{2000} appeared to be the most effective in FR-mediated gene silencing.

3.7. Antiproliferative Activity

PLK1 is a potential target in tumor therapy, because PLK1 is overexpressed in various types of human tumors [3], and its inhibition is potently antiproliferative for tumor cells. To examine whether transfection of PLK1 siRNA into KB cells with FA-PEG modified siRNA lipoplexes could selectivity inhibit tumor growth, we measured cell viability 48 h after transfection of PLK1 siRNA into KB cells. Transfection of LP-HAPC, LP-OH, and LP-OH-C lipoplexes with PLK1 siRNA largely inhibited cell growth; however, LP-HAPC-2mol%PEG$_{2000}$, LP-OH-1mol%PEG$_{2000}$, and LP-OH-C-2mol%PEG$_{2000}$ lipoplexes with PLK1 siRNA showed a decreased cytotoxic effect with the PEG-modification (Figure 8). In contrast, LP-HAPC-2mol%FA-PEG$_{2000}$, LP-OH-1mol%FA-PEG$_{2000}$, and LP-OH-C-2mol%FA-PEG$_{2000}$ lipoplexes with PLK1 siRNA strongly decreased cell proliferation, similar to Lipofectamine RNAiMax.

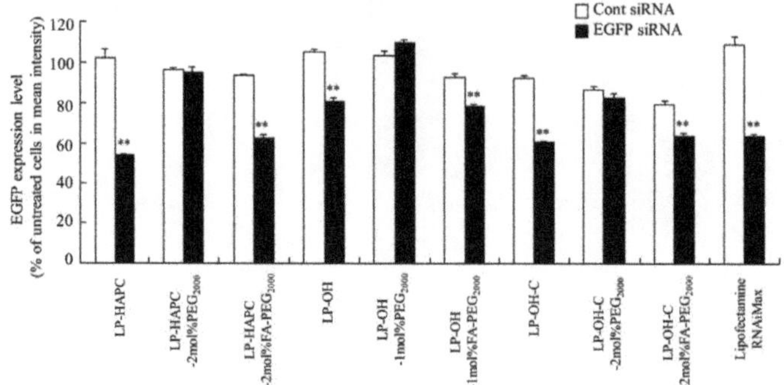

Figure 7. Effect of FA-PEG-modification of cationic liposomes on suppression of EGFP expression in KB-EGFP cells after transfection with FA-PEG-modified siRNA lipoplexes. As cationic liposomes, LP-HAPC, LP-HAPC-2mol%PEG$_{2000}$, LP-HAPC-2mol%FA-PEG$_{2000}$, LP-OH, LP-OH-1mol%PEG$_{2000}$, LP-OH-1mol%FA-PEG$_{2000}$, LP-OH-C, LP-OH-C-2mol%PEG$_{2000}$, and LP-OH-C-2mol%FA-PEG$_{2000}$ were used. The siRNA lipoplexes with Cont siRNA or EGFP siRNA were added to KB-Luc cells at 50 nM siRNA, and the EGFP expression levels were determined on the basis of EGFP-fluorescence by flow cytometry. The EGFP expression level (%) was calculated as relative to the fluorescence intensity of untransfected KB-EGFP cells. Each column represents the mean + SD (n = 3). Lipofectamine RNAiMax was used as a control. ** $p < 0.01$, compared with Cont siRNA.

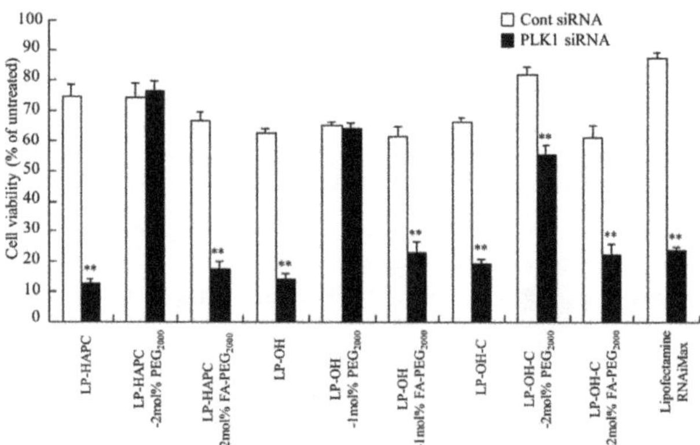

Figure 8. Effect of FA-PEG-modification of cationic liposomes on antiproliferative activities 48 h after transfection with FA-PEG-modified PLK1 siRNA lipoplexes into KB cells. As cationic liposomes, LP-HAPC, LP-HAPC-2mol%PEG$_{2000}$, LP-HAPC-2mol%FA-PEG$_{2000}$, LP-OH, LP-OH-1mol%PEG$_{2000}$, LP-OH-1mol%FA-PEG$_{2000}$, LP-OH-C, LP-OH-C-2mol%PEG$_{2000}$, and LP-OH-C-2mol%FA-PEG$_{2000}$ were used. The siRNA lipoplexes with Cont siRNA or PLK1 siRNA were added to KB-Luc cells at 50 nM siRNA. At 48 h after transfection, cell viability was measured. Each column shows the mean + SD (n = 4). Lipofectamine RNAiMax was used as a control. ** $p < 0.01$, compared with Cont siRNA.

Next, to investigate whether these cytotoxic effects by transfection of PLK1 siRNA were induced by decreased expression level of PLK1 mRNAs, we measured PLK1 mRNA levels at 24 h after transfection of PLK1 siRNA with each cationic liposome. LP-HAPC, LP-OH, and LP-OH-C lipoplexes with PLK1 siRNA significantly inhibited the expression of PLK1 mRNA, compared with Cont siRNA, which was similar to Lipofectamine RNAiMax with PLK1 siRNA (Figure 9). However, LP-HAPC-2mol%PEG$_{2000}$, LP-OH-1mol%PEG$_{2000}$, and LP-OH-C-2mol%PEG$_{2000}$ lipoplexes with PLK1 siRNA did not greatly suppress the expression of PLK1 mRNA. In contrast, transfection of LP-HAPC-2mol%FA-PEG$_{2000}$, LP-OH-1mol%FA-PEG$_{2000}$, or LP-OH-C-2mol%FA-PEG$_{2000}$ siRNA lipoplexes with PLK1 siRNA markedly decreased the expression of PLK1 mRNA in the cells. These results indicated that PLK1 siRNA transfected by FA-PEG-modified lipoplexes could specifically suppress the expression of PLK1 mRNA in the cells, and that suppression of PLK1 mRNA did affect the decrease in proliferation.

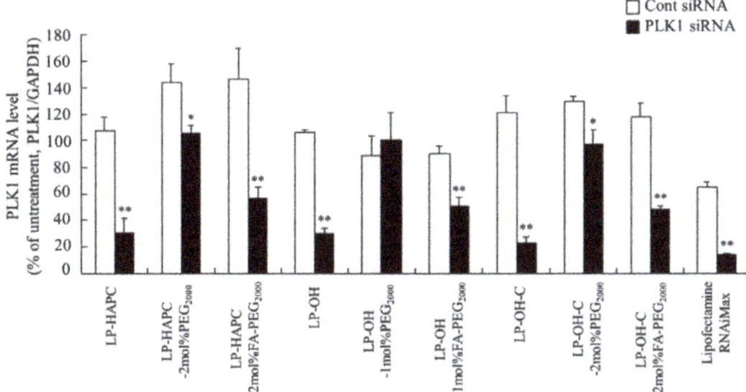

Figure 9. Effect of FA-PEG-modification of cationic liposomes on suppression of PLK1 mRNA expression by transfection with FA-PEG-modified PLK1 siRNA lipoplexes into KB cells. As cationic liposomes, LP-HAPC, LP-HAPC-2mol%PEG$_{2000}$, LP-HAPC-2mol%FA-PEG$_{2000}$, LP-OH, LP-OH-1mol%PEG$_{2000}$, LP-OH-1mol%FA-PEG$_{2000}$, LP-OH-C, LP-OH-C-2mol%PEG$_{2000}$, and LP-OH-C-2mol%FA-PEG$_{2000}$ were used. The siRNA lipoplexes with Cont siRNA or PLK1 siRNA were added to KB-Luc cells at 50 nM siRNA. At 24 h after transfection, the expression levels of PLK1 mRNA in the cells were analyzed by quantitative RT-PCR. Each result represents the mean + SD ($n = 3$). * $p < 0.05$, ** $p < 0.01$, compared with Cont siRNA.

3.8. In Vivo Gene Therapy in KB Tumor Xenografts

Among the FA-PEG-modified siRNA lipoplexes, LP-HAPC-2mol%FA-PEG$_{2000}$ lipoplexes strongly suppressed the gene silencing effect in KB cells (Figures 2A and 7). Therefore, we evaluated the anti-tumor effect by direct injection into KB tumor xenografts with LP-HAPC-2mol%FA-PEG$_{2000}$ lipoplexes of PLK1 siRNA. Injection of LP-HAPC-2mol%FA-PEG$_{2000}$ lipoplexes with Cont siRNA or PLK1 siRNA was performed a total of three times, with 2 days between each injection. The anti-tumor effect on KB tumor xenografts was evaluated by measurement of tumor volume (mm^3) and weight (mg) (Figure 10A,B). Intratumoral injections of LP-HAPC-2mol%FA-PEG$_{2000}$ lipoplexes with PLK1 siRNA inhibited tumor growth compared with the Cont siRNA, although the difference was not significant. However, intratumoral injections of LP-HAPC-2mol%PEG$_{2000}$ lipoplexes with PLK1 siRNA also inhibited growth compared with Cont siRNA, indicating that the in vivo anti-tumor effect by PEG-modified siRNA lipoplexes was not correlated with the in vitro one (Figure 8). This result suggested that the optimal formulation of PEG- or FA-PEG-modified cationic liposomes in FR-selective gene silencing effect might be different between in vitro and in vivo transfection studies.

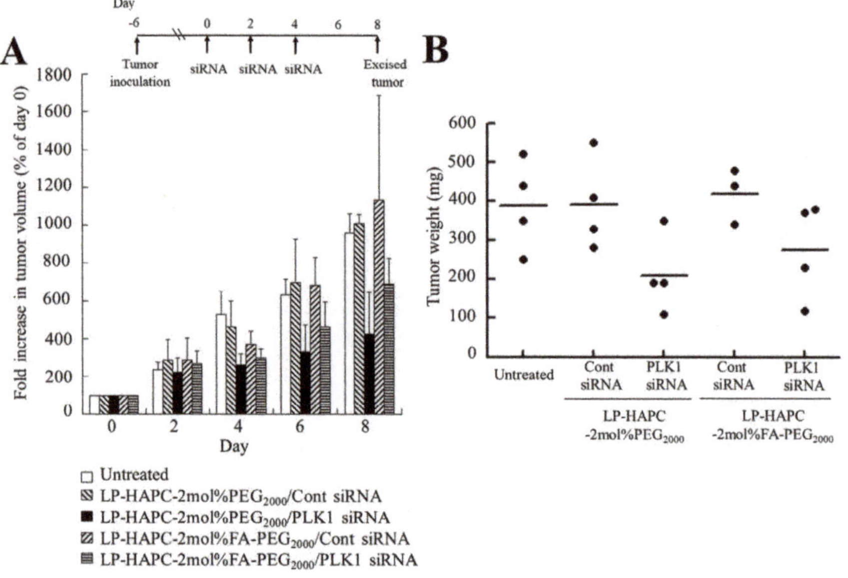

Figure 10. In vivo siRNA therapy of KB tumor xenografts with FA-PEG-modified PLK1 siRNA lipoplexes in mice. When the average volume of the xenograft tumors reached 80 mm^3 (day 0), LP-HAPC-2mol%PEG_{2000} and LP-HAPC-2mol%FA-PEG_{2000} lipoplexes with 10 μg of Cont siRNA or PLK1 siRNA were injected directly into the tumor three times (day 0, 2, and 4). Tumor volume was measured at days 0, 2, 4, 6, and 8 (**A**). Tumor volume (%) was calculated as relative to each tumor volume at the day 0. The mice were sacrificed at day 8, and then the excised tumors were weighed (**B**). Each result represents the mean + SD (n = 3–4).

4. Discussion

Several studies have investigated the application of FA-PEG-modified cationic liposomes for siRNA delivery into FR-expressing cells [20–23], although, many studies have reported on the use of FA-PEG-modified liposomes for the delivery of anticancer drugs and plasmid DNA [36,37]. For siRNA delivery with FA-PEG-modified cationic liposomes, cationic lipids such as dioctadecyldimethylammonium chloride (DODAC) [20,21], 1,2-dioleoyl-3-trimethylammonium-propane (DOTAP) [23], and 3β-(N-(N',N'-dimethylaminoethane)-carbamoyl)cholesterol (DC-Chol) [22] have been used, and FA-PEG-lipids were included at 0.5–2.5 mol% in the liposomal formulations. Previously, we reported that inclusion of 1 mol% FA-PEG_{2000}-DSPE into lipid-based cationic nanoparticles composed of OH-Chol and Tween80 could enhance siRNA delivery into tumor cells [30]. However, there have been no reports about the effect of cationic lipid type in FA-PEG-modified cationic liposomes on FR-targeting. Therefore, in this study, we used three types of cationic cholesterol derivatives for the preparation of FA-PEG-modified siRNA lipoplexes, and we examined the effect of cationic lipid type in FA-PEG-modified cationic liposomes on FR-mediated siRNA transfection in tumor cells.

Cationic cholesterol derivatives contain the following parts: a cholesteryl skeleton, a cationic head group, and a linker bound between the cholesteryl skeleton and the cationic head group. The linker between the hydrophilic and hydrophobic parts influences the gene delivery efficacy of cationic liposomes [38]. Furthermore, the introduction of a hydroxyethyl group into the cationic head group of cationic cholesterol derivatives can enhance in vitro gene transfection by cationic liposomes [39–41]. Previously, we synthesized HAPC-Chol, OH-Chol, and OH-C-Chol as cationic cholesterol derivatives with a hydroxyethyl group in the cationic head group, and demonstrated that cationic liposomes composed of their cationic cholesterol derivatives exhibited effective siRNA transfection activity [12–14,42]. Therefore, in this study, we used their cationic cholesterol derivatives, and prepared three types of FA-PEG-modified siRNA lipoplexes for the evaluation of FR-mediated siRNA transfection.

In in vitro transfection studies, FA-PEG-modification of LP-HAPC, LP-OH, and LP-OH-C could increase the gene silencing effect of siRNAs via efficient cellular uptake, compared with PEG-modification (Figures 2 and 4). In OH-Chol-based cationic liposomes, the increase in FA-PEG_{2000}-DSPE in LP-OH-FA-PEG_{2000} increased the amount of siRNA taken up by the cells (Figure 4B). However, it also decreased the gene silencing activity (Figure 2C), indicating that increased FA-PEG-modification of LP-OH could increase cellular uptake of the siRNA lipoplexes via FR, but siRNA might not be efficiently delivered into the cytoplasm or not easily released from the siRNA lipoplexes after FR-mediated endocytosis. In LP-OH-C lipoplexes, the increase in FA-PEG_{2000}-DSPE in LP-OH-C-FA-PEG_{2000} decreased the cellular uptake of the siRNA lipoplexes (Figure 4C), but it did not greatly decrease the gene silencing activity (Figure 2D), indicating that the LP-OH-C-1–3mol%FA-PEG_{2000} might deliver siRNA efficiently into cytoplasm after the endocytosis, although the increase of FA-PEG-modification in LP-OH-C decreased the cellular association. In contrast, the increase in FA-PEG_{2000}-DSPE in LP-HAPC-FA-PEG_{2000} did not greatly affect the amount of siRNA taken up by the cells (Figure 4A) and the gene silencing effect (Figure 2A), suggesting that FA-PEG-modification of LP-HAPC could mediate cellular uptake via FR, and deliver siRNA efficiently into cytoplasm after the endocytosis. These results indicated that the cellular association via FR and gene silencing efficiency by FA-PEG-modified siRNA lipoplexes were strongly affected by the cationic lipid type in cationic liposomes, and HAPC-Chol will be a better cationic lipid in FR-mediated transfection by cationic liposomes, compared with OH-Chol and OH-C-Chol. However, it was not clear why the cationic lipid type in FA-PEG-modified cationic liposomes affected cellular association via FR and the gene silencing effect. The linker groups of cationic cholesterol derivatives control the flexibility of the cationic head groups. HAPC-Chol and OH-C-Chol have a carbamate-type linker, and OH-Chol has a carboxamide-type linker. Previously, we reported that the difference in the linker group between carboxamide and carbamate in cationic cholesterol derivatives affected cellular association with siRNA nanoplexes [42]. In addition, HAPC-Chol has a slightly longer linker compared with OH-Chol and OH-C-Chol. Therefore, we speculated that the difference in the linker group of these cationic cholesterol derivatives might affect the interaction between the cationic head group on cationic liposomes and the anionic cell membrane and/or endosomal escape after transfection with FA-PEG-modified siRNA lipoplexes.

In our study, the inclusion of PEG_{2000}-DSPE or FA-PEG_{2000}-DSPE into the formulations of LP-HAPC, LP-OH, and LP-OH-C were optimal at 2 mol%, 1 mol%, and 2 mol%, respectively, for selective FR-mediated gene silencing (Figures 2 and 7). In addition, LP-HAPC-2mol%FA-PEG_{2000}, LP-OH-1mol%FA-PEG_{2000}, and LP-OH-C-2mol%FA-PEG_{2000} lipoplexes with PLK1 siRNA selectively suppressed the growth of tumor cells via down-regulation of PLK1 mRNA (Figures 8 and 9), indicating that their FA-PEG-modified siRNA lipoplexes were selectively taken up by the FR-expressing cells, and then suppressed the expression of target genes in the cells. However, in in vivo transfection, both LP-HAPC-2mol%PEG_{2000} and LP-HAPC-2mol%FA-PEG_{2000} lipoplexes with PLK1 siRNA inhibited tumor growth compared with Cont siRNA, indicating that the in vivo anti-tumor effect of PEG-modified siRNA lipoplexes was not correlated with the in vitro one. We speculated that siRNA lipoplexes after intratumoral injection were subjected to a strict environment surrounded by tumor cells, resulting in induction of non-specific uptake into tumor cells. Therefore, in in vivo transfection, PEG-modification of LP-HAPC with more PEG_{2000}-DSPE might be needed for suppression of the anti-tumor effect by PEG-modified lipoplexes with PLK1 siRNA. However, we reported previously that inclusion of 1 mol% FA-PEG_{2000}-DSPE into lipid-based cationic nanoparticles composed of OH-Chol and Tween 80 could efficiently deliver siRNAs into KB cells, and FA-PEG-modified nanoplexes of HER-2 siRNA introduced by intratumoral injection significantly inhibited the tumor growth of KB xenografts compared with

Cont siRNA, but PEG-modified nanoplexes did not [30], indicating that the in vivo anti-tumor effects by PEG-modified and FA-PEG-modified siRNA nanoplexes were well correlated with the in vitro ones. These findings suggested that the FR-selective gene silencing by PEG- and FA-PEG-modified cationic liposomes might also be affected by the liposomal formulation. Further studies must be performed to investigate liposomal formulations to improve FR-selective transfection in vivo. In this study, we selected an intratumoral injection as a route of administration for FA-PEG-modified siRNA lipoplexes; however, in future study, we will need to evaluate anti-tumor effect after intravenous injection of FA-PEG-modified siRNA lipoplexes.

5. Conclusions

In this study, we examined the effect of cationic lipids in FA-PEG-modified cationic liposomes on gene-silencing effects in tumor cells by siRNA lipoplexes. FA-PEG-modification of cationic liposomes increased in vitro gene-silencing activity regardless of cationic lipid type in cationic liposomes compared with PEG-modification; however, the cationic lipid type in FA-PEG-modified cationic liposomes affected the optimal amount of PEG_{2000}-DSPE or FA-PEG_{2000}-DSPE in the liposomal formulation. Furthermore, the in vivo anti-tumor effects by PEG-modified and FA-PEG-modified siRNA lipoplexes were not correlated with their in vitro effect, indicating that the optimized formulations of FA-PEG-modified cationic liposomes by in vitro transfection were not necessarily correlated with ones by in vivo transfection. Therefore, the in vivo optimization of FA-PEG-lipids in the liposomal formulation might be important for successful in vivo FR-mediated delivery of siRNAs by FA-PEG modified cationic liposomes.

Supplementary Materials: The following are available online at http://www.mdpi.com/1999-4923/11/4/181/s1, Figure S1. Effect of charge ratio (+:−) of siRNA lipoplexes on gene suppression in KB-Luc cells after transfection with siRNA lipoplexes.

Author Contributions: Conceptualization: Y.H.; Formal analysis: Y.H., S.S., and K.O.; Investigation: S.S., and K.O.; Project administration: Y.H.; Supervision: Y.H.; Validation: Y.H.; Writing—original draft: Y.H.; Writing—review & editing: H.O.

Funding: This research received no external funding.

Acknowledgments: We thank Yui Asami and Haruka Honma (the Department of Drug Delivery Research, Hoshi University) for their assistance with the experimental work. We thank Hiroaki Ohno (the Graduate School of Pharmaceutical Sciences, Kyoto University) for supplying OH-C-Chol.

Conflicts of Interest: The authors declare no conflict of interest.

References

1. Wilson, R.C.; Doudna, J.A. Molecular mechanisms of RNA interference. *Annu. Rev. Biophys.* **2013**, *42*, 217–239. [CrossRef]
2. Singh, A.; Trivedi, P.; Jain, N.K. Advances in siRNA delivery in cancer therapy. *Artif. Cells Nanomed. Biotechnol.* **2018**, *46*, 274–283. [CrossRef]
3. Takai, N.; Hamanaka, R.; Yoshimatsu, J.; Miyakawa, I. Polo-like kinases (PLKS) and cancer. *Oncogene* **2005**, *24*, 287–291. [CrossRef]
4. Takahashi, T.; Sano, B.; Nagata, T.; Kato, H.; Sugiyama, Y.; Kunieda, K.; Kimura, M.; Okano, Y.; Saji, S. Polo-like kinase 1 (PLK1) is overexpressed in primary colorectal cancers. *Cancer Sci.* **2003**, *94*, 148–152. [CrossRef]
5. Liu, Z.; Sun, Q.; Wang, X. PLK1, a potential target for cancer therapy. *Transl. Oncol.* **2017**, *10*, 22–32. [CrossRef]
6. Chen, X.; Mangala, L.S.; Rodriguez-Aguayo, C.; Kong, X.; Lopez-Berestein, G.; Sood, A.K. RNA interference-based therapy and its delivery systems. *Cancer Metastasis Rev.* **2018**, *37*, 107–124. [CrossRef]
7. Wang, J.; Lu, Z.; Wientjes, M.G.; Au, J.L. Delivery of siRNA therapeutics: Barriers and carriers. *AAPS J.* **2010**, *12*, 492–503. [CrossRef]
8. Zhang, S.; Zhi, D.; Huang, L. Lipid-based vectors for siRNA delivery. *J. Drug Target.* **2012**, *20*, 724–735. [CrossRef]

9. Martin, B.; Sainlos, M.; Aissaoui, A.; Oudrhiri, N.; Hauchecorne, M.; Vigneron, J.P.; Lehn, J.M.; Lehn, P. The design of cationic lipids for gene delivery. *Curr. Pharm. Des.* **2005**, *11*, 375–394. [CrossRef]
10. Xue, H.Y.; Guo, P.; Wen, W.C.; Wong, H.L. Lipid-based nanocarriers for RNA delivery. *Curr. Pharm. Des.* **2015**, *21*, 3140–3147. [CrossRef]
11. Ariatti, M. Liposomal formulation of monovalent cholesteryl cytofectins with acyclic head groups and gene delivery: A systematic review. *Curr. Pharm. Biotechnol.* **2015**, *16*, 871–881. [CrossRef] [PubMed]
12. Hattori, Y.; Takeuchi, N.; Nakamura, M.; Yoshiike, Y.; Taguchi, M.; Ohno, H.; Ozaki, K.; Onishi, H. Effect of cationic lipid type in cationic liposomes for siRNA delivery into the liver by sequential injection of chondroitin sulfate and cationic lipoplex. *J. Drug Deliv. Sci. Technol.* **2018**, *48*, 235–244. [CrossRef]
13. Hattori, Y.; Nakamura, M.; Takeuchi, N.; Tamaki, K.; Shimizu, S.; Yoshiike, Y.; Taguchi, M.; Ohno, H.; Ozaki, K.; Onishi, H. Effect of cationic lipid in cationic liposomes on siRNA delivery into the lung by intravenous injection of cationic lipoplex. *J. Drug Target.* **2019**, *27*, 217–227. [CrossRef] [PubMed]
14. Hattori, Y.; Hara, E.; Shingu, Y.; Minamiguchi, D.; Nakamura, A.; Arai, S.; Ohno, H.; Kawano, K.; Fujii, N.; Yonemochi, E. SiRNA delivery into tumor cells by cationic cholesterol derivative-based nanoparticles and liposomes. *Biol. Pharm. Bull.* **2015**, *38*, 30–38. [CrossRef]
15. Wasungu, L.; Hoekstra, D. Cationic lipids, lipoplexes and intracellular delivery of genes. *J. Control. Release* **2006**, *116*, 255–264. [CrossRef] [PubMed]
16. Liu, F.; Huang, L. Development of non-viral vectors for systemic gene delivery. *J. Control. Release* **2002**, *78*, 259–266. [CrossRef]
17. Parker, N.; Turk, M.J.; Westrick, E.; Lewis, J.D.; Low, P.S.; Leamon, C.P. Folate receptor expression in carcinomas and normal tissues determined by a quantitative radioligand binding assay. *Anal. Biochem.* **2005**, *338*, 284–293. [CrossRef]
18. Low, P.S.; Kularatne, S.A. Folate-targeted therapeutic and imaging agents for cancer. *Curr. Opin. Chem. Biol.* **2009**, *13*, 256–262. [CrossRef]
19. Xia, W.; Low, P.S. Folate-targeted therapies for cancer. *J. Med. Chem.* **2010**, *53*, 6811–6824. [CrossRef]
20. Lopes, I.; Oliveira, C.N.; Sárria, M.P.; Neves Silva, J.P.; Goncalves, O.; Gomes, A.C.; Real Oliveira, M.E. Monoolein-based nanocarriers for enhanced folate receptor-mediated RNA delivery to cancer cells. *J. Liposome Res.* **2016**, *26*, 199–210. [CrossRef]
21. Feng, C.; Wang, T.; Tang, R.; Wang, J.; Long, H.; Gao, X.; Tang, S. Silencing of the MYCN gene by siRNA delivered by folate receptor-targeted liposomes in LA-N-5 cells. *Pediatr. Surg. Int.* **2010**, *26*, 1185–1191. [CrossRef] [PubMed]
22. Xiang, B.; Dong, D.W.; Shi, N.Q.; Gao, W.; Yang, Z.Z.; Cui, Y.; Cao, D.Y.; Qi, X.R. PSA-responsive and PSMA-mediated multifunctional liposomes for targeted therapy of prostate cancer. *Biomaterials* **2013**, *34*, 6976–6991. [CrossRef]
23. Yang, T.; Li, B.; Qi, S.; Liu, Y.; Gai, Y.; Ye, P.; Yang, G.; Zhang, W.; Zhang, P.; He, X.; et al. Co-delivery of doxorubicin and bmi1 siRNA by folate receptor targeted liposomes exhibits enhanced anti-tumor effects in vitro and in vivo. *Theranostics* **2014**, *4*, 1096–1111. [CrossRef] [PubMed]
24. Hattori, Y.; Kubo, H.; Higashiyama, K.; Maitani, Y. Folate-linked nanoparticles formed with DNA complexes in sodium chloride solution enhance transfection efficiency. *J. Biomed. Nanotechnol.* **2005**, *1*, 176–184. [CrossRef]
25. Takeuchi, K.; Ishihara, M.; Kawaura, C.; Noji, M.; Furuno, T.; Nakanishi, M. Effect of zeta potential of cationic liposomes containing cationic cholesterol derivatives on gene transfection. *FEBS Lett.* **1996**, *397*, 207–209. [CrossRef]
26. Ding, W.; Hattori, Y.; Higashiyama, K.; Maitani, Y. Hydroxyethylated cationic cholesterol derivatives in liposome vectors promote gene expression in the lung. *Int. J. Pharm.* **2008**, *354*, 196–203. [CrossRef]
27. Hattori, Y.; Kikuchi, T.; Ozaki, K.; Onishi, H. Evaluation of in vitro and in vivo therapeutic antitumor efficacy of transduction of polo-like kinase 1 and heat shock transcription factor 1 small interfering RNA. *Exp. Ther. Med.* **2017**, *14*, 4300–4306. [CrossRef]
28. Hattori, Y.; Arai, S.; Kikuchi, T.; Ozaki, K.; Kawano, K.; Yonemochi, E. Therapeutic effect for liver-metastasized tumor by sequential intravenous injection of anionic polymer and cationic lipoplex of siRNA. *J. Drug Target.* **2016**, *24*, 309–317. [CrossRef]

29. Shiokawa, T.; Hattori, Y.; Kawano, K.; Ohguchi, Y.; Kawakami, H.; Toma, K.; Maitani, Y. Effect of polyethylene glycol linker chain length of folate-linked microemulsions loading aclacinomycin a on targeting ability and antitumor effect in vitro and in vivo. *Clin. Cancer Res.* **2005**, *11*, 2018–2025. [CrossRef] [PubMed]
30. Yoshizawa, T.; Hattori, Y.; Hakoshima, M.; Koga, K.; Maitani, Y. Folate-linked lipid-based nanoparticles for synthetic siRNA delivery in KB tumor xenografts. *Eur. J. Pharm. Biopharm.* **2008**, *70*, 718–725. [CrossRef]
31. Livak, K.J.; Schmittgen, T.D. Analysis of relative gene expression data using real-time quantitative PCR and the $2^{-\Delta\Delta Ct}$ method. *Methods* **2001**, *25*, 402–408. [CrossRef]
32. Hattori, Y.; Yamashita, J.; Sakaida, C.; Kawano, K.; Yonemochi, E. Evaluation of antitumor effect of zoledronic acid entrapped in folate-linked liposome for targeting to tumor-associated macrophages. *J. Liposome Res.* **2015**, *25*, 131–140. [CrossRef]
33. Ignowski, J.M.; Schaffer, D.V. Kinetic analysis and modeling of firefly luciferase as a quantitative reporter gene in live mammalian cells. *Biotechnol. Bioeng.* **2004**, *86*, 827–834. [CrossRef]
34. Bronstein, I.; Fortin, J.; Stanley, P.E.; Stewart, G.S.; Kricka, L.J. Chemiluminescent and bioluminescent reporter gene assays. *Anal. Biochem.* **1994**, *219*, 169–181. [CrossRef]
35. Corish, P.; Tyler-Smith, C. Attenuation of green fluorescent protein half-life in mammalian cells. *Protein Eng.* **1999**, *12*, 1035–1040. [CrossRef]
36. Hattori, Y.; Maitani, Y. Folate-linked lipid-based nanoparticle for targeted gene delivery. *Curr. Drug Deliv.* **2005**, *2*, 243–252. [CrossRef]
37. Chaudhury, A.; Das, S. Folate receptor targeted liposomes encapsulating anti-cancer drugs. *Curr. Pharm. Biotechnol.* **2015**, *16*, 333–343. [CrossRef]
38. Rajesh, M.; Sen, J.; Srujan, M.; Mukherjee, K.; Sreedhar, B.; Chaudhuri, A. Dramatic influence of the orientation of linker between hydrophilic and hydrophobic lipid moiety in liposomal gene delivery. *J. Am. Chem. Soc.* **2007**, *129*, 11408–11420. [CrossRef]
39. Okayama, R.; Noji, M.; Nakanishi, M. Cationic cholesterol with a hydroxyethylamino head group promotes significantly liposome-mediated gene transfection. *FEBS Lett.* **1997**, *408*, 232–234. [CrossRef]
40. Percot, A.; Briane, D.; Coudert, R.; Reynier, P.; Bouchemal, N.; Lievre, N.; Hantz, E.; Salzmann, J.L.; Cao, A. A hydroxyethylated cholesterol-based cationic lipid for DNA delivery: Effect of conditioning. *Int. J. Pharm.* **2004**, *278*, 143–163. [CrossRef]
41. Biswas, J.; Mishra, S.K.; Kondaiah, P.; Bhattacharya, S. Syntheses, transfection efficacy and cell toxicity properties of novel cholesterol-based gemini lipids having hydroxyethyl head group. *Org. Biomol. Chem.* **2011**, *9*, 4600–4613. [CrossRef]
42. Hattori, Y.; Hagiwara, A.; Ding, W.; Maitani, Y. Nacl improves sirna delivery mediated by nanoparticles of hydroxyethylated cationic cholesterol with amido-linker. *Bioorg. Med. Chem. Lett.* **2008**, *18*, 5228–5232. [CrossRef]

Article

Synergistic Anti-Angiogenic Effects Using Peptide-Based Combinatorial Delivery of siRNAs Targeting VEGFA, VEGFR1, and Endoglin Genes

Anna A. Egorova [1], Sofia V. Shtykalova [2], Marianna A. Maretina [1], Dmitry I. Sokolov [1], Sergei A. Selkov [1], Vladislav S. Baranov [1,2] and Anton V. Kiselev [1,*]

[1] D.O. Ott Research Institute of Obstetrics, Gynecology and Reproductology, 199034 Saint-Petersburg, Russia; egorova_anna@yahoo.com (A.A.E.); marianna0204@gmail.com (M.A.M.); falcojugger@yandex.ru (D.I.S.); selkovsa@mail.ru (S.A.S.); baranov@vb2475.spb.edu (V.S.B.)

[2] Department of Genetics and Biotechnology, Saint-Petersburg State University, 199034 Saint-Petersburg, Russia; sofia.shtykalova@gmail.com

* Correspondence: ankiselev@yahoo.co.uk; Tel.: +78-12-328-9809

Received: 26 April 2019; Accepted: 3 June 2019; Published: 6 June 2019

Abstract: Angiogenesis is a process of new blood vessel formation, which plays a significant role in carcinogenesis and the development of diseases associated with pathological neovascularization. An important role in the regulation of angiogenesis belongs to several key pathways such as VEGF-pathways, TGF-β-pathways, and some others. Introduction of small interfering RNA (siRNA) against genes of pro-angogenic factors is a promising strategy for the therapeutic suppression of angiogenesis. These siRNA molecules need to be specifically delivered into endothelial cells, and non-viral carriers modified with cellular receptor ligands can be proposed as perspective delivery systems for anti-angiogenic therapy purposes. Here we used modular peptide carrier L1, containing a ligand for the CXCR4 receptor, for the delivery of siRNAs targeting expression of VEGFA, VEGFR1 and endoglin genes. Transfection properties of siRNA/L1 polyplexes were studied in CXCR4-positive breast cancer cells MDA-MB-231 and endothelial cells EA.Hy926. We have demonstrated the efficient down-regulation of endothelial cells migration and proliferation by anti-VEGFA, anti-VEGFR1, and anti-endoglin siRNA-induced silencing. It was found that the efficiency of anti-angiogenic treatment can be synergistically improved via the combinatorial delivery of anti-VEGFA and anti-VEGFR1 siRNAs. Thus, this approach can be useful for the development of therapeutic angiogenesis inhibition.

Keywords: VEGFA; VEGFR1; endoglin; siRNA delivery; peptide; angiogenesis; gene silencing; migration; proliferation; endothelial cells

---

1. Introduction

Angiogenesis is the process of the formation of a new capillary network. A crucial role in the angiogenesis regulation belongs to mechanisms associated with endothelial cells, which retain their ability to divide in the adult organism [1]. Angiogenesis includes all phases of new blood vessel growth: proliferation and migration of endothelial cells, the formation of a capillary tube, and the remodulation of vascular network in organs [2,3]. The balanced functioning of this system is very important, since either the excessive vessel formation or their insufficient development leads to serious diseases. For example, intensive angiogenesis contributes to tumor growth where the formation of a branched vascular network in the tumor leads to an increase in their growth and further metastasis [4]. Therefore, approaches for the down-regulation of angiogenesis are used as elements for therapy against these diseases. Also, endothelial cells, being genetically stable, are less likely to develop drug resistance in comparison with tumor ones [5]. Taken together, antitumor agents targeting endothelial cells are supposed to be more effective than drugs targeting tumor cells. The inhibition of endothelial cell proliferation and migration may lead to the lack of structural support for tumor cells resulting in the disassembly of tumor tissues and can be used for the treatment of cancers and tumor-like diseases [6,7].

A key role in angiogenesis belongs to VEGFA (the vascular endothelial growth factor A). Its level is elevated in tissues with intensive angiogenesis and its receptors are predominantly expressed on the endothelial cells of blood vessels nearby [8,9]. VEGFA signaling has a direct impact on endothelial cells growth in vitro [10]. It prevents apoptosis of endothelial cells in vitro by inducing the expression of anti-apoptotic proteins Bcl-2 and A1 [11]. In newborn mice the suppression of VEGFA gene expression leads to apoptosis of endothelial cells in a large number of blood vessels [12]. Hypoxia is an upregulating factor for VEGFA gene expression and signaling [13,14]. For example, solid tumors under hypoxia can stimulate the production of an increased amount of VEGFA [15]. This creates conditions for the intensive development of vascular network in the growing tumors. VEGFA signaling functions through two receptor tyrosine kinases of similar structure: VEGFR1 and VEGFR2. VEGFR2 is considered to be the primary VEGFA receptor that runs angiogenesis, while VEGFA most strongly binds to the VEGFR1 receptor. VEGFR1 gene knockout mice die on the ninth day of prenatal development from disorganization and excessive growth of blood vessels. This showed that in early embryogenesis, VEGFR1 functions mainly as a decoy receptor that sequesters excess VEGFA [16]. Nevertheless, a positive modulation of angiogenesis by VEGFR-1 has been demonstrated in adults. For example, suppression of VEGFR1 led to defects in neovascularization of the eye [17]. Also, VEGFR1 expression is high in many human cancers [16,18]. VEGFR1 is suggested to serve as an alternative angiogenic pathway in the case when VEGFA is inhibited, acting in conjunction with VEGFB and PlGF ligands [19]. It gives the opportunity for VEGFR1-based inhibition of angiogenesis alternatively to the VEGFA/VEGFR2 pathway. One more alternative and promising target for the inhibition of angiogenesis is endoglin. Its expression is greatly increased in endothelial cells of blood vessels and surrounding tumors [20]. Endoglin (ENG or CD105) is a co-receptor for transforming growth factor-β (TGF-β), which participates in activating a complex signaling pathway and thus mediates the proliferation, migration, and adhesion of endothelial cells [21]. Mice with a fully inactivated endoglin die during prenatal development due to cardiac abnormalities and defects in the formation of the vascular network; their vessels stop growing and do not penetrate into the yolk sac [22]. Homozygous knockout the endoglin gene mutations in humans are also lethal. Heterozygous endoglin mutations cause hereditary hemorrhagic telangiectasia 1, which is characterized by the fragility and instability of small vessels [23]. Endoglin is preferentially expressed in the angiogenic endothelium of solid tumors, and was found to be a marker of activated endothelial cells [24]. Recently it was suggested as a promising target for antivasculogenic therapy [25,26].

A possible way for angiogenesis targeting may be the use of a gene therapy approach where small interfering RNA (siRNA) is introduced into endothelial cells with the aim to specifically inhibit the pro-angiogenic gene expression. The application of siRNA for the specific suppression of endogenous genes in cells was successfully realized by Elbashir et al. [27].

One of the most important barriers to the application of RNAi remains the necessity to create an effective siRNA delivery system. The delivery system has to provide siRNA-targeted delivery into cells, protect it from nucleases degradation, and release siRNA for its activity in the cytosol [28]. To achieve nucleic-acid-targeted delivery to endothelial cells, we and others previously suggested a new ligand-receptor pair SDF1/CXCR4 [29–32]. SDF1 (stromal cell-derived factor-1) is a ligand for the CXCR4 (chemokine receptor type 4) expressed in the endothelium of angiogenic vessels [33]. Moreover, SDF1 plays an important role during neoangiogenesis by being a main recruiter of endothelial progenitor cells [34]. The targeted delivery via CXCR4 was achieved in our previous studies by using modular peptide carriers modified with ligand derived from the N-terminus of SDF1 [31,35–39]. The developed carriers were based on cationic cysteine-flanked cross-linking peptides that can effectively bind and protect DNA and RNA fromnuclease degradation [35–37]. The anti-VEGFA siRNA-peptide polyplexes demonstrated an efficient inhibition of VEGFA expression in endothelial cells in vitro [37,39].Achieved VEGFA gene silencing by means of RNAi resulted in a significant decrease of VEGFA protein production and the rate of endothelial cell migration. Polyplexes were formed by anti-VEGFA siRNA and the most efficient peptide carrier L1 was tested using an in vivo treatment of endometriosis in a rat subcutaneous model [38]. Significantinhibition of endometriotic implants growth (55–60%) and a two-fold decrease in VEGFA gene expression were demonstrated. Anti-angiogenic effect of the polyplexes also was confirmed via immunohistochemical characterization of the endometriotic implants.

In the present study, we used L1 peptide-based polyplexes bearing siRNA against VEGFA, VEGFR1, and endoglin for the targeted suppression of angiogenesis in endothelial cells. Proliferation and migration of the transfected cells was evaluated. We analyzed the effects of VEGFA, VEGFR1, and endoglin gene silencing, either alone or in combination. Here, the tested hypothesis was that the combinatorial siRNA silencing of several angiogenic pathways may be more efficient than single gene knockdown and could result in synergistic anti-angiogenic effects in endothelial cells.

2. Materials and Methods

2.1. Cell Lines

GFP-expressing human breast cancer cell line MDA-MB 231 was kindly provided by Prof. Jessica Rosenholm, Abo Academy University, Turku, Finland. The cell line was maintained under mycoplasma-free conditions as described previously [40].

Endothelial cells EA.Hy926 (hybridoma of primary HUVEC (human umbilical vein endothelial cells) and A549 cells (human lung adenocarcinoma)) were kindly gifted by Dr. Cora-Jean C. Edgell from the University of North Carolina, USA. This cell line reproduces the main morphological, phenotypical and functional features of the endothelium [41]. The EA.Hy926 cells were maintained under mycoplasma-free conditions as described previously [36].

2.2. Peptide Synthesis and Design

L1 peptide carrier was synthesized using NPF Verta, LLC (SaintPetersburg, Russia), and stored desiccated at −20 °C. Before use, the peptide carrier was dissolved in 0.1% TFA at 2 mg/mL. The peptide purity was determined using high-performance liquid chromatography, and found to be in the range of 90–95%. L1 peptide consists of the KPVSLSYRSPSRFFESH motif connected with a DNA-binding sequence (CHRRRRRRHC) via two ε-aminocaproic acids (Ahx) [36].

2.3. siRNA Preparation of Peptide/siRNA Complexes

The sense strand of anti-VEGFA siRNA 5′-GCG GAU CAA ACC UCA CCA Att-3′ targets human VEGFA mRNA [42]. The sense strand of anti-VEGFR1 siRNA 5′-GGC CAA GAU UUG CAG AAC Utt-3′ targets human VEGFR1 mRNA [43]. The sense strand of anti-endoglin siRNA 5′-CGG UGA CGG UGA AGG UGG AAC UGA G-3′targets human endoglin mRNA [44]. The sense strand of anti-GFP siRNA 5′-CAA GCU GAC CCU GAA GUU Ctt-3′ targets GFP mRNA [45]. A non-silencing siRNA 5′-UUC UCC GAA CGU GUC ACG U- 3′served as a mock siRNA [46]. siRNAs were purchased from Syntol JSC, Moscow, Russia. siRNA/peptide complexes were prepared at 8 to 1 and 16 to 1 N/P ratios (peptide nitrogen/RNA phosphorus ratio). All positively charged amino acids were taken into account for the calculation of N/P charge ratios. The appropriate volume of the peptide carrier (2 mg/mL) was added to the siRNA solution (100 μg/mL) in Hepes-buffered mannitol (HBM) (5% *w*/*v* mannitol, 5 mM Hepes, pH 7.5) and vortexed. Then, thepolyplexes were allowed to stand at room temperature for 2 h.

X-tremeGENE liposomal transfection reagent (Roche, Mannheim, Germany) was used as a control siRNA carrier according to the manufacturer recommendations.

2.4. Cytotoxicity Assay

A total of 0.6×10^4 MDA-MB-231 and EA.Hy926 cells (at the low density) and 2.5×10^4 EA.Hy926 cells (at the high density) were seeded in 96-well plates and incubated overnight. The cytotoxicity of peptide/siRNA complexes was evaluated using Alamar blue assay (Invitrogen, Eugene, OR, USA) for cell viability after 16 h of incubation. The fluorescence was recorded on a Wallac 1420D scanning multilabel counter (Thermo Fisher Scientific Oy, Vantaa, Finland) with an excitation wavelength at 544 nm and emission wavelength at 590 nm. The relative fluorescence intensity was counted according to (F−Ff)/(Fb−Ff) × 100%, where Fb is the fluorescence intensity in untreated control and Ff is the fluorescence intensity without cells. The results are presented as mean± S.E.M of the means obtained from three independent experiments with three samples.

2.5. siRNA Transfer to MDA-MB-231 Cells

Transfection experiments were performed in triplicate. A total of 2.5×10^4 cells was seeded in 24-well plates and incubated overnight. Before transfection, the cell culture medium was replaced with medium without FBS.Anti-GFP siRNA and mock siRNA complexes were added and incubated with cells for 2.5 h. The final concentration of siRNA was 200 nM in each well and the volume of medium was 250 µL. After incubation in 1000 µL of normal culture medium for the next 48 h, cells were washed cells by 1× PBS (pH 7.2) and permeabilized with the reporter cell lysis buffer (25 mM Gly-Gly, 15 mM $MgSO_4$, 4 mM EGTA, 1 mM DTT, 1 mM PMSF; pH 7.8). GFP fluorescence in the cell extracts was measured with a Wallac 1420D scanning multilabel counter (Thermo Fisher Scientific Oy, Vantaa, Finland) at excitation wavelength of 485 nm and emission wavelength of 535 nm. The GFP fluorescence level was normalized by the total protein concentration of the cell extracts, measured using Bradford reagent (Helicon, Moscow, Russia). The data are shown as mean ± S.E.M of the means obtained from three independent experiments with three samples. Visual appearance of MDA-MB-231 cells after the transfection was registered using a Leica DM 2500 microscope (Wetzlar, Germany) with a Leica DFC345 FX camera.

2.6. siRNA Transfer to EA.Hy926 Cells

Transfection experiments in EA.Hy926 cells were performed in duplicates. The cells (15×10^4) were seeded in 24-well plates and incubated overnight. A fully supplemented cell culture medium was aspirated and replaced with medium without FBS just before the addition of siRNA complexes, followed by incubation for 4 h. The final concentration of siRNA was 200 nM per well in 1000 µL of medium. After incubation in a fully supplemented cell culture medium for the next 48 h, cells were taken for RNA extraction.

2.7. Quantitative RT-PCR

Total RNA extraction and quantitative real-time PCR analysis was performed as previously described [36,38]. The following primers were used: VEGFR1 forward primer 5′-GAGCTAAAA ATCTTGACCCACATTG-3′, reverse primer 5′-CAGTATTCAACAATCACCATCAGAG-3′; endoglin forward primer 5′-TGGTACATCTACTCGCACACGC-3′, reverse primer 5′-GGCTATGCCATGCTG CTGGTGG-3′; and endogenous reference gene β-actin was detected using forward 5′-TGCCGACAGGATGCAGAAG-3′, reverse primer 5′-GCCGATCCACACGGAGTACT-3′. The samples were measured three times and a final result was inferred by averaging the data. The values are presented as mean ± S.E.M of the means obtained from three independent experiments.

2.8. Scratch Migration Assay

The EA.Hy926 cells migration study of was performed as described previously [37,39]. siRNA/L1 complexes were prepared as described above at N/P ratios of 8/1 and 16/1 in quadruplicates (siRNA concentration was 200 and 100 nM). Also, we used a combination of different siRNA in the complexes.

Stained cells were photographed using an AxioObserver Z1 microscope (Carl Zeiss, Jena, Germany). Three random fields were registered. EA.Hy926 cell migration during the wound repair was analyzed using ImagePro Plus 6.0 software (Media Cybernetics, Bethesda, MD, USA). Number of cells (n) that migrated to the wound area was counted. Cell density (ρ) was counted in area of 17,000 μm^2. Relative number of migrated cells was computed by (n/n′)*(ρ′/ρ), where n′ is number of migrated cells in untreated control and ρ′ is the cell density in the untreated control. The results are presented as mean± S.E.M of the means obtained from five independent experiments with four samples.

2.9. Proliferation Assays

For the cell proliferation study, 0.6×10^4 EA.Hy926 cells in 100 μL DMEM-F12 medium per well were plated in a 96-well plate. siRNA/L1 complexes were prepared as described above at N/P ratios of 8/1 and 16/1 in quadruplicates (siRNA concentration was 200 and 100 nM). Also, we used a combination of different siRNA in the complexes. The cell culture medium was replaced with 50 μL of medium without FBS. siRNA/peptide complexes were added and incubated with cells for 2.5 h.Then, the cell culture medium was replaced with 100 μL of medium containing 2.5% FBS. After 72 h incubation, the cell proliferation was analyzed using Alamar blue or crystal violet staining. In the case of AlamarBlue assay, cells were incubated in a normal culture medium with 10% Alamar blue for 2 h. The fluorescence was recorded on a Wallac 1420D scanning multilabel counter (Thermo Fisher Scientific Oy, Vantaa, Finland) with an excitation wavelength at 544 nm and emission wavelength at 590 nm.In the case of crystal violet analysis, cells were stained in 100 μL of 0.2% crystal violet in 5% methanol and were then dried. Cells were dissolved in 100 μL of 50% acetic acid per well for 5 min.Absorbance was measured at 540 nm and at 630 nm. The 630 nm values were subtracted from the 450 nm values to correct for optical imperfections in the microplate. The results are shown as mean±S.E.M of the means obtained from five independent experiments with four samples.

2.10. Statistical Analysis

Statistical analysis was carried out using the Student *t*-test with the GraphPad Prism 6 software package (GraphPad Prism Inc., San Diego, CA, USA). Statistical significance was defined as * $p < 0.05$, ** $p < 0.01$, and *** $p < 0.001$.

3. Results and Discussion

RNAi targeting of different angiogenic pathways gives an opportunity for the combinatorial anti-angiogenic treatment of tumor diseases. The therapeutic RNAi can stop the growth of tumor vessels andeventually result in the disassembly of tumor tissues [47]. Here, we targeted three pathways by means of siRNA-mediated down-regulation of VEGFA, VEGFR1, and endoglin molecules. For the siRNA delivery, a previously developed original vector L1, targeting of the CXCR4 receptor was used. MDA-MB 231 and EA.Hy926 cell lines, used for the study, were shown to express CXCR4 on their surface [37,48,49]. CXCR4 is a promising receptor to be targeted for tumor-vasculature-specific delivery due to a previously found dependence between the receptor density on target cells and the efficiency of gene delivery by means of CXCR4-targeted vehicles [29–31]. This feature may additionally minimize possible off-target effects mediated by siRNA delivery in vivo via the CXCR4 receptor.

3.1. Cytotoxicity Evaluation of L1 Peptide/siRNA Complexes

Cytoxicity studies are critical for establishing of the potential of nanocarrier systems for gene therapy [50]. Cytotoxicity of the siRNA-polyplexes at two charge ratios (8/1 and 16/1) was determined using an Alamar Blue assay in MDA-MB-231 and EA.Hy926 cell lines (Figure 1). EA.Hy926 cells were seeded at high (25,000 per well) and low (6000 per well) densities according to the requirements of the migration and the proliferation analysis protocols, respectively. The low density of cells may affect results of the proliferation test after the transfection. That is why the cytotoxicity of siRNA-polyplexes in model MDA-MB-231 GFP-expressing cells, seeded at low density, was studied to choose a non-toxic siRNA-polyplexes concentration for the subsequent transfection and proliferation studies. Cytotoxicity was determined for L1/siRNA polyplexes and X-tremeGENE/siRNA lipoplexes. In the case of MDA-MB-231 cells, we used different anti-GFP siRNA concentrations (50, 100, 150, and 200 nM). We found that L1/siRNA-polyplexes at the all siRNA concentrations showed significantly lower cell toxicity than the X-tremeGENE/siRNA lipoplexes (Figure 1a). The relative fluorescence intensity after cell incubation with these polyplexes was not less than 80% to that of intact cells. These results supposed that the studied L1/siRNA-polyplexes were not involved in MDA-MB-231 cell damage and could be used in subsequent studies. In experiments with EA.Hy926 cells, we used anti-VEGFA, anti-VEGFR1, anti-endoglin, and mock siRNAs to evaluate the impact of a cationic peptide carrier on the cytotoxicity of siRNA-polyplexes. Studies with EA.Hy926 cells were conducted using 200 nM of siRNA, which was referred to our previous research on cells with high density [37]. Due to the fact that anti-endoglin siRNA has a higher molecular weight, the corresponding siRNA polyplexes consisted of more cationic L1 carrier. In order to exclude a contribution of cationic carrier on cellular damage in subsequent functional tests, we formulated mock siRNA-polyplexes with the amount of peptide carrier equivalent to anti-endoglin siRNA-polyplexes and used it as an additional control. In a preliminary experiment in endothelial cells at the low density, several siRNA concentrations were tested (50, 100, 150, and 200 nM), and no difference in cytotoxicity of the polyplexes was found (data not shown). Therefore, 200 nM of siRNA was used in subsequent experiments. In EA.Hy926 cells seeded at the low density, siRNA-polyplexes at 8/1 charge ratios were found to be non-toxic and the relative fluorescence intensity after cell incubation with these polyplexes was about 90% to that of intact cells. L1/siRNA complexes at a charge ratio of 16/1 were more toxic in comparison with 8/1 polyplexes.The relative amount of viable cells was 65–80% in comparison with that in the intact cells. However, the cytotoxicity of these polyplexes was less than that of X-tremeGENE/siRNA lipoplexes (Figure 1b).

The experiments with EA.Hy926 cells seeded with high density are showed in Figure 1c. The results of Alamar blue assay suggested that at N/P ratio of 8/1 the L1/siRNA complexes had no apparent cytotoxicity. The relative fluorescence intensity after cell incubation with siRNA-polyplexes at a 8/1 charge ratio was similar to that of intact cells. However, cytotoxicity was detected for L1/anti-endoglin siRNA and corresponding mock siRNA polyplexes formed at a N/P ratio of 16/1. It is known that the cytotoxicity of the polyplexes is mainly caused by their positive surface charge. The cationic nanocarriers interact with the cell membranes changing the membrane potential and porosity, thus eventually inducing inflammatory responses [51]. In our previous studies, a zeta potential of the L1-polyplexes formed at a 16/1 N/P ratio was shown to be highly positive (+35 mV) [37]. However, polyplexes with a positive zeta-potential were used to provide the higher transfection efficiency [52]. In the case of mock, anti-VEGFA, and anti-VEGFR1 siRNA L1-polyplexes at a 16/1 N/P ratio, the cytotoxicity was similar to that of corresponding X-tremeGENE-lipoplexes. Also, cytotoxicity exhibited by anti-endoglin siRNA-polyplexes formed at N/P ratio 16/1 was higher than in X-tremeGENE-lipoplexes. Thus, we decided to exclude the anti-endoglin siRNA-polyplexes from subsequent migration studies.

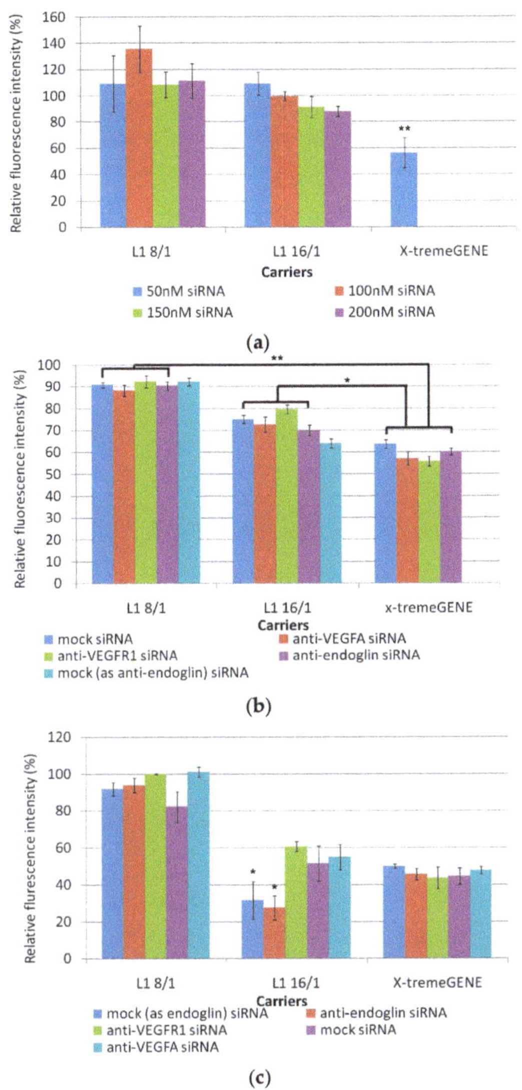

Figure 1. Cytotoxicity evaluation of the siRNA-polyplexes in MDA-MB-231 cells at the low density (**a**), EA.Hy926 cells at the low density (**b**), and EA.Hy926 cells at the high density (**c**). * $p < 0.05$, ** $p < 0.01$ when compared with X-tremeGENE-lipoplexes.

3.2. *In Vitro Transfection of MDA-MB-231 Cells*

MDA-MB-231 GFP-positive breast cancer cells were used as a convenient model for marker gene silencing in vitro. An anti-GFP siRNA transfer was performed by means of L1- and X-tremeGENE-based complexes at 8/1 and 16/1 charge ratios and with 200 nM of siRNA in 50 μL of medium (Figure 2). MDA-MB-231 cells were seeded at the low density corresponding to that in the proliferation protocol in order test whether cell density can affect the silencing efficiency. At the time of transfection, the cell growth was in log-phase. L1/mock siRNA complexes were used as a negative control and did not down-regulated GFP gene expression in comparison with non-treated cells. Meanwhile, L1/anti-GFP polyplexes caused significant decrease of GFP gene expression compared to negative control. This fact strongly confirms siRNA-induced marker gene expression silencing in the model cells. Transfection with L1/anti-GFP-siRNA complexes at N/P ratios of 8/1 and 16/1 resulted in a decrease of the relative GFP expression level of up to 22% and 17% from the intact cells, respectively. X-tremeGENE/anti-GFP-siRNA complexes also showed down-regulation of GFP gene expression up to 59%.Thus, at the low cell density conditions, we observed targeted inhibition of gene expression by means of RNAi, and the results obtained allowed us to further use RNAi in the proliferation protocol.

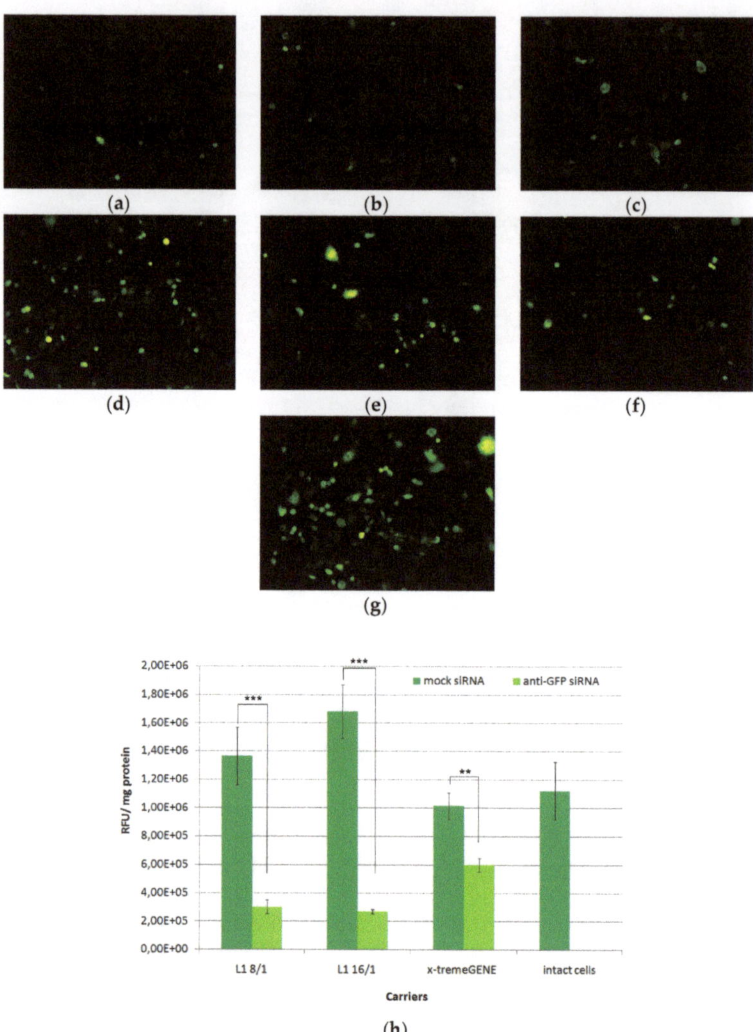

Figure 2. Silencing of the *GFP* gene expression after the treatment of MDA-MB-231 cells with L1/siRNA and X-tremeGENE/siRNA complexes. The visual appearance of MDA-MB-231 cells (magnification ×100) after treatment with (**a**) L1/anti-GFP siRNA (N/P 8/1),(**b**) L1/anti-GFP siRNA (N/P 16/1), (**c**) x-tremeGENE/anti-GFP siRNA, (**d**) L1/mock siRNA (N/P 8/1), (**e**) L1/mock siRNA (N/P 16/1), (**f**) x-tremeGENE/mock siRNA complexes (200 nM siRNA), and (**g**) intact cells. (**h**) Quantitative analysis of *GFP* gene expression, ** $p < 0.01$, *** $p < 0.001$ when compared with cells treated by mock siRNA-complexes.

3.3. *In Vitro Transfection of* EA.Hy926 *Cells*

Effects of L1 carrier-mediated siRNA delivery on gene expression were also investigated in endothelial EA.Hy926 cells. The main morphological and functional characteristics of this cell line allowed us to use it as a model of vascular endothelium and to study angiogenesis down-regulation [40]. Before performing functional tests, it was necessary to prove the specificity of VEGFA, VEGFR1, and endoglin gene expression inhibition via the RNAi mechanism. Anti-VEGFA, anti-VEGFR1, anti-endoglin, and mock siRNA were used to induce the specific silencing effects on the corresponding genes expression. L1/siRNA polyplexes were formed at 8/1 and 16/1 N/P ratios. The mock siRNA-polyplexes and free siRNA were used as negative controls, which did not show down-regulation of the gene expression (Figure 3).

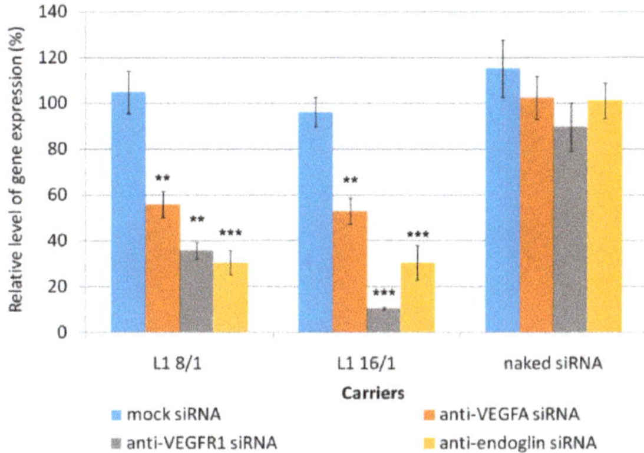

Figure 3. Silencing of VEGFA, VEGFR1, and endoglin gene expression after the treatment of EA.Hy926 cells by L1/siRNA polyplexes. ** $p < 0.01$, *** $p < 0.001$ when compared with cells treated using L1/mock siRNA polyplexes.

The cell treatment using L1/anti-VEGFA-siRNA complexes at N/P ratios of 8/1 and 16/1 resulted in a decrease of VEGFA gene expression of up to 56% and 53%, respectively. L1/anti-VEGFR1 polyplexes formed at N/P ratios of 8/1 and 16/1 demonstrated VEGFR1 knockdown of up to 36% and 11%, respectively. In the case of L1/anti-endoglin polyplexes formed at the same charge ratios down-regulated the endoglin gene expression up to 31% and 30%, respectively. The complexes with nonspecific mock siRNA did not induce any silencing compared to anti-VEGFA, anti-VEGFR1, or anti-endoglin siRNA-polyplexes. Taken together, these results suggest that the reduction in the appropriate gene expression in endothelial cells was due to a specific siRNA effect but not from the carrier toxicity.

3.4. *Inhibition of the Endothelial Cells Migration*

Migration of endothelial cells is a necessary step in the formation of new blood vessels. Endothelial cells migrate from already existing blood vessels to angiogenesis foci and interact with vascular smooth muscle cells and pericytes of the new vessels [53]. Scratch assays were performed to determine how EA.Hy926 cell migration can be affected by anti-VEGFA, anti-VEGFR1, or anti-endoglin siRNA delivery. L1/mock siRNA polyplexes were used as a negative control to demonstrate the specificity of the siRNA action. EA.Hy926 cells were transfected with siRNA-bearing polyplexes with different charge ratios and siRNA concentration. The cell monolayer was damaged and the relative number of migrated cells into the cell-free area was registered (Figure 4). The number of migrated intact cells was taken as 100% (Figure 5). It should be noted that efficient VEGFA gene silencing by L1-based polyplexes bearing 200 nM of siRNA was demonstrated previously [37]. Here, we compared the VEGFA gene silencing efficacy in EA.Hy926 cells treated with 100 and 200 nM of anti-VEGFA siRNA. The L1-based polyplexes were formed at the N/P ratio of 8/1. The decrease of cell migration after polyplex treatment with 100 nM and 200 nM of siRNA was found to be to 63% and 47%, respectively, in comparison with an appropriate mock siRNA control (Figure 5a). Actually, a key role in stimulating the blood vessels formation belongs to VEGFA. For example, VEGFA stimulates the endothelial cell migration by interacting with neuropilin-1 [54]. Therefore, a decrease in the expression of the VEGFA gene eventually leads to reduction of the cell migration.

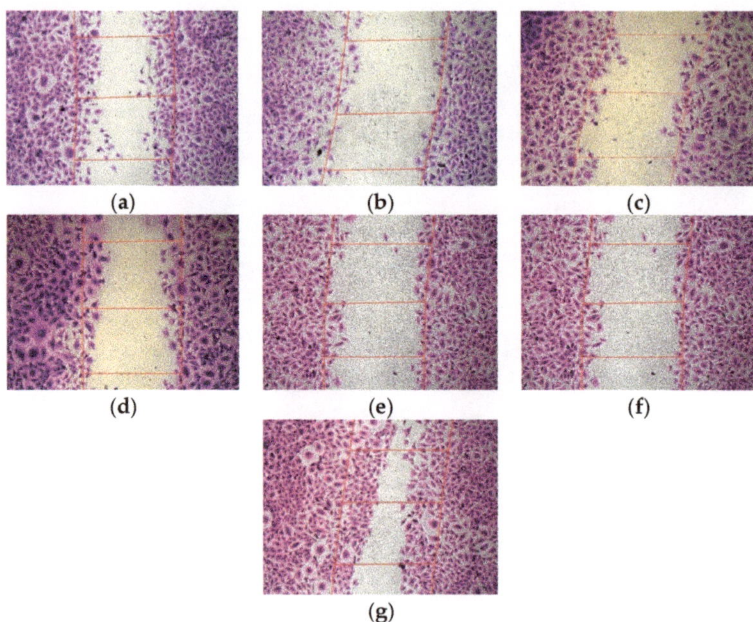

Figure 4. Appearance of migrated EA.Hy926 cells (magnification ×100) after treatment with (**a**) L1/mock siRNA,(**b**) L1/anti-VEGFA siRNA, (**c**) L1/anti-VEGFR1 siRNA, (**d**) L1/anti-endoglin siRNA, (**e**) L1/anti-VEGFA siRNA + anti-VEGFR1 siRNA, (**f**) L1/anti-VEGFA siRNA+anti-endoglin siRNA complexes (200 nM siRNA), and (**g**) intact cells.

In the case of anti-VEGFR1 siRNA treatment, the complexes were formed at the N/P ratios of 8/1 and 16/1, and with siRNA concentrations of 100 nM and 200 nM (Figure 5b). Anti-VEGFR1 siRNA complexes formed at the 16/1 N/P ratio and with 100 nM of siRNA did not contribute to a decrease in the migration activity of endothelial cells in comparison with mock siRNA-polyplexes. In contrast, significant differences between anti-VEGFR1 siRNA and mock siRNA-complexes were found for the inhibition efficiency of L1 polyplexes formed at the N/P ratio of 8/1 with 100 nM or 200 nM siRNA, and the N/P ratio of 16/1 with 200 nM siRNA. A decrease of cell migration was found to be up to 43%, 47%, and 37%, respectively, compared to appropriate mock siRNA-complexes. Therefore, in most cases, a decrease in VEGFR1 gene expression led to a reduction of endothelial cell migration. Previously, it was also demonstrated that down-regulation of VEGFR1 reduced the migration ability of endothelial cells due to the suppression of the cell actin cytoskeleton reorganization [55].

For L1/anti-endoglin polyplexes, we formed complexes at the N/P ratios of 8/1 with 100 nM and 200 nM siRNA, and 16/1 with 100 nM siRNA (Figure 5c). L1/anti-endoglin 16/1 complexes with 200 nM siRNA were excluded from the study because of their cytotoxicity (Figure 1c). Significant differences in the migration activity after the cell treatment with anti-endoglin siRNA and mock siRNA-complexes were demonstrated only when 200 nM siRNA concentrations were used. The decrease of cell migration was found to be up to 57% compared to mock siRNA-complexes. Previously, it has been demonstrated that RNAi-based inhibition of endoglin gene expression in human and mouse endothelial cells decreased their migration potential [44]. However, in a number of studies, this effect was found only after TGF-β addition into the culture medium [56]. Moreover, Lee and colleagues demonstrated an opposite effect in endoglin-negative mouse embryonic endothelial cells, which had a higher migration ability compared to wild-type cells [57]. Thus, the role of endoglin in angiogenesis should be elucidated in further studies.

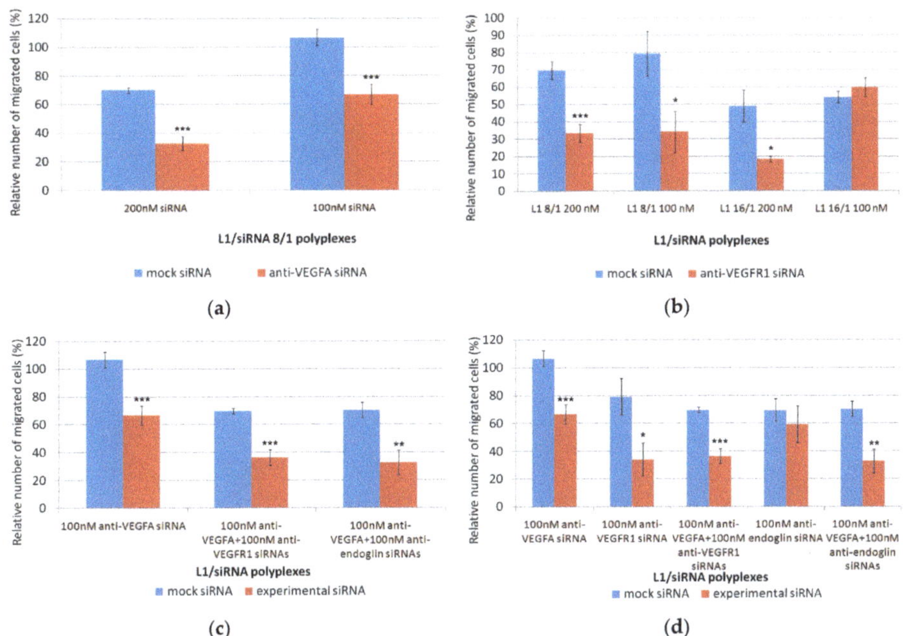

Figure 5. Number of migrated EA.Hy926 cells after treatment with (**a**) L1/anti-VEGFA siRNA, (**b**) L1/anti-VEGFR1 siRNA, (**c**) L1/ anti-endoglin siRNA, and (**d**) L1/mixed siRNA complexes. * $p < 0.05$, ** $p < 0.01$, *** $p < 0.001$ when compared with cells treated using mock siRNA.

In order to reveal a possible synergistic effect on endothelial cells migration, several siRNAs were co-delivered by L1-based polyplexes (Figure 5d). Combinatorial delivery of several siRNAs may affect the regulatory functions of the cellular miRNAs due to selective incorporation into a RISC complex [58]. However, the optimization of siRNA concentrations can weaken the competition with the miRNAs [59]. Anti-VEGFR1 or anti-endoglin siRNAs were combined with anti-VEGFA siRNA with a final concentration of 200 nM and complexed with L1 at a N/P ratio of 8/1. We did not use triple siRNA polyplexes because of cytotoxicity that could be induced by a high total concentration of siRNA (Figure 1c). Combined anti-VEGFA+anti-VEGFR1 and anti-VEGFA+anti-endoglin siRNA polyplexes decreased the relative number of migrated cells up to 52% and 47%, respectively, in comparison with control mock siRNA-complexes (Figure 5d). We did not observe acumulative effect from the combined anti-VEGFA+anti-VEGFR1 siRNA delivery. In fact, the migration suppression efficiency of anti-VEGFR1 siRNA-polyplexes alone was equal to that of combined anti-VEGFA+anti-VEGFR1 siRNA-polyplexes. In contrast, the synergistic effect of anti-VEGFA+anti-endoglin siRNA co-delivery was demonstrated and significant differences between the efficiency of these complexes and anti-VEGFA siRNA-polyplexes treatments were found ($p < 0.05$) (Figure 5d). Also, it should be noted that 100 nM anti-endoglin siRNA-polyplexes treatment did not reduce the migration of the endothelial cells (Figure 5c). The synergistic effect of combinatorial RNAi knockdown was already described in several works devoted to cancer gene therapy. For example, combinatorial siRNA targeting of EGF-Receptor and Akt2 induced tumor specific apoptosis and significantly increased survival in intracerebral glioblastoma mouse models [60]. Recently, Kamaruzman and colleagues also demonstrated the synergistic effect of combinatorial siRNA targeting both growth factor receptor and anti-apoptotic genes for therapy against breast cancer [61].

3.5. Inhibition of the Endothelial Cells Proliferation

Endothelial cells migrated from already existing blood vessels to angiogenic foci begin to proliferate under pro-angiogenic factors [62]. We used Alamar blue assay and crystal violet assay to determine whether EA.Hy926 cells proliferation was affected by anti-VEGFA, anti-VEGFR1, or anti-endoglin siRNA delivery and co-delivery. Mock siRNA-polyplexes were used as a negative control. EA.Hy926 cells were transfected with siRNA-polyplexes formed at 8/1 and 16/1 charge ratios and the siRNA concentration was 200 nM (Figure 6). It should be noted that EA.Hy926 cells were seeded at the low density in order to avoid the contact inhibition of dividing cells during the 72 h period of the proliferation assessment. Absence of the carrier-associated cytotoxicity and the successful GFP gene expression silencing after siRNA treatment of MDA-MB-231 cells seeded at the same density allowed us to perform RNAi followed by the proliferation test (Figures 1 and 2). According to the results of the proliferation tests, significant differences in the cell number after the incubation period were found between the anti-VEGFA siRNA-polyplexes and mock siRNA-polyplexes. The relative cell number after the treatment with the anti-VEGFA polyplexes at siRNA concentration of 200 nM was decreased by up to 45% and 30% from that of mock siRNA, respectively, according to the Alamar blue assay data, and up to 68% and 67%, respectively, according to crystal violet assay data (Figure 6a,b). Actually, a significant decrease in the cell proliferation was expected because VEGFA plays a critical role in the angiogenesis regulation and is important for endothelial cell physiology [63]. However, it should be noted that at a lesser anti-VEGFA siRNA concentration (100 nM), the decrease in proliferation was not observed (Figure 6g,h). This fact highlights the importance of siRNA concentration optimization to obtain VEGFA gene silencing. According to Figure 6c,d, anti-VEGFR1 siRNA polyplexes did not contribute to the decrease of endothelial cell proliferation. Differences in the relative number of EA.Hy926 cells were not observed after cell treatment, both with experimental and mock siRNA-polyplexes (Figure 6c,d). Previously, it has been demonstrated that VEGFR1 activation did not lead to the proliferation of the primary endothelial cells [64]. Also, VEGFR1 suppression did not affect the VEGFA-induced HUVEC cell line proliferation [55]. Results of the Alamar blue assay showed that in the case of L1/anti-endoglin polyplexes formed at the N/P ratio of 8/1, a significant decrease in relative cell number after the anti-endoglin siRNA treatment was found (Figure 6e). However, when the crystal violet assay was used, no differences were observed (Figure 6f). On the other hand, anti-endoglin siRNA complexes formed at the 16/1 N/P ratio did not contribute to a decrease in the proliferation of endothelial cells compared with mock siRNA-polyplexes (Figure 6e,f). Dolinsek and colleagues previously demonstrated the inhibition of endothelial cell proliferation after endoglin gene expression was suppressed by RNAi [44]. Opposite results were obtained by Pan and colleagues [65]. They showed that the level of endothelial cell proliferation in endoglin negative cells was higher than that of wild-type cells. The authors explain these results by considering TGF-β-independent signaling cascades.

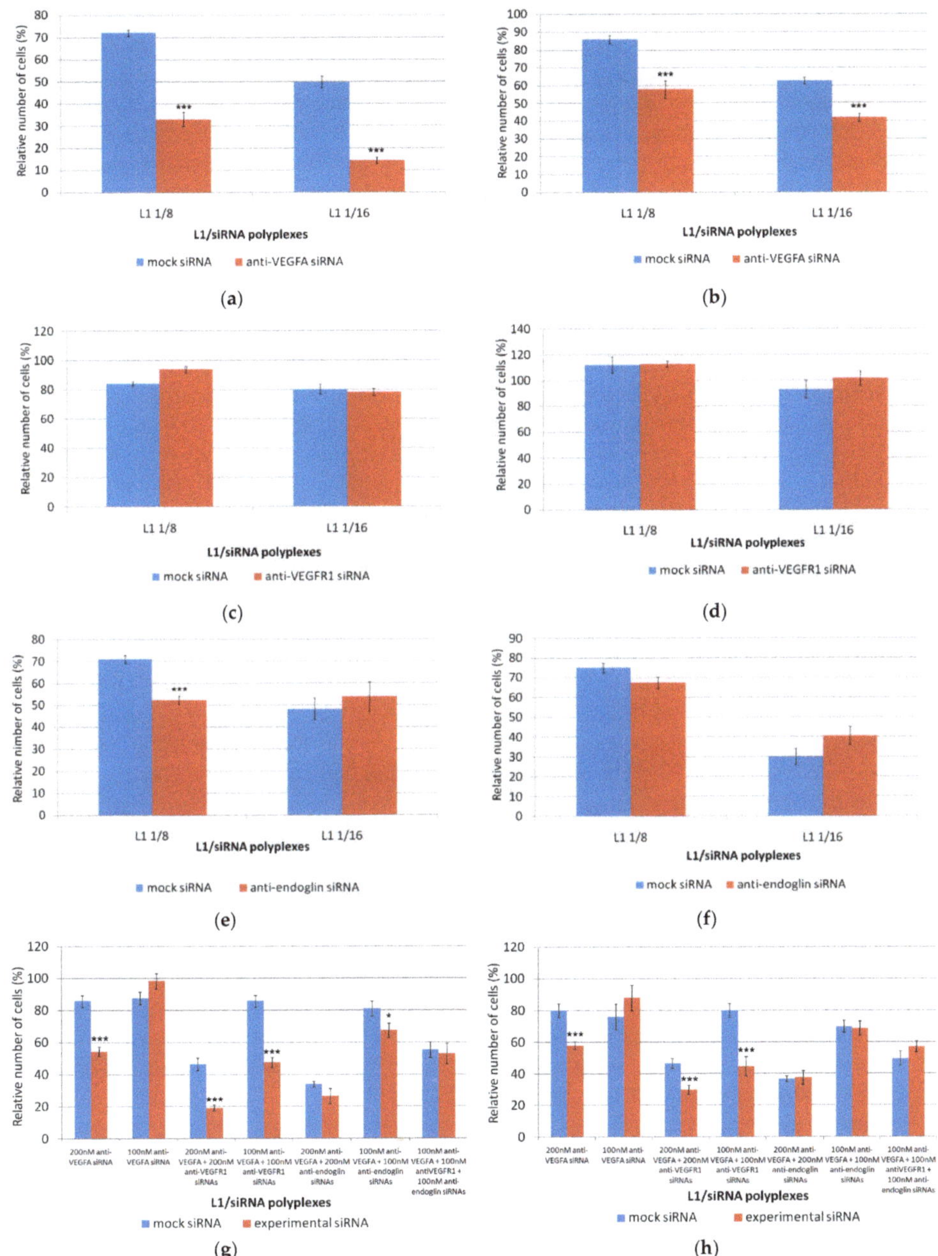

Figure 6. Number of EA.Hy926 cells after treatment with L1/anti-VEGFA siRNA (**a**,**b**), L1/anti-VEGFR1 siRNA (**c**,**d**), L1/ anti-endoglin siRNA (**e**,**f**), and L1/mixed siRNA complexes (**g**,**h**) analyzed using Alamar blue assay (**a**,**c**,**e**,**g**) and by crystal violet assay (**b**,**d**,**f**,**h**); * $p < 0.05$, *** $p < 0.001$ when compared with cells treated using mock siRNA.

Previously, it was shown that the administration of drugs that inhibit angiogenesis through only VEGFA targeting results in the elevation of VEGFR1 ligands (e.g., PlGF and VEGF-B) and eventually leads to an adaptive response and drug resistance [19]. Thus, additional targeting of an alternative angiogenesis pathway could have a greater effect on angiogenesis inhibition. As both VEGFR1 and endoglin signaling modulates angiogenesis, we investigated whether targeting both receptors and VEGFA expression might result in synergistic anti-angiogenic treatment effects in EA.Hy926 endothelial cells. To determine the synergistic effects in downregulation of the endothelial cell proliferation, several siRNAs were combined in the L1-polyplexes at an 8/1 charge ratio (Figure 6g,h). The polyplexes formed at a 16/1 charge ratio were excluded from the cell proliferation study because of an absence of difference between the two formulations in RNAi efficacy (Figures 2 and 3) and to avoid possible toxic effects of 16/1 polyplexes found in cytotoxicity experiments (Figure 1b). Anti-VEGFR1 or anti-endoglin siRNAs were mixed with anti-VEGFA siRNA at equimolar concentrations of 100 nM and 200 nM of both molecules. These polyplexes were compared with anti-VEGFA siRNA alone complexed by L1 at the concentration of 100 nM and 200 nM, respectively. We also studied triple siRNA-polyplexes consisted of L1 and 100 nM of anti-VEGFR1, anti-endoglin, and anti-VEGFA siRNAs. Similarly, the efficiency of triple combinatorial siRNA-polyplexes was compared with anti-VEGFA siRNA-polyplexes alone.

We found that the endothelial cell treatment using L1-polyplexes formed with anti-VEGFA and anti-VEGFR1 siRNAs at a concentration of 200 nM resulted in a decrease of cell proliferation up to 42% according to the Alamar blue assay results. Respective anti-VEGFA siRNA-polyplexes were less effective and decreased the cell proliferation up to 64% at the same siRNA concentration (Figure 6g). Similarly, treatment with L1-polyplexes formed with anti-VEGFA and anti-VEGFR1 siRNAs at a concentration of 100 nM, which led to a decrease of proliferation up to 55% (Figure 6g). In contrast, treatment with anti-VEGFA siRNA-polyplexes at the same siRNA concentration did not inhibit cell proliferation and was not significantly different from treatment with mock siRNA-polyplexes. Alamar blue assay results were concordant with crystal violet assay data (Figure 6h).

Thus, the combination of anti-VEGFA and anti-VEGFR1 siRNA resulted in a 1.5-fold increase of anti-angiogenic properties of L1-based polyplexes ($p < 0.001$) (Figure 6g,h). The results obtained clearly show that the combinatorial down-regulation of two VEGF-pathways led to a synergistic anti-angiogenic effect confirmed via migration and proliferation studies on the endothelial cells. Previously, it was demonstrated that combinatorial co-delivery of siRNAs against several targets could be a powerful approach to treat cancer [59,66,67]. Two strategies of combinatorial siRNA delivery were proposed. The first was copolymerization of two different siRNA sequences in a single backbone of siRNA polymer [66,67] and the second was the formation of combined siRNA-polyplexes via simultaneous complexation of multiple siRNAs with non-viral carriers [59]. Both strategies of combinatorial treatment led to synergistic anti-cancer and anti-angiogenic effects. On the over hand, combinatorial anti-VEGFA+anti-endoglin siRNAs-polyplexes and triple siRNA polyplexes mostly did not contribute to a decrease of the cell proliferation. We found that only treatment with L1-polyplexes formed with anti-VEGFA and anti-endoglin siRNAs at 100 nM concentration resulted in a slight decrease of proliferation up to 16% when analyzed using the Alamar blue assay (Figure 6g). However, this was not confirmed by the crystal violet analysis (Figure 6h).

The results of simultaneous VEGFA and endoglin suppression may be explained by considering the opposing functional activities of endoglin isoforms. Two main isoforms, membrane-anchored long endoglin (CD105-L) and soluble short endoglin (CD105-S), differ from each other regarding their cytoplasmic tails and functions [68]. CD105-L has been shown to be proangiogenic, while CD105-S exerts the opposite effect [69]. Previously, it was found that soluble CD105-S can bind some members of the TGF-β superfamily, preventing their interaction with CD105-L, and thus eventually down-regulating angiogenesis [24].The obtained results are consistent with that of Pan and colleagues who demonstrated that TGF-β-independent signaling cascades that could adversely affect cell proliferation after endoglin suppression [65]. Triple siRNA-based silencing of VEGFA, VEGFR1, and endoglin genes expression abolished anti-proliferative properties of L1-polyplexes (Figure 5g,h). The most likely explanation of this fact is the involvement of non-VEGF pro-angiogenic pathways (e.g., FGF signaling) that could be activated after significant down-regulation of the main pro-angiogenic factors (VEGFA, VEGFR1) and modulators (Endoglin) [70]. Furthermore, important information can be acquired from our data that is essential for the future development of combinatorial anti-angiogenic treatment (Table 1). According to the obtained results, the simultaneous silencing of several pro-angiogenic pathways may not be always beneficial for the efficiency of anti-angiogenic therapy as it was demonstrated by the endothelial cell treatment with triple siRNAs L1-polyplexes.

Table 1. Summary of registered anti-angiogenic effects after anti-VEGFA, anti-VEGFR1, or anti-endoglin siRNA delivery and co-delivery mediated by an L1 vector (↓ means decrease of migration/proliferation; – means no effect; ↓/– means effect detected only by Alamar Blue assay; ↓↓ means synergistic effect; NA means absence of data due to cytotoxicity).

Type of Analysis	Type of siRNA					
	Anti-VEGFA	Anti-VEGFR1	Anti-Endoglin	Anti-VEGFA + Anti-VEGFR1	Anti-VEGFA + Anti-Endoglin	Anti-VEGFA + Anti-VEGFR1 + Anti-Endoglin
Migration	↓	↓	↓	↓	↓↓	NA
Proliferation	↓	–	↓/–	↓↓	–	–

4. Conclusions

Targeting of angiogenesis by means of RNA interference is a promising and highly necessary approach for treatment of angiogenesis-related diseases induced by abnormally stimulated neovascularization such as cancer, atherosclerosis, age-related macular degeneration, endometriosis, etc. However, the need for efficient and specific siRNA delivery systems is of importance for the development of this promising approach. The efficiency of an anti-angiogenic treatment can be further improved via the combinatorial delivery of siRNAs against multiple pro-angiogenic targets. Here, we have demonstrated the efficient down-regulation of endothelial cells migration and proliferation by anti-VEGFA, anti-VEGFR1, and anti-Endoglin siRNA delivery mediated by peptide-based vector L1. Several types of single and combinatorial L1-based siRNA polyplexes have been studied and the most efficient formulation has been found. Based on our findings, we have concluded that a combinatorial treatment by L1-polyplexes formed with anti-VEGFA and anti-VEGFR1 siRNAs effectively inhibits migration and proliferation of endothelial cells and can be suggested as a useful tool for anti-angiogenic therapy.

Author Contributions: A.V.K., D.I.S., and A.A.E. designed the work; A.A.E., M.A.M., and S.V.S. performed the transfection experiments; A.A.E. and S.V.S. performed the genes expression analysis; A.A.E., M.A.M., and S.V.S. performed the functional cytological experiments; A.V.K. and A.A.E. analyzed the data; S.A.S. and V.S.B. provided funding acquisition and conceptualization of the work; S.A.S. and A.V.K. supervised the work, A.A.E. prepared the original draft; A.V.K. and D.I.S.performed the review and editing of the manuscript, and all other authors revised it critically.

Funding: This research was funded by the Russian Science Foundation, grant number 17-15-01230 (transfection/migration and proliferation studies of endothelial cells) and by the Ministry of Science and Higher Education of the Russian Federation, project number AAAA-A19-119021290033-1 (gene expression analysis and transfection studies of cancer cells). M.A.M. is supported by a President of Russian Federation scholarship (SP-822.2018.4).

Acknowledgments: Authors express their gratitude to Ksenia Markova and Anastasia Kozyreva for their assistance in the endothelial cell culturing.

Conflicts of Interest: The authors declare no conflict of interest.

References

1. Park, C.; Kim, T.M.; Malik, A.B. Transcriptional regulation of endothelial cell and vascular development. *Circ. Res.* **2013**, *112*, 1380–1400. [CrossRef] [PubMed]
2. Ferrara, N.; Kerbel, R.S. Angiogenesis as a therapeutic target. *Nature* **2005**, *438*, 967–974. [CrossRef] [PubMed]
3. Gacche, R.N.; Meshram, R.J. Angiogenic factors as potential drug target: Efficacy and limitations of anti-angiogenic therapy. *Biochim. Et Biophys. Acta—Rev. Cancer* **2014**, *1846*, 161–179. [CrossRef] [PubMed]

4. Bennett, M.R.; Sinha, S.; Owens, G.K. Vascular Smooth Muscle Cells in Atherosclerosis. *Circ. Res.* **2016**, *118*, 692–702. [CrossRef] [PubMed]
5. Kerbel, R.; Folkman, J. Clinical translation of angiogenesis inhibitors. *Nat. Rev. Cancer* **2002**, *2*, 727–739. [CrossRef] [PubMed]
6. Shubina, A.N.; Egorova, A.A.; Baranov, V.S.; Kiselev, A.V. Recent advances in gene therapy of endometriosis. *Recent Pat. Dna Gene Seq.* **2013**, *7*, 169–178. [CrossRef] [PubMed]
7. Li, T.; Kang, G.; Wang, T.; Huang, H.E. Tumor angiogenesis and anti-angiogenic gene therapy for cancer. *Oncol. Lett.* **2018**, *16*, 687–702. [CrossRef]
8. Ferrara, N. Role of vascular endothelial growth factor in regulation of physiological angiogenesis. *Am.J.Physiol.—CellPhysiol.* **2001**, *280*, C1358–C1366. [CrossRef]
9. Behrouz, R.; Malek, A.R.; Torbey, M.T. Small vessel cerebrovascular disease: The past, present, and future. *StrokeRes.Treat.* **2012**, *2012*, 839151. [CrossRef]
10. Gaengel, K.; Genové, G.; Armulik, A.; Betsholtz, C. Endothelial-mural cell signaling in vascular development and angiogenesis. *Arterioscler. Thromb. Vasc. Boil.* **2009**, *29*, 630–638. [CrossRef]
11. Liu, S.; Kong, X.; Ge, D.; Wang, S.; Zhao, J.; Su, L.; Zhang, S.; Zhao, B.; Miao, J. Identification of new small molecules as apoptosis inhibitors in vascular endothelial cells. *J. Cardiovasc. Pharmacol.* **2016**, *67*, 312–318. [CrossRef] [PubMed]
12. Gerber, H.P.; Hillan, K.J.; Ryan, A.M.; Kowalski, J.; Keller, G.A.; Rangell, L.; Wright, B.D.; Radtke, F.; Aguet, M.; Ferrara, N. VEGF is required for growth and survival in neonatal mice. *Development* **1999**, *126*, 1149–1159. [PubMed]
13. Safran, M.; Kaelin, W.G., Jr. HIF hydroxylation and the mammalian oxygen-sensing pathway. *J. Clin. Investig.* **2003**, *111*, 779–783. [CrossRef] [PubMed]
14. Morfoisse, F.; Renaud, E.; Hantelys, F.; Prats, A.-C.; Garmy-Susini, B. Role of hypoxia and vascular endothelial growth factors in lymphangiogenesis. *Mol.Cell.Oncol.* **2015**, *2*, e1024821. [CrossRef] [PubMed]
15. Luo, H.; Li, B.; Li, Z.; Cutler, S.J.; Rankin, G.O.; Chen, Y.C. Chaetoglobosin K inhibits tumor angiogenesis through downregulation of vascular epithelial growth factor-binding hypoxia-inducible factor 1α. *Anti-Cancer Drugs* **2013**, *24*, 715–724. [CrossRef] [PubMed]
16. Fischer, C.; Mazzone, M.; Jonckx, B.; Carmeliet, P. FLT1 and its ligands VEGFB and PlGF: Drug targets for anti-angiogenic therapy? *Nat. Rev. Cancer* **2008**, *8*, 942–956. [CrossRef] [PubMed]
17. Kami, J.; Muranaka, K.; Yanagi, Y.; Obata, R.; Tamaki, Y.; Shibuya, M. Inhibition of choroidal neovascularization by blocking vascular endothelial growth factor receptor tyrosine kinase. *Jpn. J. Ophthalmol.* **2008**, *52*, 91–98. [CrossRef] [PubMed]
18. Golfmann, K.; Meder, L.; Koker, M.; Volz, C.; Borchmann, S.; Tharun, L.; Dietlein, F.; Malchers, F.; Florin, A.; Büttner, R.; et al. Synergistic anti-angiogenic treatment effects by dual FGFR1 and VEGFR1 inhibition in FGFR1-amplified breast cancer. *Oncogene* **2018**, *37*, 5682–5693. [CrossRef]
19. Cao, Y. Positive and negative modulation of angiogenesis by VEGFR1 ligands. *Sci.Signal.* **2009**, *2*, re1. [CrossRef]
20. Derbyshire, E.J.; Gazdar, A.F.; King, S.W.; Thorpe, P.E.; Derbyshire, E.J.; King, S.W.; Thorpe, P.E.; Gazdar, A.F.; Vitetta, E.S.; Tazzari, P.L.; et al. Up-Regulation of Endoglin on Vascular Endothelial Cells in Human Solid Tumors: Implications for Diagnosis and Therapy. *Clin. Cancer Res.* **1995**, *1*, 1623–1634.
21. Nassiri, F.; Cusimano, M.D.; Scheithauer, B.W.; Rotondo, F.; Fazio, A.; Yousef, G.M.; Syro, L.V.; Kovacs, K.; Lloyd, R.V. Endoglin (CD105): A review of its role in angiogenesis and tumor diagnosis, progression and therapy. *Anticancer Res.* **2011**, *31*, 2283–2290. [PubMed]
22. Li, D.Y.; Sorensen, L.K.; Brooke, B.S.; Urness, L.D.; Davis, E.C.; Taylor, D.G.; Boak, B.B.; Wendel, D.P. Defective angiogenesis in mice lacking endoglin. *Science* **1999**, *284*, 1534–1537. [CrossRef] [PubMed]
23. Taskiran, C.; Erdem, O.; Onan, A.; Arisoy, O.; Acar, A.; Vural, C.; Erdem, M.; Ataoglu, O.; Guner, H. The prognostic value of endoglin (CD105) expression in ovarian carcinoma. *Int. J. Gynecol. Cancer* **2006**, *16*, 1789–1793. [CrossRef] [PubMed]
24. Kasprzak, A.; Adamek, A. Role of Endoglin (CD105) in the Progression of Hepatocellular Carcinoma and Anti-Angiogenic Therapy. *Int. J. Mol. Sci.* **2018**, *19*, 3887. [CrossRef] [PubMed]
25. Seon, B.K.; Haba, A.; Matsuno, F.; Norihiko Takahashi, M.T.; She, X.; Harada, N.; Uneda, S.; Tsujie, T.; Toi, H.; Hilda Tsai, Y.H. Endoglin-targeted cancer therapy. *Curr. Drug Deliv.* **2011**, *8*, 135–143. [CrossRef]

26. Smirnov, I.V.; Gryazeva, I.V.; Samoilovich, M.P.; Klimovich, V.B. Endoglin (CD105) - A target for visualization and anti-angiogenic therapy for malignant tumors. *Vopr. Onkol.* **2015**, *61*, 898–907.
27. Elbashir, S.M.; Harborth, J.; Lendeckel, W.; Yalcin, A.; Weber, K.; Tuschl, T. Duplexes of 21-nucleotide RNAs mediate RNA interference in cultured mammalian cells. *Nature* **2001**, *411*, 494–498. [CrossRef]
28. David, S.; Pitard, B.; Benoît, J.P.; Passirani, C. Non-viral nanosystems for systemic siRNA delivery. *Pharm. Res.* **2010**, *62*, 100–114. [CrossRef]
29. Le Bon, B.; Van Craynest, N.; Daoudi, J.M.; Di Giorgio, C.; Domb, A.J.; Vierling, P. AMD3100 Conjugates as Components of Targeted Nonviral Gene Delivery Systems: Synthesis and in Vitro Transfection Efficiency of CXCR4-Expressing Cells. *Bioconjug. Chem.* **2004**, *15*, 413–423. [CrossRef]
30. Driessen, W.H.P.; Fujii, N.; Tamamura, H.; Sullivan, S.M. Development of peptide-targeted lipoplexes to CXCR4-expressing rat glioma cells and rat proliferating endothelial cells. *Mol. Ther.* **2008**, *16*, 516–524. [CrossRef]
31. Egorova, A.; Kiselev, A.; Hakli, M.; Ruponen, M.; Baranov, V.; Urtti, A. Chemokine-derived peptides as carriers for gene delivery to CXCR4 expressing cells. *J. Gene Med.* **2009**, *11*, 772–781. [CrossRef] [PubMed]
32. Wang, Y.; Xie, Y.; Oupický, D. Potential of CXCR4 / CXCL12 Chemokine Axis in Cancer Drug Delivery. *Curr. Pharmacol. Rep.* **2016**, *2*, 1–10. [CrossRef] [PubMed]
33. Juarez, J.; Bendall, L.; Bradstock, K. Chemokines and their Receptors as Therapeutic Targets: The Role of the SDF-1 / CXCR4 Axis. *Curr. Pharm. Des.* **2005**, *10*, 1245–1259. [CrossRef]
34. Salcedo, R.; Oppenheim, J.J. Role of chemokines in angiogenesis: CXCL12/SDF-1 and CXCR4 interaction, a key regulator of endothelial cell responses. *Microcirculation* **2003**, *10*, 359–370. [CrossRef] [PubMed]
35. Kiselev, A.; Egorova, A.; Laukkanen, A.; Baranov, V.; Urtti, A. Characterization of reducible peptide oligomers as carriers for gene delivery. *Int. J. Pharm.* **2013**, *441*, 736–747. [CrossRef] [PubMed]
36. Egorova, A.; Bogacheva, M.; Shubina, A.; Baranov, V.; Kiselev, A. Development of a receptor-targeted gene delivery system using CXCR4 ligand-conjugated cross-linking peptides. *J.GeneMed.* **2014**, *16*, 336–351. [CrossRef] [PubMed]
37. Egorova, A.; Shubina, A.; Sokolov, D.; Selkov, S.; Baranov, V.; Kiselev, A. CXCR4-targeted modular peptide carriers for efficient anti-VEGF siRNA delivery. *Int.J.Pharm.* **2016**, *515*, 431–440. [CrossRef]
38. Egorova, A.; Petrosyan, M.; Maretina, M.; Balashova, N.; Polyanskih, L.; Baranov, V.; Kiselev, A. Anti-angiogenic treatment of endometriosis via anti-VEGFA siRNA delivery by means of peptide-based carrier in a rat subcutaneous model. *Gene Ther.* **2018**, *25*, 548–555. [CrossRef]
39. Egorova, A.A.; Maretina, M.A.; Kiselev, A.V. VEGFA Gene Silencing in CXCR4-Expressing Cells via siRNA Delivery by Means of Targeted Peptide Carrier. *Methods Mol. Biol.* **2019**, *1974*, 57–68.
40. Slita, A.; Egorova, A.; Casals, E.; Kiselev, A.; Rosenholm, J.M. Characterization of modified mesoporous silica nanoparticles as vectors for siRNA delivery. *Asian J. Pharm. Sci.* **2018**, *13*, 592–599. [CrossRef]
41. Edgell, C.J.S.; McDonald, C.C.; Graham, J.B. Permanent cell line expressing human factor VIII-related antigen established by hybridization. *Proc. Natl. Acad. Sci. USA* **1983**, *80*, 3734–3737. [CrossRef] [PubMed]
42. De Fougerolles, A.; Frank-Kamenetsky, M.; Manoharan, M.; Rajeev, K.G.; Hadwiger, P. *IRNA Agents Targeting VEGF, U.S. Patent No. 7,919,473*; U.S. Patent and Trademark Office: Washington, DC, USA, 2011.
43. Zhou, Z.; Zhao, C.; Wang, L.; Cao, X.; Li, J.; Huang, R.; Lao, Q.; Yu, H.; Li, Y.; Du, H.; et al. A VEGFR1 antagonistic peptide inhibits tumor growth and metastasis through VEGFR1-PI3K-AKT signaling pathway inhibition. *Am. J. Cancer Res.* **2015**, *5*, 3149–3161. [PubMed]
44. Dolinsek, T.; Markelc, B.; Bosnjak, M.; Blagus, T.; Prosen, L.; Kranjc, S.; Stimac, M.; Lampreht, U.; Sersa, G.; Cemazar, M. Endoglin silencing has significant antitumor effect on murine mammary adenocarcinoma mediated by vascular targeted effect. *Curr.GeneTher.* **2015**, *15*, 228–244. [CrossRef]
45. Raemdonck, K.; Naeye, B.; Høgset, A.; Demeester, J.; De Smedt, S.C. Prolonged gene silencing by combining siRNA nanogels and photochemical internalization. *J.Control.Release* **2010**, *145*, 281–288. [CrossRef] [PubMed]
46. Wu, J.; Qu, L.; Meng, L.; Zeng, Y.; Shou, C.; Xu, H.; Jiang, B.; Ren, T. N -α-Acetyltransferase 10 protein inhibits apoptosis through RelA/p65-regulated MCL1 expression. *Carcinogenesis* **2012**, *33*, 1193–1202.
47. Whitehead, K.A.; Langer, R.; Anderson, D.G. Knocking down barriers: advances in siRNA delivery. *Nat.Rev.Drug Discov.* **2009**, *8*, 129. [CrossRef]
48. Müller, A.; Homey, B.; Soto, H.; Ge, N.; Catron, D.; Buchanan, M.E.; McClanahan, T.; Murphy, E.; Yuan, W.; Wagner, S.N.; et al. Involvement of chemokine receptors in breast cancer metastasis. *Nature* **2001**, *410*, 50–56.

49. Wang, Z.; Ma, Y.; Yu, X.; Niu, Q.; Han, Z.; Wang, H.; Li, T. Targeting CXCR4 – CXCL12 Axis for Visualizing, Predicting, and Inhibiting Breast Cancer Metastasis with Theranostic AMD3100 – Ag2S Quantum Dot Probe. *Adv. Funct. Mater.* **2018**, *28*, 1800732. [CrossRef]
50. Vega-Villa, K.R.; Takemoto, J.K.; Yáñez, J.A.; Remsberg, C.M.; Forrest, M.L.; Davies, N.M. Clinical toxicities of nanocarrier systems. *Adv. Drug Deliv. Rev.* **2008**, *60*, 929–938. [CrossRef]
51. Vaidyanathan, S.; Anderson, K.B.; Merzel, R.L.; Jacobovitz, B.; Kaushik, M.P.; Kelly, C.N.; Van Dongen, M.A.; Dougherty, C.A.; Orr, B.G.; Banaszak Holl, M.M. Quantitative Measurement of Cationic Polymer Vector and Polymer-pDNA Polyplex Intercalation into the Cell Plasma Membrane. *ACS Nano* **2015**, *9*, 6097–6109. [CrossRef]
52. Jones, N.A.; Hill, I.R.C.; Stolnik, S.; Bignotti, F.; Davis, S.S.; Garnett, M.C. Polymer chemical structure is a key determinant of physicochemical and colloidal properties of polymer–DNA complexes for gene delivery. *Biochim. Biophys. Acta—Gene Struct. Expr.* **2000**, *1517*, 1–18. [CrossRef]
53. Herbert, S.P.; Stainier, D.Y.R. Molecular control of endothelial cell behaviour during blood vessel morphogenesis. *Nat. Rev. Mol. Cell Biol.* **2011**, *12*, 551. [CrossRef] [PubMed]
54. Herzog, B.; Pellet-Many, C.; Britton, G.; Hartzoulakis, B.; Zachary, I.C. VEGF binding to NRP1 is essential for VEGF stimulation of endothelial cell migration, complex formation between NRP1 and VEGFR2, and signaling via FAK Tyr407 phosphorylation. *Mol. Boil. Cell* **2011**, *22*, 2766–2776. [CrossRef] [PubMed]
55. Kanno, S.; Oda, N.; Abe, M.; Terai, Y.; Ito, M.; Shitara, K.; Tabayashi, K.; Shibuya, M.; Sato, Y. Roles of two VEGF receptors, Flt-1 and KDR, in the signal transduction of VEGF effects in human vascular endothelial cells. *Oncogene* **2000**, *19*, 2138–2146. [CrossRef] [PubMed]
56. Warrington, K.; Hillarby, M.C.; Li, C.; Letarte, M.; Kumar, S. Functional role of CD105 in TGF-β1 signalling in murine and human endothelial cells. *Anticancer Res.* **2005**, *25*, 1851–1864.
57. Lee, N.Y.; Ray, B.; How, T.; Blobe, G.C. Endoglin promotes transforming growth factor β-mediated Smad 1/5/8 signaling and inhibits endothelial cell migration through its association with GIPC. *J. Biol. Chem.* **2008**, *283*, 32527–32533. [CrossRef] [PubMed]
58. Castanotto, D.; Sakurai, K.; Lingeman, R.; Li, H.; Shively, L.; Aagaard, L.; Soifer, H.; Gatignol, A.; Riggs, A.; Rossi, J.J. Combinatorial delivery of small interfering RNAs reduces RNAi efficacy by selective incorporation into RISC. *Nucleic Acids Res.* **2007**, *35*, 5154–5164. [CrossRef]
59. Tiash, S.; Kamaruzman, N.I.B.; Chowdhury, E.H. Carbonate apatite nanoparticles carry siRNA(S) targeting growth factor receptor genes egfr1 and erbb2 to regress mouse breast tumor. *Drug Deliv.* **2017**, *24*, 1721–1730. [CrossRef]
60. Michiue, H.; Eguchi, A.; Scadeng, M.; Dowdy, S.F. Induction of in vivo synthetic lethal RNAi responses to treat glioblastoma. *Cancer Biol.* **2009**, *8*, 2306–2313. [CrossRef]
61. Kamaruzman, N.; Tiash, S.; Ashaie, M.; Chowdhury, E. siRNAs Targeting Growth Factor Receptor and Anti-Apoptotic Genes Synergistically Kill Breast Cancer Cells through Inhibition of MAPK and PI-3 Kinase Pathways. *Biomedicines* **2018**, *6*, 73. [CrossRef]
62. Carmeliet, P.; Jain, R.K. Molecular mechanisms and clinical applications of angiogenesis. *Nature* **2011**, *473*, 298–307. [CrossRef] [PubMed]
63. Kaufmann, P.; Mayhew, T.M.; Charnock-Jones, D.S. Aspects of human fetoplacental vasculogenesis and angiogenesis. II. Changes during normal pregnancy. *Placenta* **2004**, *25*, 114–126. [CrossRef] [PubMed]
64. Seetharam, L.; Gotoh, N.; Maru, Y.; Neufeld, G.; Yamaguchi, S.; Shibuya, M. A unique signal transduction from FLT tyrosine kinase, a receptor for vascular endothelial growth factor VEGF. *Oncogene* **1995**, *10*, 135–147. [PubMed]
65. Pan, C.C.; Bloodworth, J.C.; Mythreye, K.; Lee, N.Y. Endoglin inhibits ERK-induced c-Myc and cyclin D1 expression to impede endothelial cell proliferation. *Biochem. Biophys. Res. Commun.* **2012**, *424*, 620–623. [CrossRef] [PubMed]
66. Lee, S.J.; Yook, S.; Yhee, J.Y.; Yoon, H.Y.; Kim, M.G.; Ku, S.H.; Kim, S.H.; Park, J.H.; Jeong, J.H.; Kwon, I.C.; et al. Co-delivery of VEGF and Bcl-2 dual-targeted siRNA polymer using a single nanoparticle for synergistic anti-cancer effects in vivo. *J. Control. Release* **2015**, *220*, 631–641. [CrossRef] [PubMed]
67. Jang, M.; Han, H.D.; Ahn, H.J. A RNA nanotechnology platform for a simultaneous two-in-one siRNA delivery and its application in synergistic RNAi therapy. *Sci. Rep.* **2016**, *6*, 32363. [CrossRef]

68. Hu, J.; Guan, W.; Liu, P.; Dai, J.; Tang, K.; Xiao, H.; Qian, Y.; Sharrow, A.C.; Ye, Z.; Wu, L.; et al. Endoglin Is Essential for the Maintenance of Self-Renewal and Chemoresistance in Renal Cancer Stem Cells. *Stem Cell Rep.* **2017**, *9*, 464–477. [CrossRef] [PubMed]
69. Pérez-Gómez, E.; Eleno, N.; López-Novoa, J.M.; Ramirez, J.R.; Velasco, B.; Letarte, M.; Bernabéu, C.; Quintanilla, M. Characterization of murine S-endoglin isoform and its effects on tumor development. *Oncogene* **2005**, *24*, 4450–4461.
70. Clarke, J.M.; Hurwitz, H.I. Understanding and targeting resistance to anti-angiogenic therapies. *J. Gastrointest. Oncol.* **2013**, *4*, 253–263.

Article

Intracellular Delivery of siRNAs Targeting AKT and ERBB2 Genes Enhances Chemosensitization of Breast Cancer Cells in a Culture and Animal Model

Tahereh Fatemian [1], Hamid Reza Moghimi [2] and Ezharul Hoque Chowdhury [1,*]

1 Jeffry Cheah School of Medicine and Health Sciences, Monash University Malaysia, Jalan Lagoon Selatan, 46150 Bandar Sunway, Malaysia

2 School of Pharmacy, Shahid Beheshti University of Medical Sciences, 19839-63113 Tehran, Iran

* Correspondence: md.ezharul.hoque@monash.edu; Tel.: +60-3-5514-4978

Received: 30 June 2019; Accepted: 26 August 2019; Published: 3 September 2019

Abstract: Pharmacotherapy as the mainstay in the management of breast cancer suffers from various drawbacks, including non-targeted biodistribution, narrow therapeutic and safety windows, and also resistance to treatment. Thus, alleviation of the constraints from the pharmacodynamic and pharmacokinetic profile of classical anti-cancer drugs could lead to improvements in efficacy and patient survival in malignancies. Moreover, modifications in the genetic pathophysiology of cancer via administration of small nucleic acids might pave the way towards higher response rates to chemotherapeutics. Inorganic pH-dependent carbonate apatite (CA) nanoparticles were utilized in this study to efficiently deliver various classes of therapeutics into cancer cells. Co-delivery of drugs and genetic materials was successfully attained through a carbonate apatite delivery device. On 4T1 cells, siRNAs against AKT and ERBB2 plus paclitaxel or docetaxel resulted in the largest increase in anti-cancer effects compared to CA/paclitaxel or CA/docetaxel. Therefore, these ingredients were selected for further in vivo investigations. Animals receiving injections of CA/paclitaxel or CA/docetaxel loaded with siRNAs against AKT and ERBB2 possessed significantly smaller tumors compared to CA/drug-treated mice. Interestingly, synergistic interactions in target protein knock down with combinations of CA/AKT/paclitaxel, CA/ERBB2/docetaxel were documented via western blotting.

Keywords: siRNA; drug delivery; nanoparticle; carbonate apatite; ERBB2; AKT; breast cancer

1. Introduction

Combinations of individual therapeutics indicated against malignancies might result in superior outcomes in resolving the complexity of cancer cells. Among the tackled complexities is an acquired or intrinsic resistance of the cancer cells to treatment. In fact, combining different therapeutic strategies might provide various benefits as follows: (a) Maximized therapeutic efficacy without increased overall toxicity to the host due to different mechanisms of action; (b) prevention of the development of resistance to single agents; (c) covering the heterogeneous tumor cell population with different drug sensitivity profiles; and (d) possible synergy between therapeutics, resulting in increased anticancer efficacy [1].

The underlying mechanisms of resistance might involve genetic alterations, which could be depleted via RNA interference technologies, such as small interfering RNAs (siRNAs). Phenotype modifications through suppression of mRNA transcripts by the usage of siRNA, might render the cancer cells more responsive to the accompanying chemotherapeutic agents. In the co-delivery of nucleic acids and small molecule drugs, protective vectors would assist in boosted cellular uptake, lysosomal escape, and protection against serum nucleases together with limited off-target effects.

Nano-based formulations have the potential to be versatile and multifunctional, enabling the co-delivery of multiple agents entrapped in the nanoparticles' structure to gain synergistic anticancer effects or multi-functions.

Co-delivery of drugs and siRNAs has been documented via application of different carrier systems [2–4]. In one study, doxorubicin and Bcl-2-targeted siRNA were applied via a polyethylenimine-coated graphene oxide (PEI-GO) vehicle. Here, the GO moiety adsorbs the therapeutics while PEI enhances cell membrane penetration [2]. In another example, a triblock polymer loaded with VEGF siRNA and paclitaxel delivered augmented efficacy against paclitaxel [3]. *N*-succinyl chitosan-poly-L-lysine-palmitic acid (NSC-PLL-PA) has been utilized in the synthesis of a triblock copolymer for co-delivery of P-glycoprotein-targeted siRNA and doxorubicin at pH 7.4. These therapeutics are released upon degradation of the carrier in the acidic pH of lysosomes [4].

In fact, numerous examples of synthetic dual siRNA and drug-delivery vehicles have been reported in the literature, with varying formulations and results [5–8]. However, carriers consisting of protein or peptide origins have demonstrated superiority due to biodegradability, complex structure details and functional groups, feasible adjustments in the synthesis process, and also compatibility in the construction of multifunctional hybrid materials [9].

As published data demonstrate, a lipoproteoplex has been developed for the purpose of dual delivery of siRNA and doxorubicin, with the ability to condense siRNA and encapsulate the small-molecule chemotherapeutic, doxorubicin. The lipoproteoplex demonstrates improved doxorubicin loading, resulting in a substantial decrease in MCF-7 cell viability, plus effective transfection of GAPDH (60% knockdown) in MCF-7 breast cancer cells [9].

Inorganic carbonate apatite is a recently developed nanocarrier synthesized via calcium phosphate precipitation in the presence of bicarbonate. The controlled crystal growth dynamic leads to the formation of particles with a size ranging between 50 and 300 nm. These carriers encompass optimal features of efficient endocytosis, fast dissolution rate in endosomal acidic pH, and effective release of loaded therapeutics. In more detail, addition of 3 mM of Ca in the synthesis of carbonate apatite results in the formation of nanoparticles with a size less than 50 nM. In presence of 10 nM of paclitaxel, NPs reach a maximum size of around 170 nm. Markedly, paclitaxel-loaded carbonate apatite demonstrates a $20.71 \pm 4.34\%$ loading efficiency [10]. Carbonate apatite nanoparticles have been utilized for co-delivery of anti-cancer drugs and various siRNAs, resulting in improved outcomes [11,12].

Members of the human epidermal growth factor receptor (HER) family are among the key factors in regulating the response of breast cancer cells to chemotherapy. As another mediator in the complex pathway of oncogenesis, AKT, a serine–threonine protein kinase, has been heavily studied.

AKT signaling has been shown to be regulated by enhanced HER signals, mutational activation of Ras leading to AKT activation via the PI3K pathway, or the mutational inactivation of the PTEN phosphatase resulting in diminished AKT activity. Secondary to these alterations, AKT kinase activity would be augmented. Thereafter, increased AKT kinase activity on its own and in the absence of any upregulation in AKT protein concentrations may have a broader effect on oncogenesis as well as the cellular response to cancer therapy [13].

Thus, various components of these inter-related signaling cascades have been targeted via siRNA application in this study to obtain an accurate assessment of their role in the response of the cancer cells to chemotherapy agents.

2. Materials and Methods

2.1. Materials

Dulbecco's modified eagle medium (DMEM), calcium chloride dehydrate ($CaCl_2.2H_2O$), sodium bicarbonate ($NaHCO_3$), dimethyl sulphoxide (DMSO), 3-(4,5-dimethylthiazol-2-yl)-2,5-diphenyl tetrazolium bromide (MTT), phosphoric acid solution (H_3PO_4), trifluroacetice acid (TFA; CF_3COOH), Ethylene Diamine Tetraacetic Acid (EDTA), and anti-cancer drugs docetaxel (Doc) and paclitaxel (Pac) were purchased from Sigma-Aldrich (St Louis, MO, USA). DMEM powder, fetal bovine serum (FBS), trypsin ethylene diamine tetraacetate (trypsin-EDTA), 4-(2-hydroxyethyl)-1-piperazineethanesulfonic acid (HEPES), and penicillin-streptomycin were obtained from Gibco BRL (Carlsbad, CA, USA). All functionally validated siRNAs used in this study (listed in Table 1) were obtained from Qiagen and dissolved in RNase-free water provided by the company to obtain 10-μM stock solution. MCF-7, 4T1 and MDA-MB-231 cells were originally from ATCC.

Table 1. List of validated siRNAs (Qiagen, MD, USA) that were used in this study.

siRNA	Target Sequence	Targeted Gene	Validation Cell Line	% knockdown
Hs-AKT1-5	AATCACACCACCTGACCAAGA	AKT (protein kinase B)	HeLa S3	90
Hs_MAPK1_l0	AAGTTCGAGTAGCTATCAAGA	Mitogen-activated protein kinase	HeLa S3	90
Hs-ROS1-5	AAGGTAATTGCTCTAACTTTA	ROS Proto-Oncogene 1, Receptor tyrosine kinase	HeLa	87
Hs-ERBB2-14	AACAAAGAAATCTTAGACGAA	Receptor tyrosine-protein kinase ERBB-2 or human epidermal growth factor receptor 2 (HER2)	MCF-7	92

2.2. *Cell Culture*

MCF-7 and MDA-MB-231 as human breast cancer cell lines and 4T1 as a mouse breast cancer cell line were cultured in 75-cm^2 tissue culture flasks (Nunc, Orlando, FL, USA) using DMEM supplemented with 10% FBS, 1% penicillin and streptomycin, and 1% HEPES at 37 °C in a humidified 5% CO_2-containing atmosphere.

2.3. *Generation and Characterization of Carbonate Apatite*

Carbonate apatite nanoparticles were manufactured via addition of Ca^{2+} from 1 M $CaCl_2$ stock solution to the bicarbonate-buffered cell culture medium (DMEM, pH adjusted to 7.5, containing 44 mM HCO^{3-}), which already contained the third reactant (0.9 mM phosphate), followed by incubation at 37 °C for 30 min. As a result, microscopically visible carbonate apatite particles were formed through precipitation following nucleation in a supersaturated solution. Turbidity determination and size and zeta potential measurement of the variously formulated nanoparticles were employed for characterization of the resulting products [10].

2.4. *Complexation of Drugs and siRNAs with Carbonate Apatite*

Various concentrations of drugs and siRNAs (Table 1) were added in DMEM media (44 mM bicarbonate, pH 7.5) containing particular concentrations of $CaCl_2$ and subjected to a 30-min incubation at 37 °C to allow formation of complexes.

2.5. *In Vitro Viability Assay*

Cytotoxicity of carbonate apatite nanoparticles alone and also differently loaded NPs on human and murine breast cancer cell lines was assessed by MTT assay. Briefly, the cells from the exponential growth phase were seeded in 24-well plates (Griener, Frickenhausen, Germany) (approximately 50,000 cells/well) in DMEM with 10% FBS at 37 °C with 5% CO_2. After 24 h, cells were exposed to various treatments for a consecutive period of 48 h. Two days later the viability was assessed by adding 50 μL MTT solution (5 mg/mL in phosphate buffered solution (PBS)) to each well and incubating for 4 h in dark. Then, the medium was removed and 300 μL DMSO was added to each well to dissolve the purple formazan crystals. Formazan quantification in the form of optical density (OD) was performed at test and reference wavelengths of 595 nm and 630 nm by a plate reader (benchmark plus, Bio Rad).

Cell viability was determined using the following formula:

$$\text{Cell viability (\%) (CV)} = \frac{OD_{(treated)} - OD_{(reference)}}{OD_{(untreated)} - OD_{(reference)}} \times 100.$$

The reference was the optical density of DMSO only in the applied wavelengths.

Each experiment was done in triplicate and results are expressed as mean ± SD of % of cell viability. Subsequently, an increase in cytotoxic effect of the loaded NPs was calculated as follows:

$$\text{Increase in toxicity (\%)} = CV_{\text{ baseline treatment}} - CV_{\text{ complete treatment}},$$

where $CV_{\text{ baseline treatment}}$ and $CV_{\text{ complete treatment}}$ represent the cell viability resulting from the baseline treatment and complete treatment, respectively. In all bar charts (Supplementary Figures) displaying cell viability values, each two adjacent bars were compared together and an increase in toxicity was calculated for all different concentrations and expressed as mean ± SD.

2.6. Sodium Dodecyl Sulfate Polyacrylamide Gel Electrophoresis (Sds-Page) and Western Blot

Treated cells in each well of a 24-well plate were lysed by addition of 200 µL of whole cell IP-lysis buffer and protein sample was collected. Then, 5 µL solution was used for estimation of total protein content using the BSA assay kit according to instruction provided by the manufacturer (Quick-start Bradford protein assay kit, Bio Rad, Hercules, CA, USA).

Samples of the cell lysate containing equal amounts of total protein (e.g., 10 µg) were mixed with 10 µL of 10× loading dye and heated for 5 min at 95 °C and then resolved by SDS-PAGE using stain free mini protean SFX gels (10 wells) in 1X running buffer. In total, 7 µL of precision plus protein standards-dual color were used as a molecular weight marker to confirm the molecular weight of the proteins in the samples. The protein samples were transferred from gel to the 0.2-µm polyvinylidene difluoride (PVDF) membranes and attached using a trans-blot turbo transfer system (Bio Rad). Membranes were blocked in 5% skimmed milk in 1X TBST for an hour at room temperature.

The membrane was probed with indicated primary antibody (Table 2) overnight at 4 °C. Unbound primary antibodies were washed using 1X TBST buffer for 5 times, 5 min each, with gentle agitation. Blots were probed with horseradish peroxide conjugated secondary antibody (anti-rabbit IgG, 1:3000) for 1 h at room temperature. TBST was again used to remove excess secondary antibody by 5 wash cycles each 5 min long with gentle agitation. Clarity Western Enhanced Chemiluminescence (ECL) substrate (Bio-Rad) was applied onto the membrane in the dark for 5 min and the signals on the membrane were visualized via a Bio-Rad Gel documentation system.

Table 2. Information for the primary antibodies used for western blot in this study.

Name	Manufacturer	Molecular Weight	Clonality	Used Dilution
AKT (pan) (C67E7) Rabbit mAb	Cell signaling (Danvers, MA, USA)	60 KDa	Monoclonal	1:1000
p 44/42 MAPK (Erk 1/2) (137F5) Rabbit mAb	Cell signaling (Danvers, MA, USA)	42/44 KDa	Monoclonal	1:1000
ERBB2 (HER2)	Thermo Fisher scientific (Waltham, MA, USA)	200 KDa	Polyclonal	1:1000
GAPDH (glyceraldehyde-3-phosphate dehydrogenase)	Cell signaling (Danvers, MA, USA)	37 KDa	Monoclonal	1:3000

2.7. Formulation of Particles for In Vivo Study

The injectable nanoparticles were formulated in 100 µL of freshly prepared bicarbonated (44 mM) DMEM media to which $CaCl_2$ was added. Samples were then incubated at 37 °C for 30 min followed by maintenance on ice to prevent aggregation during injection. In drugcontaining samples, 1.25 mg/kg of Pac and 1 mg/kg of Doc were used prior to incubation. In case of using siRNAs, 50 nM of each siRNA was added to the media prior to incubation. The resulting therapeutics were used for iv treatment of animals.

2.8. 4T1-Induced Breast Cancer Murine Model

Female Balb/c mice with the age of 6 to 8 weeks and body weight of 15 to 20 g were used in this study.

Animals were maintained in a 12:12 light:dark condition and provided with food ab libitum and water. All experiments were performed in complete adherence to the regulations of Monash University Animal Welfare Committee. The details of the animal study were approved by Monash Animal Ethics Committee on 3 August 2012 with the project identification code of MARP/2012/087 under the title of "Delivery of anti-cancer drugs to breast cancer cells using nanoparticles". Tumor induction was performed via subcutaneous injection of 4T1 cells (in 100 μL PBS) on the mammary fat pad of mice. Injection day was considered as day 1 of the animal study. The development of tumor was regularly assessed through manual examination of the injection site. Randomization and treatment were carried out when the volume of the tumor reached an average of 13.20 ± 2.51 mm^3 (Table 3). Treatment was administered via intravenous injection through the right or left caudal vein. Duration of study was 30 days, which involved close monitoring of the animals together with recording their body weights and tumor outgrowth every other day. The following formula was used for calculation of the tumor volume:

$$\text{Tumor volume }(\text{mm}^3) = \frac{(\text{Length} \times \text{Width}^2)}{2}$$

Table 3. Treatment groups used for in vivo study.

Group	Regimen
Untreated	-
CA	7 μL of 1 M $CaCl_2$ in DMEM
CA	4 μL of 1 M $CaCl_2$ in DMEM
Pac	1.25 mg/kg paclitaxel in DMEM
Doc	1 mg/kg docetaxel in DMEM
CA/Pac	1.25 mg/kg paclitaxel and 7 μL of 1 M $CaCl_2$ in DMEM
CA/Pac	1.25 mg/kg paclitaxel and 4 μL of 1 M $CaCl_2$ in DMEM
CA/Doc	1 mg/kg docetaxel and 4 μL of 1 M $CaCl_2$ in DMEM
CA/AKT	50 nM of AKT siRNA and 4 μL of 1 M $CaCl_2$ in DMEM
CA/ERBB2	50 nM of ERBB2 siRNA and 4 μL of 1 M $CaCl_2$ in DMEM
CA/AKT/ERBB2/Pac	1.25 mg/kg paclitaxel and 50 nM of ERBB2 and AKT siRNA and 4 μL of 1 M $CaCl_2$ in DMEM
CA/AKT/ERBB2/Doc	1 mg/kg docetaxel and 50 nM of ERBB2 and AKT siRNA and 4 μL of 1 M $CaCl_2$ in DMEM

2.9. Statistics

For determining statistical significance of quantifications, student's *t*-test was used; all data are presented as mean ± SD. Data was considered significant for p values < 0.05.

3. Results and Discussion

3.1. Cytotoxicity of siRNA-Loaded NPs on MCF-7 and 4T1 Cells

To explore the efficacy of carbonate apatite in the delivery of siRNAs into the cells and also evaluate the effect of single gene knock down on the viability of cancer cells, carbonate apatite nanoparticles harboring single siRNA were tested on MCF-7 and 4T1 cells. Here, 4 mM of $CaCl_2$ was applied in the presence of various amounts of single siRNAs to produce treatment formulations.

Binding of siRNA to NPs, in the presence of increasing concentrations of calcium and fixed siRNA amounts, is saturable and reaches a plateau level, in spite of the escalating pattern of particle size. Higher amounts of calcium together with limited endogenous phosphate (0.9 mM) accounts for an increased particle size due to aggregation and not larger individual particles. Limited surface area of the particles secondary to aggregation results in restricted siRNA binding [14].

Remarkably, CA capacity in the efficient delivery of siRNAs into their action site is displayed by the considerable enhancement in the cytotoxic effects for almost all amounts of siRNAs complexed with CA compared to free siRNA. The most killing effect on MCF-7 and 4T1 seems to be achieved by muting the HER2 or ROS1 cascade (Table 4) (Supplementary Figures S1 and S2), indicating the strong impact of ERBB2 or ROS signaling on the survival of both cell lines.

Table 4. Cytotoxicity (%) enhancement on 4T1 and MCF-7 cells treated with various siRNAs (1 pM, 10 pM, 100 pM, 1 nM, and 10 nM) incorporated into an apatite structure formed with 3 mM of $CaCl_2$. The data is presented as mean ± SD compared to free siRNA.

Cell line	Treatment	siRNA Concentration				
		1 pM	10 pM	100 pM	1 nM	10 nM
4T1	CA/ERBB2	16.44 ± 5.77	33.24 ± 9.14	26.70 ± 3.93	28.09 ± 6.28	19.80 ± 1.13
	CA/AKT	14.11 ± 5.41	−4.58 ± 5.01	3.42 ± 1.23	−3.27 ± 4.48	−3.96 ± 2.98
	CA/MAPK	10.68 ± 3.69	−0.72 ± 0.87	17.59 ± 1.04	5.36 ± 1.15	25.45 ± 4.64
	CA/ROS1	13.85 ± 6.99	14.68 ± 3.87	27.46 ± 1.17	16.65 ± 1.44	35.56 ± 2.81
MCF-7	CA/ERBB2	07.49 ± 8.01	24.56 ± 3.19	13.82 ± 3.34	19.29 ± 3.91	8.63 ± 1.67
	CA/ROS1	12.52 ± 2.79	13.69 ± 3.07	23.97 ± 2.13	28.88 ± 2.29	29.70 ± 3.96

In another research, carbonate apatite nanoparticles have been confirmed as efficient carriers for electrostatically associated siRNA inside the cells, releasing bound siRNA from endosomes to the cytosol through pH-responsive self-dissolution. This was concluded based on the significant increase in the fluorescence intensity of the intracellular components from the cells treated with CA/fluorescent siRNA compared to untreated and CA-treated cells [14].

Moreover, application of different concentrations of 'Allstars Negative Control siRNA' (1 pM to 10 nM) loaded into a carbonate apatite structure resulted in no alteration of breast cancer cells' viability [11].

However, the duration of response of the cancer cells to single-oncogene inhibitors, even the most potent oncogenes, is mostly short due to the development of resistance or upregulation of compensatory mechanisms in survival and proliferation signaling. Therefore, sustained and significant outcomes are mainly achieved with combinations of various classes of anti-cancer therapies. In case of no initial response to treatment or development of drug resistance, combinatory regimens are of great benefit to induce therapeutic response. In fact, putting off the causal cascades of resistance by one agent might result in re-sensitization of the cells to other therapeutics in the combination.

Markedly, pharmacokinetics and pharmacodynamics interactions between the agents in combination therapy might develop in case of a shared action site, uptake, and metabolism or elimination routes.

Delivery of drugs and siRNAs by means of carbonate apatite nanoparticles into cells was further studied. The aim was to evaluate the impact of silencing different signaling pathways on the response of cancer cells to the cytotoxic effects of drugs. Different concentrations of paclitaxel, docetaxel, mitomycin C, and topotecan (10 pM, 100 pM, and 1 nM) were used together with 1 pM of siRNAs against AKT, ERBB2, MAPK, and ROS1 and 3 mM Ca. Based on extensive experiments and with the aim of exploring possible synergistic effects, the lowest effective doses of therapeutics were applied. This was to perhaps achieve a therapeutic effect with lower doses and hence less side effects.

Notably, application of anti HER2 siRNA together with any of the four drugs caused positive changes in cytotoxicity against 4T1 cells (Supplementary Figures S3–S6). Additionally, co-delivery of Pac and ROS1 siRNA or Doc and MAPK siRNA seemed to sensitize 4T1 cells to the drug to a great extent.

Response of MDA-MB-231 cells to chemotherapy in case of single pathway silencing was also explored. Paclitaxel and docetaxel were used in complexes of carbonate apatite using 3 mM Ca and 10 pM, 100 pM, and 1 nM of the drug together with 1 nM of each siRNA. As the cell viability data reveals, Pac efficacy benefits from silencing ERBB2 or ROS1 pathways in terms of a highest increase in cytotoxicity on MDA-MB-231 cells (Supplementary Figure S7). Whereas, AKT or ERBB2 pathways display a significant role in response to docetaxel in MDA-MB-231 cells, as with knock down of these two oncogenes, all concentrations of Doc exert a substantial higher cytotoxicity (Supplementary Figure S8).

As the next step, simultaneous silencing of AKT and ERBB2 oncogenes was implemented to check if these pathways have any impact on the response to classical anti-cancer drugs.

Carbonate apatite was formed with 3 mM of $CaCl_2$ and 1 pM of AKT and ERBB2 siRNA together with 10 pM, 100 pM, and 1 nM of paclitaxel, docetaxel, mitomycin C, and topotecan. Based on cell viability results (Supplementary Figure S9), knock down of AKT and ERBB2 seems to have the greatest impact on the response to paclitaxel or docetaxel in 4T1 cells. The highest enhancement in cytotoxicity on 4T1 is equal to 19.97 ± 1.73% for Pac and 15.16 ± 3.55% for Doc, resulting from simultaneous blockade of AKT and ERBB2 signaling. Thus, these combinations were applied for in vivo studies since the animal tumor model was developed by local injection of 4T1 cells in mice.

On MDA-MB-231 cells, application of siRNAs for simultaneous silencing of two pathways led to augmented cytotoxicity of paclitaxel as well. This effect was higher with usage of AKT and ROS1 siRNA or co-delivery of AKT and MAPK siRNAs in the presence of Pac (Supplementary Figure S10).

Tables 5–7 demonstrate the enhancement of the cytotoxic effect of classical anti-cancer drugs resulting from the silencing of various pathways in MCF-7, 4T1, and MDA-MB-231 cells, respectively.

Table 5. Effect of silencing various pathways on cytotoxicity of classical anti-cancer drugs in MCF-7 cells. Data is presented as mean ± SD compared to CA/drug.

Treatment on MCF-7 Cells	Drug Concentration		
	10 pM	100 pM	1 nM
CA/Pac/AKT	10.93 ± 1.05	8.33 ± 0.84	3.31 ± 0.45
CA/Pac/ERBB2	7.25 ± 1.91	4.11 ± 0.35	3.40 ± 1.81
CA/Pac/MAPK	−10.79 ± 2.12	1.53 ± 0.55	2.06 ± 0.35
CA/Pac/ROS1	4.19 ± 1.41	3.14 ± 0.74	3.64 ± 0.19
CA/Doc/AKT	5.14 ± 1.22	10.45 ± 1.33	11.69 ± 2.18
CA/Doc/ERBB2	8.95 ± 2.28	10.25 ± 0.99	14.28 ± 1.04
CA/Doc/MAPK	3.22 ± 0.91	4.27 ± 0.08	3.67 ± 1.72
CA/Doc/ROS1	1.06 ± 0.59	2.83 ± 0.71	1.25 ± 0.53

Table 6. Effect of silencing various pathways on cytotoxicity of classical anti-cancer drugs in 4T1 cells. Data is presented as mean ± SD compared to CA/drug.

Treatment on 4T1 Cells	Drug Concentration		
	10 pM	100 pM	1 nM
CA/Pac/AKT	−15.93 ± 0.95	8.70 ± 2.14	−2.68 ± 0.54
CA/Pac/ERBB2	12.93 ± 0.36	9.15 ± 0.14	2.42 ± 2.84
CA/Pac/MAPK	−13.79 ± 1.18	2.53 ± 0.20	2.16 ± 0.11
CA/Pac/ROS1	10.09 ± 1.03	2.04 ± 0.85	7.24 ± 0.07
CA/Doc/AKT	−5.04 ± 0.77	−9.43 ± 0.51	1.30 ± 0.62
CA/Doc/ERBB2	5.85 ± 1.08	7.21 ± 1.17	2.20 ± 1.95
CA/Doc/MAPK	3.16 ± 0.21	5.07 ± 0.98	2.67 ± 0.76
CA/Doc/ROS1	2.76 ± 0.58	1.81 ± 0.97	0.05 ± 0.42

Table 6. *Cont.*

Treatment on 4T1 Cells	Drug Concentration		
	10 pM	100 pM	1 nM
CA/Mito/AKT	0.20 ± 0.71	4.97 ± 1.15	1.92 ± 0.28
CA/Mito/ERBB2	2.13 ± 3.64	5.11 ± 0.85	1.71 ± 0.38
CA/Mito/MAPK	1.26 ± 1.23	0.23 ± 1.45	7.77 ± 0.95
CA/Mito/ROS1	0.72 ± 0.92	0.52 ± 0.16	4.20 ± 1.12
CA/Topo/AKT	0.02 ± 1.64	3.09 ± 0.42	5.28 ± 3.07
CA/Topo/ERBB2	4.54 ± 2.27	5.85 ± 0.47	0.86 ± 2.51
CA/Topo/MAPK	6.54 ± 0.55	0.55 ± 0.51	4.74 ± 0.88
CA/Topo/ROS1	0.58 ± 0.25	0.03 ± 0.36	0.52 ± 0.81
CA/Pac/AKT/ERBB2	19.97 ± 1.73	5.48 ± 2.09	11.63 ± 2.23
CA/Doc/AKT/ERBB2	7.87 ± 1.82	1.45 ± 0.37	15.16 ± 3.55
CA/Mito/AKT/ERBB2	6.04 ± 0.28	0.59 ± 0.47	4.60 ± 0.82
CA/Topo/AKT/ERBB2	−10.41 ± 0.66	−4.73 ± 1.52	−10.77 ± 0.71

Table 7. Effect of silencing various pathways on cytotoxicity of classical anti-cancer drugs in MDA-MB-231 cells. Data is presented as mean ± SD compared to CA/drug.

Treatment on MDA-MB-231 Cells	Drug Concentration		
	10 pM	100 pM	1 nM
CA/Pac/AKT	4.60 ± 1.53	−8.84 ± 0.92	7.55 ± 1.43
CA/Pac/ERBB2	12.03 ± 2.44	0.59 ± 1.32	12.03 ± 3.57
CA/Pac/MAPK	8.96 ± 2.16	0.35 ± 0.36	8.25 ± 1.28
CA/Pac/ROS1	11.08 ± 2.09	0.35 ± 0.09	7.90 ± 1.83
CA/Doc/AKT	3.66 ± 0.24	8.97 ± 2.94	16.16 ± 4.52
CA/Doc/ERBB2	20.05 ± 4.61	7.97 ± 3.06	8.02 ± 1.49
CA/Doc/MAPK	−1.06 ± 0.58	6.14 ± 2.73	6.49 ± 1.91
CA/Doc/ROS1	0.59 ± 0.34	10.74 ± 3.16	7.90 ± 1.52
CA/Pac/AKT/ERBB2	7.07 ± 2.22	3.71 ± 0.85	13.29 ± 1.12
CA/Pac/AKT/MAPK	18.62 ± 3.64	14.42 ± 2.14	3.37 ± 3.47
CA/Pac/AKT/ROS1	11.26 ± 1.05	−0.12 ± 0.93	22.40 ± 3.44
CA/Pac/ERBB2/MAPK	1.32 ± 0.07	−1.44 ± 0.12	12.22 ± 2.33
CA/Pac/ERBB2/ROS1	−0.48 ± 0.52	−9.70 ± 0.49	5.51 ± 1.86
CA/Pac/MAPK/ROS1	9.58 ± 1.24	−4.79 ± 0.93	6.23 ± 2.22

According to the cell viability data displayed in Table 5, silencing AKT or ERBB2 regulates the response of MCF-7 cells to paclitaxel or docetaxel substantially. Thus, the effect of these therapeutic agents on the protein expression was further evaluated via Western blotting.

Induction of cell death followed by application of therapeutically loaded carbonate apatite could be associated with involvement of the apoptotic pathway, based on escalation of caspase-7-mediated signaling [12].

Silencing of AKT or HER2 signaling to expedite treatment outcomes in cancer is backed up by a solid body of knowledge. Expression of HER2 in MCF7 cells has been shown to increase under certain conditions [15,16].

According to recent research, docetaxel induces either apoptosis accompanied by survivin upregulation or necrosis and a lower rate of survivin upregulation in various breast cancer cell lines, based on cells receptor expression profile and the molecular phenotype. Additionally, inhibition of p-AKT was shown to revert survivin upregulation and also induce docetaxel-dependent apoptosis [17].

In another study, inhibition of NRF2 in ERBB2-overexpressing ovarian carcinoma cells was shown to suppress ERBB2 expression, which led to a decrease in phospho-AKT and enhanced p27 protein together with increased sensitivity of these cells to docetaxel cytotoxicity and apoptosis [18].

There are also other published data associating increased paclitaxel-induced apoptosis with inhibition of AKT in ovarian cancer cells. While paclitaxel-resistant cells demonstrated higher levels of p-AKT compared to paclitaxel-sensitive cells, inhibition of AKT increased paclitaxel therapeutic efficacy in both cell lines [19].

Moreover, combination of shRNA against AKT1 with paclitaxel exerted synergistic anti-cancer effects, thus inhibiting the growth of human breast cancer MDA-MB-231 and MCF-7 cells, and breast cancer MDA-MB-231 cell xenografts in mice as well. The combination therapy demonstrated enhanced anti-cancer effects through inhibition of AKT1 signaling and induction of apoptosis [20].

Extensive in vitro and in vivo research has associated ERBB2 overexpression with resistance of breast cancer cells to certain chemotherapeutic agents, namely docetaxel and paclitaxel. Furthermore, in clinical studies, herceptin, the anti-ERBB2 antibody, enhanced the antitumor activity of paclitaxel and doxorubicin against ERBB2-overexpressing human breast cancer xenografts, and the paclitaxel response rate of patients with ERBB2-overexpressing breast cancers was significantly higher among patients receiving paclitaxel plus herceptin than those receiving paclitaxel alone [21]. Additionally, co-administration of a PI3K inhibitor with cytotoxic or targeted anticancer agents, such as carboplatin, paclitaxel, or erlotinib, led to increased tumor growth inhibition over the corresponding single agents [22].

Moreover, high AKT activity has been shown to be responsible for the enhanced resistance of ERBB2-overexpressing cancer cells toward chemotherapeutic agents, and inhibition of AKT activation by peptide aptamer resulted in the restoration of regular sensitivity of breast cancer MCF7 cells towards paclitaxel [23].

In another research, siRNA-based knockdown of HER-2 conferred increased sensitivity to paclitaxel in endometrial cancer cells, attenuating the induction of p-AKT on paclitaxel stimulation, which was cancelled by inactivating AKT by the introduction of a dominant-negative form [24].

3.2. *Effect of CA/siRNA/Drug at the Protein Level*

As cell viability assays revealed, silencing AKT or ERBB2 exerts the maximum effect on the enhancement of cytotoxicity of Pac or Doc on MCF-7. Thus, the underlying mechanism of reduced viability following administration of these treatments was investigated via western blotting. This was performed to confirm that the applied siRNA is effective in the knock down of the target protein synthesis and also to observe the changes in protein expression leading to enhanced chemo sensitivity.

AKT expression following treatment of MCF-7 cells with free AKT siRNA and CA-complexed AKT siRNA was assessed and compared with untreated cells. As the bands on the blot and densitometry analysis reveal (Figure 1), complexation of siRNA with carbonate apatite leads to significantly decreased expression of AKT compared to free AKT siRNA and untreated cells (p value < 0.05). This confirms the efficacy of CA in the successful delivery of siRNA and knock down of the target protein.

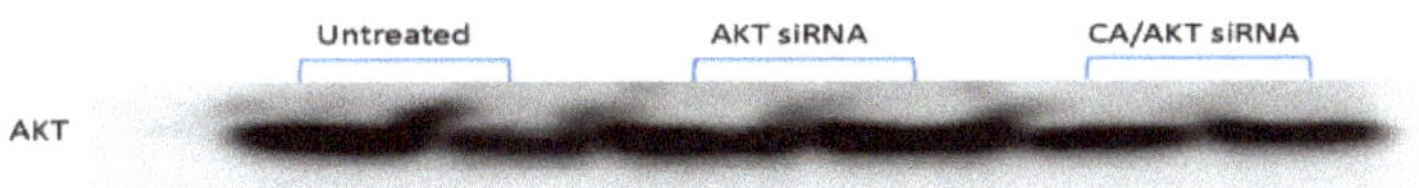

Figure 1. AKT protein expression in MCF-7 cells. Cells were treated with media (untreated), free AKT siRNA, and CA/AKT siRNA formed with 3 mM of $CaCl_2$ and 1 nM siRNA for 44 h. Cellular lysates were resolved by SDS-PAGE and transferred in PVDF membrane followed by incubation with primary antibodies raised in rabbit against AKT. HRP-conjugated goat anti-rabbit secondary antibody was used to detect the chemiluminescent signals.

The same experiment design was utilized with MAPK siRNA. According to Figure 2, free MAPK siRNA does not change the expression level of MAPK protein whereas CA/MAPK significantly knocks down production of the target protein (p value < 0.05).

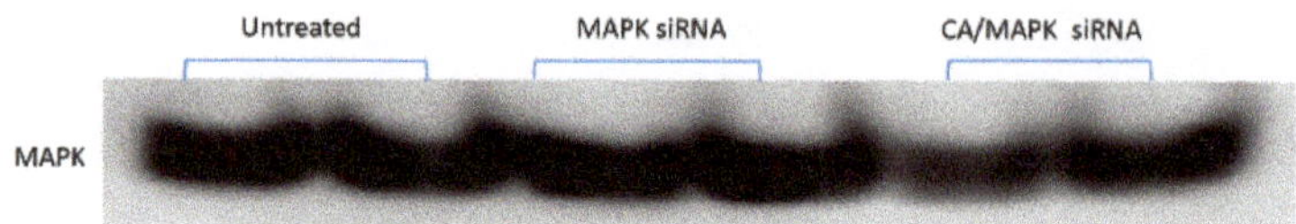

Figure 2. MAPK protein expression in MCF-7 cells. Cells were treated with media (untreated), free MAPK siRNA, and CA/MAPK siRNA formed with 3 mM of $CaCl_2$ and 1 nM siRNA for 44 h. Cellular lysates were resolved by SDS-PAGE and transferred into PVDF membrane followed by incubation with primary antibodies raised in rabbit against AKT. HRP-conjugated goat anti-rabbit secondary antibody was used to detect the chemiluminescent signals.

Next, changes in protein expression resulting from the combinations of drugs and siRNAs, as highlighted in the cell viability data, were studied. MCF-7 cell lysate treated with free and CA-bound AKT siRNA, CA/AKT/Pac and CA/Pac were loaded on the gels. According to the resulting bands (Figure 3), the lowest level of AKT protein was obtained by CA/AKT/Pac treatment versus CA/Pac or CA/AKT. This implies the development of synergistic interactions to suppress survival pathways in the presence of classical anti-cancer drugs together with genetic downregulation of survival proteins, which in turn might lead to enhanced sensitivity of the cells to the anti-cancer medication.

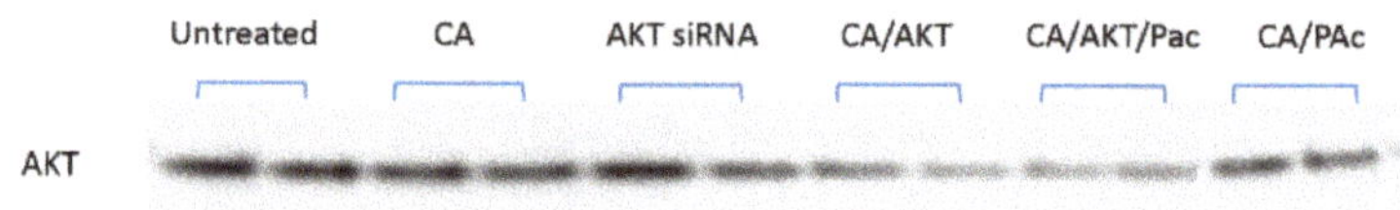

Figure 3. AKT protein expression in MCF-7 cells. Cells were treated with media (untreated), carbonate apatite, free AKT siRNA, CA/AKT siRNA, CA/AKT/Pac, and CA/Pac. Ingredients include 3 mM of $CaCl_2$, 1 nM siRNA, and 1 nM Pac. Cellular lysates were run in SDS-PAGE and transferred into PVDF membrane and incubated with primary antibodies raised in rabbit against AKT. HRP-conjugated goat anti-rabbit secondary antibody was used to detect the chemiluminescent signals.

The same synergistic pattern, but with a lower potency, was observed with application of CA/AKT/Doc in the downregulation of AKT protein against CA/AKT or CA/Doc (Figure 4).

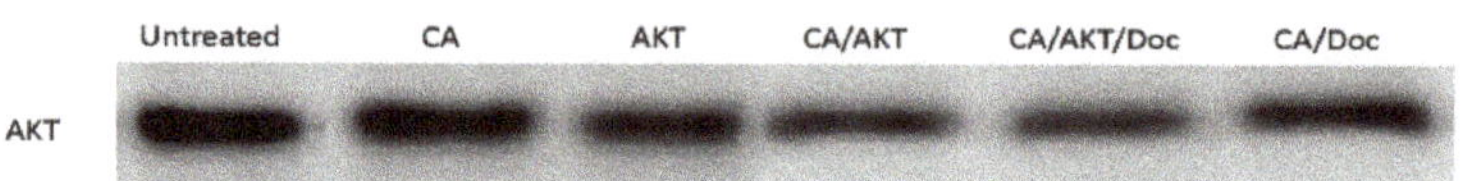

Figure 4. AKT protein expression in MCF-7 cells. Cells were treated with media (untreated), carbonate apatite, free AKT siRNA, CA/AKT siRNA, CA/AKT/Doc, and CA/Doc. Ingredients include 3 mM of $CaCl_2$, 1 nM siRNA, and 1 nM Doc. Cellular lysates were run in SDS-PAGE and transferred into PVDF membrane and incubated with primary antibodies raised in rabbit against AKT. HRP-conjugated goat anti-rabbit secondary antibody was used to detect the chemiluminescent signals.

Based on a substantial enhancement in efficacy, the impact of different combinations of ERBB2 siRNA with paclitaxel or docetaxel on the expression level of ERBB2 protein was also examined. As the blot in Figure 5 shows, the highest knock down was achieved by co-delivery of CA-bound Doc and ERBB2 siRNA followed by CA/Pac/ERBB2. Again, the synergy in the therapeutic efficacy of the drug together with siRNA is present since the impact of the drug or siRNA as a single agent is considerably weaker.

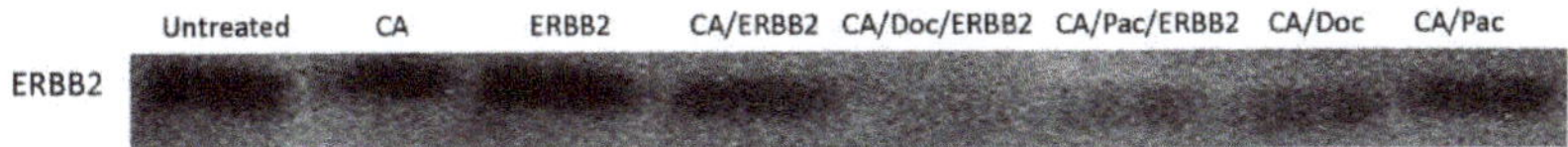

Figure 5. ERBB2 protein expression in MCF-7 cells. Cells were treated with media (untreated), carbonate apatite, free ERBB2 siRNA, CA/ERBB2 siRNA, CA/Doc/ERBB2, CA/Pac/ERBB2, CA/Doc, and CA/Pac. Ingredients include 3 mM of $CaCl_2$, 1 nM siRNA, and 1 nM drug. Cellular lysates were run in SDS-PAGE and transferred into PVDF membrane and incubated with primary antibodies raised in rabbit against ERBB2. HRP-conjugated goat anti-rabbit secondary antibody was used to detect the chemiluminescent signals.

3.3. *In Vivo Efficacy of CA/Drug/siRNA*

According to cell viability data on 4T1 cells and the enhancement in the cytotoxicity of drugs, co-delivery of AKT and ERBB2 siRNAs together with Pac or Doc resulted in the highest increase in the drug's efficacy. Thus, these two combinations were applied in animal study.

For comparison of the in vivo efficacy of the therapeutics in the first batch, animals were treated on day 8 and 11. Formulations were prepared by mixing 4 mM Ca, 1.25 mg/kg Pac, 50 nM AKT siRNA plus 50 nM ERBB2 siRNA in 100 µL of HCO_3-DMEM. There were no significant changes in the pattern of body weight change and also general signs and symptoms of the animals among different groups of the studies.

As the t-test results reveal, CA-AKT-ERBB2-Pac treatment significantly reduced the tumor volume on day 14 and 16 compared to Ca/Pac. Moreover, the group treated with CA/ERBB2 displayed significantly smaller tumors compared to the CA group on day 12 and 14. The effect of Pac complexed with CA on tumor regression was significant on day 14 versus free paclitaxel therapy (Figure 6).

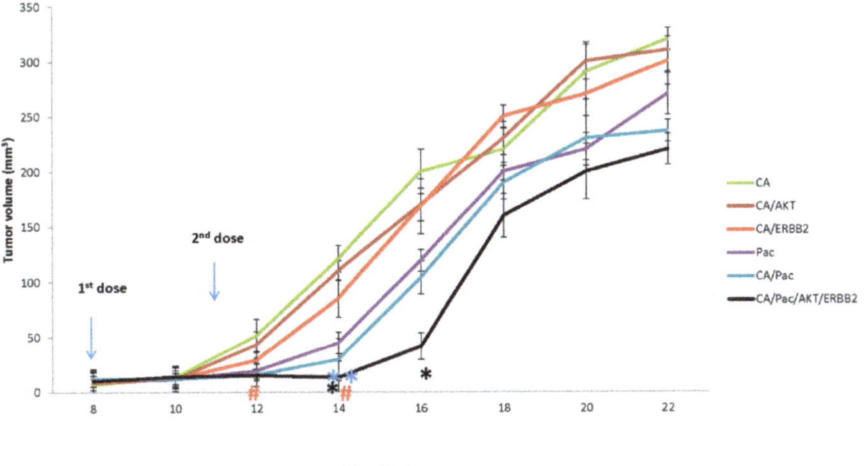

Figure 6. Effect of silencing AKT and ERBB2 pathways on in vivo efficacy of paclitaxel. Mice were purchased from Razi Research Institute, Tehran, Iran. Approximately 10^6 4T1 cells were inoculated subcutaneously on the mammary pad of mice. Based on tumor volume calculations, mice were randomized and treated intravenously through tail-vein injection on day 8 and 11. The therapeutics included 100 µL of carbonate apatite entailing 4 µL of 1M $CaCl_2$ and 50 nM of AKT and ERBB2 siRNA plus 1.25 mg/kg Pac. Body weight and tumor outgrowth were monitored every other day. Data is represented as mean ± SD, $n = 6$ and values are significant when * p value < 0.05 for CA/Pac/AKT/ERBB2 vs. CA/Pac, # p value < 0.05 for CA/ERBB2 compared to CA and ** p value < 0.05 for CA/Pac against Pac.

In another batch, injections on day 10 and 13 encompassed preparations of 4 mM Ca, 1 mg/kg Doc, 50 nM AKT siRNA, and 50 nM ERBB2 siRNA in 100 μL bicarbonated DMEM (Figure 7).

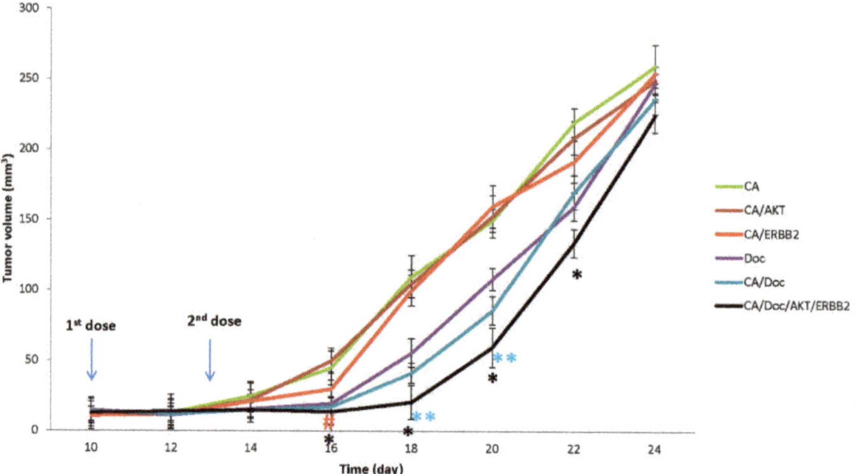

Figure 7. Effect of silencing AKT and ERBB2 pathways on in vivo efficacy of docetaxel. Mice were purchased from Razi Research Institute, Tehran, Iran. Approximately 10^6 4T1 cells were inoculated subcutaneously on the mammary pad of mice. Based on tumor volume calculations, mice were randomized and treated intravenously through tail-vein injection on day 10 and 13. The therapeutics included 100 μL of carbonate apatite entailing 4 μL of 1M $CaCl_2$ and 50 nM of AKT and ERBB2 siRNA plus 1mg/kg Doc. Body weight and tumor outgrowth were monitored every other day. Data is represented as mean ± SD, $n = 6$ and values are significant when * p value < 0.05 for CA/Doc/AKT/ERBB2 vs CA/Doc, # p value < 0.05 for CA/ERBB2 compared to CA and ** p value < 0.05 for Ca/Doc against Doc.

According to the p values calculated in the t-test, treatment of animals with CA/Doc/AKT/ERBB2 resulted in significantly smaller tumor volumes on days 16, 18, 20, and 22 compared to CA/Doc. Whereas CA/ERBB2 displayed a significant anti-tumor efficacy only on day 16 compared to CA, and CA/AKT was not effective in tumor regression. Animals treated with CA/Doc had significantly smaller tumors on day 18 and 20 compared to Doc-treated animals.

ERBB2 overexpression has been linked to elevated levels of inhibitory phosphorylation of Cdc2 and suppression of paclitaxel-induced cell death in breast cancer cells via deregulation of the G2/M cell cycle checkpoint. This provides a mechanistic rationale for the association between ERBB2 overexpression and paclitaxel resistance. Interestingly, overexpression of a subunit of PI3k in ovarian cancer cells has been shown to confer paclitaxel resistance. Additionally, selective inhibition of the PI3k pathway could restore the efficacy of paclitaxel in those cells [25]. Thus, alterations in the expression and activity levels of key components of these signaling networks regulating cellular proliferation and survival may confer paclitaxel resistance. Further, circumvention of this type of resistance might be achieved via application of selective inhibitors of these proteins to increase drug sensitivity.

Simultaneous targeting of AKT and ERBB2 to achieve improvements in eradication of oncogenesis has been brought up in various studies.

Overexpression or activation of HER members and also AKT has been linked to limited benefits of treatment in patients, leading to poor prognosis in breast cancer. In fact, functionally relevant alterations in AKT1 could be a putative mediator of tumor progression and drug resistance. Moreover, amplification or gain-of-function mutations of ERBB2 can account for hyperactivation of the AKT cascade in breast cancer cells. Crosstalk with heterologous receptors and amplification of HER2 signaling, amplifications of the PI3K/AKT pathway, and de-repression and/or activation of compensatory survival pathways through increased PI3K/AKT signaling are among the resistance mechanisms against anti-HER2 therapy. Factors associated with resistance to ERBB2-targeted agents have been invariably associated with a reactivation of the PI3K/AKT signaling cascade [26,27]. However, defects in HER2/PI3K/AKT axes and their impact on regulation of the response of cancer cells to treatment need further investigation.

In an experimental design, HER2-positive MCF7 cells showed a PI3K-dependent increase in AKT activity together with increased resistance to a panel of five chemotherapeutic agents with known different mechanisms of action (paclitaxel, doxorubicin, 5-fluorouracil, etoposide, and camptothecin). Selective inhibition of PI3K or AKT in these HER2-overexpressing MCF7 cells reduced the levels of phosphorylated (activated) AKT and sensitized the cells to the chemotherapeutic agents. It has been further confirmed that expression of a constitutively active AKT vector alone in MCF7 cells caused similarly increased resistance of the cells to the chemotherapeutic agents [13]. Therefore, the HER2/PI3K/AKT pathway is confirmed to play a causal role in the resistance of breast cancer cells against several therapeutics, and targeting components of this pathway might resolve the resistance and enhance the efficacy of the treatments. In view of that, the in vitro and in vivo results of this study are in alignment with the documented role of AKT and ERBB2 in drug resistance.

Strikingly, the efficacy of simultaneous knock down of AKT and ERBB2 in augmenting the response of cancer cells to docetaxel was reported for the first time in this study.

Markedly, the strategic location of AKT and its activation by multiple upstream signal transduction pathways makes it a better target than its upstream targets, such as HER2, Ras, or PI3K, in sensitizing cancer cells to chemotherapy or radiotherapy. Thus, there could be clinical benefits from an appropriate combination of conventional chemotherapeutic drugs with a new generation of signal transduction inhibitors that inhibit the HER/PI3K/AKT pathway for the treatment of breast cancer.

A summary of the ERBB2 and AKT signaling cascades regulating cell's proliferation, survival, and resistance to apoptosis and treatment is illustrated in Figure 8. This is for clarification of the enhanced efficacy of Pac or Doc attained by simultaneous blockade of AKT and ERBb2 cascades and the impact on the response of the cells to the drugs.

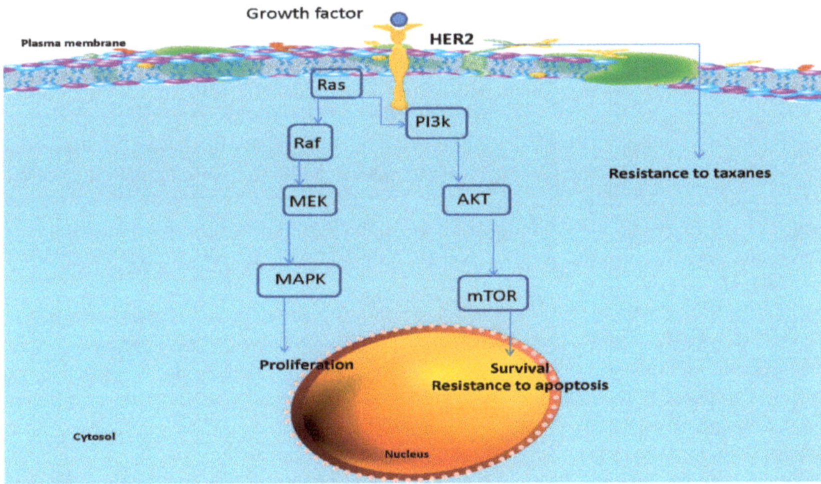

Figure 8. Summary of the ERBB2 and AKT signaling cascades regulating the cell's proliferation, survival, and resistance to apoptosis and treatment.

As revealed through the extensive experiments on breast cancer cells and animal models plus protein expression studies, simultaneous knock down of AKT1 and ERBB2 expression in the presence of paclitaxel or docetaxel leads to a substantial increase in cellular response to the chemotherapeutic agent both in culture and an animal model. Synergistic interactions in target protein knock down have been documented via western blotting in the cells treated with these combinations, which would shed light on the underlying mechanisms of the enhanced sensitivity of the cells to treatment. Taken together, with a favorable in vitro profile and also superior in vivo efficacy, co-administration of genetic materials and classical chemotherapeutics could propose a promising platform for improved strategies against cancer cells in upcoming practice.

Supplementary Materials: The following are available online at http://www.mdpi.com/1999-4923/11/9/458/s1, Figure S1. Cell viability assessment in 4T1 cells treated with siRNA loaded carbonate apatite. Figure S2. Carbonate apatite facilitated delivery of ERBB2 (A) and ROS1 (B) siRNA to MCF-7 cells. Figure S3. Effect of single pathway silencing on cytotoxicity of paclitaxel on 4T1 cells. Figure S4. Effect of single pathway silencing on cytotoxicity of docetaxel on 4T1 cells. Figure S5. Effect of single pathway silencing on cytotoxicity of mitomycin C on 4T1 cells. Figure S6. Effect of single pathway silencing on cytotoxicity of topotecan on 4T1 cells. Figure S7. Effects of single pathway silencing on paclitaxel cytotoxicity in MDA-MB-231. Figure S8. Effects of single pathway silencing on docetaxel cytotoxicity. Figure S9. Effects of silencing AKT and ERBB2 oncogenes on drugs cytotoxicity in 4T1. Figure S10. Cell viability assay on MDA-MB-231 cells treated with carbonate apatite complexed with paclitaxel and two siRNAs.

Author Contributions: T.F. carried out experiments and data analysis under the supervision of H.R.M. and E.H.C., H.R.M. provided resources while E.H.C. originally designed the project and provided the resources.

Funding: This research was supported by MOSTI Science fund (Project ID: 02-02-10-SF0299) of Malaysia.

Conflicts of Interest: The authors declare no conflict of interest.

References

1. Holohan, C.; Van Schaeybroeck, S.; Longley, D.B.; Johnston, P.G. Cancer drug resistance: An evolving paradigm. *Nat. Rev. Cancer* **2013**, *13*, 714–726. [CrossRef] [PubMed]
2. Zhang, L.; Lu, Z.; Zhao, Q.; Huang, J.; Shen, H.; Zhang, Z. Enhanced Chemotherapy Efficacy by Sequential Delivery of siRNA and Anticancer Drugs Using PEI-Grafted Graphene Oxide. *Small* **2011**, *7*, 460–464. [CrossRef] [PubMed]
3. Zhu, C.; Jung, S.; Luo, S.; Meng, F.; Zhu, X.; Park, T.G.; Zhong, Z. Co-delivery of siRNA and paclitaxel into cancer cells by biodegradable cationic micelles based on PDMAEMA–PCL–PDMAEMA triblock copolymers. *Biomaterials* **2010**, *31*, 2408–2416. [CrossRef] [PubMed]
4. Zhang, C.; Zhu, W.; Liu, Y.; Yuan, Z.; Yang, S.; Chen, W.; Li, J.; Zhou, X.; Liu, C.; Zhang, X. Novel polymer micelle mediated co-delivery of doxorubicin and P-glycoprotein siRNA for reversal of multidrug resistance and synergistic tumor therapy. *Sci. Rep.* **2016**, *6*, 23859. [CrossRef] [PubMed]
5. Meng, H.; Liong, M.; Xia, T.; Li, Z.; Ji, Z.; Zink, J.I.; Nel, A.E. Engineered Design of Mesoporous Silica Nanoparticles to Deliver Doxorubicin and P-Glycoprotein siRNA to Overcome Drug Resistance in a Cancer Cell Line. *Am. Chem. Soc. Nano* **2010**, *4*, 4539–4550. [CrossRef] [PubMed]
6. Taratula, O.; Kuzmov, A.; Shah, M.A.; Garbuzenko, O.B.; Minko, T. Nanostructured lipid carriers as multifunctional nanomedicine platform for pulmonary co-delivery of anticancer drugs and siRNA. *J. Control. Release* **2013**, *171*, 349–357. [CrossRef]
7. Meng, H.; Mai, W.X.; Zhang, H.; Xue, M.; Xia, T.; Lin, S.; Wang, X.; Zhao, Y.; Ji, Z.; Zink, J.I.; et al. Codelivery of an Optimal Drug/siRNA Combination Using Mesoporous Silica Nanoparticles to Overcome Drug Resistance in Breast Cancer in Vitro and in Vivo. *Am. Chem. Soc. Nano* **2013**, *7*, 994–1005. [CrossRef]
8. Elzoghby, A.O.; Samy, W.M.; Elgindy, N.A. Protein-based nanocarriers as promising drug and gene delivery systems. *J. Control. Release* **2012**, *161*, 38–49. [CrossRef]
9. Liu, C.F.; Chen, R.; Frezzo, J.A.; Katyal, P.; Hill, L.K.; Yin, L.; Srivastava, N.; More, H.T.; Renfrew, P.D.; Bonneau, R.; et al. Efficient Dual siRNA and Drug Delivery Using Engineered Lipoproteoplexes. *Biomacromolecules* **2017**, *18*, 2688–2698. [CrossRef]
10. Fatemian, T.; Chowdhury, E.H. Cytotoxicity Enhancement in Breast Cancer Cells with Carbonate Apatite-Facilitated Intracellular Delivery of Anti-Cancer Drugs. *Toxics* **2018**, *6*, E12. [CrossRef]
11. Tiash, S.; Chua, M.J.; Chowdhury, E.H. Knockdown of ROS1 gene sensitizes breast tumor growth to doxorubicin in a syngeneic mouse model. *Int. J. Oncol.* **2016**, *48*, 2359–2366. [CrossRef] [PubMed]

12. Tiash, S.; Chowdhury, E.H. siRNAs targeting multidrug transporter genes sensitize breast tumor to doxorubicin in a syngeneic mouse model. *J. Drug Target.* **2018**, *27*, 325–337. [CrossRef] [PubMed]
13. Knuefermann, C.; Lu, Y.; Liu, B.; Jin, W.; Liang, K.; Wu, L.; Schmidt, M.; Mills, G.B.; Mendelsohn, J.; Fan, Z. HER2/PI-3K/Akt activation leads to a multidrug resistance in human breast adenocarcinoma cells. *Oncogene* **2003**, *22*, 3205–3212. [CrossRef] [PubMed]
14. Tiash, S.; Kamaruzman, N.I.B.; Chowdhury, E.H. Carbonate apatite nanoparticles carry siRNA(s) targeting growth factor receptor genes, EGFR1 and ERBB2 to regress mouse breast tumor. *Drug Deliv.* **2017**, *24*, 1721–1730. [CrossRef] [PubMed]
15. Knowlden, J.M.; Hutcheson, I.R.; Jones, H.E.; Madden, T.; Gee, J.M.; Harper, M.E.; Barrow, D.; Wakeling, A.E.; Nicholson, R.I. Elevated levels of epidermal growth factor receptor/c-erbB2 heterodimers mediate an autocrine growth regulatory pathway in tamoxifen-resistant MCF-7 cells. *Endocrinology* **2003**, *144*, 1032–1044. [CrossRef]
16. Kumar, R.; Mandal, M.; Lipton, A.; Harvey, H.; Thompson, C.B. Overexpression of HER2 modulates bcl-2, bcl-XL, and tamoxifen-induced apoptosis in human MCF-7 breast cancer cells. *Clin. Cancer Res.* **1996**, *2*, 1215–1219.
17. De Iuliis, F.; Salerno, G.; Giuffrida, A.; Milana, B.; Taglieri, L.; Rubinacci, G.; Giantulli, S.; Terella, F.; Silvestri, I.; Scarpa, S. Breast cancer cells respond differently to docetaxel depending on their phenotype and on survivin upregulation. *Tumor Biol.* **2016**, *37*, 2603–2611. [CrossRef]
18. Manandhar, S.; Choi, B.H.; Jung, K.A.; Ryoo, I.G.; Song, M.; Kang, S.J.; Choi, H.G.; Kim, J.A.; Park, P.H.; Kwak, M.K. NRF2 inhibition represses ErbB2 signaling in ovarian carcinoma cells: Implications for tumor growth retardation and docetaxel sensitivity. *Free Radic. Biol. Med.* **2012**, *52*, 1773–1785. [CrossRef]
19. Kim, S.H.; Juhnn, Y.S.; Song, Y.S. Akt Involvement in Paclitaxel Chemoresistance of Human Ovarian Cancer Cells. *Ann. N. Y. Acad. Sci.* **2007**, *1095*, 82–89. [CrossRef]
20. Guo, D.D.; Hong, S.H.; Jiang, H.L.; Kim, J.H.; Minai-Tehrani, A.; Kim, J.E.; Shin, J.Y.; Jiang, T.; Kim, Y.K.; Choi, Y.J.; et al. Synergistic effects of Akt1 shRNA and paclitaxel-incorporated conjugated linoleic acid-coupled poloxamer thermosensitive hydrogel on breast cancer. *Biomaterials* **2012**, *33*, 2272–2281. [CrossRef]
21. Hung, D.Y.; Chie, M. Overexpression of ErbB2 in cancer and ErbB2-targeting strategies. *Oncogene* **2000**, *19*, 6115–6121.
22. Echeverria, C.G.; Sellers, W.R. Drug discovery approaches targeting the PI3K/Akt pathway in cancer. *Oncogene* **2008**, *27*, 5511–5526. [CrossRef] [PubMed]
23. Kunz, C.; Borghouts, C.; Buerger, C.; Groner, B. Peptide Aptamers with Binding Specificity for the Intracellular Domain of the ErbB2 Receptor Interfere with AKT Signaling and Sensitize Breast Cancer Cells to Taxol. *Mol. Cancer Res.* **2006**, *4*, 983–998. [CrossRef] [PubMed]
24. Mori, N.; Kyo, S.; Nakamura, M.; Hashimoto, M.; Maida, Y.; Mizumoto, Y.; Takakura, M.; Ohno, S.; Kiyono, T.; Inoue, M. Expression of HER-2 affects patient survival and paclitaxel sensitivity in endometrial cancer. *Br. J. Cancer* **2010**, *103*, 889–898. [CrossRef] [PubMed]
25. Orr, G.A.; Verdier-Pinard, P.; McDaid, H.; Horwitz, S.B. Mechanisms of Taxol resistance related to microtubules. *Oncogene* **2003**, *22*, 7280–7295. [CrossRef] [PubMed]
26. Zhao, Y.H.; Liu, H.; Liu, Z.; Ding, Y.; Ledoux, S.P.; Wilson, G.L.; Voellmy, R.; Lin, Y.; Lin, W.; Nahta, R.; et al. Overcoming trastuzumab resistance in breast cancer by targeting dysregulated glucose metabolism. *Cancer Res.* **2011**, *71*, 4585–4597. [CrossRef]
27. Shattuck, D.L.; Miller, J.K.; Carraway, K.L.; Sweeney, C. Met receptor contributes to trastuzumab resistance of Her2-overexpressing breast cancer cells. *Cancer Res.* **2008**, *68*, 1471–1477. [CrossRef]

Article

Polymeric Nanoparticles Based on Tyrosine-Modified, Low Molecular Weight Polyethylenimines for siRNA Delivery

Alexander Ewe [1], Sandra Noske [1,2], Michael Karimov [1] and Achim Aigner [1,*]

[1] Rudolf-Boehm-Institute for Pharmacology and Toxicology, Faculty of Medicine, Clinical Pharmacology, Leipzig University, 04107 Leipzig, Germany; alexander.ewe@medizin.uni-leipzig.de (A.E.); sandra.noske@medizin.uni-leipzig.de (S.N.); michael.karimov@medizin.uni-leipzig.de (M.K.)

[2] Faculty of Chemistry, Technical University Kaiserslautern, 67663 Kaiserslautern, Germany

* Correspondence: achim.aigner@medizin.uni-leipzig.de; Tel.: +49-(0)341-9724661

Received: 29 September 2019; Accepted: 8 November 2019; Published: 12 November 2019

Abstract: A major hurdle for exploring RNA interference (RNAi) in a therapeutic setting is still the issue of in vivo delivery of small RNA molecules (siRNAs). The chemical modification of polyethylenimines (PEIs) offers a particularly attractive avenue towards the development of more efficient non-viral delivery systems. Here, we explore tyrosine-modified polyethylenimines with low or very low molecular weight (P2Y, P5Y, P10Y) for siRNA delivery. In comparison to their respective parent PEI, they reveal considerably increased knockdown efficacies and very low cytotoxicity upon tyrosine modification, as determined in different reporter and wildtype cell lines. The delivery of siRNAs targeting the anti-apoptotic oncogene survivin or the serine/threonine-protein kinase PLK1 (polo-like kinase 1; PLK-1) oncogene reveals strong inhibitory effects in vitro. In a therapeutic in vivo setting, profound anti-tumor effects in a prostate carcinoma xenograft mouse model are observed upon systemic application of complexes for survivin or PLK1 knockdown, in the absence of in vivo toxicity. We thus demonstrate the tyrosine-modification of (very) low molecular weight PEIs for generating efficient nanocarriers for siRNA delivery in vitro and in vivo, present data on their physicochemical and biological properties, and show their efficacy as siRNA therapeutic in vivo, in the absence of adverse effects.

Keywords: poly(ethylene) imine; PEI; RNA; siRNA delivery; tyrosine-modification; tumor xenograft

---

1. Introduction

The induction of RNA interference (RNAi) by small interfering RNAs (siRNAs) [1] has great potential, allowing for the downregulation of pathophysiologically over-expressed genes on the post-transcriptional level. Since siRNAs can be specifically directed against virtually any target gene, including otherwise "undruggable" genes, they offer novel avenues towards customized therapies against many diseases, for example, the treatment of cancer (see, e.g., [2] for review). Today, siRNAs can be easily designed and chemically synthesized to target a gene of interest in a sequence-specific manner. By harnessing the cellular enzyme machinery, siRNAs are incorporated in the RNA-induced silencing complex (RISC) and, upon activation, the siRNA guide strand binds to its complementary target mRNA, leading to mRNA cleavage, degradation, and thus long-lasting gene knockdown. This is a unique characteristic of RNAi which cannot be achieved with small molecules or anti-sense oligonucleotides [3,4].

While it is thus necessary and sufficient to deliver siRNAs to the cell for inducing RNAi-mediated gene knockdown, the physicochemical properties of siRNAs, i.e., relatively high molecular weight, high polyanionic charge density, hydrophilic properties, and their instability/susceptibility to nuclease degradation, largely prevent their use in an unmodified form for therapeutic approaches [5]. Beyond chemical modifications of the siRNA, delivery systems are required [6,7]. Apart from siRNA conjugates specifically suited for liver uptake, a variety of different cationic lipids, polymeric materials, and other nanoscale systems have been developed to adsorb, electrostatically interact with or incorporate siRNA and mediate cellular uptake [8–12]. For in vitro application several compounds have been commercialized as transfection reagents and are able to induce a potent gene knockdown at low siRNA concentrations in cell culture [13]. However, they are generally not suited for in vivo applications and clinical translation is still challenging, and thus require the development of novel delivery systems. On the other hand, the feasibility of siRNA therapeutics has been shown when the first siRNA drug, ONPATTRO® (Patisiran), was approved by the US Food and Drug Administration (FDA) and European Medicines Agency (EMA) in 2018. ONPATTRO® is a lipid-based siRNA nanoparticle designed to downregulate transthyretin in the liver for treatment of the polyneuropathy of hereditary transthyretin-mediated amyloidosis (hATTR) [14,15]. Despite this success, it might be assumed that one delivery platform cannot address all diseases, especially when it comes to target tissues other than the liver [16].

Polymeric reagents are another well studied class of materials for the delivery of nucleic acids (for review, see [17,18]). A highly investigated cationic polymer is polyethylenimine (PEI), which is able to efficiently deliver nucleic acids in vitro and in vivo. PEIs are available in branched and linear topologies over a wide range of molecular weights (0.8–100 kDa) (for review, see [19,20]). While higher molecular weight PEIs show biological activity but are associated with toxicity, low molecular weight PEIs are more biocompatible but essentially lack transfection efficacy, especially in the case of smaller oligonucleotides like siRNA. Branched PEIs are more efficient for the delivery of siRNA than their linear counterparts [21–23]. Since the commercially available 25 kDa branched PEI shows some efficacy but is also attributed to toxic effects, the lower molecular weight PEI F25-LMW (branched; 4–10 kDa) has been introduced for siRNA delivery in vitro and in vivo [24]. Still, further improvements are warranted and various approaches have been explored to optimize transfection efficacies of low molecular weight PEIs and/or to decrease the toxicity of high molecular weight PEIs. These include the covalent modification of PEI with polyethylene glycol (PEG) [25–27], the cross-linking of small PEIs with hydrolysable linkers, e.g., esters, ketals, or reductive cleavable disulfide groups leading to the reversible formation of higher molecular weights ([28–31]; see [32] for review), the copolymerization/grafting with other polymers like polycaprolactone (PCL), starch or polyvinylalcohol (PVA) [33–35], the modification with lipophilic molecules, e.g., fatty acids, alkanes or cholesterol ([36,37], see [38] for review), or the combination of PEI-based complexes with liposomes [39–42].

The chemical modification of PEI with amino acids offers another very promising approach for nucleic acid delivery in vivo. In previous studies, the grafting of arginine, lysine, or leucine onto PEI showed increased β-galactosidase activity in tumor-bearing mice upon i.v. injection of modified PEI/pDNA complexes [43]. In another report, a library of twenty amino acids conjugated to a poly(amidoamine) (PAMAM G5) dendrimer was characterized for pDNA delivery in vitro and identified cationic and hydrophobic amino acids as most effective [44]. The triple modification of PAMAM G5 with arginine, histidine, and phenylalanine revealed synergistically enhanced pDNA reporter gene expression in vitro and in vivo as compared to their mono-functionalized derivatives [45]. Amino acid modifications have also been shown to improve the delivery of siRNA. Derivatives of a branched 25 kDa PEI with leucine, tryptophan, phenylalanine, or tyrosine were evaluated in vitro and identified the tyrosine-modified PEI as efficient for siRNA transfection [46]. The same high molecular weight 25 kDa branched PEI modified with *N*-acetyl-leucine was used for the delivery of miR34a, significantly improving bone regeneration in an orthodontic in vivo model [47]. The incorporation of histidine and tyrosine also improved in vitro siRNA transfection efficacy in the case of oligoamino amides, a new class of polymers prepared by solid phase synthesis, and decreased the EG5 mRNA expression upon i.v. injection of siEG5 nanoparticles in tumor bearing mice [48,49].

It becomes apparent from these studies that for the successful modification of cationic polymers with amino acids a balanced hydrophobicity and hydrophilicity is of great importance. This may then also allow for exploring lower molecular weight PEIs which are insufficiently active in the unmodified form, especially for the delivery of small nucleic acid molecules like siRNAs, but show significantly improved bioactivities upon introducing modifications. In a recent study, we selected a low molecular 10 kDa branched PEI for tyrosine modification. This new PEI derivative (P10Y) showed excellent in vitro siRNA transfection efficacies and increased complex stability upon serum contact. Moreover, P10Y/siRNA complexes decreased tumor growth in an aggressively growing melanoma xenograft mouse model [50,51]. Based on these results, we extended this approach towards even smaller 2 kDa and 5 kDa branched PEIs. Initial data indicated high knockdown efficacies of these small tyrosine-modified derivatives compared to their otherwise inactive parent PEIs [51]. In this study, we comprehensively explore these small tyrosine-PEIs in various cell lines, for targeting different reporter genes as well as endogenous genes. We identify P5Y as particularly active, and thus also studied this most efficient derivative in a tumor xenograft mouse model.

2. Materials and Methods

2.1. Cells and Cell Culture

Cell culture plastics and consumables were from Sarstedt (Nümbrecht, Germany). Cell culture media were from Sigma-Aldrich (Taufkirchen, Germany) and fetal calf serum was purchased from Biochrom (Berlin, Germany). The cell lines HCT116 (colorectal carcinoma), H441 (lung adenocarcinoma), PC3 (prostate carcinoma), Saos-2 (osteosarcoma), and MV3 (melanoma) were obtained from ATCC/LGC Promochem (Wesel, Germany), and G55T2 (glioblastoma) was a kind gift from Katrin Lamszus (University Clinic Hamburg-Eppendorf, Hamburg, Germany). All cell lines were cultivated in a humid atmosphere at 37 °C and 5% CO_2. HCT116, PC3, G55T2 cells were grown in IMDM medium, H441 and MV3 cells were grown in RPMI 1640 medium, and Saos-2 cells were grown in McCoy's 5a medium. All media were supplemented with 10% FCS and 2 mM alanyl-glutamine.

Reporter cell lines were stably transduced by lentiviral transduction. The EGFP/luciferase plasmid pCCLc-MNDU3-Luciferase-PGK-EGFP-WPRE was a gift from Fernando Fierro (University of California, Davis, CA, USA; Addgene plasmid # 89608) [52]) and the luciferase plasmid pLenti PGK V5-LUC Puro (w543-1) was a gift from Eric Campeau and Paul Kaufman (University of Massachusetts, Worchester, MA, USA; Addgene plasmid # 19360) [53]. Cell lines were cultivated in the above mentioned media and stable reporter cells were selected by either using puromycin (luciferase plasmid) or sorted by FACS (EGFP/luciferase plasmid).

2.2. Synthesis of Tyrosine-Modified PEIs

Branched polyethylenimines were obtained from the following suppliers: 2 kDa PEI (Sigma-Aldrich, Taufkirchen, Germany), 5 kDa PEI (a kind gift from BASF, Ludwigshafen, Germany), and 10 kDa PEI (Polysciences, Eppelheim, Germany). *N*-Boc-tyrosine-OH, *N*-hydroxysuccinimide, EDC·HCl were from Carbolution Chemicals (Saarbrücken, Germany). Dry *N*,*N*-Dimethylformamide (DMF) was from VWR (Darmstadt, Germany), Trifluoroacetic acid (TFA), ethanol, and methanol were from Carl Roth (Karlsruhe, Germany).

For the tyrosine modification, *N*-Boc (*tert*-butyloxycarbonyl) protected tyrosine (0.65 g, 2.3 mmol) and *N*-hydroxysuccinimide (0.27 g, 2.3 mmol) were dissolved in 3 mL dry DMF in a glass vial, followed by addition of EDC·HCl (0.45 g, 2.3 mmol), and stirred for 4 h under a nitrogen atmosphere at RT. In a second vial, PEI (0.2 g, 4.65 mmol in ethylenimine) was dissolved in 3 mL dry DMF, and the pre-activated tyrosine mixture was added and further stirred under nitrogen for 3 d. The reaction mixture was purified by dialysis (1 kDa MWCO regenerated cellulose, Serva, Heidelberg, Germany) against methanol for 10 h with intermediate solvent replacement to remove by-products. Next, the methanol was removed in vacuo and the viscous polymer was dissolved in 3 mL TFA, prior to stirring overnight for Boc-deprotection. Excess TFA was removed by co-evaporation with ethanol. Finally, the crude polymer was dissolved in 0.1 M HCl and excessively purified by dialysis against 0.05 M HCl for 24 h, then against water for 48 h. Lyophilization yielded the tyrosine-modified PEIs as yellowish fluffy powders. The degree of functionalization (~25–30% based on ethylenimine monomer) was confirmed by ^{1}H-NMR (Mercury plus, 300 MHz, Varian Agilent Technologies, Santa Clara, CA, USA) as described previously [50,51].

2.3. Polyplex Preparation

The polyplexes were prepared based on polymer/siRNA mass ratios. Typically, the P5Y- and P10Y-based complexes were prepared at a mass ratio of 2.5 unless indicated otherwise. For P2Y, a mass ratio of 10 was used. For standard transfection experiments in a 24 well plate, 30 pmol (0.4 μg) siRNA per well was diluted in 12.5 μL HN buffer (150 mM NaCl, 10 mM HEPES (4-(2-hydroxyethyl)-1-piperazineethanesulfonic acid), pH 7.4). In a separate tube, the calculated amount of the polymer was diluted in 12.5 μL HN buffer and the siRNA solution was added to the polymer solution, thoroughly mixed by pipetting up and down and incubated for 30 min at RT. For serum stability studies, the prepared polyplexes were additionally incubated with different volumes of FCS.

For in vivo experiments, the polyplexes were prepared as above, with slight modifications. Per dose for injection, 10 μg siRNA was diluted in 75 μL HN buffer and 25 μg (mass ratio 2.5) of the polymer (P5Y) was diluted in 75 μL 5% (*w/v*) glucose, 10 mM HEPES pH 7.4. The complexation was performed as described above.

2.4. Polyplex Characterization

The hydrodynamic diameters and zeta potentials were measured by dynamic light scattering (DLS) and phase analysis light scattering (PALS) with a Brookhaven ZetaPALS system (Brookhaven Instruments, Holtsville, NY, USA). Polyplexes containing 5 μg siRNA in 250 μL were prepared as described above and diluted to 1.7 mL with millipore water. The data were analyzed using the manufacturer's software, with applying the viscosity and refractive index of pure water at 25 °C. For the size determination, polyplexes were measured in five runs with a run duration of 1 min per experiment. Zeta potentials were analyzed in ten runs, with each run containing ten cycles using the Smoluchowski model. Additionally, polyplex sizes were analyzed by nanoparticle tracking (NTA) using a NanoSight LM10 (Malvern, Wiltshire, UK) equipped with a 640 nm sCMOS camera, software NTA 3.0, and a siRNA concentration of 1 μg/mL.

Complex stabilities were determined by a heparin displacement assay. The complexes based on the different tyrosine-modified PEIs were prepared at their optimal mass ratios as described above, prior to further incubation in FCS (50%, *v/v*) for 1 h. The complexes were then subjected to the polyanion heparin at different concentrations (units), as indicated in Figure S3C. A total of 25 μL of the mixture containing 0.2 μg siRNA was analyzed by 1.5% agarose gel electrophoresis.

2.5. Cell Transfection

For standard transfection experiments, the cells were seeded at a density of 35,000 cells per well of a 24 well plate in 0.5 mL fully supplemented medium. The next day, the cells were transfected with polyplexes containing 30 pmol/0.4 μg siRNA per 24 well as described above in the presence of FCS and without further medium change. For siRNA targets and sequences see Table S1.

2.6. Determination of Knockdown Efficacies and Complex Uptake

Luciferase activities were determined 72 h after transfection using the Beetle-Juice Kit (PJK, Kleinblittersdorf, Germany). The medium was aspirated and the cells were lysed with 300 μL Luciferase Cell Culture Lysis Reagent (Promega, Mannheim, Germany) for 30 min at RT. In a test tube,

10 μL lysate was mixed with 25 μL substrate and the luminescence was immediately measured in a luminometer (Berthold, Bad Wildbad, Germany).

The EGFP expression was determined by flow cytometry after 72 h. The cells were trypsinized, centrifuged for 3 min at 3000 rpm, and resuspended in 0.5 mL PBS, 1% BSA. The cells were measured in an Attune® Acoustic Focusing Cytometer (Applied Biosystems, Foster City, CA, USA). Finally, 20,000 events in the vital gate were analyzed using the Attune® software (V2.1.0).

For uptake experiments, complexes were prepared as described above (Section 2.4), but using a fluorophore-labeled siRNA-Atto488. PC3 wildtype cells were transfected, after 24 h prepared as described above with an additional PBS washing step and analyzed by flow cytometry.

To visualize the complex uptake by confocal microscopy, PC3 wildtype cells were seeded at a density of 200,000 cells onto glass coverslips in a 6 well plate. The next day, the cells were transfected with complexes of the different tyrosine-modified PEIs containing Alexa Fluor ® 647-labeled siRNA (2 μg siRNA/well). After 36 h, the siRNA uptake and intracellular distribution was analyzed by confocal microscopy (TCS SP8, Leica, Wetzlar, Germany).

2.7. Cell Proliferation and Viability Assays

For proliferation assays, cells were seeded at a density of 2000 cells (PC3 and Saos-2) or 1000 cells (MV3) per well of a 96 well plate in 100 μL medium on the day before transfection. For endpoint cell viability assays, PC3 or G55T2 cells were seeded at a density of 7000 cells per 96 well in 100 μL medium. The number of metabolically active cells after transfection was determined by using a colorimetric assay (WST-8 Cell Counting Kit-8; Dojindo Molecular Technologies EU, Munich, Germany). Briefly, after aspirating the medium, 50 μL of a 1:10 dilution of WST-8 in serum-free medium was added to the cells and incubated at 37 °C for 1 h. The absorbance was measured at 450/620 nm in a plate reader.

Acute cell damage caused by the polyplexes was determined by measuring the extent of lactate dehydrogenase (LDH) release, using the Cytotoxicity Detection Kit (Roche, Mannheim, Germany) according to the manufacturer's manual. Briefly, conditioned medium from transfection experiments (negative control siRNA) was collected after 24 h. For the determination of the maximum LDH release, cells were lysed with Triton X-100 at a final concentration of 2%, and conditioned medium from untreated cells served as negative control. In a 96 well plate, 50 μL sample medium was mixed with 50 μL reagent mix and incubated for 30 min in the dark. The reaction was stopped with 50 μL 1 M acetic acid and the absorption at 490/620 nm was measured using a plate reader. For blank value correction, fresh fully supplemented medium was mixed with the reagent and subtracted from all values.

2.8. RNA Preparation and Quantitative RT-PCR (RT-qPCR)

Total RNA was isolated using a combined TRI reagent and silica column protocol. Per 24 well, cells were lysed in 500 μL TRI reagent (TriFast, VWR, Darmstadt, Germany) and incubated for 5 min at RT. The lysate was transferred into a 1.5 mL tube and 200 μL chloroform was added, mixed well by shaking and incubated for 5 min at RT. The samples were centrifuged at 12,000 g for 15 min and the upper aqueous phase (~300 μL) was pipetted into a new 1.5 mL tube containing 300 μL 100% ethanol. After mixing, the sample was loaded onto the column (Zymo-Spin IC, Zymo Research, Freiburg, Germany), prior to centrifugation at 8000 g for 1 min. The flow-through was discarded. The column was washed twice with 450 μL 3 M sodium acetate pH 5.2, and once with 70% ethanol by centrifugation at 8000 g for 1 min; the collection tube was emptied after each step. For the drying step, the collection tube was replaced by a new one, and the columns were centrifuged at 12,000 g for 2 min. For RNA elution, 15 μL DEPC-water, pre-warmed to 65 °C, was added onto the column and incubated for 2 min, before finally eluting the RNA by centrifugation at 8000 g for 2 min. RNA concentrations and purities were measured in a NanoDrop 2000c (Thermo Fisher, Schwerte, Germany).

The total RNA was reverse transcribed with the RevertAid™ H Minus First strand cDNA synthesis Kit (Thermo Fisher, Waltham, MA, USA). In brief, 1 μL random hexamer primer was added to 1 μg RNA diluted in 10 μL DEPC-treated water, followed by an incubation step for 5 min at 65 °C and cooling to 4 °C. Then, 4 µL 5× reaction buffer, 2 µL 10 mM dNTP mix, 2.5 µL DEPC-treated ddH_2O, and 0.5 µL RevertAid™ H Minus M-MuLV Reverse Transcriptase (200 U/µL) were added. The cDNA synthesis mixture was incubated at 25 °C for 10 min, 42 °C for 60 min, and heat denatured at 70 °C for 10 min.

For quantitative real time PCR, a StepOnePLUS Real-Time PCR System (Applied Biosystems, Foster City, CA, USA) was used, applying the ΔΔCt method [54]. The cDNA was 1:10 diluted with DEPC-water and 4 µL were mixed with 5 µL 2x PerfeCTa SYBR® Green FastMix ROX (Quantabio, Beverly, MA, USA) and 1 µL 5 µM forward and reverse primer mix (see Table S2 for primer sequences). The qPCR were run with a pre-incubation at 95 °C for 2 min, followed by 45 amplification cycles (95 °C for 15 s, 55 °C for 15 s, 72 °C 15 s).

2.9. *Hemolysis*

To determine the hemolytic activity of the complexes, whole blood from healthy mice was diluted in Ringer's solution and erythrocytes were purified by several centrifugation and washing steps at 5000 rpm for 5 min, until the supernatant became clear. After the final washing step, the cells were taken up into physiological NaCl solution. Then, 1×10^6 cells in 50 µL were mixed with 50 µL complexes containing different siRNA amounts and incubated for 1 h at 37 °C. For absorption measurements, 25 µL of each sample were transferred into a 96 well plate in triplicates. Erythrocytes incubated with buffer served as negative control and erythrocytes treated with Triton X-100 to a final concentration of 1% served as positive control (= 100%). The absorbance of the samples was measured at 550 nm using a plate reader.

To evaluate erythrocyte aggregation, 50 µL of the purified cells (~1×10^6 cells/mL) were mixed with 50 µL aqueous solution containing 7.5 µg P5Y/3 µg siCtrl, or 750 kDa branched PEI (1 and 3 µg) as positive control. After incubation for 2 h at 37 °C, the samples were mounted onto glass slides and examined microscopically.

2.10. *In Vivo Tumor Therapy*

Athymic nude mice (Foxn1nu, Charles River Laboratories, Sulzfeld, Germany) were kept at 23 °C in a humidified atmosphere, 12 h light/dark cycle, with standard rodent chow and water ad libitum. Animal studies were performed according to the national regulations and approved by the local authorities (Landesdirektion Sachsen, approval NO. TVV 38/16, date: 20 January 2017). Tumor xenografts were established by injecting 3×10^6 PC3 cells in 150 µL PBS subcutaneously (s.c.) into both flanks of mice. When tumors reached a size of ~100 mm^3, the mice were randomized into different treatment groups (6–9 tumors/group). The polyplexes were prepared as described above and amounts corresponding to 10 µg siRNA were i.p. injected every 2–3 days over 14 days.

For in vivo luciferase knockdown experiments, 3×10^6 HCT116-Luc cells were s.c. injected for the induction of tumor xenografts described above. Mice were treated five times every 2–3 days with P5Y/siRNA complexes (siLuc2 = siCtrl and siLuc3 = Luciferase specific siRNA). Tumors were excised and lysed with Luciferase Cell Culture Lysis Reagent (app. 100 mg tumor tissue/1 mL buffer), prior to homogenization using an ULTRA-TURRAX®. Homogenates were repeatedly centrifuged at 13,000 rpm for 5 min and the supernatants were transferred into fresh tubes until the lysate became clear. Luminescence was determined as described above. Relative light units (RLU) were normalized for total protein concentration of the lysates using the BCA assay (Pierce Thermo Fisher, Schwerte, Germany) according to the manufacturer's protocol.

For the determination of blood serum markers, healthy nude mice were i.p. injected with P5Y/siLuc3 complexes containing 10 µg siRNA every 2–3 days, with a total of four times over 8 days. Three hours after the last injection, the blood was collected. Untreated mice served as negative control. The serum was diluted 1:20 with water and serum levels of various analytes were determined using an AU480 (Beckman Coulter, Krefeld, Germany).

For the analysis of the immunostimulatory cytokines TNF-α and INF-γ, P5Y/siLuc3 complexes (10 μg siRNA) were i.v. injected twice within 24 h into immunocompetent C57BL/6 mice. Four hours after the last injection, the blood was collected. Mice treated with lipopolysaccharides (LPS) 50 μg in 150 μL (single injection) served as positive control and untreated mice served as negative control. The serum levels of TNF-α and INF-γ were measured using ELISA kits (PreproTech, Hamburg, Germany) following the manufacturer's instructions.

2.11. Statistics

Statistical analyses were performed by Student's *t*-test or One-way ANOVA, and significance levels are * = $p < 0.05$, ** = $p < 0.01$, *** = $p < 0.001$, and # = not significant, with at least $n = 3$.

3. Results

3.1. Identification of Optimal Complexes for In Vitro Transfection

Based on the superior siRNA complexation and transfection efficacy of 10 kDa PEI upon its tyrosine-modification, we analyzed the performance of even smaller PEIs, which had been shown previously to be inactive [51], upon tyrosine engraftment. We hypothesized that even in the case of 5 kDa or 2 kDa PEIs, the tyrosine modification may result in polymers capable of efficient siRNA complexation and delivery, thus leading to target gene knockdown and RNAi-mediated tumor cell-inhibitory effects in various cellular assays. The tyrosine was covalently coupled onto branched PEIs by using standard NHS/EDC coupling chemistry as outlined in Figure 1A. The degree of tyrosine modification was ~30% compared to ethylenimine monomers, determined by ^{1}H-NMR analysis as described ([50,51]; see Figure S1). Transfection experiments with complexes based on the tyrosine-modified 5 kDa PEI (P5Y) revealed >90% knockdown of serine/threonine-protein kinase PLK1 (polo-like kinase 1; PLK-1) when transfecting a specific siPLK1 as compared to a negative control siRNA (siCtrl; Figure 1B, left). Complexes based on P10Y or P5Y were even more efficient than their higher molecular weight 25 kDa counterpart (P25Y; data not shown), while P2Y/siRNA complexes comprising the very low molecular weight 2 kDa PEI were less efficient. Similar results were obtained after transfection of an siRNA targeting the anti-apoptotic protein survivin, with P5Y/siRNA and P10Y/siRNA complexes again resulting in a 80–90% knockdown (Figure 1B, right) and thus performing better than their P25Y/siRNA counterparts (not shown). P2Y-based nanoparticles showed somewhat reduced activity, even when using higher N/P ratios. The knockdown of the oncogenes/proto-oncogenes PLK-1 and survivin also exerted profound tumor cell-inhibitory effects in anchorage-dependent proliferation assays (Figure 1C). When using P5Y or P10Y, the transfection of 6 pmol/well siRNA was sufficient to largely abolish cell proliferation. In contrast, in the case of P2Y/siRNA complexes, even double amounts (12 pmol) were incapable of inhibiting cell growth. Only the transfection of a very potent ubiquitin siRNA, which essentially abolished tumor cell viabilities in the P5Y and P10Y transfection groups, was strong enough to partially inhibit cell growth when complexed with P2Y. For the very efficient P10Y and P5Y, the siRNA concentration could be further reduced to 3 pmol siRNA with still showing strong inhibitory effects (Figure S2A). In addition, this reduction revealed differences in the potencies of the siRNAs against different target genes.

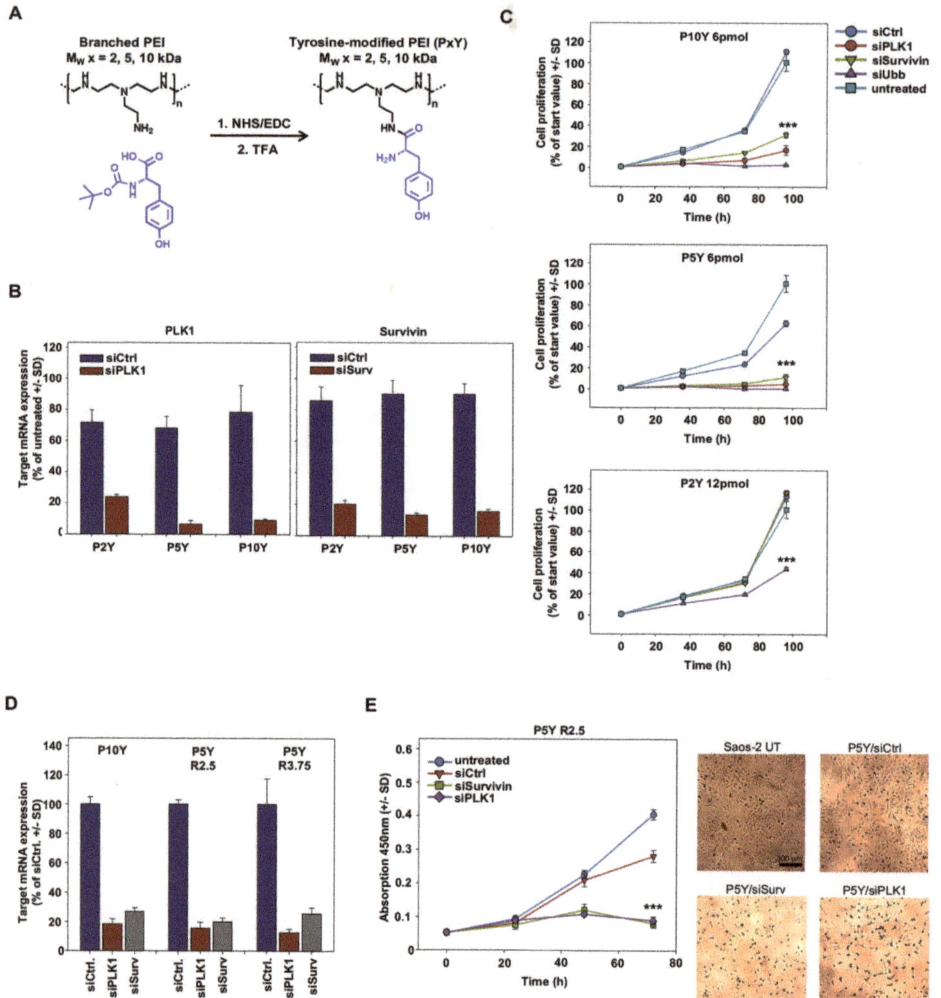

Figure 1. (**A**) Reaction scheme of tyrosine-coupling onto branched polyethylenimines (PEIs). Abbreviations: TFA, Trifluoroacetic acid; NHS, *N*-Hydroxysuccinimid; EDC, 1-Ethyl-3-(3-dimethylaminopropyl) carbodiimid (**B**) Knockdown efficacies of various tyrosine-modified PEI/siRNA complexes in PC3 cells targeting the oncogenes polo-like kinase 1 (PLK1) (left) or survivin (right), quantitated by RT-qPCR. (**C**) Determination of anchorage-dependent proliferation of PC3 cells upon transfection with the tyrosine-modified PEI/siRNA complexes using different anti-proliferative siRNAs at the amounts indicated in the figures. SiRNA-mediated cell inhibition is compared to negative-control transfected or untreated cells. (**D**) Knockdown upon transfecting Saos-2 cells with siRNAs against PLK1 or survivin, as determined by RT-qPCR. Increasing mass ratios from 2.5 to 3.75 does not further improve the knockdown efficacy. (**E**) Determination of anchorage-dependent proliferation of Saos-2 cells upon transfection with P5Y/siRNA complexes targeting PLK1 or survivin. The statistical significance indicates the difference to negative-control transfected cells. Left: quantitation based on WST-8 measurements; right: original pictures.

To further analyze possible reasons for these observed differences in biological activity, we determined intracellular levels of fluorophore-labeled siRNAs upon transfection of PC3 cells complexed with the different tyrosine-modified PEIs. The comparison of P5Y/atto488-siRNA complexes with their P10Y counterparts used for transfection in the presence of fetal calf serum (FCS) revealed a ~30% decrease in siRNA fluorescence when switching from P5Y to P10Y, while siRNA delivery in P2Y complexes was poor (Figure S3A). This was also confirmed in confocal microscopy, demonstrating atto488-siRNA signals to be largely absent upon transfection with P2Y (Figure S3B, upper panel). In contrast, profound fluorescence was observed in the case of P5Y and P10Y, with somewhat stronger signals in the case of P5Y/siRNA complexes (Figure S3B, center and lower panels). Beyond complex uptake, these differences may be based on variations in complex stabilities. Interestingly, a heparin displacement assay revealed even enhanced P5Y/siRNA complex stability in the presence vs. in the absence of serum, as indicated by the shift towards higher heparin concentrations required for siRNA release from the complex upon its pre-incubation in FCS (Figure S3C, upper panels). Under the same conditions, siRNA release from P2Y/siRNA complexes was already observed at lower heparin concentrations, indicating poor complex stability. In stark contrast, siRNA bands from heparin displacement of P10Y/siRNA complexes were weaker, independent of the heparin concentration, and thus indicating incomplete siRNA release even under very stringent conditions. Taken together, this suggests differences in siRNA delivery, complex stability, and siRNA release as underlying reasons for molecular weight-dependent differences between the tyrosine-modified PEIs. Very high efficacies of tyrosine-modified branched PEIs were observed up to as low as 5 kDa (P5Y), while only an even further reduction of the molecular weight (P2Y) eventually led to reduced transfection efficacies. This prompted us to focus in particular on P5Y as the tyrosine-modified PEI with the lowest molecular weight and highest activity, which was even slightly above P10Y.

The very profound siRNA transfection efficacies of P5Y were also confirmed in other cell lines like Saos-2 osteosarcoma cells (Figure 1D). A 1.5-fold increase of the polymer/siRNA mass ratio did not lead to enhanced knockdown, indicating that the very low polymer/siRNA mass ratio of 2.5 was already sufficient for maximum complex activity (Figure 1D). In contrast, knockdown efficacies were found to be dependent on the selection of the siRNA and the target gene, as indicated by the particularly profound >90% knockdown in the case of an siRNA targeting glyceraldehyde-3-phosphate dehydrogenase (GAPDH). As before in PC3 cells, an almost complete abolishment of cell proliferation was observed upon transfecting Saos-2 cells with P5Y/siPLK-1 or P5Y/siSurv complexes (Figure 1E). The microscopic evaluation of the wells also revealed a complete (siSurv) or almost complete (siPLK-1) loss of viable cells after 72 h (Figure 1E, right panel). In all experiments, P5Y-complexed negative control siRNA (siCtrl) was transfected in parallel, with the comparison to untransfected cells indicating non-specific transfection effects to be largely absent. As seen before in PC3 cells, an increase in the polymer/siRNA ratio by 1.5-fold did not further enhance biological efficacies. However, the P5Y/siCtrl curve indicated some non-specific inhibitory effects at this ratio due to excess polymer (Figure S2B). Specific growth inhibition upon P5Y/siRNA-mediated survivin knockdown was also observed in MV3 melanoma cells (Figure S2C), and GAPDH target gene reduction similar to the results shown above were also found in other cell lines (H441 lung adenocarcinoma cells, PC3 prostate carcinoma cells, and G55T2 glioblastoma cells; Figure S2D).

3.2. *Characterization of P5Y/siRNA Complex Properties*

For physicochemical characterization, complexes with different polymer/siRNA mass ratios were subjected to Zetasizer and Nanosight measurements. The hydrodynamic diameters determined by dynamic light scattering (DLS) were around 200–400 nm, with a slight trend towards smaller complexes with lower P5Y/siRNA mass ratios (Figure 2A).

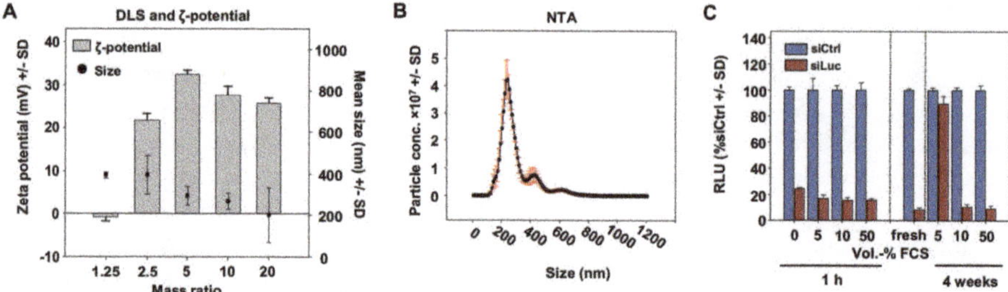

Figure 2. (**A**) Determination of the hydrodynamic diameters and zeta potentials of P5Y/siRNA complexes at different mass ratios as indicated in the figure. (**B**) Size measurement of P5Y/siRNA complexes by nanoparticle tracking (NTA) at the optimal mass ratio of 2.5. (**C**) Luciferase knockdown efficacies in H441-Luc cells for P5Y/siRNA complexes pre-incubated with increasing amounts of FCS (*v/v*) prior to transfection for 1 h (left) or upon storage for 4 weeks at 4 °C.

The diameter for the optimal mass ratio of 2.5 was additionally characterized by nanoparticle tracking (NTA) of which the main peak was at 200 nm and a second small peak at 400 nm (Figure 2B). The measurement of the zeta potentials revealed a slightly negative value for the lowest mass ratio of 1.25 indicating non complexed siRNA, and strongly increased to +20 mV for the mass ratio 2.5. The zeta potential further increased to ~30 mV at mass ratio 5 and reached a plateau of ~25 mV for mass ratios above 10. In many nanoparticle formulations, serum stability has been identified as one major issue limiting biological activity. Our transfection experiments, however, were exclusively performed in serum-containing media (10% FCS). Moreover, the presence of serum even proved beneficial: While PEI-based complexes tend to readily aggregate in aqueous solutions [42], our P5Y/siRNA complexes could be stored over weeks in the presence of 10% or even 50% FCS without losing activity, while 5% FCS was insufficient (Figure 2C).

3.3. High Biocompatibility/Absence of Toxicity of P5Y/siRNA Complexes

High biological activity may well be associated with increased cytotoxicity or other adverse effects, which were therefore assessed next. Lactate dehydrogenase (LDH) release assays, however, revealed no increase of LDH levels over background (untransfected cells). This was true for both, P5Y/siRNA and P10Y/siRNA complexes and was found in two reporter cell lines (PC3-Luc-EGFP and G55T2-Luc-EGFP cells; Figure 3A,D).

Again, the transfection led to profound target gene knockdown, indicated by 80–90% decreased EGFP activities upon siEGFP transfection as determined in flow cytometry (Figure 3B,E). Concomitantly, cell viabilities remained unaffected by P5Y/siRNA or P10Y/siRNA transfection, even after 48 h and independent of siRNA amounts, which could be increased from 4 pmol to 12 pmol without appreciable decrease in metabolic activity (Figure 3C,F). In line with the absence of adverse effects on cell membrane integrity (LDH release assay), a hemoglobin release assay in erythrocytes revealed no adverse effects. Within an almost 10-fold range of different complex amounts, no hemoglobin release over background was observed (Figure 3G). Similarly, the P5Y/siRNA complexes were tested for their erythrocyte aggregation potential. Identical complex amounts were incubated with red blood cells and analyzed under the microscope (Figure 3H). While the positive control a 750 kDa PEI led to strong aggregates at 1 µg, the P5Y complexes did not damage the red blood cells up to 3 µg complexed siRNA.

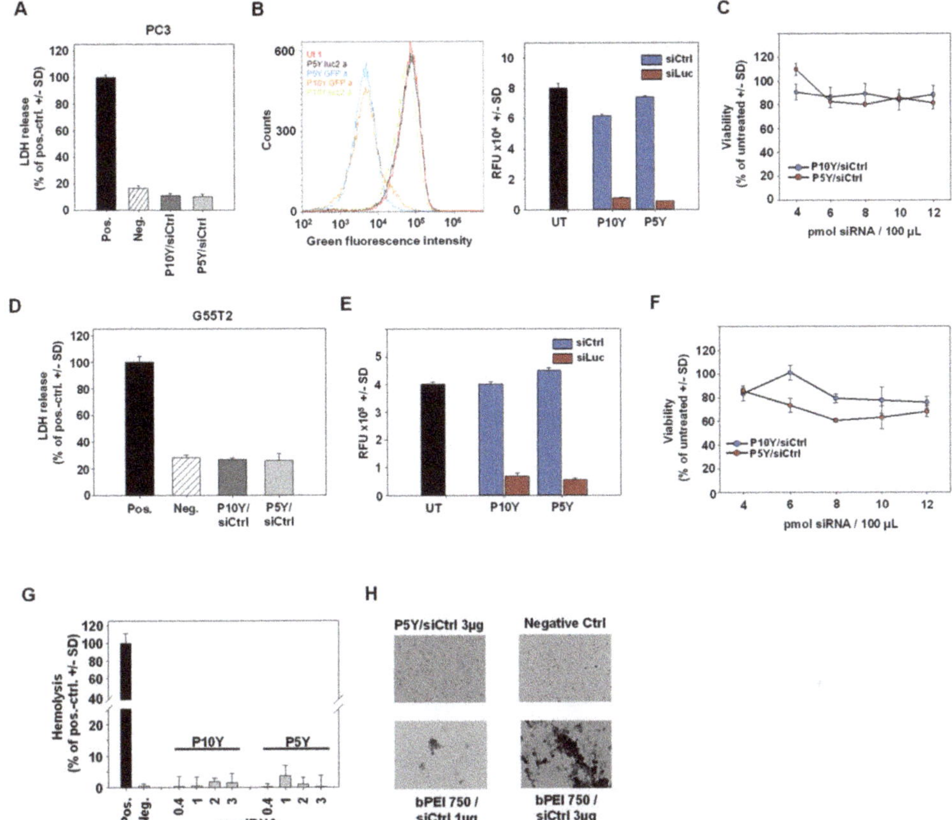

Figure 3. (**A**,**D**) Absence of early cytotoxic effects, as determined in lactate dehydrogenase (LDH) release assays for P10Y/siCtrl and P5Y/siCtrl complexes in PC3 (**A**) and G55T2 (**D**) cells. LDH release was quantitated 24 h after transfection of 30 pmol siCtrl/polymer per 24 well. (**B**,**E**) Knockdown of the Enhanced Green Fluorescent Protein (EGFP) reporter gene in stably expressing PC3-EGFP/Luc (**B**) and G55T2-EGFP/Luc (**E**) cells. Cells were transfected with P10Y and P5Y siRNA complexes and analyzed after 72 h by flow cytometry. (**C**,**F**) Cell viabilities of PC3 cells (**C**) and G55T2 cells (**F**) 48 h after transfection of P10Y/siCtrl and P5Y/siCtrl complexes at mass ratio 2.5 with increasing siRNA-concentrations. (**G**) Hemolysis assay for P5Y/siRNA complexes, demonstrating the absence of erythrocyte damage over a wider range of different concentrations. (**H**) Erythrocyte aggregation assay, further confirming the biocompatibility of P5Y/siRNA complexes as shown for 3 µg complexed siRNA (upper left; all lens magnifications: 4×). A total of 750 kDa bPEI complexes served as positive control, which led to a dose-dependent, profound aggregation of the red blood cells.

3.4. Therapeutic In Vivo Application of P5Y/siRNA Complexes Leads to Profound Anti-Tumor Effects

Pivotal in the development of nanoparticles for siRNA delivery is their applicability in vivo. In previous biodistribution assays, we showed that PEI/siRNA or P10Y/siRNA complexes become systemically available upon i.p. injection, and thus are able to reach organs/tissues remote from the injection site including tumors [50,55]. These studies were now extended towards P5Y/siRNA complexes. In a first experiment, P5Y/siRNA complexes were tested for reporter gene knockdown in mice bearing s.c. HCT116-Luc tumor xenografts. Prior to the in vivo application, the P5Y complexes were first evaluated in vitro for their potential to transfect this cell line. The luciferase expression was decreased to 60% at the lowest siRNA concentration of 15 pmol and further to 25% of control levels at 30 pmol siRNA (Figure 4A).

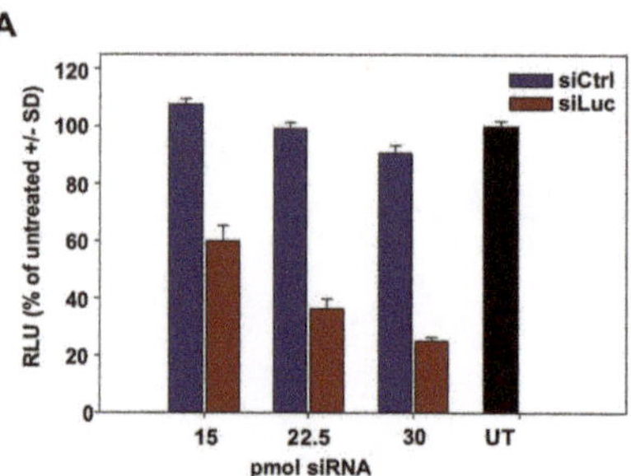

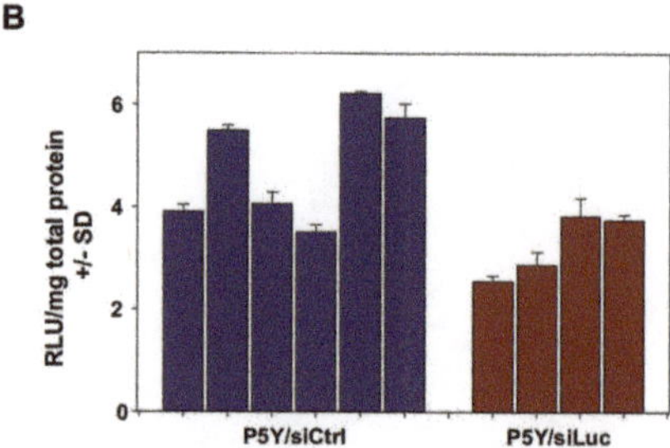

Figure 4. (**A**) Evaluation of luciferase knockdown efficacies of P5Y/siRNA complexes for in vivo use, in stably expressing HCT116-Luc cells over a range of different siRNA concentrations as indicated in the figure. (**B**) Luciferase activity of lysed HCT116-Luc tumors after in vivo treatment with P5Y/siRNA complexes. Bars represent the luciferase activities normalized for total protein concentrations from individual tumors of siCtrl-treated (blue) and siLuc-treated (red) mice.

Next, mice were injected with 5×10^6 tumor cells and, upon establishment of tumor xenografts, treated by repeated systemic application of 10 µg P5Y-complexed siRNA (3× every second day, intraperitoneal injection). Upon termination of the experiment at day 6 after treatment start, tumors were harvested and luciferase activities were analyzed from tumor lysates. Some variations were found between individual samples, especially in the negative control group. Still, the comparison of luciferase activities between the specific (P5Y/siLuc3) and the non-specific negative control (P5Y/siCtrl) group revealed a ~25% reduction of luciferase expression already after three injections (Figure 4B), and thus confirm that the treatment of mice with P5Y/siRNA complexes leads indeed to the knockdown of a given target gene in the tumors.

Switching to a more relevant therapy model, PC3 cells were s.c. injected into mice. Upon establishment of tumor xenografts of ~100 mm^3 in size and with clear growth kinetics, mice were randomized into different groups. Specific treatment relied on the P5Y-complexed siRNAs targeting the oncogenic proteins PLK-1 or survivin. For P10Y complexation, only PLK1 was selected as specific siRNA, while untreated mice or mice treated with siCtrl-containing complexes served as negative controls. The comparison of the latter groups revealed the absence of non-specific effects of repeated treatments (3×/week), as indicated by identical tumor growth curves (Figure 5A,B). In contrast, the treatment with the P5Y/siPLK-1 or P5Y/siSurv (Figure 5A) or P10Y/siSurv complexes (Figure 5B) led to profound antitumor effects.

During the experiment and upon its termination, mice were also screened and tested for possible adverse effects due to the repeated i.p. administration of P5Y/siRNA complexes. Mice showed no alterations in behavior, no obvious weight loss, and no other signs of adverse effects. In addition, possible toxic effects were determined by measuring important serum parameters. Healthy mice were repeatedly i.p. injected with P5Y complexes comprising 10 µg siCtrl four times over eight days. The serum analyses revealed no increase in liver enzymes (ASAT, ALAT), blood glucose (Gluc), urea, albumin, or LDH (Figure 6A). Other side effects may include the unwanted stimulation of the innate immune system. To test for this, immunocompetent mice were treated twice within 24 h with P5Y/siCtrl complexes. In this experiment, i.v. injection was preferred over i.p. administration to achieve instantaneous 100% bioavailability, and serum levels were analyzed at 4 h after the last injection, as described previously [50]. As shown in Figure 6B, no increase in TNFα or IFNγ levels were observed. We thus conclude that P5Y/siRNA complexes are efficient and safe in vitro and in vivo.

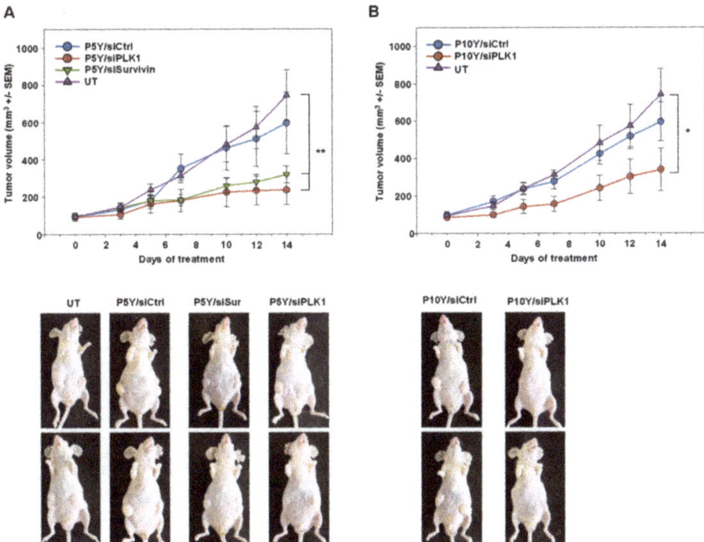

Figure 5. Therapy studies in PC3 tumor bearing mice using P5Y/siRNA (**A**) or P10Y/siRNA (**B**) complexes targeting either the oncogenes PLK1 or survivin. Mice were intraperitoneally treated with the complexes equivalent to 10 µg siRNA every 2–3 days. Tumor growth curves (upper panels) demonstrate the growth inhibiting effects of the specifically treated groups as compared to a negative control treated or untreated group. Lower panels: representative examples of mice.

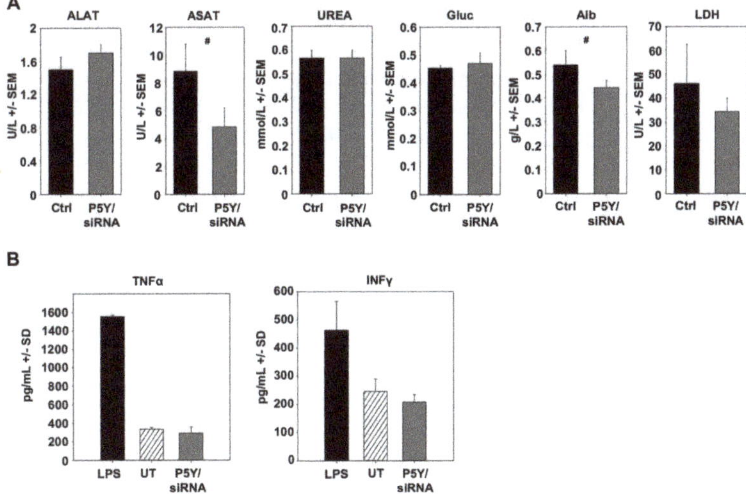

Figure 6. (**A**) Determination of blood serum markers for hepatotoxicity, cardiac or skeletal muscle damage, and kidney dysfunction. Healthy mice were repeatedly i.p. injected with P5Y/siCtrl complexes and blood serum levels were compared to untreated mice. ALAT: Alanine aminotransferase; ASAT: Aspartate aminotransferase; Gluc: Glucose; Alb: Albumin; LDH: Lactate deghydrogenase. (**B**) Absence of immunostimulation of P5Y/siCtrl complexes. Complexes were i.v. injected into immunocompetent mice twice within 24 h and blood was collected 4 h after the last injection. Treatment with lipopolysaccharides (LPS) served as positive control. #, not significant.

4. Discussion

PEIs and other cationic polymers are promising with regard to nucleic acid delivery, but they are often associated with toxicity. This is particularly true for higher molecular weight PEIs, while their lower molecular weight counterparts suffer from poor efficacy [21–23]. The toxicity of high molecular weight PEIs occurs at later stages. An early toxic effect is largely due to a disturbance of the cell membrane after contact with the complexes, leading to increased LDH release and induction of necrosis [56]. The late-stage cytotoxicity was reported to be caused by a damage of the mitochondrial membrane and subsequent activation of caspases and cytochrome C release [57–59]. While the latter findings have been frequently reported, the genotoxic potential of PEI is controversially discussed. Both, the presence and absence of DNA-damaging potential has been reported [60–63] and, possibly explaining these discrepancies, may be dependent on the tested concentrations and selected cell lines.

The fact that PEIs offer relatively easy access to chemical modification is a particularly attractive approach for further improvement of physicochemical and biological properties of the polymer and its complexes. Tyrosine may be considered as among the best candidate amino acids for chemical modification, due to its ability to balance between hydrophilicity (primary amines necessary for nucleic acid complexation) and hydrophobicity (favoring cellular internalization and endosomal escape [44,46]). Many chemical modifications of low molecular weight PEIs in order to increase their bioactivity rely on the cross-linking to create larger PEI derivatives, the coupling of lipidic groups to facilitate micellar structures for improved nucleic acid binding, or modifications with cationic groups to increase the electrostatic interactions (see Introduction). The aromatic amino acid tyrosine plays a key role in protein/DNA interactions. Several studies have reported that tyrosine is essential for the binding to nucleic acids by hydrogen bonding and by π–π-stacking [64,65], which can provide an explanation for the substantial increase in very low molecular weight PEI bioactivity upon tyrosine modification. Generally, phenol-containing molecules are an interesting motif to positively affect a given nucleic acid delivery system. For example, the pre-incubation of siRNA with EGCG, a polyphenol from green tea, and subsequent complexation with low molecular weight polyamines ≤5 kDa (PEI, PLL, PAMAM) strongly increased the knockdown efficacy [22].

Serum stability is an important factor for nucleic acid delivery systems, and we demonstrate high serum compatibility of our complexes based on tyrosine-modified PEIs (see Results and [50]). This may be attributed to the hydroxyl group of tyrosine. In contrast to phenylalanine, tyrosine shows negligible interactions with bovine serum albumin (BSA), as shown previously by the incubation of thymine-dityrosine or thymine-diphenylalanine with BSA solutions [66]. Still, P5Y/siRNA complex stability was even slightly enhanced upon incubation in serum (see Figure S3C). The stabilizing effect of hydroxyl groups was demonstrated in another study, by the modification of PEI with a hydroxyl-containing cyclic carbonate (EHDO). This modification improved serum-tolerance and reduced cytotoxicity as compared to the parent 25 kDa bPEI [67]. Similar results were reported for the Tris-modification of 25 kDa bPEI for pDNA and siRNA delivery [68].

Indeed, we could demonstrate in this study that tyrosine-modification allows for using very low, otherwise inactive PEIs. High complexation efficacies even at very low mass ratios are particularly relevant in the case of small oligonucleotides like siRNAs, considering the lack of binding cooperativity and their high rigidity [69]. Our results are in line with other studies demonstrating low toxicity and high biocompatibility of other (non-PEI) tyrosine-based polymers [70], as well as the safety of tyrosine in humans [71]. The beneficial effects of tyrosine modification thus allowed for using very small amounts of (derivatized) non-toxic low molecular weight PEIs whose bioactivity and biocompatibility critically relies on the grafting with tyrosine. Optimal complex stabilities, which need to be sufficiently high for efficient siRNA complexation/protection and sufficiently low for efficient intracellular siRNA release from the complex, are additional requirements. The potential relevance of the approach of using tyrosine-modified PEIs for translation into the clinic is emphasized by the fact that these nanoparticles were found to be efficacious also in vivo.

Supplementary Materials: The following are available online at http://www.mdpi.com/1999-4923/11/11/600/s1, Figure S1: Degrees of tyrosine grafting and original ^{1}H-NMR spectra, Figure S2: Additional data to Figure 1, Figure S3A,B: Cellular siRNA uptake, Figure S3C: complex stabilities, Table S1: siRNA sequences, Table S2: RT-qPCR primer sequences.

Author Contributions: Conceptualization, A.E. and A.A.; methodology, A.E., S.N., and M.K.; investigation, A.E., S.N., and M.K.; writing—original draft preparation, A.E. and A.A.; writing—review and editing, A.E. and A.A.; supervision, A.A.; project administration, A.A.; funding acquisition, A.A.

Funding: This research was funded by the Deutsche Forschungsgemeinschaft (DFG), grant number AI 24/21-1, and the Deutsche Krebshilfe (grant number 111616) to A.A.

Acknowledgments: The authors are grateful to Markus Böhlmann and Anne-Kathrin Krause for expert mouse maintenance, the core unit "fluorescence technologies" (Kathrin Jäger) for cell sorting and Gabi Oehme for technical assistance in tissue culture experiments. Aileen Heinze is acknowledged for proofreading the manuscript.

Conflicts of Interest: The authors declare no conflict of interest.

References

1. Fire, A.; Xu, S.; Montgomery, M.K.; Kostas, S.A.; Driver, S.E.; Mello, C.C. Potent and specific genetic interference by double-stranded RNA in Caenorhabditis elegans. *Nature* **1998**, *391*, 806–811. [CrossRef] [PubMed]
2. Setten, R.L.; Rossi, J.J.; Han, S.P. The current state and future directions of RNAi-based therapeutics. *Nat. Rev.* **2019**, *18*, 421–446. [CrossRef] [PubMed]
3. Bartlett, D.W.; Davis, M.E. Insights into the kinetics of siRNA-mediated gene silencing from live-cell and live-animal bioluminescent imaging. *Nucleic Acids Res.* **2006**, *34*, 322–333. [CrossRef] [PubMed]
4. Dahlman, J.E.; Barnes, C.; Khan, O.; Thiriot, A.; Jhunjunwala, S.; Shaw, T.E.; Xing, Y.; Sager, H.B.; Sahay, G.; Speciner, L.; et al. In vivo endothelial siRNA delivery using polymeric nanoparticles with low molecular weight. *Nat. Nanotechnol.* **2014**, *9*, 648–655. [CrossRef]
5. Whitehead, K.A.; Langer, R.; Anderson, D.G. Knocking down barriers: advances in siRNA delivery. *Nat. Rev.* **2009**, *8*, 129–138.
6. Khvorova, A.; Watts, J.K. The chemical evolution of oligonucleotide therapies of clinical utility. *Nat. Biotechnol.* **2017**, *35*, 238–248. [CrossRef]
7. Miele, E.; Spinelli, G.P.; Miele, E.; Di Fabrizio, E.; Ferretti, E.; Tomao, S.; Gulino, A. Nanoparticle-based delivery of small interfering RNA: challenges for cancer therapy. *Int. J. Nanomed.* **2012**, *7*, 3637–3657.
8. Basarkar, A.; Singh, J. Nanoparticulate systems for polynucleotide delivery. *Int. J. Nanomed.* **2007**, *2*, 353–360.
9. Merdan, T.; Kopecek, J.; Kissel, T. Prospects for cationic polymers in gene and oligonucleotide therapy against cancer. *Adv. Drug Deliv. Rev.* **2002**, *54*, 715–758. [CrossRef]
10. Ojea-Jimenez, I.; Tort, O.; Lorenzo, J.; Puntes, V.F. Engineered nonviral nanocarriers for intracellular gene delivery applications. *Biomed. Mater.* **2012**, *7*, 054106. [CrossRef]
11. Balazs, D.A.; Godbey, W. Liposomes for use in gene delivery. *J. Drug Deliv.* **2011**, *2011*, 326497. [CrossRef] [PubMed]
12. Zhi, D.; Zhang, S.; Cui, S.; Zhao, Y.; Wang, Y.; Zhao, D. The headgroup evolution of cationic lipids for gene delivery. *Bioconjug. Chem.* **2013**, *24*, 487–519. [CrossRef] [PubMed]
13. Jensen, K.; Anderson, J.A.; Glass, E.J. Comparison of small interfering RNA (siRNA) delivery into bovine monocyte-derived macrophages by transfection and electroporation. *Vet. Immunol. Immunopathol.* **2014**, *158*, 224–232. [CrossRef] [PubMed]
14. Hu, B.; Weng, Y.; Xia, X.H.; Liang, X.J.; Huang, Y. Clinical advances of siRNA therapeutics. *J. Gene Med.* **2019**, *21*, e3097. [CrossRef]
15. Adams, D.; Gonzalez-Duarte, A.; O'Riordan, W.D.; Yang, C.C.; Ueda, M.; Kristen, A.V.; Tournev, I.; Schmidt, H.H.; Coelho, T.; Berk, J.L.; et al. Patisiran, an RNAi Therapeutic, for Hereditary Transthyretin Amyloidosis. *N. Engl. J. Med.* **2018**, *379*, 11–21. [CrossRef]
16. Lorenzer, C.; Dirin, M.; Winkler, A.M.; Baumann, V.; Winkler, J. Going beyond the liver: progress and challenges of targeted delivery of siRNA therapeutics. *J. Control. Release* **2015**, *203*, 1–15. [CrossRef]
17. Ulkoski, D.; Bak, A.; Wilson, J.T.; Krishnamurthy, V.R. Recent advances in polymeric materials for the delivery of RNA therapeutics. *Expert Opin. Drug Deliv.* **2019**, *16*, 1149–1167. [CrossRef]

18. Peng, L.; Wagner, E. Polymeric Carriers for Nucleic Acid Delivery: Current Designs and Future Directions. *Biomacromolecules* **2019**, *20*, 3613–3626. [CrossRef]
19. Hobel, S.; Aigner, A. Polyethylenimines for siRNA and miRNA delivery in vivo. *Wiley Interdiscip. Reviews. Nanomed. Nanobiotechnol.* **2013**, *5*, 484–501. [CrossRef]
20. Pandey, A.P.; Sawant, K.K. Polyethylenimine: A versatile, multifunctional non-viral vector for nucleic acid delivery. *Mater. Sci. Eng. Cmaterials Biol. Appl.* **2016**, *68*, 904–918. [CrossRef]
21. Kwok, A.; Hart, S.L. Comparative structural and functional studies of nanoparticle formulations for DNA and siRNA delivery. *Nanomed. Nanotechnol. Biol. Med.* **2011**, *7*, 210–219. [CrossRef] [PubMed]
22. Shen, W.; Wang, Q.; Shen, Y.; Gao, X.; Li, L.; Yan, Y.; Wang, H.; Cheng, Y. Green Tea Catechin Dramatically Promotes RNAi Mediated by Low-Molecular-Weight Polymers. *ACS Cent. Sci.* **2018**, *4*, 1326–1333. [CrossRef] [PubMed]
23. Shim, M.S.; Kwon, Y.J. Acid-responsive linear polyethylenimine for efficient, specific, and biocompatible siRNA delivery. *Bioconjug. Chem.* **2009**, *20*, 488–499. [CrossRef] [PubMed]
24. Werth, S.; Urban-Klein, B.; Dai, L.; Hobel, S.; Grzelinski, M.; Bakowsky, U.; Czubayko, F.; Aigner, A. A low molecular weight fraction of polyethylenimine (PEI) displays increased transfection efficiency of DNA and siRNA in fresh or lyophilized complexes. *J. Control. Release* **2006**, *112*, 257–270. [CrossRef] [PubMed]
25. Williford, J.M.; Archang, M.M.; Minn, I.; Ren, Y.; Wo, M.; Vandermark, J.; Fisher, P.B.; Pomper, M.G.; Mao, H.Q. Critical Length of PEG Grafts on lPEI/DNA Nanoparticles for Efficient in Vivo Delivery. *ACS Biomater. Sci. Eng.* **2016**, *2*, 567–578. [CrossRef] [PubMed]
26. Malek, A.; Czubayko, F.; Aigner, A. PEG grafting of polyethylenimine (PEI) exerts different effects on DNA transfection and siRNA-induced gene targeting efficacy. *J. Drug Target.* **2008**, *16*, 124–139. [CrossRef]
27. Malek, A.; Merkel, O.; Fink, L.; Czubayko, F.; Kissel, T.; Aigner, A. In vivo pharmacokinetics, tissue distribution and underlying mechanisms of various PEI(-PEG)/siRNA complexes. *Toxicol. Appl. Pharmacol.* **2009**, *236*, 97–108. [CrossRef]
28. Li, S.; Wang, Y.; Zhang, J.; Yang, W.H.; Dai, Z.H.; Zhu, W.; Yu, X.Q. Biodegradable cross-linked poly(amino alcohol esters) based on LMW PEI for gene delivery. *Mol. Biosyst.* **2011**, *7*, 1254–1262. [CrossRef]
29. Wang, C.; Yuan, W.; Xiao, F.; Gan, Y.; Zhao, X.; Zhai, Z.; Zhao, X.; Zhao, C.; Cui, P.; Jin, T.; et al. Biscarbamate Cross-Linked Low-Molecular-Weight Polyethylenimine for Delivering Anti-chordin siRNA into Human Mesenchymal Stem Cells for Improving Bone Regeneration. *Front. Pharmacol.* **2017**, *8*, 572. [CrossRef]
30. Bonner, D.K.; Zhao, X.; Buss, H.; Langer, R.; Hammond, P.T. Crosslinked linear polyethylenimine enhances delivery of DNA to the cytoplasm. *J. Control. Release* **2013**, *167*, 101–107. [CrossRef]
31. Breunig, M.; Hozsa, C.; Lungwitz, U.; Watanabe, K.; Umeda, I.; Kato, H.; Goepferich, A. Mechanistic investigation of poly(ethylene imine)-based siRNA delivery: disulfide bonds boost intracellular release of the cargo. *J. Control. Release* **2008**, *130*, 57–63. [CrossRef] [PubMed]
32. Jiang, H.L.; Islam, M.A.; Xing, L.; Firdous, J.; Cao, W.; He, Y.J.; Zhu, Y.; Cho, K.H.; Li, H.S.; Cho, C.S. Degradable Polyethylenimine-Based Gene Carriers for Cancer Therapy. *Top. Curr. Chem.* **2017**, *375*, 34. [CrossRef] [PubMed]
33. Feldmann, D.P.; Xie, Y.; Jones, S.K.; Yu, D.; Moszczynska, A.; Merkel, O.M. The impact of microfluidic mixing of triblock micelleplexes on in vitro / in vivo gene silencing and intracellular trafficking. *Nanotechnology* **2017**, *28*, 224001. [CrossRef]
34. Goyal, R.; Tripathi, S.K.; Vazquez, E.; Kumar, P.; Gupta, K.C. Biodegradable poly(vinyl alcohol)-polyethylenimine nanocomposites for enhanced gene expression in vitro and in vivo. *Biomacromolecules* **2012**, *13*, 73–83. [CrossRef] [PubMed]
35. Yamada, H.; Loretz, B.; Lehr, C.M. Design of starch-graft-PEI polymers: an effective and biodegradable gene delivery platform. *Biomacromolecules* **2014**, *15*, 1753–1761. [CrossRef] [PubMed]
36. Wu, P.; Luo, X.; Wu, H.; Yu, F.; Wang, K.; Sun, M.; Oupicky, D. Cholesterol Modification Enhances Antimetastatic Activity and siRNA Delivery Efficacy of Poly(ethylenimine)-Based CXCR4 Antagonists. *Macromol. Biosci.* **2018**, *18*, e1800234. [CrossRef] [PubMed]
37. Khan, O.F.; Kowalski, P.S.; Doloff, J.C.; Tsosie, J.K.; Bakthavatchalu, V.; Winn, C.B.; Haupt, J.; Jamiel, M.; Langer, R.; Anderson, D.G. Endothelial siRNA delivery in nonhuman primates using ionizable low-molecular weight polymeric nanoparticles. *Sci. Adv.* **2018**, *4*, eaar8409. [CrossRef]

38. Thapa, B.; Plianwong, S.; Remant Bahadur, K.C.; Rutherford, B.; Uludag, H. Small hydrophobe substitution on polyethylenimine for plasmid DNA delivery: Optimal substitution is critical for effective delivery. *Acta Biomater.* **2016**, *33*, 213–224. [CrossRef]
39. Schafer, J.; Hobel, S.; Bakowsky, U.; Aigner, A. Liposome-polyethylenimine complexes for enhanced DNA and siRNA delivery. *Biomaterials* **2010**, *31*, 6892–6900. [CrossRef]
40. Ewe, A.; Aigner, A. Nebulization of liposome–polyethylenimine complexes (lipopolyplexes) for DNA or siRNA delivery: Physicochemical properties and biological activity. *Eur. J. Lipid Sci. Technol.* **2014**, *116*, 1195–1204. [CrossRef]
41. Ewe, A.; Panchal, O.; Pinnapireddy, S.R.; Bakowsky, U.; Przybylski, S.; Temme, A.; Aigner, A. Liposome-polyethylenimine complexes (DPPC-PEI lipopolyplexes) for therapeutic siRNA delivery in vivo. *Nanomed. Nanotechnol. Biol. Med.* **2016**, *13*, 209–218. [CrossRef] [PubMed]
42. Ewe, A.; Schaper, A.; Barnert, S.; Schubert, R.; Temme, A.; Bakowsky, U.; Aigner, A. Storage stability of optimal liposome-polyethylenimine complexes (lipopolyplexes) for DNA or siRNA delivery. *Acta Biomater.* **2014**, *10*, 2663–2673. [CrossRef] [PubMed]
43. Navath, R.S.; Menjoge, A.R.; Wang, B.; Romero, R.; Kannan, S.; Kannan, R.M. Amino acid-functionalized dendrimers with heterobifunctional chemoselective peripheral groups for drug delivery applications. *Biomacromolecules* **2010**, *11*, 1544–1563. [CrossRef] [PubMed]
44. Wang, F.; Hu, K.; Cheng, Y. Structure-activity relationship of dendrimers engineered with twenty common amino acids in gene delivery. *Acta Biomater.* **2016**, *29*, 94–102. [CrossRef] [PubMed]
45. Wang, F.; Wang, Y.; Wang, H.; Shao, N.; Chen, Y.; Cheng, Y. Synergistic effect of amino acids modified on dendrimer surface in gene delivery. *Biomaterials* **2014**, *35*, 9187–9198. [CrossRef]
46. Creusat, G.; Zuber, G. Self-assembling polyethylenimine derivatives mediate efficient siRNA delivery in mammalian cells. *Chembiochem A Eur. J. Chem. Biol.* **2008**, *9*, 2787–2789. [CrossRef]
47. Yu, W.; Zheng, Y.; Yang, Z.; Fei, H.; Wang, Y.; Hou, X.; Sun, X.; Shen, Y. N-AC-l-Leu-PEI-mediated miR-34a delivery improves osteogenic differentiation under orthodontic force. *Oncotarget* **2017**, *8*, 110460–110473. [CrossRef]
48. He, D.; Muller, K.; Krhac Levacic, A.; Kos, P.; Lachelt, U.; Wagner, E. Combinatorial Optimization of Sequence-Defined Oligo(ethanamino)amides for Folate Receptor-Targeted pDNA and siRNA Delivery. *Bioconjug. Chem* **2016**, *27*, 647–659. [CrossRef]
49. Lee, D.J.; He, D.; Kessel, E.; Padari, K.; Kempter, S.; Lachelt, U.; Radler, J.O.; Pooga, M.; Wagner, E. Tumoral gene silencing by receptor-targeted combinatorial siRNA polyplexes. *J. Control. Release* **2016**, *244*, 280–291. [CrossRef]
50. Ewe, A.; Przybylski, S.; Burkhardt, J.; Janke, A.; Appelhans, D.; Aigner, A. A novel tyrosine-modified low molecular weight polyethylenimine (P10Y) for efficient siRNA delivery in vitro and in vivo. *J. Control. Release* **2016**, *230*, 13–25. [CrossRef]
51. Ewe, A.; Hobel, S.; Heine, C.; Merz, L.; Kallendrusch, S.; Bechmann, I.; Merz, F.; Franke, H.; Aigner, A. Optimized polyethylenimine (PEI)-based nanoparticles for siRNA delivery, analyzed in vitro and in an ex vivo tumor tissue slice culture model. *Drug Deliv. Transl. Res.* **2017**, *7*, 206–216. [CrossRef] [PubMed]
52. Beegle, J.; Lakatos, K.; Kalomoiris, S.; Stewart, H.; Isseroff, R.R.; Nolta, J.A.; Fierro, F.A. Hypoxic preconditioning of mesenchymal stromal cells induces metabolic changes, enhances survival, and promotes cell retention in vivo. *Stem. Cells* **2015**, *33*, 1818–1828. [CrossRef] [PubMed]
53. Campeau, E.; Ruhl, V.E.; Rodier, F.; Smith, C.L.; Rahmberg, B.L.; Fuss, J.O.; Campisi, J.; Yaswen, P.; Cooper, P.K.; Kaufman, P.D. A versatile viral system for expression and depletion of proteins in mammalian cells. *PLoS ONE* **2009**, *4*, e6529. [CrossRef] [PubMed]
54. Livak, K.J.; Schmittgen, T.D. Analysis of relative gene expression data using real-time quantitative PCR and the 2(-Delta Delta C(T)) Method. *Methods* **2001**, *25*, 402–408. [CrossRef]
55. Hobel, S.; Koburger, I.; John, M.; Czubayko, F.; Hadwiger, P.; Vornlocher, H.P.; Aigner, A. Polyethylenimine/small interfering RNA-mediated knockdown of vascular endothelial growth factor in vivo exerts anti-tumor effects synergistically with Bevacizumab. *J. Gene Med.* **2010**, *12*, 287–300.
56. Moghimi, S.M.; Symonds, P.; Murray, J.C.; Hunter, A.C.; Debska, G.; Szewczyk, A. A two-stage poly(ethylenimine)-mediated cytotoxicity: implications for gene transfer/therapy. *Mol. Ther.* **2005**, *11*, 990–995. [CrossRef]

57. Monnery, B.D.; Wright, M.; Cavill, R.; Hoogenboom, R.; Shaunak, S.; Steinke, J.H.G.; Thanou, M. Cytotoxicity of polycations: Relationship of molecular weight and the hydrolytic theory of the mechanism of toxicity. *Int. J. Pharm.* **2017**, *521*, 249–258. [CrossRef]
58. Hall, A.; Larsen, A.K.; Parhamifar, L.; Meyle, K.D.; Wu, L.P.; Moghimi, S.M. High resolution respirometry analysis of polyethylenimine-mediated mitochondrial energy crisis and cellular stress: Mitochondrial proton leak and inhibition of the electron transport system. *Biochim. Biophys. Acta* **2013**, *1827*, 1213–1225. [CrossRef]
59. Yamamoto, T.; Tsunoda, M.; Ozono, M.; Watanabe, A.; Kotake, K.; Hiroshima, Y.; Yamada, A.; Terada, H.; Shinohara, Y. Polyethyleneimine renders mitochondrial membranes permeable by interacting with negatively charged phospholipids in them. *Arch. Biochem. Biophys.* **2018**, *652*, 9–17. [CrossRef]
60. Gholami, L.; Sadeghnia, H.R.; Darroudi, M.; Kazemi Oskuee, R. Evaluation of genotoxicity and cytotoxicity induced by different molecular weights of polyethylenimine/DNA nanoparticles. *Turk. J. Biol.* **2014**, *38*, 380–387. [CrossRef]
61. Castan, L.; Jose da Silva, C.; Ferreira Molina, E.; Alves Dos Santos, R. Comparative study of cytotoxicity and genotoxicity of commercial Jeffamines(R) and polyethylenimine in CHO-K1 cells. *J. Biomed. Mater. Research. Part. Bapplied Biomater.* **2018**, *106*, 742–750. [CrossRef] [PubMed]
62. Beyerle, A.; Long, A.S.; White, P.A.; Kissel, T.; Stoeger, T. Poly(ethylene imine) nanocarriers do not induce mutations nor oxidative DNA damage in vitro in MutaMouse FE1 cells. *Mol. Pharm.* **2011**, *8*, 976–981. [CrossRef] [PubMed]
63. Hall, A.; Lachelt, U.; Bartek, J.; Wagner, E.; Moghimi, S.M. Polyplex Evolution: Understanding Biology, Optimizing Performance. *Mol. Ther.* **2017**, *25*, 1476–1490. [CrossRef] [PubMed]
64. Plyte, S.E.; Kneale, G.G. The role of tyrosine residues in the DNA-binding site of the Pf1 gene 5 protein. *Protein Eng.* **1991**, *4*, 553–560. [CrossRef]
65. Huang, J.; Zhao, Y.; Liu, H.; Huang, D.; Cheng, X.; Zhao, W.; Taylor, I.A.; Liu, J.; Peng, Y.L. Substitution of tryptophan 89 with tyrosine switches the DNA binding mode of PC4. *Sci. Rep.* **2015**, *5*, 8789. [CrossRef]
66. Roviello, G.N.; Oliviero, G.; Di Napoli, A.; Borbone, N.; Piccialli, G. Synthesis, self-assembly-behavior and biomolecular recognition properties of thyminyl dipeptides. *Arab J. Chem* 2018. [CrossRef]
67. Luo, X.H.; Huang, F.W.; Qin, S.Y.; Wang, H.F.; Feng, J.; Zhang, X.Z.; Zhuo, R.X. A strategy to improve serum-tolerant transfection activity of polycation vectors by surface hydroxylation. *Biomaterials* **2011**, *32*, 9925–9939. [CrossRef]
68. Dong, X.; Lin, L.; Chen, J.; Guo, Z.; Tian, H.; Li, Y.; Wei, Y.; Chen, X. A serum-tolerant hydroxyl-modified polyethylenimine as versatile carriers of pDNA/siRNA. *Macromol. Biosci.* **2013**, *13*, 512–522. [CrossRef]
69. Bolcato-Bellemin, A.L.; Bonnet, M.E.; Creusat, G.; Erbacher, P.; Behr, J.P. Sticky overhangs enhance siRNA-mediated gene silencing. *Proc. Natl. Acad. Sci. USA* **2007**, *104*, 16050–16055. [CrossRef]
70. Bourke, S.L.; Kohn, J. Polymers derived from the amino acid L-tyrosine: polycarbonates, polyarylates and copolymers with poly(ethylene glycol). *Adv. Drug Deliv. Rev.* **2003**, *55*, 447–466. [CrossRef]
71. Baldrick, P.; Richardson, D.; Wheeler, A.W. Review of L-tyrosine confirming its safe human use as an adjuvant. *J. Appl. Toxicol. Jat* **2002**, *22*, 333–344. [CrossRef] [PubMed]

Article

Efficient Delivery of Therapeutic siRNA by Fe_3O_4 Magnetic Nanoparticles into Oral Cancer Cells

Lili Jin [1], Qiuyu Wang [1], Jiayu Chen [2], Zixiang Wang [1], Hongchuan Xin [3] and Dianbao Zhang [2,*]

[1] School of Life Science, Liaoning University, Shenyang 110036, China; lilijin@lnu.edu.cn (L.J.); qiuyuwang@lnu.edu.cn (Q.W.); zxwlnu@163.com (Z.W.)

[2] Department of Stem Cells and Regenerative Medicine, Key Laboratory of Cell Biology, National Health Commission of China, and Key Laboratory of Medical Cell Biology, Ministry of Education of China, China Medical University, Shenyang 110122, China; chenjiayu@cmu.edu.cn

[3] Qingdao Institute of Bioenergy and Bioprocess Technology, Chinese Academy of Sciences, Qingdao 266101, China; xinhc@qibebt.ac.cn

* Correspondence: zhangdianbao@gmail.com

Received: 22 August 2019; Accepted: 15 November 2019; Published: 17 November 2019

Abstract: The incidence of oral cancer is increasing due to smoking, drinking, and human papillomavirus (HPV) infection, while the current treatments are not satisfactory. Small interfering RNA (siRNA)-based therapy has brought hope, but an efficient delivery system is still needed. Here, polyethyleneimine (PEI)-modified magnetic Fe_3O_4 nanoparticles were prepared for the delivery of therapeutic siRNAs targeting B-cell lymphoma-2 (BCL2) and Baculoviral IAP repeat-containing 5 (BIRC5) into Ca9-22 oral cancer cells. The cationic nanoparticles were characterized by transmission electronic microscopy (TEM), scanning electronic microscopy (SEM), dynamic light scattering (DLS), and vibrating sample magnetometer (VSM). By gel retardation assay, the nanoparticles were found to block siRNA in a concentration-dependent manner. The cellular uptake of the nanoparticle/siRNA complexes under a magnetic field was visualized by Perl's Prussian blue staining and FAM labeling. High gene silencing efficiencies were determined by quantitative real-time PCR and western blotting. Furthermore, the nanoparticle-delivered siRNAs targeting BCL2 and BIRC5 were found to remarkably inhibit the viability and migration of Ca9-22 cells, by cell counting kit-8 assay and transwell assay. In this study, we have developed a novel siRNA-based therapeutic strategy targeting BCL2 and BIRC5 for oral cancer.

Keywords: magnetic nanoparticle; iron oxide; siRNA delivery; BCL2; BIRC5/survivin; oral cancer

1. Introduction

The incidence of oral cancer has increased due to risk factors such as tobacco, alcohol, and human papillomavirus (HPV), resulting in nearly 180 thousand deaths worldwide in 2018 [1,2]. The clinical treatment of oral cancer mainly depends on surgery, radiotherapy, chemotherapy, and several targeted drugs, but the prognosis is poor [3,4]. Therefore, it is necessary to develop novel therapeutic strategies to overcome the limitations of current therapies for oral cancer.

Recent progress in nanotechnology-based gene therapy has brought hopes for cancer treatment [5]. RNA interference (RNAi) is a sequence-specific post-transcriptional gene silencing process in eukaryotes [6]. RNAi could be triggered by microRNA (miRNA) and small interfering RNA (siRNA), which could be designed to target almost any gene [7]. It is exploited by researchers for loss-of-function studies and holds promise for the development of therapeutic gene silencing [8]. The first siRNA drug Onpattro (patisiran) targeting transthyretin (TTR) has been approved by the U.S. Food and Drug Administration (FDA) in 2018, for the treatment of peripheral nerve disease polyneuropathy in adults.

This major progress, with the elucidation of more and more disease-related target genes, has greatly stimulated the research and development of siRNA drugs. However, how to effectively deliver siRNA drugs is a bottleneck to clinical practice [6,9]. The currently developed siRNA delivery strategies mainly include siRNA conjugation, lipid-based, and polymer-based delivery systems [10,11]. In our previous studies, the prepared polyethyleneimine (PEI)-coated Fe_3O_4 nanoparticles exhibited siRNA protection and delivery capacities for mesenchymal stem cells and glioblastoma cells [12,13]. However, the feasibility of these type of nanoparticles for delivering siRNA to oral cancer cells remains unknown, and the uniformity and efficiency of delivery needs to be improved.

An increasing number of cancer target genes has been reported in recent years. BCL2 (B-cell lymphoma-2) is a gene that is overexpressed in many cancers to escape cell death [14]. It is a promising cancer therapeutic target, but there are few targeting agents with clinical significance [14–16]. For oral cancers, BCL2 were proved to be important in cancer progression and chemoradiation resistance [17]. Inhibition of BCL2 in oral cancer cells inhibited proliferation and induced apoptosis, and also augmented the inhibitory effects of cisplatin in vitro and in vivo [18]. BIRC5 (Baculoviral IAP repeat-containing 5, also named survivin) is a conserved gene essential for cell proliferation. It is expressed in proliferating cells and upregulated in most cancers, the inhibition of BIRC5 leads to apoptosis or sensitization to chemotherapy and radiotherapy [19,20]. BIRC5 is rarely mutated in oral cancer samples and upregulated compared to non-cancerous tissue [21]. In this study, we prepared Fe_3O_4 nanoparticles and design siRNAs targeting BCL2 and BIRC5, aiming to explore the efficient delivery of therapeutic siRNA into oral cancer cells by Fe_3O_4 nanoparticles, which might provide a novel strategy for the future therapy of oral cancer.

2. Materials and Methods

2.1. Synthesis and Characterization of Nanoparticles

The magnetic nanoparticles were synthesized based on the oxidative hydrolysis method [22]. Briefly, $FeSO_4$ and PEI dissolved in H_2SO_4 solution were dripped into KNO_3 and NaOH solutions under nitrogen bubbling in a triple neck flask. After precipitation, the PEI–Fe_3O_4 nanoparticles were obtained by ultrafiltration (100 KDa, UFC910096, Millipore, Beijing, China). The particle size and morphology were analyzed by transmission electronic microscopy (TEM) and scanning electronic microscopy (SEM) [12]. The particle size was analyzed by ImageJ software. The zeta potential and hydrodynamic size of nanoparticles were analyzed by dynamic light scattering (DLS, Mastersizer 2000, Malvern, Worcestershire, UK). The amino group density was determined by the conductivity meter, and the elemental content was analyzed by energy dispersive spectroscopy (EDS). The magnetization of nanoparticles was measured by the vibrating sample magnetometer (VSM, Lakeshore7404, Westerville, OH, USA).

2.2. Gel Retardation Assay

The binding capacity of the nanoparticles to siRNA was analyzed by gel retardation assay. In general, 1 µg siRNA was mixed with 0, 0.5, 1, 1.5, and 2 µg nanoparticles in Opti-MEM Reduced Serum Medium (51985034, Gibco, Shanghai, China) and incubated at room temperature for 10 min. The mixtures were subjected to 2% agarose gel electrophoresis at 100 V for 20 min. The gels were stained with Biosafe nucleic acid dye (170-3001, Tanon, Shanghai, China) and imaged under Tanon-1600 imaging system (Tanon).

2.3. Cell Culture

Human oral cancer cell Ca9-22 and CAL 27 (Procell, Wuhan, China) cells were cultured in Dulbecco's Modified Eagle Medium (DMEM) with high glucose (SH30022.01, Hyclone, Beijing, China) supplemented with 10% Certified Fetal Bovine Serum (FBS) (04-001-1A, Bioind, Kibbutz Beit-Haemek,

Israel) and 1% Penicillin Streptomycin (SV30010, Hyclone) at 37 °C in a humidified atmosphere containing 5% CO_2.

2.4. Cell Transfection

The Ca9-22 cells were seeded on 24-well culture plates and incubated for 12 h before transfection. Normally, 0.6 μg of nanoparticles and 0.2 μg of siRNA were diluted with 20 μL of Opti-MEM Reduced Serum Medium separately and then mixed together. The mixtures were incubated at room temperature for 10 min and added to the cells. The cells were incubated under the magnetic field for 30 min (Mag0201, Nanoeast, Nanjing, China) and then under normal conditions. The transfection of siRNA using Lipofectamine 3000 Transfection Reagent (L3000015, Invitrogen, Shanghai, China) was carried out according to the manufacturer's instructions, serving as a positive control. FAM-siRNAs were used for observation of the cellular uptake of siRNA by fluorescence microscope (Observer A1, Carl Zeiss, Oberkochen, Germany). All the siRNA duplexes were chemically synthesized by GenePharma (Shanghai, China) and the sequences were as follows: siBCL2—5′-GGGAGAACAGGGUACGAUATT-3′; siBIRC5—5′-GAAGCAGUUUGAAGAAUUATT-3′; NC (non-targeting siRNA serving as a negative control) 5′-UCCGAACGUGUCACGUTT-3′.

2.5. Perl's Prussian Blue Staining

The cellular uptake of the nanoparticles was visualized by Perl's Prussian blue staining. The cells were cultured in 24-well plates and transfected as indicated. After incubation for 12 h, the cells were fixed with 4% Paraformaldehyde Fix Solution (P0099, Beyotime) for 15 min. The cells were washed with PBS and stained with 5% potassium ferrocyanide in 10% hydrochloric acid for 30 min at room temperature. The working solution was made by mixing equal volumes of 10% potassium ferrocyanide and 20% hydrochloric acid solution just before use. The cells were observed under a microscope (Olympus CKX41, Tokyo, Japan).

2.6. Quantitative Real-time PCR

The mRNA levels of genes were analyzed by quantitative real-time PCR. Total RNA was prepared using RNAiso Plus (9108, Takara, Dalian, China) according to the manufacturer's instructions. The RNA concentration and quality were determined by NanoDrop 2000C spectrophotometer (Thermo Fisher Scientific, Wilmington, DE, USA). Equal amounts total RNA from samples were subjected to reverse transcription separately using PrimeScript RT reagent Kit with gDNA Eraser (Perfect Real Time) (RR047A, Takara). The TB Green Premix Ex Taq II (Tli RNaseH Plus) (RR820A, Takara) was used for quantitative real-time PCR in a 7500 Real-Time PCR Systems (Applied Biosystems, Foster City, CA, USA) following the standard protocol provided by the manufacturer. The relative mRNA levels were calculated by the $\Delta\Delta$Ct method and GAPDH served as an internal control. All the primers were chemically synthesized by Sangon Biotech (Shanghai, China). The sequences of primers were listed in Table 1.

Table 1. Primer sequences for quantitative real-time PCR.

Gene	Primer Sequence	Product Length (bp)
BCL2	forward: 5′-GATAACGGAGGCTGGGATGC-3′ reverse: 5′-CAGGGCCAAACTGAGCAGAG-3′	105
BIRC5	forward: 5′-TGAGAACGAGCCAGACTTGG-3′ reverse: 5′-GTTCCTCTATGGGGTCGTCA-3′	86
GAPDH [13]	forward: 5′-GCACCGTCAAGGCTGAGAAC-3′ reverse: 5′-TGGTGAAGACGCCAGTGGA-3′	138

2.7. Western Blotting

The protein levels were determined by the western blotting assay. Total protein lysis was prepared using the RIPA Lysis Buffer (P0013B, Beyotime) and quantified by the BCA Protein Assay Kit (T9300A, Takara). The protein samples for western blotting were prepared using SDS-PAGE Sample Loading Buffer (P0015L, Beyotime). Equal amounts of total proteins were loaded onto 10% SDS-PAGE (P0012AC, Beyotime) and separated by electrophoresis. The separated proteins were transferred onto a PVDF membrane (IPVH00010, Millipore, Shanghai, China) and blocked by 5% skim milk (232100, BD Bioscience, San Jose, CA, USA). The membranes were incubated with primary antibodies Bcl-2 Rabbit Polyclonal Antibody (1:1000, AF0060, Beyotime), Anti-Survivin Rabbit pAb (1:1000, GB11177, Servicebio, Wuhan, China) and GAPDH Mouse Monoclonal antibody (1:5000, 60004-1-Ig, Proteintech, Wuhan, China) at 4 °C overnight. Then HRP-conjugated Affinipure Goat Anti-Mouse IgG (1:5000, SA00001-1, Proteintech) and HRP-conjugated Affinipure Goat Anti-Rabbit IgG (1:5000, SA00001-2, Proteintech) were used to probe the membrane at room temperature for 1 h. The protein bands were visualized using Amersham ECL Prime Western Blotting Detection Reagent (RPN2232, GE Healthcare, Princeton, NJ, USA) and imaged by Tanon-5200 chemiluminescence detection system (Tanon).

2.8. Cell Viability Assay

The cell viability was analyzed using Cell Counting Kit-8 assay (MA0218, Meilunbio, Dalian, China). In brief, the cells were cultured in 24-well plates and transfected as indicated, and then seeded into 96-well plates. A 10 μL of CCK-8 solution was added to each well and incubated at 37 °C for 1 h. The absorbance at 450 nm was detected using an iMARK microplate reader (Bio-Rad, Hercules, CA, USA) with a reference wavelength of 630 nm.

2.9. Transwell Migration Assay

The migration capacity of cells was assessed using the transwell migration assay. The CA9-22 cells were transfected as indicated for 24 h and then seeded with serum-free culture medium into the upper chamber of Transwell (3422, Coring, Corning, NY, USA). The complete medium was added into the lower chamber. After incubation for 12 h, the cells were fixed with 4% Paraformaldehyde Fix Solution for 15 min. The cells on the upper side of the membranes were removed with a cotton swab. The migrated cells were stained with Crystal Violet Staining Solution (C0121, Beyotime) and visualized using a microscope (Olympus CKX41).

2.10. Cell Cycle Analysis

The cell cycle progression was determined using the Cell Cycle and Apoptosis Analysis Kit (C1052, Beyotime). The cells were collected and fixed in ice-cold 70% ethanol overnight. The fixed cells were washed with PBS and stained with PI in Staining Buffer supplemented with RNase A at 37 °C for 30 min in the dark. Then the stained cells were analyzed by flow cytometry FACSCalibur (BD Bioscience).

2.11. Statistical Analysis

The experiments were carried out in triplicates and the data were presented as mean ± standard deviation (SD). Statistical significance was determined by one-way analysis of variance (ANOVA) following post-hoc multiple comparisons. $p < 0.05$ was considered to be statistically significant.

3. Results

3.1. Nanoparticle Characterization

TEM and SEM were carried out to characterize the prepared nanoparticles. As shown in Figure 1a, the synthesized nanoparticles presented a dense spherical morphology. Based on the TEM image, the diameter of nanoparticles was analyzed and the average diameter was 7.95 nm (Figure 1b). The SEM image in Figure 1c further confirmed the uniform morphology and good dispersion of the nanoparticles. The hydrodynamic diameter of the nanoparticles was detected by DLS, the average diameter was 26.12 nm (Figure 1d). By EDS energy-mapping, the element nitrogen was proven to contribute 10.6% of the dry weight of the nanoparticles, indicating PEI accounted for about 33.6% (Figure 1e). The amino group density analyzed by the conductivity meter was 1482 nmol/g. The average zeta potential analyzed by DLS was +46.5 mV, further quantifying the positive charge of the nanoparticles. The magnetization curve obtained by VSM showed saturation magnetization as 52.7 emu/g, without a hysteresis loop (Figure 1f). These data proved that the prepared Fe_3O_4 nanoparticles were small, superparamagnetic, and positively charged.

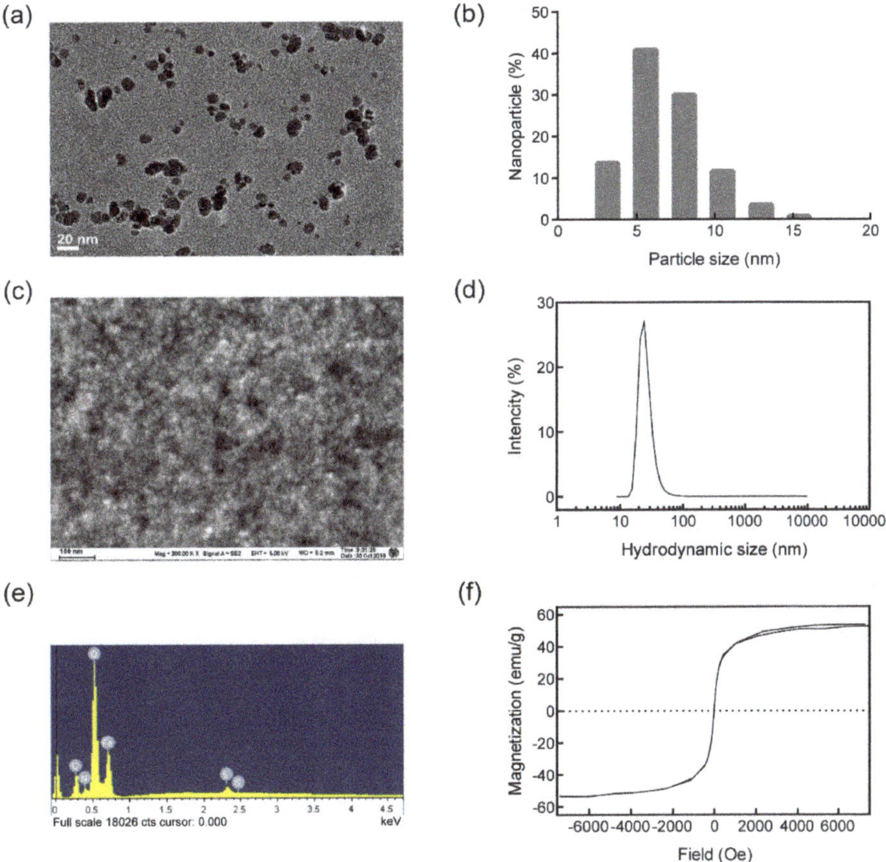

Figure 1. Characterization of the prepared nanoparticles. (**a**) The transmission electron microscope (TEM) image of the nanoparticles, and the bar indicates 20 nm; (**b**) the diameter distribution of the nanoparticles; (**c**) the scanning electron microscopy (SEM) image of the nanoparticles, and the bar indicates 100 nm; (**d**) the distribution of the hydrodynamic diameter of the nanoparticles; (**e**) the element composition; (**f**) the magnetization curve of the nanoparticles.

3.2. Cellular Upkake of Nanoparticle/siRNA

The capacity of nanoparticles to form complexes with siRNA ex vitro is essential for siRNA delivery into cells, and it is also key to optimize the delivery parameters. The gel retardation assay was performed to assess the interaction between the nanoparticles and siRNA. After 1 μg of siRNA was incubated with 0, 0.5, 1, 1.5, and 2 μg of nanoparticles and subjected to agarose gel electrophoresis. When the siRNA formed complexes with the nanoparticles, the complexes were so large that they would not move under the electric field in the agarose gel and remain in the loading holes. The observed bands indicated the free siRNA; it presented concentration-dependent gel retardation in both siBCL2 (Figure 2a) and siBIRC5 (Figure 2b) gel images. When 2 μg of nanoparticles were incubated with 1 μg of siRNA, all the siRNA could not run in the agarose gel electrophoresis, suggesting that siRNA totally formed complexes with nanoparticles when the weight ratio was more than 2.0. In other words, when the weight ratio was less than 2.0, the ability of nanoparticles to load siRNA was saturated or nearly saturated. Most of the positive charges of nanoparticles were neutralized by negatively charged siRNA, it was not conducive for the complexes to approach the surface of cell membranes, which was negatively charged. Thus, more nanoparticles were essential for siRNA delivery into the cells. For siRNA delivery, 0.6 μg of nanoparticles and 0.2 μg of siRNA were used for Ca9-22 cells in each well of 24-well plates. The cells were incubated under the magnetic field for 30 min and then under normal conditions for 12 h. The cellular uptake of the nanoparticle/siRNA was detected by Perl's Prussian blue staining and FAM-labeled siRNAs. As shown in Figure 2c, almost all the cells were stained blue in nanoparticle-delivered FAM-siBCL2 and siBIRC5 groups. Under a fluorescence microscope, green fluorescence was observed in these two groups and lipo (lipofectamine 3000)/FAM-NC group (positive control). These results indicated an efficient siRNA delivery by the nanoparticles into Ca9-22 cells.

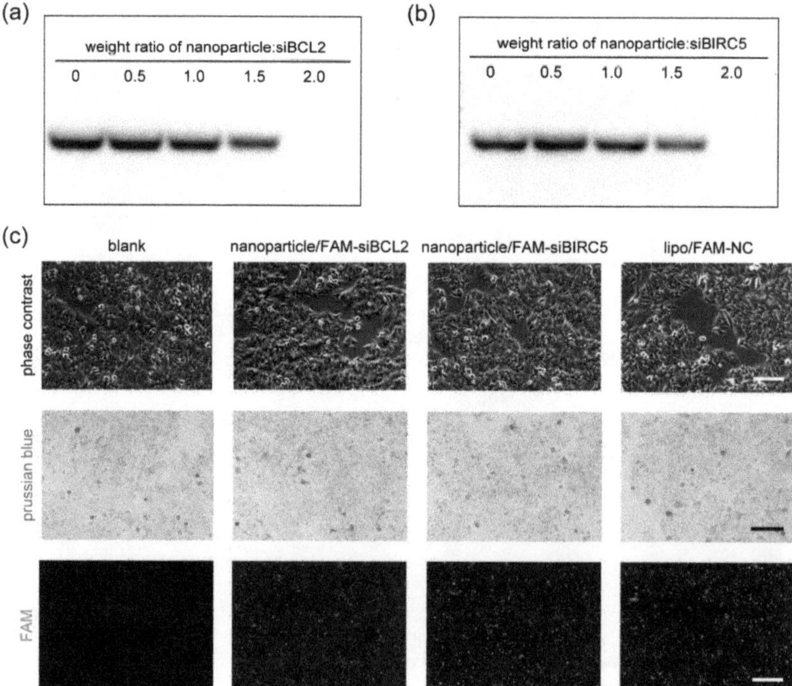

Figure 2. The cellular uptake of the siRNA mediated by the nanoparticles. (**a**) Gel retardation assay for the interaction between nanoparticles and siBCL2; (**b**) gel retardation assay for the interaction between nanoparticles and siBIRC5; (**c**) the delivery of siRNAs into Ca9-22 cells by the nanoparticles visualized by Perl's Prussian blue staining and FAM-labeled siRNAs. Lipofectamine 3000 (lipo) was served as a positive control for siRNA transfection. Bar indicates 100 μm.

3.3. Gene-Silencing Efficiency

To determine the gene-silencing efficiency of nanoparticle-delivered siRNA in oral cancer cells, 0.6 μg of nanoparticles and 0.2 μg of siRNA were used for Ca9-22 and CAL 27 cells in each well of 24-well plates, the cells were incubated under the magnetic field for 30 min and then under normal conditions for 48 h. Total RNA was extracted and analyzed using quantitative real-time PCR. As shown in Figures 3a and 4a, the mRNA level of BCL2 was significantly reduced by nanoparticle-delivered siBCL2 to 18% in Ca9-22 cells and 56% in CAL 27 cells, compared with the nanoparticle+NC groups. The silencing of BCL2 was further verified in the protein level using western blotting (Figures 3c and 4c). For BIRC5 silencing, similar results were observed in Figure 3b,d and Figure 4b,d. The BIRC5 in Ca9-22 was significantly silenced by nanoparticle-delivered siBIRC5 and proven by quantitative real-time PCR and western blotting in both mRNA and protein levels. Lipofectamine 3000 (lipo) was served as a positive control for siRNA transfection. These data proved the satisfied siRNA delivery efficiency of the nanoparticles in Ca9-22 and CAL 27 cells, at least for BCL2 and BIRC5 silencing.

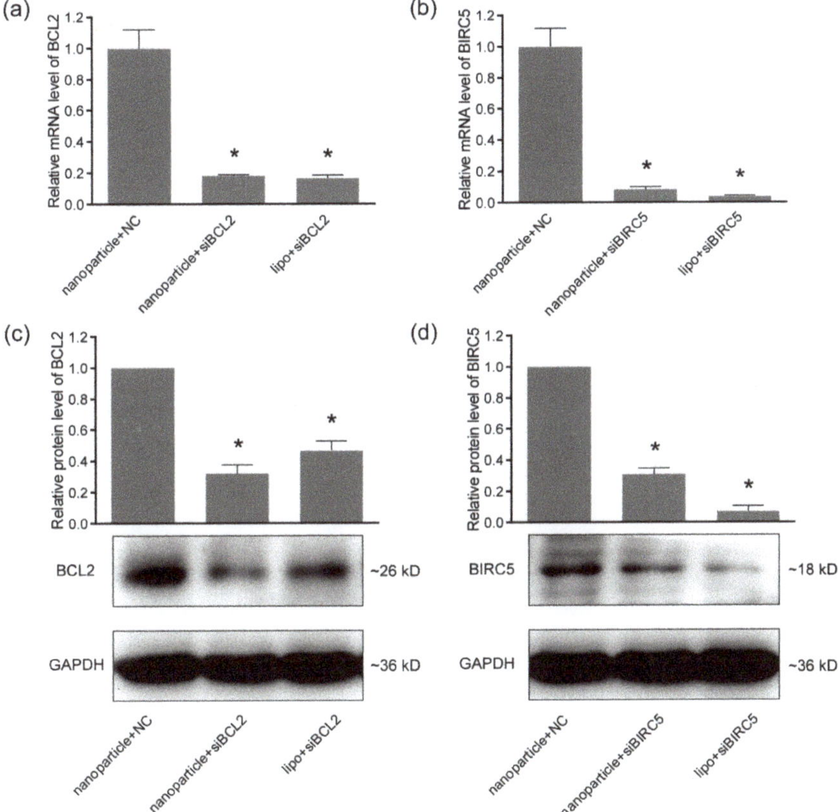

Figure 3. The gene silencing efficiencies of siRNA delivered by nanoparticles in Ca9-22 cells. (**a**) The mRNA levels of BCL2 in Ca9-22 cells detected by quantitative real-time PCR; (**b**) the mRNA levels of BIRC5 detected by quantitative real-time PCR; (**c**) the protein levels of BCL2 analyzed by western blotting; (**d**) the protein levels of BIRC5 analyzed by western blotting. GAPDH was served as an internal control. Lipofectamine 3000 (lipo) was served as a positive control for siRNA transfection. * $p < 0.05$ compared with nanoparticle+NC (negative control) group.

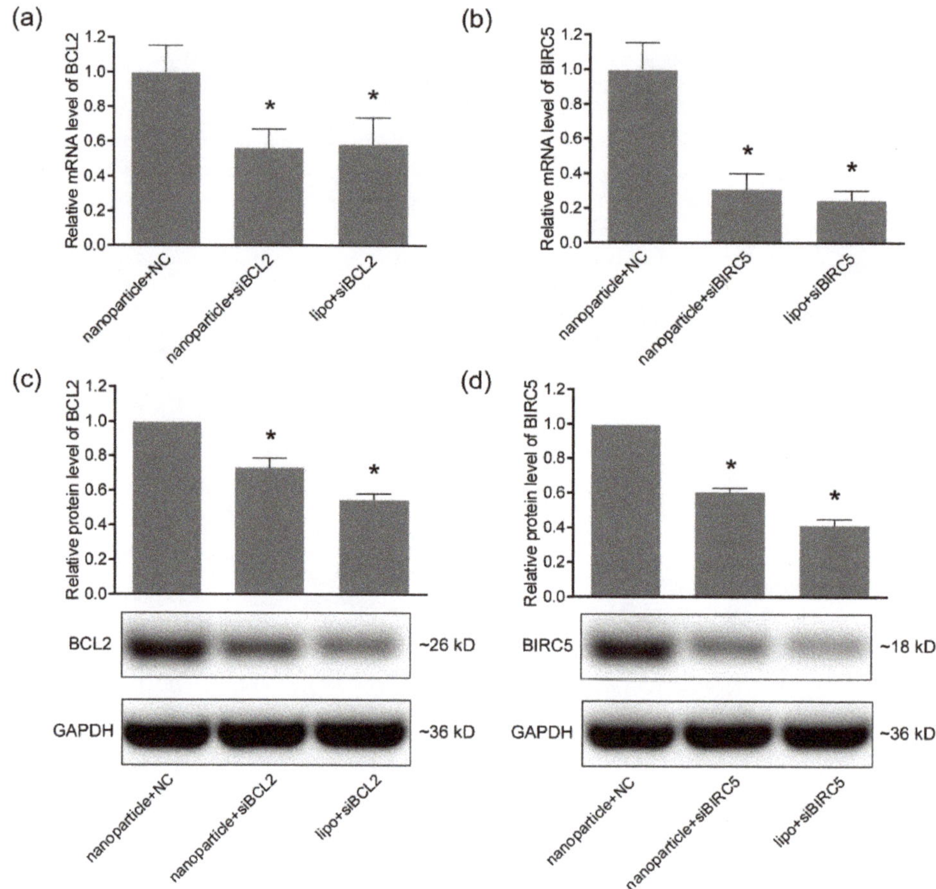

Figure 4. The gene silencing efficiencies of siRNA delivered by nanoparticles in CAL 27 cells. (**a**) The mRNA levels of BCL2 in CAL 27 cells detected by quantitative real-time PCR; (**b**) the mRNA levels of BIRC5 detected by quantitative real-time PCR; (**c**) the protein levels of BCL2 analyzed by western blotting; (**d**) the protein levels of BIRC5 analyzed by western blotting. GAPDH was served as an internal control. Lipofectamine 3000 (lipo) was served as a positive control for siRNA transfection. * $p < 0.05$ compared with nanoparticle+NC (negative control) group.

3.4. Anti-Tumor Activity

To evaluate the anti-tumor activity of the nanoparticle-delivered therapeutic siRNA targeting BCL2 and BIRC5, the siRNAs were delivered to Ca9-22 cells by the nanoparticles as described in Section 3.3. By the CCK-8 assay, cell viability showed significant reduction in the nanoparticle+siBCL2 and nanoparticle+siBIRC5 groups, compared with the nanoparticle+NC group (Figure 5a). The treatment with nanoparticle+NC also showed no significant difference versus blank groups, indicating the safety of the nanoparticles at working concentrations (0.6 μg per well of 24-well plate). Our previous data showed that these types of nanoparticles were toxic to cells at high concentrations [23]. Thus, the nanoparticles need to be used at appropriate concentrations to avoid cytotoxicity. The migration capacity of tumor cells is critical for cancer metastasis [24], and the Ca9-22 cell migration was determined using the transwell assay. As shown in Figure 5b,c, the numbers of migrated cells were remarkably reduced by the nanoparticle-delivered siBCL2 and siBIRC5. Furthermore, cell cycle distribution was analyzed by flow cytometry to verify the inhibitory effects. G1 phase cell cycle arrest was observed in nanoparticle+siBCL2 group and G2 phase arrest in nanoparticle+siBIRC5 group (Figure 5d). These results demonstrated the effective delivery of therapeutic siRNA to Ca9-22 cells by the nanoparticles, indicating a novel potential therapeutic strategy for oral cancer therapy.

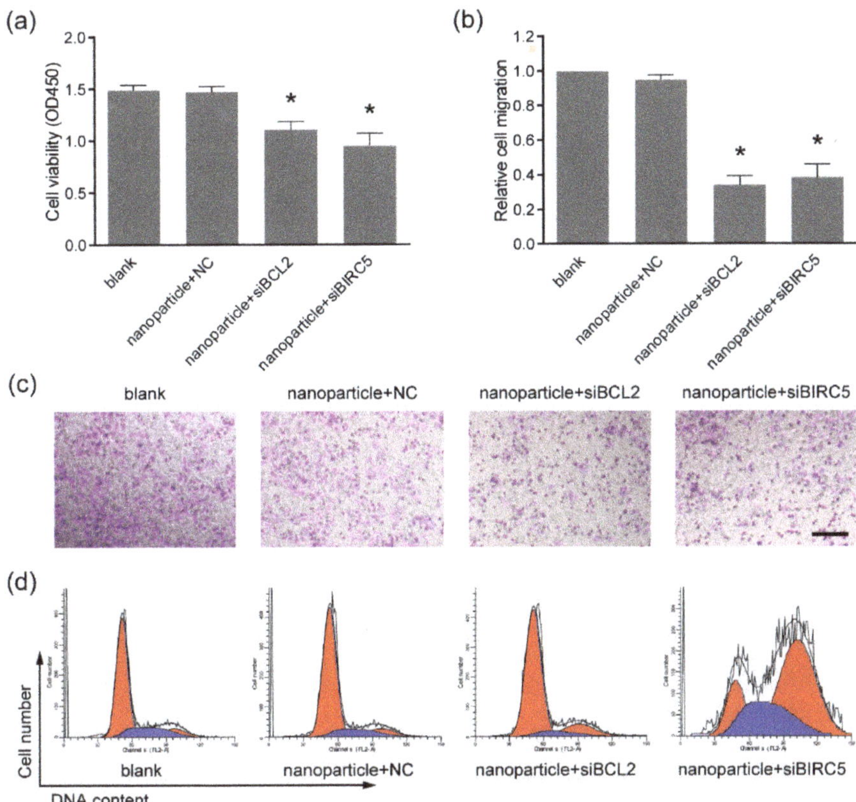

Figure 5. Cell viability and migration were inhibited by the siRNA delivered by nanoparticles. (**a**) The cell viability examined by cell counting kit-8 assay (CCK-8). * $p < 0.05$ compared with nanoparticle+NC group; (**b**) the cell migration analyzed by the transwell assay; (**c**) the representative images of migrated cells in the transwell assay. Bar indicates 100 μm. (**d**) The cell cycle distribution analyzed by PI staining followed by flow cytometry. G1 phase arrest and G2 phase arrest were observed upon nanoparticle-delivered siBCL2 and siBIRC5, respectively.

4. Discussion

Investigations for molecular mechanisms during oral cancer occurrence and development provide a large number of candidate target genes, but most of which have not been targeted for therapy until now [4]. RNAi-based gene therapy expands the scope of targeted therapies, due to rational drug design instead of high-throughput screening and the increased target space including non-druggable targets and non-coding RNA [5,9]. Meanwhile, the stability, bioavailability, and the delivery across biomembranes still limit siRNA drug development [25–28]. In previous studies, cationic polymer polyethylenimine-modified nanoparticles were used for siRNA delivery to oral cancer cells. PEI-modified mesoporous silica nanoparticles were used for codelivery of doxorubicin and MDR1-siRNA to overcome multidrug resistance for oral squamous carcinoma treatment [29].

Polyethylene glycol-polyethyleneimine-chlorin e6 (PEG-PEI-Ce6) nanoparticles delivered Wnt-1 siRNA to the cytoplasm of KB cells enhanced the cancer cell-killing effect by photodynamic therapy (PDT) [30]. Here, a type of PEI-modified Fe_3O_4 nanoparticle was prepared for siRNA delivery into Ca9-22 oral cancer cells. The zeta potential of the cationic nanoparticle was +46.5 mV, providing satisfied siRNA adsorption capacity and dispersibility, as verified by SEM, TEM, and gel retardation. The gel retardation data indicated that the siRNA targeting BCL2 and BIRC5 could be completely blocked by the nanoparticles of twice siRNA mass, achieved by the high PEI content and amino group density. During siRNA delivery, the PEI coating on the Fe_3O_4 core provided a positive charge, which is essential for siRNA capture and cellular uptake. For siRNA delivery into Ca9-22 cells, the weight ratio of the nanoparticles to the siRNAs was 3:1 to retain the positive charge of the particles, thereby facilitating delivery into the cells. To enhance the delivery efficiency further, the superparamagnetism of the nanoparticles was utilized by applying an external magnetic field when transfection. As presented in the images obtained by Prussian blue staining and FAM labeling, cellular uptake of the nanoparticles and siRNAs was uniform with high-efficiency for both siBCL2 and siBIRC5, which was comparable with commercial in vitro transfection regent Lipofectamine 3000. The superparamagnetic property could be utilized for magnetofection and MIR imaging in future studies.

The siRNA adsorbed by nanoparticles needs to be released after entering cells to play the role of gene silencing by forming an RNA-induced silencing complex (RISC). To verify the effectiveness of the nanoparticle-delivered siRNA in oral cancer cells, designed siRNAs targeting BCL2 and siBIRC5 were used separately. The gene silencing efficiency was verified by quantitative real-time PCR and western blotting at mRNA and protein levels. Both siBCL2 and siBIRC5 achieved high silencing potencies, indicating the efficient siRNA delivery by the Fe_3O_4 nanoparticles into Ca9-22 cells. While the silencing efficiency was lower in oral cancer CAL 27 cells. The gene silencing efficacy of RNAi relies on siRNA design, delivery strategy, and cell properties [31–33]. The same siRNA sequences and delivery methods were used for the two cell lines. The positive control group treated with lipo also showed lower silencing efficiency in CAL 27 cells. Therefore, it was speculated that CAL 27 cells might be more difficult to receive transfection or the activation of the RNAi pathway was lower, leading to the lower efficiency of gene silencing. These phenomena drive us to further optimize the delivery scheme and improve the structure of nanoparticles in future research in more cells and in vivo experiments. In order to further verify the anti-cancer effects of the delivered siRNAs, cell viabilities and migration were proved to be remarkably inhibited by either siBCL2 or siBIRC5. The molecular weight and charge of the synthesized siRNAs are similar, so the delivery system based on the Fe_3O_4 nanoparticles would be suitable for more reasonably designed siRNA delivery and would not be limited to cancer therapy. In summary, we have successfully developed a novel therapeutic strategy for oral cancer, as well as a simple and universal siRNA delivery system for more applications.

Author Contributions: Conceptualization, L.J., Q.W., and D.Z.; investigation, L.J., J.C., and Z.W.; writing—original draft preparation, D.Z. and H.X.; writing—review and editing, L.J. and Q.W.

Funding: This research was funded by the National Natural Science Foundation of China, grant number 31071916 and 81703102, the Scientific Research Program for Public Welfare of Liaoning Province, grant number 20170022, and the Foundation of China Medical University (XZR20160022).

Conflicts of Interest: The authors declare no conflict of interest.

References

1. Bray, F.; Ferlay, J.; Soerjomataram, I.; Siegel, R.L.; Torre, L.A.; Jemal, A. Global cancer statistics 2018: GLOBOCAN estimates of incidence and mortality worldwide for 36 cancers in 185 countries. *CA Cancer J. Clin.* **2018**, *68*, 394–424. [CrossRef] [PubMed]
2. Ferlay, J.; Colombet, M.; Soerjomataram, I.; Mathers, C.; Parkin, D.M.; Pineros, M.; Znaor, A.; Bray, F. Estimating the global cancer incidence and mortality in 2018: GLOBOCAN sources and methods. *Int. J. Cancer* **2019**, *144*, 1941–1953. [CrossRef] [PubMed]

3. Cheraghlou, S.; Schettino, A.; Zogg, C.K.; Judson, B.L. Changing prognosis of oral cancer: An analysis of survival and treatment between 1973 and 2014. *Laryngoscope* **2018**, *128*, 2762–2769. [CrossRef] [PubMed]
4. Li, C.C.; Shen, Z.; Bavarian, R.; Yang, F.; Bhattacharya, A. Oral Cancer: Genetics and the Role of Precision Medicine. *Dent. Clin. N. Am.* **2018**, *62*, 29–46. [CrossRef] [PubMed]
5. Marcazzan, S.; Varoni, E.M.; Blanco, E.; Lodi, G.; Ferrari, M. Nanomedicine, an emerging therapeutic strategy for oral cancer therapy. *Oral Oncol.* **2018**, *76*, 1–7. [CrossRef] [PubMed]
6. Setten, R.L.; Rossi, J.J.; Han, S.P. The current state and future directions of RNAi-based therapeutics. *Nat. Rev. Drug Discov.* **2019**, *18*, 421–446. [CrossRef]
7. Kim, D.H.; Rossi, J.J. Strategies for silencing human disease using RNA interference. *Nat. Rev. Genet.* **2007**, *8*, 173–184. [CrossRef]
8. Hu, B.; Weng, Y.; Xia, X.H.; Liang, X.J.; Huang, Y. Clinical advances of siRNA therapeutics. *J. Gene Med.* **2019**, *21*, e3097. [CrossRef]
9. Yin, H.; Kanasty, R.L.; Eltoukhy, A.A.; Vegas, A.J.; Dorkin, J.R.; Anderson, D.G. Non-viral vectors for gene-based therapy. *Nat. Rev. Genet.* **2014**, *15*, 541–555. [CrossRef]
10. Dong, Y.; Siegwart, D.J.; Anderson, D.G. Strategies, design, and chemistry in siRNA delivery systems. *Adv. Drug Dev. Rev.* **2019**. [CrossRef]
11. Bochicchio, S.; Dapas, B.; Russo, I.; Ciacci, C.; Piazza, O.; De Smedt, S.; Pottie, E.; Barba, A.A.; Grassi, G. In vitro and ex vivo delivery of tailored siRNA-nanoliposomes for E2F1 silencing as a potential therapy for colorectal cancer. *Int. J. Pharm.* **2017**, *525*, 377–387. [CrossRef] [PubMed]
12. Wang, R.; Degirmenci, V.; Xin, H.; Li, Y.; Wang, L.; Chen, J.; Hu, X.; Zhang, D. PEI-Coated Fe_3O_4 Nanoparticles Enable Efficient Delivery of Therapeutic siRNA Targeting REST into Glioblastoma Cells. *Int. J. Mol. Sci.* **2018**, *19*, 2230. [CrossRef] [PubMed]
13. Zhang, D.; Wang, J.; Wang, Z.; Wang, R.; Song, L.; Zhang, T.; Lin, X.; Shi, P.; Xin, H.; Pang, X. Polyethyleneimine-Coated Fe3O4 Nanoparticles for Efficient siRNA Delivery to Human Mesenchymal Stem Cells Derived from Different Tissues. *Sci. Adv. Mater.* **2015**, *7*, 1058–1064. [CrossRef]
14. Delbridge, A.R.; Grabow, S.; Strasser, A.; Vaux, D.L. Thirty years of BCL-2: Translating cell death discoveries into novel cancer therapies. *Nat. Rev. Cancer* **2016**, *16*, 99–109. [CrossRef]
15. Radha, G.; Raghavan, S.C. BCL2: A promising cancer therapeutic target. *Biochim. Biophys. Acta* **2017**, *1868*, 309–314. [CrossRef]
16. Knight, T.; Luedtke, D.; Edwards, H.; Taub, J.W.; Ge, Y. A delicate balance—The BCL-2 family and its role in apoptosis, oncogenesis, and cancer therapeutics. *Biochem. Pharmacol.* **2019**, *162*, 250–261. [CrossRef]
17. Alam, M.; Kashyap, T.; Pramanik, K.K.; Singh, A.K.; Nagini, S.; Mishra, R. The elevated activation of NFkappaB and AP-1 is correlated with differential regulation of Bcl-2 and associated with oral squamous cell carcinoma progression and resistance. *Clin. Oral Investig.* **2017**, *21*, 2721–2731. [CrossRef]
18. Xiong, L.; Tang, Y.; Liu, Z.; Dai, J.; Wang, X. BCL-2 inhibition impairs mitochondrial function and targets oral tongue squamous cell carcinoma. *Springerplus* **2016**, *5*, 1626. [CrossRef]
19. Wheatley, S.P.; Altieri, D.C. Survivin at a glance. *J. Cell Sci.* **2019**, *132*. [CrossRef]
20. Martinez-Garcia, D.; Manero-Ruperez, N.; Quesada, R.; Korrodi-Gregorio, L.; Soto-Cerrato, V. Therapeutic strategies involving survivin inhibition in cancer. *Med. Res. Rev.* **2019**, *39*, 887–909. [CrossRef]
21. Troiano, G.; Guida, A.; Aquino, G.; Botti, G.; Losito, N.S.; Papagerakis, S.; Pedicillo, M.C.; Ionna, F.; Longo, F.; Cantile, M.; et al. Integrative Histologic and Bioinformatics Analysis of BIRC5/Survivin Expression in Oral Squamous Cell Carcinoma. *Int. J. Mol. Sci.* **2018**, *19*, 2664. [CrossRef] [PubMed]
22. Calatayud, M.P.; Riggio, C.; Raffa, V.; Sanz, B.; Torres, T.E.; Ibarra, M.R.; Hoskins, C.; Cuschieri, A.; Wang, L.; Pinkernelle, J.; et al. Neuronal cells loaded with PEI-coated Fe3O4 nanoparticles for magnetically guided nerve regeneration. *J. Mater. Chem. B* **2013**, *1*, 3607–3616. [CrossRef]
23. Zhang, D.; Wang, R.; Wang, Z.; Dong, J.; Shi, P.; Zhang, T.; Lin, X.; Pang, X. In Vitro Toxicity of PEI-Coated Fe3O4 Nanoparticles in HaCaT Cells. *Mater. Focus* **2014**, *3*, 145–148. [CrossRef]
24. Denais, C.M.; Gilbert, R.M.; Isermann, P.; McGregor, A.L.; te Lindert, M.; Weigelin, B.; Davidson, P.M.; Friedl, P.; Wolf, K.; Lammerding, J. Nuclear envelope rupture and repair during cancer cell migration. *Science* **2016**, *352*, 353–358. [CrossRef] [PubMed]
25. Gomes-da-Silva, L.C.; Fonseca, N.A.; Moura, V.; Pedroso de Lima, M.C.; Simoes, S.; Moreira, J.N. Lipid-based nanoparticles for siRNA delivery in cancer therapy: Paradigms and challenges. *Acc. Chem. Res.* **2012**, *45*, 1163–1171. [CrossRef]

26. Selvam, C.; Mutisya, D.; Prakash, S.; Ranganna, K.; Thilagavathi, R. Therapeutic potential of chemically modified siRNA: Recent trends. *Chem. Biol. Drug Des.* **2017**, *90*, 665–678. [CrossRef]
27. Kaur, K.; Rath, G.; Chandra, S.; Singh, R.; Goyal, A.K. Chemotherapy with si-RNA and Anti-Cancer Drugs. *Curr Drug Deliv.* **2018**, *15*, 300–311. [CrossRef]
28. Maheshwari, R.; Tekade, M.; Gondaliya, P.; Kalia, K.; D'Emanuele, A.; Tekade, R.K. Recent advances in exosome-based nanovehicles as RNA interference therapeutic carriers. *Nanomedicine* **2017**, *12*, 2653–2675. [CrossRef]
29. Wang, D.; Xu, X.; Zhang, K.; Sun, B.; Wang, L.; Meng, L.; Liu, Q.; Zheng, C.; Yang, B.; Sun, H. Codelivery of doxorubicin and MDR1-siRNA by mesoporous silica nanoparticles-polymerpolyethylenimine to improve oral squamous carcinoma treatment. *Int. J. Nanomed.* **2018**, *13*, 187–198. [CrossRef]
30. Ma, C.; Shi, L.; Huang, Y.; Shen, L.; Peng, H.; Zhu, X.; Zhou, G. Nanoparticle delivery of Wnt-1 siRNA enhances photodynamic therapy by inhibiting epithelial-mesenchymal transition for oral cancer. *Biomater. Sci.* **2017**, *5*, 494–501. [CrossRef]
31. Liu, T.; Wang, L.; Xin, H.; Jin, L.; Zhang, D. Delivery Systems for RNA Interference-Based Therapy and Their Applications Against Cancer. *Sci. Adv. Mater.* **2020**, *12*, 75–86. [CrossRef]
32. Reynolds, A.; Leake, D.; Boese, Q.; Scaringe, S.; Marshall, W.S.; Khvorova, A. Rational siRNA design for RNA interference. *Nat. Biotechnol.* **2004**, *22*, 326–330. [CrossRef] [PubMed]
33. Sioud, M.; Sorensen, D.R. Cationic liposome-mediated delivery of siRNAs in adult mice. *Biochem. Biophys. Res. Commun.* **2003**, *312*, 1220–1225. [CrossRef] [PubMed]

Article

Design of New Polyaspartamide Copolymers for siRNA Delivery in Antiasthmatic Therapy

Emanuela Fabiola Craparo [1], Salvatore Emanuele Drago [1], Nicolò Mauro [1,2], Gaetano Giammona [1] and Gennara Cavallaro [1,*]

1 Lab of Biocompatible Polymers, Department of Biological, Chemical and Pharmaceutical Sciences and Technologies (STEBICEF), University of Palermo, via Archirafi 32, 90123 Palermo, Italy; emanuela.craparo@unipa.it (E.F.C.); salvatoreemanuele.drago@unipa.it (S.E.D.); nicolo.mauro@unipa.it (N.M.); gaetano.giammona@unipa.it (G.G.)

2 Fondazione Umberto Veronesi, Piazza Velasca 5, 20122 Milano, Italy

* Correspondence: gennara.cavallaro@unipa.it; Tel.: +39-091-2389-1931

Received: 3 December 2019; Accepted: 18 January 2020; Published: 22 January 2020

Abstract: Here, a novel protonable copolymer was realized for the production of polyplexes with a siRNA (inhibitor of STAT6 expression in asthma), with the aim of a pulmonary administration. The polycation was synthesized by derivatization of α,β-poly(N-2-hydroxyethyl)D,L-aspartamide (PHEA) with 1,2-Bis(3-aminopropylamino)ethane (bAPAE) in proper conditions to obtain a PHEA-*g*-bAPAE graft copolymer with a derivatization degree in amine (DD_{bAPAE}%) equal to 35 mol%. The copolymer showed a proper buffering behavior, i.e., ranging between pH 5 and 7.4, to potentially give the endosomal escape of the obtained polycations. In effect, an in vitro experiment demonstrated the effect on biological membranes of the copolymer on bronchial epithelial cells (16-HBE) strongly dependent on the pH of the medium, i.e., higher at pH 5. bAPAE-based copolymers were further obtained with an increasing pegylation degree, i.e., equal to 1.9, 2.7, and 4.4 mol%, respectively. All the obtained copolymers were able to complex siRNA at a N/P ratio that decreases as the pegylation degree increases. At the same time, the tendency of polyplexes to aggregate and the capability to interact with mucin also decreases as the pegylation in the copolymer increases. Gene silencing experiments on 16-HBE showed that these copolymers have a significant role in improving the intracellular transport of naked siRNA, where the presence of PEG does not seem to hinder the cellular uptake of polyplexes. The latter obtained at polymer/siRNA weight ratio (R) equal to 10 with PHEA-*g*-PEG(C)-*g*-bAPAE also seems to be not susceptible to the presence of mucin, avoiding the polyanionic exchange of complexed siRNA, thus showing adequate behavior to be used as an effective vector for siRNA.

Keywords: siRNA; STAT6; polyaspartamide; pegylation; polyamine; polyplexes; asthma

1. Introduction

Although it is not a deadly disease, asthma can be highly debilitating, resulting in irreversible lung damage [1]. The international guidelines of the Global Initiative for Asthma (GINA) define asthma as a pathological state characterized by chronic inflammation of the airways and reversible limitation of air flow. However, persistent lung inflammation could cause airway obstruction or hyper-reactivity, so that when the treatment is inadequate, airway re-modelling can occur and the obstruction becomes irreversible [1]. The inflammatory cascade in allergic asthma, which involves the lung infiltration of eosinophils, T lymphocytes, mast cells, and other inflammatory cells, is schematically summarized in Figure 1.

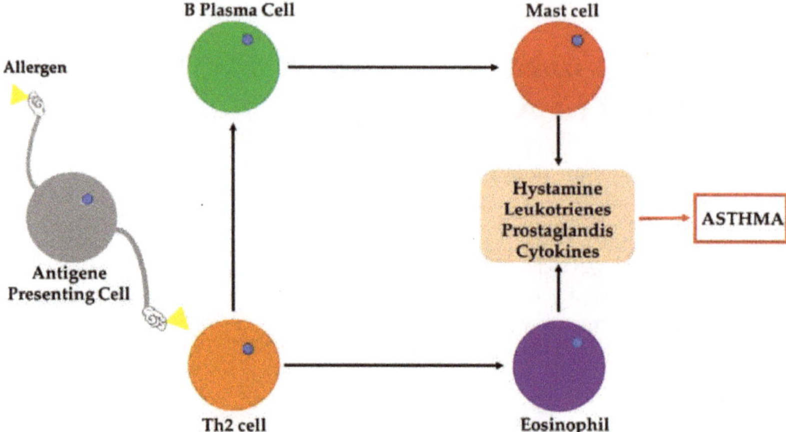

Figure 1. Schematic representation of the inflammatory cascade in allergic asthma.

The conventional drug therapy includes both bronchodilators (such as β-2 adrenergic agonists and anticholinergics) to control symptoms, and corticosteroids to reduce the inflammatory process. Inhaled corticosteroids represent an effective first-line treatment in mild asthma, but they have therapeutic limitations in patients with moderate to severe asthma due to several side effects, although the local administration allows to significantly reduce the administered dose [2,3].

The conventional therapies, therefore, are often associated with a series of limitations that have pushed the research towards the identification of new biological targets for treatment of the pathology. It has been seen that in individuals suffering from asthma there is overexpression of numerous genes and proteins, such as Signal Transducer and Activator of Transcription 6 (STAT6) [4], Plasminogen Activator Inhibitor-1 (PAI-1) [5], and Spleen Tyrosine Kinase (Syk) [6]. In particular, STAT6 regulates the T helper type 2 (Th_2) immune response, PAI-1 is associated with asthma severity because of its role in airway remodeling, while Syk is implicated as central immune modulator promoting allergic airway inflammation; thus, their inhibition could be an effective therapeutic approach in asthma.

For these reasons, an alternative therapeutic approach of pathological states caused by an increase in the expression of some genes, as in the case of asthma, could be gene therapy, through the administration of small interference RNA (siRNA). These consists of small double-stranded RNA fragments capable of triggering the degradation of a specific mRNA [7] and therefore ultimately capable of blocking the synthesis of certain proteins. Recently, the true efficacy of siRNA directed against particular targets, such as STAT6, PAI-1, and Syk for the treatment of asthma was demonstrated through appropriate in vitro and in vivo studies [4–6].

However, despite the potential of this genetic material, it is well known that it cannot be administered as it is, also if given directly in situ, but requires the use of particular vectors capable of conveying it into the body [7,8]. The use of siRNA as therapeutic agents necessarily requires the use of a vector, which by neutralizing the negative charge may allow them to enter cells, as well as increasing their stability against enzymatic degradation by nucleases. Among gene vectors, polymeric materials have many advantages, as they can carry large quantities of genetic material and can be chemically derivatized to obtain systems specifically oriented towards particular target tissues [9,10]. The cationic character of such polymers, necessary for the establishment of interactions with negative-charged gene material to form polyplexes, is conferred by protonable amino groups at physiological or neutral pH. However, other peculiarities are often required of the polymeric material, therefore a starting material easily modifiable by simple chemical reactions is highly sought after.

Considering the above, the aim of the present experimental work was to realize a novel protonable polymeric derivative able to form a stable electrostatic complex with the chosen siRNA, and to delivery it, through the inhalation route, at the bronchial level for the realization of an innovative formulation for the management of asthma. We have chosen to work with the α,β-poly-N-2-hydroxyethyl-DL-aspartamide (PHEA) as the starting polymeric material [11], being a highly soluble in water, biocompatible, non-immunogenic, non-antigenic polymer, already used for the development of highly performing polymeric gene vectors [12,13], as well as many other drug carriers [14–17]. Moreover, as molecule to conjugate to the PHEA backbone, in order to give the cationic behavior, we have chosen an oligoamine, the 1,2-Bis(3-aminopropylamino)ethane (bAPAE), that could give a good complexing capability, an improvement of cell internalization, associated with a good cytocompatibility [18]. As genetic material has been chosen a therapeutic siRNA able to reduce the expression of STAT6, that is one of the most important transcription factors that regulate the production of Th2 cytokines and effector functions mediated by Th2 cytokines [19–22], which seems to have a major role in the mechanism that initiates an asthmatic attack.

In addition to complexing ability to the genetic material, the copolymer forming the siRNA complex should possess additional characteristics, after administration by the pulmonary route, to further diffuse trough the mucus layer present at the airway level, until reach the bronchial epithelial cells. The latter are the main target for the siRNA delivery. To confer penetrating mucus capacity to the system it is possible to increase the superficial hydrophilicity of the system, thus reducing interactions with the protein chains of the mucin; for this reason, the PHEA backbone was also conjugated with a proper amount of poly(ethyleneglycole) (PEG) [23,24]. Therefore the potential of this new polyaspartamide copolymers as material able to complex and delivery a specific siRNA for antiasthmatic therapy was tested.

2. Materials and Methods

2.1. Materials

Triethylamine (TEA), Bis(4-nitrophenyl)carbonate (BNPC), anhydrous *N,N'*-dimethylformamide (a-DMF), 1,2-Bis(3-aminopropylamino)ethane (bAPAE), *O*-(2-Aminoethyl)-*O'*-methyl poly(ethylene glycol) 2000 (H_2N-PEG_{2000}) (0.4 mmol NH_2/g), disuccinimidylcarbonate (DSC), dichloromethane, aceton, diethylether, 2,4,6-Trinitrobenzenesulfonic acid (TNBS), agarose, ethidium bromide, mucine from porcin stomach were purchased from Sigma-Aldrich (Milan, Italy). All used reagents were of analytic grade.

Duplexed siRNA were purchased from Biomers.net (Ulm, Germany). The gene target sequences (5'→3') are: CAGUUCCGCCACUUGCCAA (sense), UUGGCAAUGGCGGAACUG (antisense).

α,β-poly(N-2-hydroxyethyl)-D,L-aspartamide (PHEA) was synthetized via polysuccinimide (PSI) reaction with ethanolamine in DMF solution, and purified according to a previously reported procedure [11].

^{1}H-NMR (300 MHz, D_2O, 25°C, TMS): δ 2.71 (m, $2H_{PHEA}$, –COCHC$\mathbf{H_2}$CONH–), δ 3.24 (m, $2H_{PHEA}$, –NHC$\mathbf{H_2}$$CH_2$O–), δ 3.55 (m, $2H_{PHEA}$, –NHCH_2C$\mathbf{H_2}$OH), δ 4.59 [m, $1H_{PHEA}$, –NHC**H**(CO)CH_2–].

2.2. Copolymer Synthesis

2.2.1. General Procedure for the Derivatization and Characterization of α,β-poly(*N*-2-hydroxyethyl)D,L-aspartamide with 1,2-Bis(3-aminopropylamino)ethane (PHEA-*g*-bAPAE)

Derivatization of PHEA with 1,2-Bis(3-aminopropylamino)ethane (bAPAE) was carried out by using Bis(4-nitrophenyl) carbonate (BNPC) as coupling agent. Two hundred milligrams of PHEA (1.26 mmol of repeating units (RU)) were dissolved in 4 mL of a-DMF; after complete solubilization, 230 mg of solid BNPC was added. The solution was stirred at 40 °C for 4 h. Simultaneously, 922.71 μL of bAPAE was dissolved in 7 mL of a-DMF. After activation time, the resulting polymeric solution was added dropwise and slowly to bAPAE solution. The reaction was carried out under and continuous stirring at 25 °C for 20 h. The amounts of each reagent were properly determined accordingly to

R_1 = (mmol of BNPC/mmol of functionalizable RU on PHEA) = 0.6 and R_2 = (mmol of bAPAE/mmol of functionalizable RU on PHEA) = 4.

After this time, the polymer was isolated from reaction mixture by precipitation in mixture 2:1 *v/v* diethyl ether/dichloromethane and the supernatant was removed by centrifugation at 4 °C for 8 min, at 9800 rpm. The obtained solid product was washed with acetone, until the pH of a mixture between the washing acetone with water (vol:vol 1:1) was neutral. Then, the obtained product was dried under vacuum. The solid residue was dissolved in double distilled water and then the solution was purified by dialysis (SpectraPor Dialysis Tubing, at MWCO 25 kDa), for two days against basic water (NaOH) and for other three days against bidistilled water, subsequently the solution was freeze-dried and stored for further characterization. PHEA-*g*-bAPAE graft copolymer was obtained with a yield of 80 wt % based on the starting PHEA.

^{1}H-NMR (300 MHz, D_2O pD 5, 25 °C, TMS): δ 1.70–2.20 (m, $4H_{bAPAE}$, $-NHCH_2CH_2CH_2NHCH_2CH_2NHCH_2CH_2CH_2NH-$), δ 2.73 (m, $2H_{PHEA}$, $-COCHCH_2CONH-$), δ 3,12 (m, $8H_{bAPAE}$, $-NHCH_2CH_2CH_2NHCH_2CH_2NHCH_2CH_2CH_2NH_2$), δ 3.23 (m, $2H_{PHEA}$, $-NHCH_2CH_2O-$), 3,38 (m, $4H_{bAPAE}$, $-CONHCH_2CH_2-$, $-CH_2CH_2CH_2NH_2$), δ3.54 (m, $2H_{PHEA}$, $-NHCH_2CH_2OH$), δ 3.60 (m, $4H_{PEG}$, $-[OCH_2CH_2O]_{44}-$), δ 4.02 (m, $2H_{PHEA}$, $-NHCH_2CH_2OCO-$), δ 4.62 (m, $1H_{PHEA}$, $-NHCH(CO)CH_2-$). The content of amine was also determined by TNBS assay [12].

2.2.2. General Procedure for the Derivatization and Characterization of PHEA with methoxy polyethylene glycol amine (H_3CO-PEG-NH_2)

Derivatization of PHEA with different amount of H_3CO-PEG-NH_2, was carried out by using N,N'-disuccinimidyl carbonate (DSC) as coupling agent [23]. Five hundred milligrams of PHEA (6.32 mmol of RU) was dissolved in 10 mL of a-DMF at 40 °C and then a proper amount of triethylamine (TEA), as catalyst, and DSC were added; subsequently, the reaction mixture was left at 40 °C for 4 h.

After the activation time, the latter dispersion of DSC-activated PHEA was added drop-wise to increasing volumes of H_3CO-PEG-NH_2 dispersions in a-DMF, at a concentration of 50 mg/mL. Then, the obtained mixture reactions were left at 25 °C for 18 h. The amounts of TEA, DSC and PEG were added according to the following moles ratios, as reported in Table 1.

Table 1. Molar ratio and mL of poly(ethyleneglycole) (PEG) solution (50 mg/mL) used for synthesis.

Copolymers	R_3	R_4	R_5	PEG Solution (mL)	PEG Weight Amount (mg)
PHEA-*g*-PEG(A)	0.03	0.04	1	4	200
PHEA-*g*-PEG(B)	0.075	0.1	1	9.5	475
PHEA-*g*-PEG(C)	0.12	0.16	1	15	750

R_3 = (mmol of aminoPEG/mmol of functionalizable RU on PHEA)
R_4 = (mmol of DSC/mmol of functionalizable RU on PHEA)
R_5 = (mmol of TEA/mmol of DSC)

After this time, each polymer was isolated from reaction mixture by precipitation diethyl ether and the supernatant was removed by centrifugation at 4 °C for 8 min, at 9800 rpm. The obtained solid product was washed with acetone one time and then, the obtained product was dried under vacuum. The solid residue was dissolved in double distilled water and then the solution was purified by dialysis (SpectraPor Dialysis Tubing, at MWCO 25 kDa), subsequently freeze-dried and stored for further characterization. PHEA-*g*-PEG graft copolymers were obtained with a yield of 220 wt % based on the starting PHEA. Three different PHEA-*g*-PEG graft copolymers in terms of PEG grafted on the PHEA backbone, that were named PHEA-*g*-PEG(A) and PHEA-*g*-PEG(B) and PHEA-*g*-PEG(C).

^{1}H-NMR (300 MHz, D_2O, 25 °C, TMS): δ 2.71 (m, $2H_{PHEA}$, $-COCHCH_2CONH-$), δ 3.24 (m, $2H_{PHEA}$, $-NHCH_2CH_2O-$), δ3.55 (m, $2H_{PHEA}$, $-NHCH_2CH_2OH$), δ 3.60 (m, $4H_{PEG}$, $-[OCH_2CH_2O]_{44}-$), δ 4.59 (m, $1H_{PHEA}$, $-NHCH(CO)CH_2-$).

2.2.3. General Procedure for the Derivatization and Characterization of PHEA-*g*-PEG with bAPAE

Derivatization of PHEA-*g*-PEG with bAPAE was carried out by using BNPC as coupling agent. 232 mg of PHEA-*g*-PEG(A), 278 mg of PHEA-*g*-PEG(B), or 329 mg of PHEA-*g*-PEG(C) (corresponding to 1.26 mmol of functionalizable RU) was dissolved in 4 mL of a-DMF; after complete solubilization, 230 mg of solid BNPC was added. The solution was stirred at 40 °C for 4 h. Simultaneously, 922.71 µL of bAPAE was dissolved in 7 mL of a-DMF. The reagents were added accordingly to R_1 = (mmol of BNPC/mmol of functionalizable RU on PHEAPEG) = 0.6 and R_7 = (mmol of bAPAE/mmol of functionalizable RU on PHEAPEG) = 4.

After activation time, the resulting polymeric solution was added dropwise and slowly to bAPAE solution. The reaction was carried out under continuous stirring at 25 °C for 20 h. After this time, the polymer was isolated from reaction mixture by precipitation in mixture 2:1 *v/v* diethyl ether/dichloromethane and the supernatant was removed by centrifugation at 4 °C for 8 min, at 9800 rpm. The obtained solid product was washed with acetone, until the pH of the washing surnatant was neutral. Then, the obtained product was dried under vacuum. The solid residue was dissolved in double distilled water and then the solution was purified by dialysis (SpectraPor Dialysis Tubing, at MWCO 25 kDa), subsequently the solution was freeze-dried and stored for further characterization. PHEA-*g*-PEG-*g*-bAPAE graft copolymers were obtained with a yield of 80 wt% based on the starting PHEA-*g*-PEG.

^{1}H-NMR (300 MHz, D_2O pD 5, 25 °C, TMS): δ 1.70–2.20 (m, $4H_{bAPAE}$, $-NHCH_2CH_2CH_2NHCH_2CH_2NHCH_2CH_2CH_2NH-$), δ 2.73 (m, $2H_{PHEA}$, $-COCHCH_2CONH-$), δ 3.12 (m, $8H_{bAPAE}$, $-NHCH_2CH_2CH_2NHCH_2CH_2NHCH_2CH_2CH_2NH_2$), δ 3.23 (m, $2H_{PHEA}$, $-NHCH_2CH_2O-$), 3,38 (m, $4H_{bAPAE}$, $-CONHCH_2CH_2-$, $-CH_2CH_2CH_2NH_2$), δ3.54 (m, $2H_{PHEA}$, $-NHCH_2CH_2OH$), δ 3.60 (m, $4H_{PEG}$, $-[OCH_2CH_2O]_{44}-$), δ 4.02 (m, $2H_{PHEA}$, $-NHCH_2CH_2OCO-$), δ 4.62 (m, $1H_{PHEA}$, $-NHCH(CO)CH_2-$).

2.3. *Determination of the Amine Content*

The content of amine-terminated side chains was also determined by TNBS assay. A stock solution of PHEA-*g*-bAPAE or PHEA-*g*-PEG(A)-*g*-bAPAE or PHEA-*g*-PEG(B)-*g*-bAPAE or PHEA-*g*-PEG(C)-*g*-bAPAE (5 mg/mL) was prepared in a borate buffer (0.1M $Na_2B_4O_7 \cdot H_2O$, pH 9.3). An aliquot of this solution (50 µL) was added to a cuvette containing 900 µL of borate buffer and 50 µL of 0.03 M TNBSA solution. After 120 min incubation, absorbance at λ 500 nm was measured and compared with that estimated for the reaction of H_2N-PEG-OCH_3 ($-NH_2$ in the range between 0.01 and 0.001 mmol/mL) with TNBSA.

2.4. *Size Exclusion Chromatography*

Weight-average molecular weight ($\overline{M}_w$), polydispersity index ($\overline{M}_w/\overline{M}_n$), of each copolymer was determined by a size exclusion chromatography (SEC) analysis, performed using Tosho Bioscience TSK-Gel G4000 PWXL and G3000 PWXL columns(Sursee, Switzerland) connected to an Agilent 1260 Infinity Multi-Detector GPC/SEC system(Santa Clara, United States), and a refractive index detector. Analyses were performed with buffer citrate/phosphate 0.15 M + 0.1 M NaCl pH 5 as eluent with a flow of 1 mL/min and poly(ethylene oxide) standard (40 kDa) to obtain the calibration curve. The column temperature was set at 30 °C.

2.5. *Potentiometric Titration of PHEA-g-bAPAE Graft Copolymer*

2.5.1. Qualitative Titration of PHEA-*g*-bAPAE Copolymer

To determine the relative buffering capacity of PHEA-*g*-bAPAE copolymer, potentiometric acid–base titrations were performed. Typically, 6 mg of copolymer was dissolved in 30 mL in 0.1 N NaCl, used as ionic strength stabilizer, and the pH was adjusted to nearly 10.0 using 0.1 N sodium hydroxide. Then the mixture was titrated by gradually adding 20 µL of 0.1 N HCl until reaching pH 3.

Titrations of comparable amounts of PHEA, and bAPAE, calculated considering the derivatization degree of the PHEA-*g*-bAPAE copolymer, at the same concentration present in PHEA-bAPAE, were also studied.

2.5.2. Determination of the pKa Values of PHEA-*g*-bAPAE Copolymer by Potentiometric Titration

30 mg of PHEA-*g*-bAPAE were dissolved in 0.1 N degassed NaCl (30 mL), used as ionic strength stabilizer, and termostated at 25 °C under argon atmosphere. The solution was then titrated using 0.05 N HCl until pH 3 under inhert conditions. Backward titrations were performed using 0.05 N NaOH. For all titrations an AMEL 631 differential electrometer was used, which was calibrated against a set of multiple standard buffers ($2.50 \pm 0.01 \leq pH \leq 10.00 \pm 0.01$). The pKa values of the amine groups were extrapolated using the De Levie method of acid–base chemical equilibria for polyelectrolytes.

2.6. Biological Studies

2.6.1. Cell Culture

In this study, an immortalized normal bronchial epithelial cell line (16-HBE) was used (furnished by Istituto Zoo-profilattico of Lombardia and Emilia Romagna). 16-HBE cells were maintained in a humidified atmosphere of 5% CO_2 in air at 37 °C, cultured as adherent monolayers in Dulbecco's Modified Eagle's medium (DMEM) (EuroClone), supplemented with 10% fetal bovine serum (FBS) (Gibco), 2 mM L-glutamine (EuroClone), 100 U/mL penicillin, 100 g/mL streptomycin, and 0.6 g/mL amphotericin B (Sigma–Aldrich, Milan, Italy).

2.6.2. Membrane Destabilization Study

16-HBE cells were plated on a 48-well plate at a cell density of 10.000 cells/well in DMEM containing 10% FBS. After 24 h of incubation, the medium was removed and then the cells were incubated with 100 µL of DPBS (pH 7.4) or 20 mM MES (pH 5.5, 130 mM NaCl) containing PHEA-*g*-bAPAE (0.5 mg/mL) for 20 min at 37 °C; in the same condition, blank analysis was also performed. After this time, 100 µL of Tripan Blue 0.2% was added to each well. After 1 min of incubation, the supernatant was removed and cells were observed with miscroscope (Axio Cam MRm, Zeiss, Oberkochen, Germany). All analysis was performed in triplicate. Simultaneously, 16-HBE cells were plated on a 24-well plate at a cell density of 100,000 cells/well in DMEM containing 10% FBS. After 24 h of incubation, the medium was removed and then the cells were incubated with 100 µL of PBS (pH 7.4) or 20 mM MES (pH 5.5, 130 mM NaCl) containing PHEA-*g*-bAPAE (0.5 mg/mL) for 20 min at 37 °C; in the same condition, blank analysis was also performed. After this time 500 µL of Trypsin was added in each well and after 5 min the content of 3 well, treated with the same condition, was reunited in a small centrifuge tube and mixed with 200 µL of Trypan Blue 0.2%. After 1 min the supernatant was removed by centrifugation at 2000 rpm and the residual pellet was washed with 2 mL of DPBS. After washing, the supernatant was removed by centrifugation at 2000 rpm and the pellet in each tube was treated with 250 µL of TRITON X100 1% and plated on a 96-well plate. After 1 h of incubation, the absorbance at 580 nm was read using a Microplate reader (Multiskan Ex, Thermo Labsystems, Vantaa, Finland). All analysis was performed in duplicate.

2.6.3. Cell Viability Assay

Cell viability was assessed by a MTS assay on 16-HBE cells, using a commercially available kit (Cell Titer 96 Aqueous One Solution Cell Proliferation assay, Promega) containing 3-(4,5-dimethylthiazol-2-yl)-5-(3-carboxymethoxyphenyl)-2-(4-sulphophenyl)-2H-tetrazolium (MTS) and phenazine ethosulfate. 16-HBE cells were plated on a 96-well plate at a cell density of 25,000 cells/well in DMEM containing 10% FBS. After 24 h of incubation, the medium was removed and then the cells were incubated with 200 µL per well with an aqueous dispersion (DMEM containing 10% FBS) of each copolymer at concentrations between 2 mg/mL to 0.0625 mg/mL. All polymers dispersions were sterilized by filtration using 220 nm filter. After 24 and 48 h incubation, the polymers dispersions were removed and each plate was washed with sterile DPBS; after this, cells in each well were incubated with with 100 μL of fresh DMEM and 20 μL of a MTS solution and plates were incubated for 2 h at 37 °C. The absorbance at 490 nm was read using a Microplate reader (Multiskan Ex, Thermo Labsystems, Finland). Relative cell viability (percentage) was expressed as (Abs490 treated cells/Abs490control cells) × 100, on the basis of three experiments conducted in multiple of six. Cells incubated with the medium were used as negative control.

2.7. Complexation Study

Complexation study were evaluated by gel retardation assay and by measurements of size and potential. Polyplex were formed by adding a volume of the copolymer dispersion at different concentrations to the same volume of siRNA solution at a fixed concentration, in order to obtain different polymer/siRNA weight ratios (R); the mixture was mixed by gently pipetting, followed by 30 min incubation at room temperature, before analysis. For gel retardation assay, siRNA/copolymer polyplexes were formed in nuclease free Hepes buffer 10 mM, at pH7.4, containing glucose 5% (*w/v*). siRNA concentration was 0.1 mg/mL and polymer/siRNA weight ratios (R) were: 0, 1, 2, 2.5, 3, 3.5, 4, and 5. Ten microliters of each sample were then loaded on a 1.5% agarose gel, containing 70 mL ethidium bromide and run at 100 V in trisacetate/EDTA (TAE) buffer at pH 8 for 30 min. The gels were then visualized against an UV trans-illuminator and photographed using a digital camera. For dynamic light scattering studies (DLS), siRNA/copolymer polyplexes were formed in nuclease free Hepes buffer 10 mM, at pH 7.4. siRNA concentration was 0.05 mg/mL and R were 0, 1, 2, 2.5, 3, 4, 5, 7, and 10. DLS measurements were performed on 50 μL of sample at 25 °C with a Malvern Zetasizer NanoZS instrument fitted with a 532 nm laser at a fixed scattering angle of 173°, using the Dispersion Technology Software 7.02. For potential, siRNA/copolymer polyplexes were formed in nuclease free Hepes buffer 10 mM, at pH7.4. siRNA concentration was 0.2 mg/mL and R were 0, 1, 2, 2.5, 3, 4, 5, 7, and 10. Four hundred microliters of each sample was diluted with Hepes buffer until 900 μL befeore measure. potential measurements were performed by aqueous electrophoresis measurements, recorded at 25 °C using the same apparatus for DLS measurement. The potential values (mV) were calculated from the electrophoretic mobility using the Smoluchowski relationship.

2.8. Polyplex Stability in Presence of Mucin

2.8.1. Polyanionic Exchange

The stability of polyplexes to polyanionic exchange was determined after polyplexes incubation with mucin dispersion. Polyplexes were prepared as described before in gel retardation assay; the resulting polyplexes (30 μL), were mixed with 5 μL of mucin dispersion (7 mg/mL), in order to have a final mucin concentration of 1 mg/mL, and samples were incubated at room temperature for 2 or 5 h. Gel electrophoresis was then performed as described in complexation study.

2.8.2. Turbidimetric Assay

Measurements of interactions between polyplexes and mucin was determined by turbidimetry. 50 μL of polyplexes, prepared as described in gel retardation assay, were mixed with 50 μL of mucin dispersion at the concentration of 2 mg/mL in Hepes buffer 10 mM pH 7.4. After incubation at 37 °C, the turbidity was measured each 50 min until 6 h approximately. The absorbance at the λ of 500 nm was recorded by Microplate reader (Multiskan Ex, Thermo Labsystems, Finland).

2.9. Gene Silencing Assay

The evaluation of the gene silencing capacity was evaluated by ELISA test, using a IL-4 Human ELISA Kit kits from Life Technologies. 16-HBE cells were plated on a 96-well plate at a cell density of 25,000 cells/well in DMEM containing 10% FBS. After 24 h of incubation, the medium was removed and then the cells were incubated with 200 μL of a polyplexes dispersion (for each well 0.01 nmol of siRNA was used) at different R (3, 5, 10) for 48 h; after this time supernatant was removed ed the cells were incubated with 200 μL of LPS 500 ng/mL for 6 h. After this time cells was washed with DPBS and treated following the protocol provided. For this study polyplexes were formed in OPTIMEM medium using a siRNA 0,1 μM. LPS, copolymers, and siRNA solution were sterilized by filtration using 220 nm filter before analysis.

3. Results and Discussion

3.1. Polymer Synthesis and Characterization

An ideal carrier for achieving the delivery of genetic material into a target tissue must be able, after in vivo administration, to interact with specific cells and to release the genetic material in the cytosol of target cells, overcoming both cellular membrane or the endosomal–lysosomal membrane [7,25]. Synthetic polycations represents in principle valid candidates in this field, thanks to the fact that can be realized with proper structural and functional properties able to confer specific characteristics that a vector of genetic material should have [7]. For this reason, the researchers explored the possibility of producing protonable copolymers with various functionalities in order to confer different properties to a single macromolecule.

Here, a novel polycation derivative of α,β-poly(N-2-hydroxyethyl)-D,L-aspartamide (PHEA) was produced by grafting on the PHEA backbone the 1,2-Bis(3-aminopropylamino)ethane (bAPAE), obtaining the PHEA-*g*-bAPAE graft copolymer. The grafting of bAPAE molecules on PHEA backbone allow to realise a copolymer carrying on the side chains with protonable amines conferring the capability to complex the genetic material by electrostatic interactions.

The reaction involved the activation of free PHEA hydroxyl groups with bis-nitrophenyl carbonate (BNPC), chosing the stoichiometry of reagents in order to obtain deficiency of BNPC over hydroxyl groups of repeating units (6:10). Using this strategy we obtained a suitable amount of activated groups able to further react with amine functions of bAPAE. However, being a polyamine, a huge excess bAPAE was employed is the second step of the reaction, thus avoiding crosslinking owing to multiple nucleofilic attack of side chains. In these experimental conditions, a derivatization degree in bAPAE (DD_{bAPAE}) of PHEA-*g*-bAPAE graft copolymer of about 35 mol% was obtained. The latter was calculated by ^{1}H-NMR analysis by using the ratio between the integral of the signals corresponding to 4H of bAPAE (at δ 1.70 and 2.20 ppm), to the integral of the signal corresponding to 2H of PHEA repeating unit (at δ 2.73 ppm); it was confirmend by the colorimetric TNBS assay, that gives a DD% value superimposable to that obtained by ^{1}H-NMR analysis [12]. The occurring of the conjugation of 1,2-Bis(3-aminopropylamino)ethane (bAPAE) was demonstrated also by the appearance of signal at about δ 4.1, related to the CH_2 of the side chain of functionalized PHEA repeat unit ($NHCH_2\mathbf{CH_2}OCO$-) near to the OCONH bond. The scheme or reaction is reported in Figure 2a.

The obtained copolymer was further characterised by SEC analysis in terms of weight average molecular weight ($\overline{M}_w$) and polydispersity index ($\overline{M}_w/\overline{M}_n$)(for SEC chromatogram, see Figure S1 in Supplementary Materials), and obtained values are reported in Table 2, together with DD_{Bapae} value.

Table 2. Weight-average molecular weight ($\overline{M}_w$), polydispersity index ($\overline{M}_w/\overline{M}_n$), and chemical composition of obtained copolymers.

Copolymers	Molecular Weight		Degree of Derivatization (DD)	
	$\overline{M}_w$ (g/mol)	$\overline{M}_w/\overline{M}_n$	DD_{PEG}	DD_{bAPAE}
PHEA	67500	1.24	—	—
PHEA-*g*-PEG(A)	82,410	1.2	1.9	—
PHEA-*g*-PEG(B)	95,360	1.3	2.7	—
PHEA-*g*-PEG(C)	110,800	1.3	4.4	—
PHEA-*g*-bAPAE	20,921	1.41	—	34
PHEA-*g*-PEG(A)-*g*-bAPAE	25,400	1.32	1.9	35
PHEA-*g*-PEG(B)-*g*-bAPAE	32,100	1.35	2.7	36
PHEA-*g*-PEG(C)-*g*-bAPAE	34,700	1.41	4.4	33

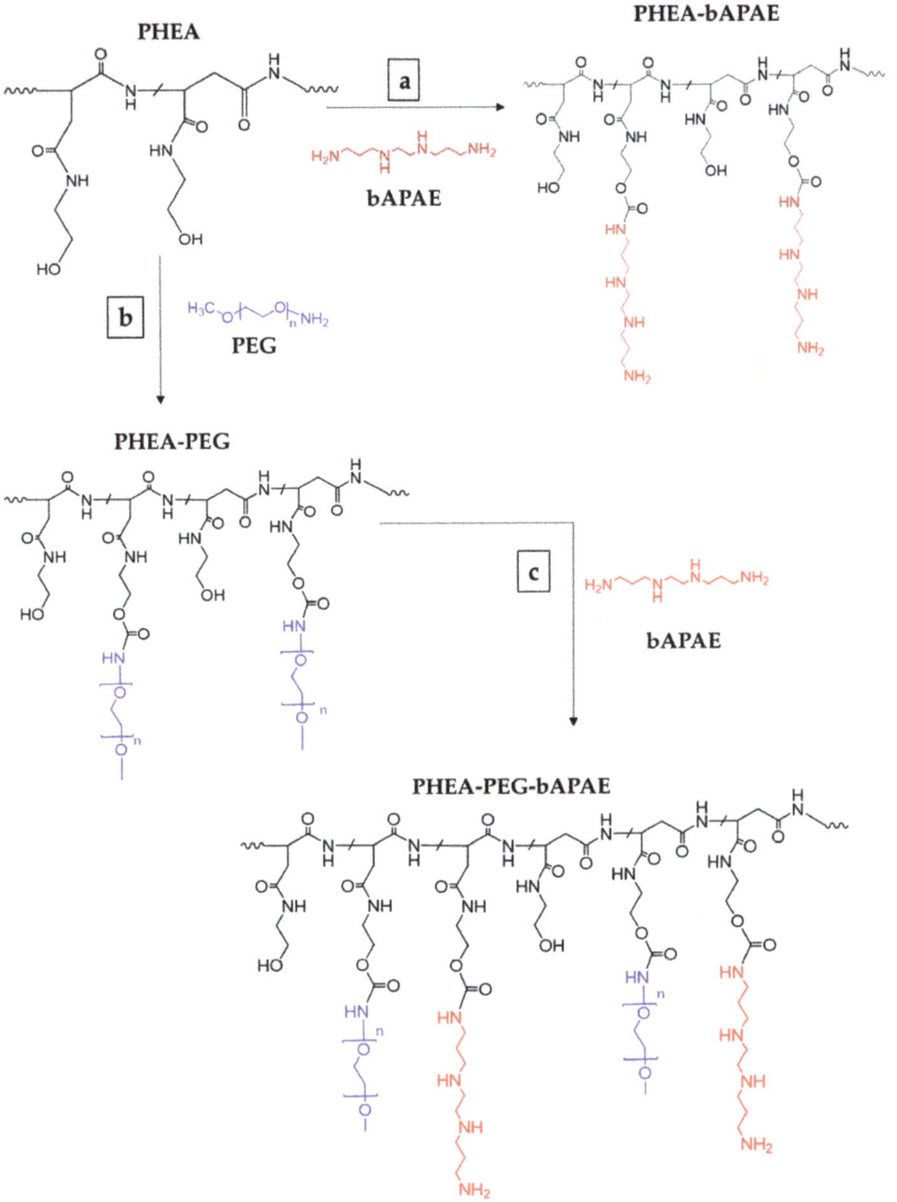

Figure 2. The synthetic route of (**a**) PHEA-*g*-bAPAE and (**b**) PHEA-*g*-PEG and (**c**) PHEA-*g*-PEG-*g*-bAPAE graft copolymers ($n = 44$). Reagents and conditions: a) a-DMF, BNPC, 4 h at 40 °C, 20 h at 25 °C; b) a-DMF, DSC, 4 h at 40 °C, 18 h at 25 °C; b) a-DMF, BNPC, 4 h at 40 °C, 20 h at 25 °C.

As previously evidenced for other reactions of amine with PHEA, the $\overline{M}_w$ undergoes a rather drastic reduction, due to the experimental condition to achieve high degree of functionalizations [26]. This fact could be explained with the use of a high amount of amine (four times compared to the RU of PHEA) for carrying out the functionalization reaction of PHEA with bAPAE, that could break some amide bound in the main chain. However authors can not exclude that also the modification of the copolymer conformation respect to that parent polymer can be the reason of the detection of lower molecular weight of copolymers by SEC. However, the use of materials with low $\overline{M}_w$ is often preferred as polymeric carriers for complexing genetic material [7].

The functionalization of PHEA with bAPAE was done in order to: (a) Make the polymer susceptible to pH changes, that is, modulable in terms of charges to be used for complexing genetic material; and (b) confer it a buffering behavior, that is also a main characteristic request to a genetic vector in order to increase the possibility of giving rise to endosome/lysosome escape, once internalized by cells, due to the so-called proton spoge effect [7].

The buffering behavior of PHEA-*g*-bAPAE graft copolymer was investigated by an acid–base titration. The titration profile was also obtained for PHEA or bAPAE, in aqueous dispersions, at concentrations present in PHEA-*g*-bAPAE dispersion. Data are reported in Figure 3.

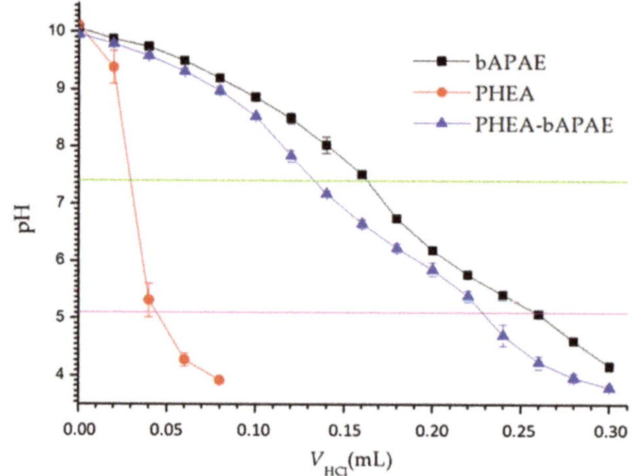

Figure 3. Acid–base titration profiles (pH versus acid volume) of PHEA-*g*-bAPAE (0.2 mg/mL) graft copolymer, PHEA (0.2 mg/mL) and bAPAE moieties (0.047 mg/mL).

As is evidenced in the graphic, PHEA-*g*-bAPAE copolymer shows a buffering capability in the pH interval from 7.4 (extracellular and cytoplasmatic pH) to 5.1 (endosomal/lysosomal pH), important for endosomal escaping with proton sponge effect [7,27,28]. Moreover, the PHEA-*g*-bAPAE titration profile show a buffering behavior quite superimposable to that obtained by the bAPAE alone. On the contrary, the titration curve of PHEA dispersion showed rapid reduction in pH value, suggesting (as expected) no buffering capacity. This result indicates that the conjugation of the amine confers to the polymeric backbone the buffering capability of the amine itself.

A further potentiometric titration was performed in order to determine the K_a constants; titration plot of backward was elaborated using Origin software according to the De Levie method of acid–base equilibria for polyprotic acid and bases, taking into account the activity corrections through the Davies expression (1) [29–32].

$$\log y = -0,5\left(\frac{\sqrt{I}}{1+\sqrt{I}} - 0,3I\right) \tag{1}$$

where y is the activity coefficient and I is the ionic strength.

For this titration, the fitting function obtained is the following (2):

$$V_B = -\frac{V_0[C_0(3\alpha_3 + 2\alpha_2 + \alpha_1) + \Delta] + V_A(\Delta - C_A)}{\Delta + C_B} \tag{2}$$

$$\Delta = C_{H^+} - \frac{K_W}{C_{H^+} \cdot y^2} + C_B \quad (3)$$

where V_B and C_B are respectively the volume and molarity of NaOH for the backword titration, V_0 is the volume of NaCl using for dissolving PHEA-*g*-bAPAE, C_0 is the analyte molarity, V_A and C_A are respectively the volume and molarity of HCl used for the forward titration, while $\alpha_3, \alpha_2, \alpha_1$ are the protonation degree, K_W is the dissociation constant of water and C_{H^+} is the H^+ concentration.

The titration profile and the fitting are reported in Figure 4.

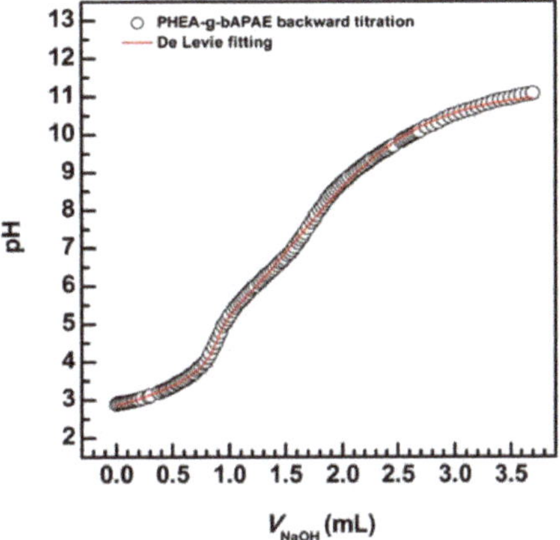

Figure 4. Backward acid–base titration (NaOH volume versus pH) of PHEA-*g*-bAPAE and De Levie fitting curve.

Obtained pK_{a1}, pK_{a2}, and pK_{a3} values from the curve fitting analysis were found, respectively, equal to 4.8, 7.9, and 10.9.

As shown in Figure 5, at pH 7.4 the diprotonated species (L^{+2}) is the mostly present, while, as the pH value of the medium decreases, the amount of the triprotonated species (L^{+3}) begins to increase, until becoming the mostly present species at a pH value approximately of 4.5. This behavior could determine a different effect of the copolymer on the biological membranes dependin on the pH of the medium, i.e., on cytosolic membranes at pH 7.4 and on endosomal membranes at pH 5 (more or less), demonstrated in our next study.

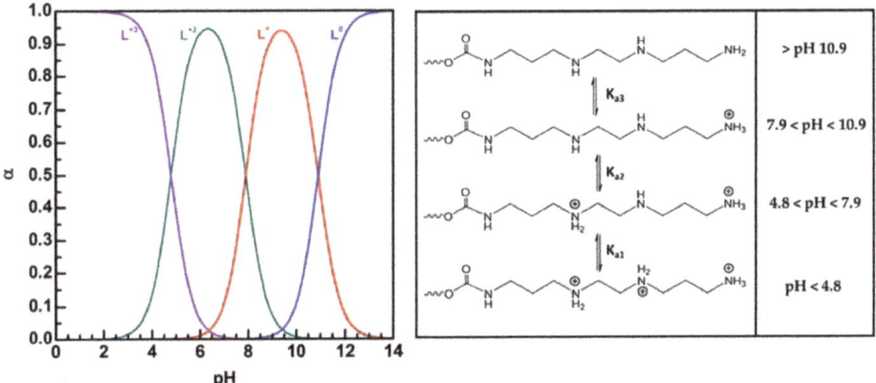

Figure 5. Speciation curves obtained for Plot α versus pH for each α of PHEA-bAPAE PHEA-*g*-bAPAE and protonation state of PHEA-bAPAE PHEA-*g*-bAPAE at different pH values.

3.1.1. Membrane Destabilization Study

Once proved its buffering behaviour, the copolymer effect on cell membranes was evaluated in detail as a function of the pH of the medium, i.e., depending on the amount of protonate amines on its structure, by mimicking the cytosol (pH 7.4) and the lysosomal compartment (pH 5).

Therefore, an in vitro study was carried out by using human bronchial cells (16-HBE) as biological membrane model and by incubating these cells in the presence of PHEA-*g*-bAPAE graft copolymer in an aqueous dispersion at two different pH values of the medium: 7.4 and 5. Trypan blue was used as colorant to distinguish healthy cells from those that are dead or in apoptotic state under the microscope, as it is capable of entering cells only when the membranes are destabilized. After incubation for 20 min, cell were treated with Trypan blue and images were recorded and reported in Figure 6.

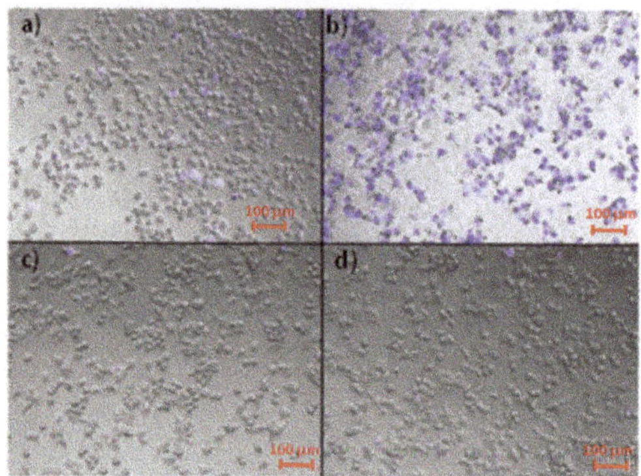

Figure 6. Microscope pictures of 16-HBE treated in different ways: (**a**) cells treated with PHEA-*g*-bAPAE in DPBS pH 7.4. (**b**) cells treated with PHEA-*g*-bAPAE in isotonic MES 20 mM pH 5, (**c**) cells treated with DPBS pH 7.4, (**d**) cells treated with isotonic MES 20 mM pH5.

As shown in figure, considerable cell membrane destabilization occurs only when cell are treated with PHEA-*g*-bAPAE at pH 5, that is in the protonated state, as demonstrated by the fact that trypan blue has free access to all cells. This phenomenon is due to the fact that at pH 5, most of the polymeric species in solution are in the diprotonated state (60%) and in the triprotonated state (40%), while at pH 7.4 most of the polymeric species in solution are in the diprotonated state (75%) and in the monoprotonated state (25%), as showed in Figure 5.

To confirm this result, a quantification of the trypan blue amount was performed by ultraviolet spectrophotometry and obtained values are reported in Figure 7; data shown the maximum absorbance when cells are treated with PHEA-*g*-bAPAE, at pH 5. Even if the quantification does not reflect the same behaviour seen in the microscope images, the results are in any case in agreement with what has been hypothesized, that is the destabilizing effect of the protonated PHEA-*g*-bAPAE graft copolymer on the cellular membrane as a function of pH.

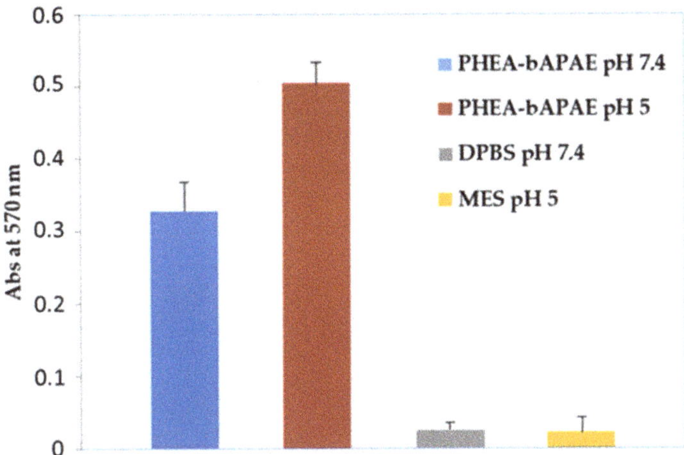

Figure 7. Destabilization assay: absorbance values at 570 nm (trypan blue) in 16-HBE cell dispersion after incubation with PHEA-*g*-bAPAE copolymer at different pH values.

3.1.2. Pegylation

The possibility of using as starting polymer a material that can be easily modified by conjugation with proper molecules is certainly an advantage in cases where it is necessary to take into account the peculiarity of the material to be transported and the barriers to be overcomed after in vivo administration. Here, the PHEA-*g*-bAPAE copolymer was thought to complex siRNA to form polyplexes to be administered by inhalation, then locally to the lung.

Polyethylenglycols (PEG) represent the proper material to influence the properties of the resulting conjugate, thanks to their unique properties such as hydrophilicity and biocompatibility [23]. In most of cases, pegylation of polyplexes provides reduction of toxicity, shielding and stealth properties. However, these improvements are often associated to as a reduction of siRNA condensation, which usually need in using higher amount of polymer [7].

To potentially minimize the tendency to aggregation of polyplexes and interactions with mucus components of lungs, i.e., mucin, PHEA was functionalised with polyethylene glycol (PEG) to obtain PHEA-*g*-PEG graft copolymer. It is well known from previously studies that the mucus-penetrating capability of colloidal carriers through the mucus layer is related to the pegylation degree (DD_{PEG}) [23]. PHEA-*g*-PEG-*g*-bAPAE graft copolymer was synthesized by two reaction steps. In the first step, PHEA was left to react with methoxy(polyethylene glycol) amine (CH_3O-PEG-NH_2) molecules by using disuccinimidyl carbonate (DSC) as coupling agent; while, in the second step, recovered PHEA-*g*-PEG was left to react with bAPAE by using BNPC as coupling agent. The schematic representation of both steps are reported in Figure 1b,c.

In order to obtain copolymers with a different pegylation degree and to evaluate the effect of the PEG amounts on the interactions with mucin of the resulting polyplexes with siRNA, three different theoretical molar ratio values between CH_3O-PEG-NH_2 molecules and PHEA RU (R_1 = 0.03, 0.075, and 0.12) were used to carry out the reaction, thus obtaining three different PHEA-*g*-PEG graft copolymers with increasing derivatization degrees (DD_{PEG} mol %): 1.9, 2.7, and 4.4 mol %, named respectively PHEA-*g*-PEG(A), PHEA-*g*-PEG(B) and PHEA-*g*-PEG(C).

The DD values were calculated by ^{1}H-NMR analysis as the ratio between the integral of the signals corresponding to protons on PEG (at δ 3.60 ppm), to the integral of those corresponding to 2H of PHEA repeating unit (at δ 3.24 ppm).

All the obtained PHEA-*g*-PEG copolymers were characterised in terms of $\overline{M}_w$ and $\overline{M}_w/\overline{M}_n$ (for SEC chromatograms, see Figure S1 in Supplementary Materials), and obtained values are reported in

Table 2, together with chemical composition expressed as DD_{PEG}. The increase in the copolymer $\overline{M}_w$ is in accordance with the theoretical value calculated considering thr starting PHEA $\overline{M}_w$ value and the DD in PEG of each copolymer.

In the second step, PHEA or PHEA-*g*-PEG(A)-(C) graft copolymers were further functionalized with bAPAE by using BNPC, obtaining a DD_{bAPAE} of each PHEA-*g*-PEG-*g*-bAPAE graft copolymers of about 35 mol%. The latter was calculated by ^{1}H-NMR analysis as reported before for PHEA-*g*-bAPAE graft copolymer. The typical ^{1}H-NMR spectrum of a PHEA-*g*-PEG-*g*-bAPAE graft copolymer is reported in Figure 8.

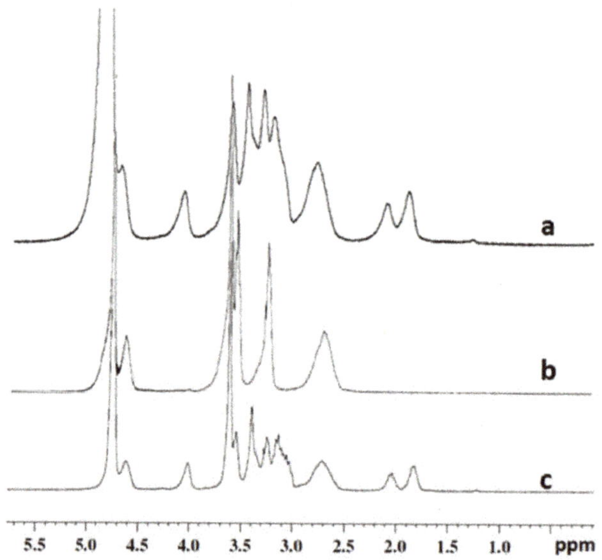

Figure 8. The typical ^{1}H-NMR spectrum of **a**) PHEA-*g*-bAPAE (D_2O at pD < 5), **b**) PHEA-*g*-PEG(C) (D_2O), and **c**) PHEA-*g*-PEG(C)-*g*-bAPAE (D_2O at pD < 5).

The obtained DD mol% was also confirmend by the colorimetric TNBS assay, that gives a DD% value superimposable to that obtained by ^{1}H-NMR analysis. All the obtained copolymers were characterised in terms of $\overline{M}_w$ and $\overline{M}_w/\overline{M}_n$ (for SEC chromatogram, see Figure S1 in Supplementary Materials), and obtained values are reported in Table 2, together with chemical composition espressed as DD_{PEG} and DD_{bAPAE} mol%. In this case, the reduction in the $\overline{M}_w$ of each PHEA-*g*-PEG starting copolymer is in agreement with what described before for PHEA following the reaction with the amine bAPAE. Moreover, the differences between experimental and theoretical $\overline{M}_w$ values of PHEA-*g*-PEG-*g*-bAPAE graft copolymers could be attributed to conformational modifications of obtained copolymers in the aqueous medium depending on the amount of linked PEG.

3.1.3. Cell Viability Assay

Considering the potential application of the PHEA-*g*-bAPAE and/or PHEA-*g*-PEG-*g*-bAPAE graft copolymers as a starting material to realise a formulation for pulmonary administration of siRNA, cytocompatibility of all copolymers was evaluated by the MTS assay on 16-HBE cells at different concentrations, after 24 and 48 incubation. Results are shown in Figure 9a,b.

As can be seen, also after 48 h incubation, all copolymers showed a good cytocompatibility at all tested concentrations, showing a cell viability higher than 80% compared to the control experiment, where cells are incubated only with DMEM medium.

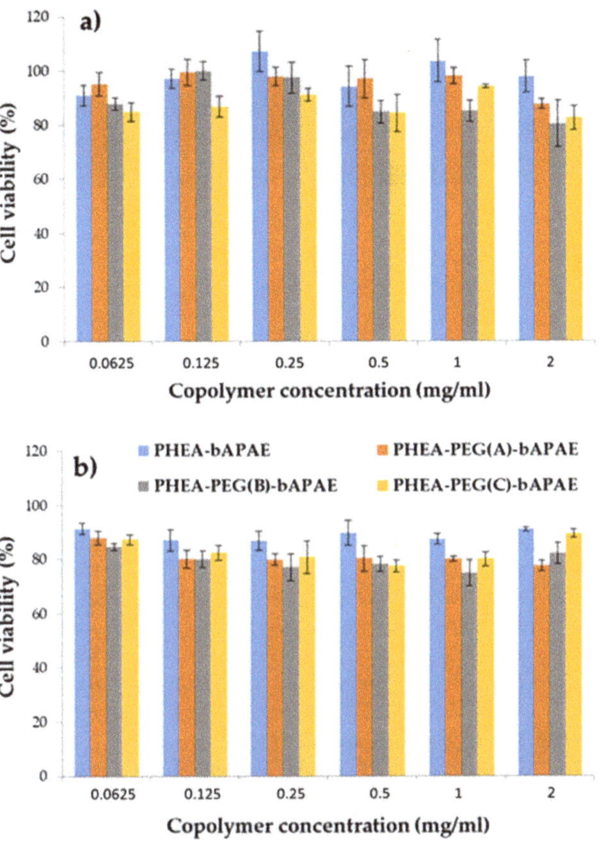

Figure 9. 16-HBE viability assay after (**a**) 24 and (**b**) 48 h of incubation with all copolymers.

3.2. Complexation Studies and Characterization of Obtained Polyplexes

The obtained copolymers, PHEA-*g*-bAPAE, PHEA-*g*-PEG(A)-*g*-bAPAE, PHEA-*g*-PEG(B)-*g*-bAPAE, and PHEA-*g*-PEG(C)-*g*-bAPAE, possess in their chemical structure an equal amount of amine (about 35 mol%) and increasing PEG chains (about 0, 1.9, 2.7, and 4.4 mol%) linked on the PHEA backbone, that could reduce the capability to the polyplexes to interact with mucus components but at the same time, the capability to electrostatically interact with siRNA and to be internalized by cells. For this reason, all four were used for the subsequent evaluation of the siRNA complexing capacity. As therapeutic siRNA, a siRNA able to reduce the expression of STAT6 was chosen, that regulate the production of Th2 cytokines and effector functions mediated by Th2 cytokines [4], and that which seems to have a major role in the mechanism that initiates an asthmatic attack.

To understand if each synthetized copolymer can electrostatically bind negatively charged siRNA molecules, complexation studies were performed. To do this, a mixing of two equal volumes of two dispersions, on containing a fixed concentration of siRNA and the other one containing increasing concentration of each copolymer was done, in order to obtain different polymer/siRNA weight ratios (R) ranging between 1 and 5. To evaluate the formation of stable complexes, an electrophoresis analysis on agarose gel was performed; results are reported in Figure 10.

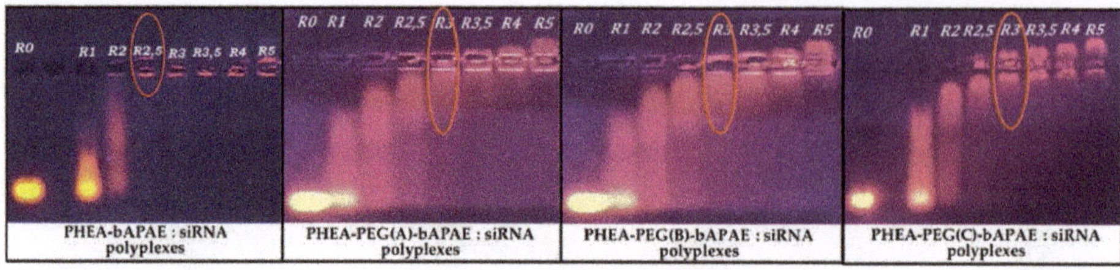

Figure 10. Agarose gel electrophoresis of polyplexes obtained in HEPES 10 mM at various graft copolymer to siRNA weight ratios (R) ranging between 1 and 5.

As can be seen in Figure 10, each PHEA-*g*-PEG-*g*-bAPAE copolymer was able to retard the electrophoresis run of siRNA molecules starting from a polymer/siRNA weight ratio of 3; on the other hand, PHEA-*g*-bAPAE was able to retard the electrophoresis run of siRNA molecules starting from a polymer/siRNA weight ratio of 2,5. This difference probably is due to the fact that the same amounts of PHEA-*g*-bAPAE and PHEA-*g*-PEG-*g*-bAPAE copolymers contain a different amounts of amine groups, that are higher in PHEA-*g*-bAPAE due to the absence of PEG chains in the polymeric backbone. In effect, if the N/P ratio is considered, it means that the unpegylated copolymer stop the siRNA run at N/P equal to 3.5, while for PHEA-*g*-PEG(A)-*g*-bAPAE, PHEA-*g*-PEG(B)-*g*-bAPAE PHEA-*g*-PEG(C)-*g*-bAPAE graft copolymers correspond to a N/P ratio of 3.5, 3.4, and 3, respectively. This result means that the pegylated copolymers were able to retard the electrophoresis run of siRNA molecules at lower N/P ratios than PHEA-*g*-bAPAE graft copolymer, and that increasing the PEG amount improves this behaviour. In all cases, the R need to stop the mobility of siRNA is quite low if compared to other synthetic copolymers proposed as non viral vectors for siRNA terapy, as reported elsewhere [7].

In order to confirm these results, the mean size, PDI and potential values of obtained polyplexes at different R values (ranging between 1 and 10) were measured and obtained data reported in Figure 11 for each synthesized copolymer.

As expected, for PHEA-*g*-bAPAE based polyplexes, a negative potential is recorded for the lowest weight ratio; the potential increases then, up to the point of reversing itself, as a higher quantity of copolymer is mixed with siRNA. Regarding to size profile, higher dimensions of the polyplexes are measured as they present a potential near neutrality, which decrease when the potential of the polyplex increases; this phenomenon is due to the fact that near a zero potential, phenomena of aggregation between polyplexes occur, while when the potential deviates from neutrality, the phenomena of repulsion cause a reduction of this aggregation. For polyplexes obtained with pegylated copolymers, a slightly different behaviour is recorded. First, regarding the potential, there are no high potential values even for the greatest weight ratios, but a stasis is observed in the region of neutrality; this phenomenon gradually increases with the increase in the amount of PEG present in the polymeric backbone. This behaviour is probably due to the ability of the PEG chains to form a shell that shields the surface charges. But at the same time, this shell causes a reduction of the phenomena of aggregation; in fact, there are no large and quite static dimensions, especially for the complexes obtained by PHEA-*g*-PEG(C)-*g*-bAPAE, where the PEG amount is higher. Thus, the presence of PEG in the copolymer allows to obtain polyplexes with siRNA that does not aggregate while maintaining a charge close to neutrality.

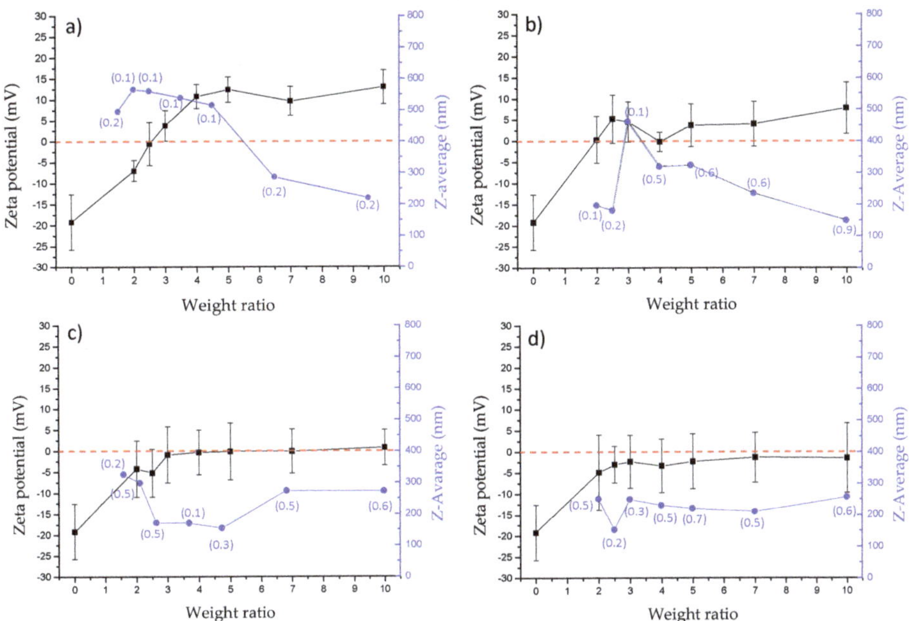

Figure 11. Mean size (blue line), PDI (value enclosed in brakets) and potential (black line) of polyplexes formed by siRNA and: (**a**) PHEA-*g*-bAPAE; (**b**) PHEA-*g*-PEG(A)-*g*-bAPAE; (**c**) PHEA-*g*-PEG(B)-*g*-bAPAE; (**d**) PHEA-*g*-PEG(C)-*g*-bAPAE graft copolymers, at weight ratios (R) ranging between 0 and 10 (data are reported as means SD, $n = 3$).

3.3. Interaction Studies of Polyplexes with Mucin

Being the copolymer/siRNA complexes in view to be administered by the inhalation route, it was necessary to assess whether these polyplexes interact with the mucin and if the presence of PEG in the copolymer structure can effectively influence these interactions, given that the mucus layer represents the main barrier that the inhaled particles must overcome.

The first study was carried out considering that, given the polyanionic nature of the mucin (due to the presence of sialic acid residues), a polyanionic exchange between siRNA and mucin may occur. Thus, a study was carried out by evaluating the electrophoretic mobility of siRNA in the polyplexes in the presence of mucin in the dispersion medium. In particular, the study was carried out in the presence of mucin at a concentration of 1 mg/mL, for 2 and 5 h. Obtained images for R = 5 and 10 are reported in Figure 12.

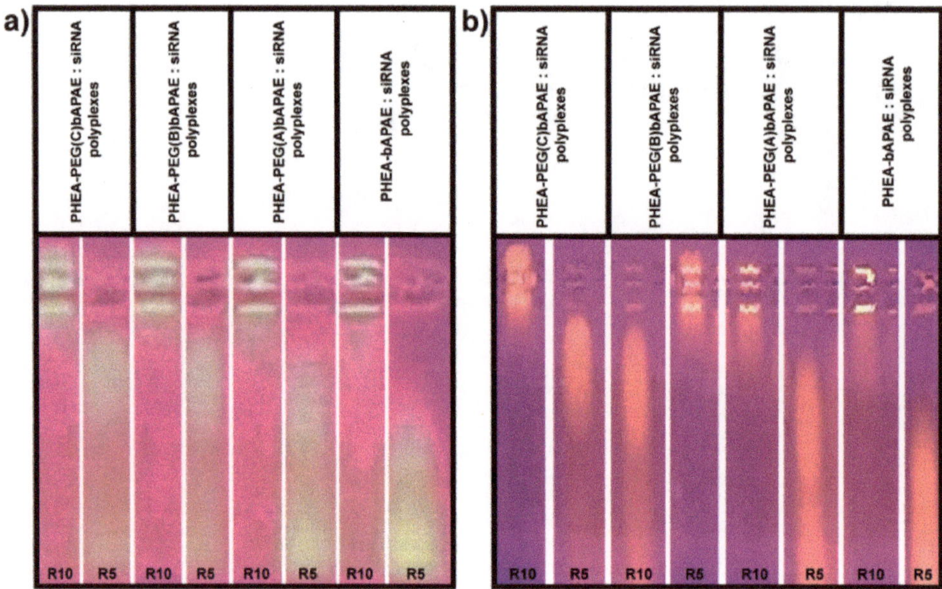

Figure 12. Gel electrophoresis in the presence of mucin, after (**a**) 2 h and (**b**) 5 h of incubation, at R = 5 and 10.

As showed in Figure 12, all the polyplexes give a polyanionic exchange between siRNA and mucin, so that a higher weight ratio (R = 10) needs to stop the electrophoretic run of siRNA respect to that request in the absence of mucin, as reported in Figure 10.

Therefore, the mucin competes with the siRNA for the electrostatic interactions with the copolymer, so that a higher amount of copolymer is request to form polyplexes. In order to obtain more informations regarding the specific interaction of each copolymer forming the polyplex with mucin, a turbidimetric assay was carried out at three different R values (3, 5, and 10) as a function of incubation time, considering that if polyplexes-mucin interactions occur, it involves a reduction of transmittance of the dispersion. Data are reported in Figure 13a–d, as trasmittance % as a function of incubation time.

As can be seen, polyplexes obtained with PHEA-*g*-bAPAE shown a muco-adhesive behaviour, especially as the weight ratio value R increases. The dependence from R is also showed by the polyplexes obtained with the pegylated copolymers. On the other hand, for the pegylated polyplexes, the interaction capability with mucin decreases as the amount of PEG in the polymeric backbone increases to all selected R values. The polyplexes obtained with PHEA-*g*-PEG(C)-*g*-bAPAE seems to be not susceptible to the presence of mucin, showing high trasmittance values at all the incubation times also at higher concentration (corresponding to R = 10). The latter also represents the minimum weight ratio to avoid the polyanionic exchange of complexed siRNA with mucin, thus showing adequate behavior to be used as an effective vector for siRNA.

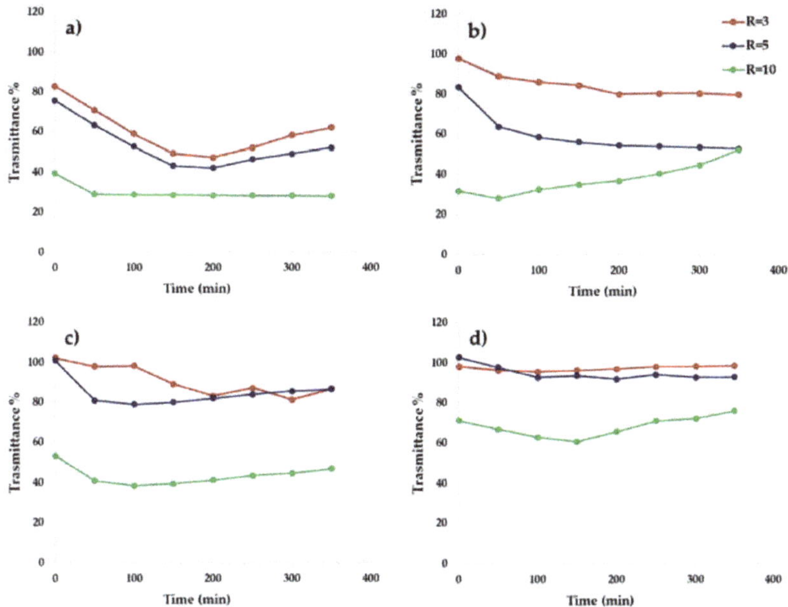

Figure 13. Turbidimetric analysis. Trasmittance at 500 nm of a dispersion containing mucin in the presence of polyplexes between siRNA and: (**a**) PHEA-*g*-bAPAE; (**b**) PHEA-*g*-PEG(A)-*g*-bAPAE; (**c**) PHEA-*g*-PEG(B)-*g*-bAPAE; (**d**) PHEA-*g*-PEG(C)-*g*-bAPAE graft copolymers, at R equal to 3, 5, and 10.

3.4. Gene Silencing Assay

Once demonstrated that these polyplexes are stable the capability to act as an effective carrier for siRNA and to allow the cellular internalization was evaluated. In particular, the gene silencing capacity of obtained polyplexes was evaluated by in vitro ELISA test upon 16-HBE cells. Cells were treated with: a) Naked siRNA and b) polyplexes obtained with each copolymer and at three different R, and then exposed to an inflammatory agent, such as LPS. Relative IL-4 production (percentage) was expressed as (Abs treated cells/Abs positive control cells) × 100. Obtained values are reported in Figure 14.

Obtained results show that cell incubated with naked siRNA produce an IL amount non significantly different from that obtained with control cells. When cells are incubated in the presence of polyplexes, there is a reduction in the expression of IL-4 in all cases, even if not in a massive way. However, this experiment showed that polymeric carrier has a fundamental role in cellular transport and uptake of siRNA. Another important result is given by the fact that presence of PEG does not seem to hinder the cellular uptake of polyplexes, as found for other pegylated non viral polymeric vectors for siRNA [7]. However, at R = 10, among the pegylated polyplexes, the capability to allow the intracellular delivery of siRNA is inversally related to the PEG amount in a significant way, while at the other R no significant differences are observed.

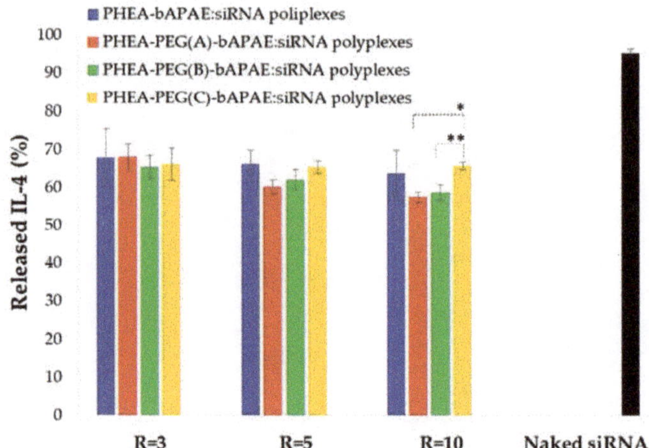

Figure 14. IL-4 (%) released by 16-HBE cells after incubation with polyplexes or with naked siRNA for 48 hrs, compared to positive control. (*$p < 0.01$; **$p < 0.001$).

4. Conclusions

The discovery of new targets for the treatment of pathologies has gone hand in hand with the need for new carriers able to allow an adequate release to the site of action and to obtain the maximum effectiveness on the treated cells. Gene therapy using non-viral vectors is very popular today, especially because it is safer than using viral vectors, although less efficient. Therefore, the realization of synthetic polymeric carriers has allowed us to design a polymer and its realization by adding perfectly functional components to its structure.

In this work, a protonable copolymer was designed for the production of polyplexes with a siRNA for the treatment of asthma. It was obtained by using PHEA as starting polymer, that is highly functionalisable, and as molecule to be conjugated in a side chain to complex siRNA, we have chosen a oligoamin bAPAE. The latter was bonded in a suitable quantity by means of a easy to reproduce reaction. Since it was thought to administer the polyplex directly to the lung, the carrier has been appropriately functionalized with variable amounts of PEG, since it is already known that the latter molecule is able to modulate the hydrophilicity, shield the charge of the material, and thus confer the ability to penetrate through a mucus layer.

All the obtained copolymers, pegylated or not, were able to complex the selected siRNA at quite low values of weight ratios, giving nanosized complexes with a tendence to the aggregation invertially proportional to the pegylation degree. These polyplexes seem to be destabilized by mucin, that gives polyanionic exchange with siRNA by direct electrostatic interactions of each copolymer with mucin. However, these interactions decrease as a function of the amount of linked PEG, resulting very low for the highest pegylated polyplexes based on the PHEA-*g*-PEG(C)-*g*-bAPAE copolymer. Moreover, the complexation of siRNA with all the synthesized copolymers significantly increases the ability of complexed siRNA to be internalized by the 16 HBE cells. Therefore, the PHEA-*g*-PEG(C)-*g*-bAPAE copolymer possesses interesting potential as a pulmonary siRNA carrier, being not interacting with mucins and able to give neutral polyplexes with quite static dimensions in a wide range of weight ratio.

Furthermore, the performance of PHEA-*g*-PEG(C)-*g*-bAPAE copolymer, as siRNA delivery system, is going to be improved by grafting on the copolymer structure other functional moieties such as cell penetrating peptides, able to improve cell internalization of the siRNA/copolymer polyplexes and/or cross-linking agents able to increase polyplexe stability in biological environment.

Supplementary Materials: The following are available online at http://www.mdpi.com/1999-4923/12/2/89/s1, Figure S1: SEC chromatograms of: a) PHEA-g-bAPAE (black), PHEA-g-PEG(A)-g-bAPAE (blue), PHEA-g-PEG(B)-g-bAPAE (red), PHEA-g-PEG(c)-g-bAPAE (green); b) PHEA-g-PEG(A) (black), PHEA-g-PEG(B) (red), PHEA-g-PEG(C) (green).

Author Contributions: G.C. and G.G. conceived and designed the experiments and organized the manuscript writing. S.E.D. and E.F.C. performed the experiments, analyzed the data and wrote the paper. N.M. analyzed and fitted titration data. All authors have read and agreed to the published version of the manuscript.

Funding: This research received no external funding.

Acknowledgments: Authors thank ATeN Center of University of Palermo—Laboratory of Preparation and Analysis of Biomaterials, for the support in the Size Exclusion Chromatography analysis.

Conflicts of Interest: The authors declare no conflict of interest.

References

1. Shakshuki, A.; Agu, R.U. Improving the Efficiency of Respiratory Drug Delivery: A Review of Current Treatment Trends and Future Strategies for Asthma and Chronic Obstructive Pulmonary Disease. *Pulm. Ther.* **2017**, *3*, 267–281. [CrossRef] [PubMed]
2. Anderson, S.D. Repurposing drugs as inhaled therapies in asthma. *Adv. Drug Deliv. Rev.* **2018**, *133*, 19–33. [CrossRef] [PubMed]
3. Kanemitsu, Y.; Matsumoto, H.; Izuhara, K.; Tohda, Y.; Kita, H.; Horiguchi, T.; Kuwabara, K.; Tomii, K.; Otsuka, K.; Fujimura, M.; et al. Increased periostin associates with greater airflow limitation in patients receiving inhaled corticosteroids. *J. Allergy Clin. Immunol.* **2013**, *132*, 305–312. [CrossRef] [PubMed]
4. Miyake, T.; Miyake, T.; Sakaguchi, M.; Nankai, H.; Nakazawa, T.; Morishita, R. Prevention of Asthma Exacerbation in a Mouse Model by Simultaneous Inhibition of NF-κB and STAT6 Activation Using a Chimeric Decoy Strategy. *Mol. Ther. Nucleic Acids* **2018**, *10*, 159–169. [CrossRef]
5. Cho, S.H.; Jo, A.; Casale, T.; Jeong, S.J.; Hong, S.J.; Cho, J.K.; Holbrook, J.T.; Kumar, R.; Smith, L.J. Soy isoflavones reduce asthma exacerbation in asthmatic patients with high PAI-1–producing genotypes. *J. Allergy Clin. Immunol.* **2019**, *144*, 109–117. [CrossRef]
6. Tabeling, C.; Herbert, J.; Hocke, A.C.; Lamb, D.J.; Wollin, S.L.; Erb, K.J.; Boiarina, E.; Movassagh, H.; Scheffel, J.; Doehn, J.M.; et al. Spleen tyrosine kinase inhibition blocks airway constriction and protects from Th2-induced airway inflammation and remodeling. *Allergy Eur. J. Allergy Clin. Immunol.* **2017**, *72*, 1061–1072. [CrossRef]
7. Cavallaro, G.; Sardo, C.; Craparo, E.F.; Porsio, B.; Giammona, G. Polymeric nanoparticles for siRNA delivery: Production and applications. *Int. J. Pharm.* **2017**, *525*, 313–333. [CrossRef]
8. Park, J.; Park, J.; Pei, Y.; Xu, J.; Yeo, Y. Pharmacokinetics and biodistribution of recently-developed siRNA nanomedicines. *Adv. Drug Deliv. Rev.* **2016**, *104*, 93–109. [CrossRef]
9. David, S.; Pitard, B.; Benoît, J.P.; Passirani, C. Non-viral nanosystems for systemic siRNA delivery. *Pharmacol. Res.* **2010**, *62*, 100–114. [CrossRef]
10. Cooper, B.M.; Putnam, D. Polymers for siRNA delivery: A Critical Assessment of Current Technology Prospects for Clinical Application. *ACS Biomater. Sci. Eng.* **2016**, *2*, 1837–1850. [CrossRef]
11. Giammona, G.; Carlisi, B.; Palazzo, S. Reaction of α,β-poly(N-hydroxyethyl)-DL-aspartamide with derivatives of carboxylic acids. *J. Polym. Sci. Part A Polym. Chem.* **1987**, *25*, 2813–2818. [CrossRef]
12. Cavallaro, G.; Farra, R.; Craparo, E.F.; Sardo, C.; Porsio, B.; Giammona, G.; Perrone, F.; Grassi, M.; Pozzato, G.; Grassi, G.; et al. Galactosylated polyaspartamide copolymers for siRNA targeted delivery to hepatocellular carcinoma cells. *Int. J. Pharm.* **2017**, *525*, 397–406. [CrossRef] [PubMed]
13. Gioia, S.D.; Sardo, C.; Belgiovine, G.; Triolo, D.; d'Apolito, M.; Castellani, S.; Carbone, A.; Giardino, I.; Giammona, G.; Cavallaro, G.; et al. Cationic polyaspartamide-based nanocomplexes mediate siRNA entry and down-regulation of the pro-inflammatory mediator high mobility group box 1 in airway epithelial cells. *Int. J. Pharm.* **2015**, *491*, 359–366. [CrossRef] [PubMed]
14. Craparo, E.F.; Cavallaro, G.; Bondì, M.L.; Giammona, G. Preparation of polymeric nanoparticles by photo-crosslinking of an acryloylated polyaspartamide in w/o microemulsion. *Macromol. Chem. Phys.* **2004**, *205*, 1955–1964. [CrossRef]

15. Pitarresi, G.; Palumbo, F.S.; Calabrese, R.; Craparo, E.F.; Giammona, G. Crosslinked hyaluronan with a protein-like polymer: Novel bioresorbable films for biomedical applications. *J. Biomed. Mater. Res.-Part A* **2008**, *84*, 413–424. [CrossRef]
16. Craparo, E.F.; Teresi, G.; Licciardi, M.; Bondí, M.L.; Cavallaro, G. Novel composed galactosylated nanodevices containing a ribavirin prodrug as hepatic cell-targeted carriers for HCV treatment. *J. Biomed. Nanotechnol.* **2013**, *9*, 1107–1122. [CrossRef]
17. Licciardi, M.; Di Stefano, M.; Craparo, E.F.; Amato, G.; Fontana, G.; Cavallaro, G.; Giammona, G. PHEA-graft-polybutylmethacrylate copolymer microparticles for delivery of hydrophobic drugs. *Int. J. Pharm.* **2012**, *433*. [CrossRef]
18. Suma, T.; Miyata, K.; Ishii, T.; Uchida, S.; Uchida, H.; Itaka, K.; Nishiyama, N.; Kataoka, K. Enhanced stability and gene silencing ability of siRNA-loaded polyion complexes formulated from polyaspartamide derivatives with a repetitive array of amino groups in the side chain. *Biomaterials* **2012**, *33*, 2770–2779. [CrossRef]
19. Darcan-Nicolaisen, Y.; Meinicke, H.; Fels, G.; Hegend, O.; Haberland, A.; Kühl, A.; Loddenkemper, C.; Witzenrath, M.; Kube, S.; Henke, W.; et al. Small Interfering RNA against Transcription Factor STAT6 Inhibits Allergic Airway Inflammation and Hyperreactivity in Mice. *J. Immunol.* **2009**, *182*, 7501–7508. [CrossRef]
20. Rippmann, J.F.; Schnapp, A.; Weith, A.; Hobbie, S. Gene silencing with STAT6 specific siRNAs blocks eotaxin release in IL-4/TNFα stimulated human epithelial cells. *FEBS Lett.* **2005**, *579*, 173–178. [CrossRef]
21. Lomonossoff, G. Modified Plant Virus Particles and Uses Therefor. U.S. Patent US20120015899A1, 19 January 2012.
22. Wu, Z.; Liu, J.; Hu, S.; Zhu, Y.; Li, S. Serine/threonine kinase 35, a target gene of stat3, regulates the proliferation and apoptosis of osteosarcoma cells. *Cell. Physiol. Biochem.* **2018**, *45*, 808–818. [CrossRef] [PubMed]
23. Craparo, E.F.; Porsio, B.; Sardo, C.; Giammona, G.; Cavallaro, G. Pegylated Polyaspartamide–Polylactide-Based Nanoparticles Penetrating Cystic Fibrosis Artificial Mucus. *Biomacromolecules* **2016**, *17*, 767–777. [CrossRef]
24. Craparo, E.F.; Porsio, B.; Mauro, N.; Giammona, G.; Cavallaro, G. Polyaspartamide-Polylactide Graft Copolymers with Tunable Properties for the Realization of Fluorescent Nanoparticles for Imaging. *Macromol. Rapid Commun.* **2015**, *36*, 1409–1415. [CrossRef]
25. Montana, G.; Bondì, M.L.; Carrotta, R.; Picone, P.; Craparo, E.F.; San Biagio, P.L.; Giammona, G.; Di Carlo, M. Employment of cationic solid-lipid nanoparticles as RNA carriers. *Bioconjug. Chem.* **2007**, *18*, 302–308. [CrossRef] [PubMed]
26. Craparo, E.F.; Triolo, D.; Pitarresi, G.; Giammona, G.; Cavallaro, G. Galactosylated micelles for a ribavirin prodrug targeting to hepatocytes. *Biomacromolecules* **2013**, *14*, 1838–1849. [CrossRef] [PubMed]
27. Behr, J.P. The proton sponge: A trick to enter cells the viruses did not exploit. *Chimia* **1997**, *51*, 34–36.
28. Patel, S.; Kim, J.; Herrera, M.; Mukherjee, A.; Kabanov, A.V.; Sahay, G. Brief update on endocytosis of nanomedicines. *Adv. Drug Deliv. Rev.* **2019**, *144*, 90–111. [CrossRef]
29. Mauro, N.; Fiorica, C.; Varvarà, P.; Di Prima, G.; Giammona, G. A facile way to build up branched high functional polyaminoacids with tunable physicochemical and biological properties. *Eur. Polym. J.* **2016**, *77*, 124–138. [CrossRef]
30. Ferruti, P.; Mauro, N.; Falciola, L.; Pifferi, V.; Bartoli, C.; Gazzarri, M.; Chiellini, F.; Ranucci, E. Amphoteric, prevailingly cationic L-arginine polymers of poly(amidoamino acid) structure: Synthesis, acid/base properties and preliminary cytocompatibility and cell-permeating characterizations. *Macromol. Biosci.* **2014**, *14*, 390–400. [CrossRef]
31. De Levie, R. *How to Use Excel ® in Analytical Chemistry and in General Scientific Data Analysis*; Cambridge University Press: Cambridge, UK, 2001; ISBN 9780511808265.
32. Scialabba, C.; Sciortino, A.; Messina, F.; Buscarino, G.; Cannas, M.; Roscigno, G.; Condorelli, G.; Cavallaro, G.; Giammona, G.; Mauro, N. Highly Homogeneous Biotinylated Carbon Nanodots: Red-Emitting Nanoheaters as Theranostic Agents toward Precision Cancer Medicine. *ACS Appl. Mater. Interfaces* **2019**, *11*, 19854–19866. [CrossRef]

MDPI
St. Alban-Anlage 66
4052 Basel
Switzerland
Tel. +41 61 683 77 34
Fax +41 61 302 89 18
www.mdpi.com

Pharmaceutics Editorial Office
E-mail: pharmaceutics@mdpi.com
www.mdpi.com/journal/pharmaceutics

www.ingramcontent.com/pod-product-compliance
Lightning Source LLC
LaVergne TN
LVHW070158170726
843515LV00007B/1947

* 9 7 8 3 0 3 9 3 6 2 0 0 4 *